Lecture Notes in Computer Science 11914

More information about this series at http://www.springer.com/series/7407

Mohsen Ghaffari · Mikhail Nesterenko ·
Sébastien Tixeuil · Sara Tucci ·
Yukiko Yamauchi (Eds.)

Stabilization, Safety, and Security of Distributed Systems

21st International Symposium, SSS 2019
Pisa, Italy, October 22–25, 2019
Proceedings

Editors
Mohsen Ghaffari
ETH Zurich
Zurich, Switzerland

Mikhail Nesterenko
Kent State University
Kent, OH, USA

Sébastien Tixeuil
Sorbonne University
Paris, France

Sara Tucci
CEA LIST
Gif-sur-Yvette, France

Yukiko Yamauchi
Kyushu University
Fukuoka, Japan

ISSN 0302-9743   ISSN 1611-3349   (electronic)
Lecture Notes in Computer Science
ISBN 978-3-030-34991-2   ISBN 978-3-030-34992-9   (eBook)
https://doi.org/10.1007/978-3-030-34992-9

LNCS Sublibrary: SL1 – Theoretical Computer Science and General Issues

This Springer imprint is published by the registered company Springer Nature Switzerland AG
The registered company address is: Gewerbestrasse 11, 6330 Cham, Switzerland

Preface

The papers in this volume were presented at the 21st International Symposium on Stabilization, Safety, and Security of Distributed Systems (SSS 2019), held during October 22–25, 2019, in Pisa, Italy.

SSS is an international forum for researchers and practitioners in the design and development of distributed systems with a focus on systems that are able to provide guarantees on their correctness, performance, and/or security in the face of an adverse operational environment. Research in distributed systems is now at a crucial point in its evolution, marked by the importance and variety of dynamic distributed systems such as robotic networks, large-scale sensor networks, mobile ad hoc networks, blockchains and peer-to-peer networks, cloud computing, and many others. Moreover, new applications such as grid and Web services, distributed command and control, and a vast array of decentralized computations in a variety of disciplines have driven the need to ensure that distributed computations are self-stabilizing, safe, secure, and efficient.

SSS started as the Workshop on Self-Stabilizing Systems (WSS), the first two of which were held in Austin in 1989 and in Las Vegas in 1995. Starting in 1995, the workshop was held biennially; it was held in Santa Barbara (1997), Austin (1999), and Lisbon (2001). As interest grew and the community expanded, in 2003 the title of the forum was changed to the Symposium on Self-Stabilizing Systems (SSS). SSS was organized in San Francisco in 2003 and in Barcelona in 2005. As SSS broadened its scope and attracted researchers from other communities, significant changes were made in 2006. It became an annual event, and the name of the conference was changed to the International Symposium on Stabilization, Safety, and Security of Distributed Systems (SSS). From then, SSS conferences were held in Dallas (2006), Paris (2007), Detroit (2008), Lyon (2009), New York (2010), Grenoble (2011), Toronto (2012), Osaka (2013), Paderborn (2014), Edmonton (2015), Lyon (2016), Boston (2017), and Tokyo (2018). This edition marks the 30th year of the conference.

This year the program was organized into four tracks reflecting major trends related to distributed systems: (A) Foundations of Distributed Computing (chaired by Mohsen Ghaffari), (B) Moving and Computing (chaired by Yukiko Yamauchi), (C) Theoretical and Practical Aspects of Self-stabilizing Systems (chaired by Mikhail Nesterenko), and (D) Security and Privacy (chaired by Sara Tucci). We received 45 submissions from 20 countries. Each submission was reviewed by at least three Program Committee members with the help of external reviewers. Out of the submitted papers, 21 were selected for presentation as regular papers. The symposium also included seven brief announcements. Selected extended papers from the symposium will be published in a special issue of the journal *Information and Computation*. The committee also selected the following papers to be awarded:

- **Best Paper**: Armando Castaneda, Pierre Fraigniaud, Ami Paz, Sergio Rajsbaum, Matthieu Roy, and Corentin Travers: "Synchronous t-Resilient Consensus in Arbitrary Graphs."

– **Best Student Paper**: Shota Nagahama, Fukuhito Ooshita, and Michiko Inoue: "Ring Exploration of Myopic Luminous Robots with Visibility More than One."

On behalf of the Program Committee, we would like to thank all the authors who submitted their work to SSS.

We sincerely acknowledge the tremendous time and effort that the Program Committee members invested in the symposium. We are grateful to the external reviewers for their valuable and insightful comments and to EasyChair for largely simplifying the reviewing process and the preparation of the proceedings.

We also thank the SSS Steering Committee for invaluable advice. We gratefully acknowledge the Organizing Committee chaired by Giuseppe Prencipe and the publicity chair Doina Bein for their time and effort that greatly contributed to the success of this symposium. This conference was supported by the "Gruppo Nazionale per il Calcolo Scientifico" (GNCS – INdAM).

September 2019

Mohsen Ghaffari
Mikhail Nesterenko
Sébastien Tixeuil
Sara Tucci
Yukiko Yamauchi

Organization

Program Committee

Yehuda Afek	Tel-Aviv University, Israel
Dan Alistarh	IST Austria, Austria
Leonardo Aniello	University of Southampton, UK
James Aspnes	Yale, USA
Hagit Attiya	Technion, Israel
Alysson Bessani	Universidade de Lisboa, Portugal
Lélia Blin	Sorbonne University, France
Borzoo Bonakdarpour	Iowa State University, USA
Francois Bonnet	Tokyo Institute of Technology, Japan
Silvia Bonomi	Sapienza University of Rome, Italy
Quentin Bramas	University of Strasbourg, France
Miguel Correia	Universidade de Lisboa, Portugal
Shantanu Das	Aix-Marseille University, France
Sylvie Delaet	University Paris-Saclay, France
Stephan Devismes	University of Grenoble, France
Swan Dubois	Sorbonne University, Inria, France
Kokoris Kogias Eleftherios	Ecole Polytechnique Fédérale de Lausanne, Switzerland
Sándor Fekete	TU Braunschweig, Germany
Paola Flocchini	University of Ottawa, Canada
Dianne Foreback	University of Akron, USA
Rati Gelashvili	University of Toronto, Canada
Chryssis Georgiou	University of Cyprus, Cyprus
Mohsen Ghaffari	ETH Zurich, Switzerland
Sukumar Ghosh	University of Iowa, USA
Seth Gilbert	National University of Singapore, Singapore
Niv Gilboa	Ben Gurion University, Israel
Vincent Gramoli	The University of Sydney, Australia
Magnus Halldorsson	Reykjavik University, Iceland
Sayaka Kamei	Hiroshima University, Japan
Rüdiger Kapitza	TU Braunschweig, Germany
Evangelos Kranakis	Carleton University, Canada
Mario Larangeira	Tokyo Institute of Technology, Japan
Christoph Lenzen	MPI for Informatics, Germany
Euripides Markou	University of Thessaly, Greece
Toshimitsu Masuzawa	Osaka University, Japan
Othon Michail	The University of Liverpool, UK
Zarko Milosevic	Interchain Foundation, Switzerland
Achour Mostefaoui	University of Nantes, France
Alfredo Navarra	University of Perugia, Italy

Mikhail Nesterenko	Kent State University, USA
Nicolas Nicolaou	KIOS Research and Innovation Center of Excellence, University of Cyprus, Cyprus
Fukuhito Ooshita	NAIST, Japan
Sriram Pemmaraju	The University of Iowa, USA
Seth Pettie	University of Michigan, USA
Maria Potop-Butucaru	Sorbonne University, France
Walter Rudametkin	University of Lille, Inria, France
Christian Scheideler	University of Paderborn, Germany
Elad Schiller	Chalmers University of Technology, Sweden
Gokarna Sharma	Kent State University, USA
Weidong Shi	University of Houston, USA
Sébastien Tixeuil	Sorbonne University, France
Corentin Travers	LaBRI, University of Bordeaux, France
Sara Tucci	CEA, France
Koichi Wada	HOSEI University, Japan
Jennifer Welch	Texas A&M University, USA
Yukiko Yamauchi	Kyushu University, Japan

Additional Reviewers

Cicerone, Serafino
Connor, Matthew
Dufoulon, Fabien
Durand, Anaïs
Fraigniaud, Pierre
Gasieniec, Leszek
Giachoudis, Nikos
Hinnenthal, Kristian
Hood, Kendric
Katayama, Yoshiaki
Kshemkalyani, Ajay
Lahav, Ori

Lamani, Anissa
Pai, Shreyas
Palios, Leonidas
Rabie, Mikael
Rieck, Christian
Rosenbaum, Will
Rovedakis, Stephane
Schmidt, Arne
Shibata, Masahiro
Skretas, George
Theofilatos, Michail
Trahan, Jerry

Sponsors

Keynote Talks

Moving and Computing
in Time-Varying Graphs

Paola Flocchini

University of Ottawa, Canada
paola.flocchini@uottawa.ca

Abstract. The study of mobile entities operating in discrete environments or in continuous spaces (*Moving and Computing*) has been quite extensive in the past two decades (for a recent account of the research in the area, see [16] and chapters therein).

The investigations in discrete spaces have been carried out under a variety of models and assumptions, depending on the environment where the entities operate (the graph topology), the type of communication mechanisms they use (e.g., face to face, whiteboards, wireless), the degree of synchronization of the network (e.g., fully synchronous, semi-synchronous, asynchronous), the level of knowledge the entities have about the environment (e.g., topological knowledge, sense of direction, size of the agents' team), the amount of persistent memory of the entities (e.g., no memory, constant memory, unbounded memory), etc.

In spite of all these differences, until recently a common assumption has been the fact that the environment where the entities move is *static*, that is the topology of the graph does not change during the computation.

Researchers in distributed computing have recently started to investigate classical problems by mobile agents moving on *time-varying graphs* (TGV) [4], that is graphs where the structure of the links keeps changing and where these topological changes are not sporadic or anomalous, but rather inherent in the nature of the network.

Moving and computing in TVGs has been studied especially in the context of *graph exploration*, both in centralized settings (e.g., [10–12, 19, 22]) as well as in distributed ones [2, 3, 8, 14, 15, 17, 18, 20, 21]. Other mobile agents problems have been studied as well, e.g., *gathering* [9], *patrolling* [5], *grouping* [6], *dispersion* [1], and *cops and robbers* [13]; for a recent survey, see [7].

In this talk, I review recent results on distributed computations by mobile agents operating in time-varying graphs, indicating open questions, unexplored problems, and future research directions.

References

1. Agarwalla, A., Augustine, J., Moses, Jr. W.K., Madhav, S., Sridhar, A.K.: Deterministic dispersion of mobile robots in dynamic rings. In: 19th International Conference on Distributed Computing and Networking (ICDCN), pp. 1–4 (2018)

This work was supported in part by an NSERC Discovery Grant and the University Research Chair.

2. Bournat, M., Datta, A.K., Dubois, S.: Self-stabilizing robots in highly dynamic environments. Theor. Comput. Sci. **772**, 88–110 (2019)

3. Bournat, M., Dubois, S., Petit, F.: Computability of perpetual exploration in highly dynamic rings. In: 37th International Conference on Distributed Computing Systems (ICDCS), pp. 794–804 (2017)

4. Casteigts, A., Flocchini, P., Quattrociocchi, W., Santoro, N.: Time-varying graphs and dynamic networks. Int. J. Parallel Emergent Distrib. Syst. **27**(5), 387–408 (2012)

5. Das, S., Di Luna, G.A., Gasieniec, L.A.: Patrolling on dynamic ring networks. In: Catania, B., Královič, R., Nawrocki, J., Pighizzini, G. (eds.) SOFSEM 2019. LNCS, vol. 11376, pp. 150–163. Springer, Cham (2019). https://doi.org/10.1007/978-3-030-10801-4_13

6. Das, S., Di Luna. G., Pagli. L., Prencipe. G.: Compacting and grouping mobile agents on dynamic rings. In: Gopal, T., Watada, J. (eds.) TAMC 2019. LNCS, vol. 11436, pp. 114–133. Springer, Cham (2019). https://doi.org/10.1007/978-3-030-14812-6_8

7. Di Luna, G.A.: Mobile Agents on Dynamic Graphs. Chap. 20 of [16], pp. 549–584 (2019)

8. Di Luna, G.A., Dobrev, S., Flocchini, P., Santoro, N.: Live exploration of dynamic rings. In: IEEE 36th International Conference on Distributed Computing Systems (ICDCS), pp. 570–579 (2016)

9. Di Luna, G.A., Flocchini, P., Pagli, L., Santoro, N., Viglietta, G.: Gathering in dynamic rings. Theor. Comput. Sci. (2019)

10. Erlebach, T., Hoffmann, M., Kammer, F.: On temporal graph exploration. In: Halldórsson, M., Iwama, K., Kobayashi, N., Speckmann, B. (eds.) ICALP 2015. LNCS, vol. 9134, pp. 444–455 . Springer, Heidelberg (2015). https://doi.org/10.1007/978-3-662-47672-7_36

11. Erlebach, T., Kammer, F., Luo, K., Sajenko, A., Spooner, J.T.: Two moves per time step make a difference. In: 46th International. Colloquium on Automata, Languages, and Programming (ICALP), pp. 1–14 (2019)

12. Erlebach, T., Spooner, J.T.: Faster exploration of degree-bounded temporal graphs. In: 43rd International Symposium on Mathematical Foundations of Computer Science (MFCS 2018), pp. 1–13 (2018)

13. Erlebach, T., Spooner, J.T.: A game of cops and robbers on graphs with periodic edge-connectivity, CoRR abs/1908.06828 (2019)

14. Flocchini, P., Kellett, M., Mason, P., Santoro, N.: Searching for black holes in subways. Theory Comput. Syst. **50**(1), 158–184 (2012)

15. Flocchini, P., Mans, B., Santoro, N.: On the exploration of time-varying networks. Theor. Comput. Sci. **469**, 53–68 (2013)

16. Flocchini, P., Prencipe, G., Santoro, N. (eds.): Distributed Computing by Mobile Entities. Springer (2019)

17. Gotoh, T., Flocchini, P., Masuzawa, T., Santoro, N.: Tight bounds on exploration of temporal graphs. Manuscript (2019)

18. Gotoh, T., Sudo, Y., Ooshita, F., Kakugawa, H., Masuzawa, T.: Group exploration of dynamic tori. In: IEEE 38th International Conference on Distributed Computing Systems (ICDCS), pp. 775–785 (2018)

19. Ilcinkas, D., Klasing, R., Wade, A.M.: Exploration of constantly connected dynamic graphs based on cactuses. In: Halldórsson, M.M. (eds.) SIROCCO 2014. LNCS, vol. 8576, pp. 250–262. Springer, Cham (2014). https://doi.org/10.1007/978-3-319-09620-9_20

20. Ilcinkas, D., Wade, A.M.: On the power of waiting when exploring public transportation systems. In: 15th International Conference on Principles of Distributed Systems (OPODIS), pp. 451–464 (2011)

21. Ilcinkas, D., Wade, A.M.: Exploration of the T-interval-connected dynamic graphs: the case of the ring. Theory Comput. Syst. **62**(4), 1144–1160 (2018)

22. Michail, O., Spirakis, P.: Traveling salesman problems in temporal graphs. Theor. Comput. Sci. **634**, 1–23 (2016)

What Can Be Computed Asynchronously

Petr Kuznetsov

LTCI, Télécom Paris, Institut Polytechnique Paris
petr.kuznetsov@telecom-paris.fr

Abstract. When we devise a computing system, it makes sense to assume as little as possible about the environment in which the system is expected to run. For example, *asynchronous* distributed systems do not rely on timing assumptions, which makes them extremely robust with respect to communication disruptions and computational delays. It is, however, notoriously difficult and sometimes even impossible to make such systems *fault-tolerant*.

In this talk, we try to understand what can and what cannot be computed in an asynchronous and fault-tolerant manner. We focus on the problem of implementing a *replicated service*, where a collection of servers maintain *replicas* of the service state and respond to the clients requests.

The folklore CAP theorem [2, 3] states that no replicated service can combine (strong) consistency, availability, and partition-tolerance. To encompass scenarios in which partitions cannot be avoided, the notion of *eventual* consistency has been introduced [12] and classes of objects that allow for asynchronous eventually consistent implementations have been studied [11]. We show, however, that a *universal* eventually consistent replicated service which, intuitively, can be used to obtain an eventually consistent distributed version of any sequential service, cannot be implemented without nontrivial synchrony assumptions [4].

In the case when partitions are excluded, e.g., by assuming that a majority of replicas are correct, building a strongly consistent universal construction is equivalent to solving *consensus* [9, 10]. Intuitively, consensus is used here to ensure that concurrent operations on the replicated service are *totally ordered*. As no fault-tolerant asynchronous solution to consensus exists [6], one may ask if there is an *asynchronous universal* abstraction, analogous to consensus in partially synchronous systems. We argue that *lattice agreement* [1, 5] can be seen as such an abstraction. We observe that many important applications tolerate specific *lattice* partial orders on their operations and, therefore, can be implemented asynchronously using lattice agreement. As an example, we consider the *asset transfer* problem that lies at the core of modern cryptocurrencies [7]. Finally, we discuss how *randomized* asynchronous algorithms can circumvent consistency and complexity barriers of deterministic ones [8].

References

1 Attiya, H., Herlihy, M., Rachman, O.: Atomic snapshots using lattice agreement. Distrib. Comput. **8**(3), 121–132 (1995)

2. Brewer, E.A.: Towards robust distributed systems (abstract). In: PODC, p. 7 (2000)

3. Castro, M., Liskov, B.: Practical byzantine fault tolerance and proactive recovery. ACM Trans. Comput. Syst. (TOCS) **20**(4), 398–461 (2002)
4. Dubois, S., Guerraoui, R., Kuznetsov, P., Petit, F., Sens, P.: The weakest failure detector for eventual consistency. In: PODC, pp. 375–384 (2015)
5. Faleiro, J., Rajamani, S., Rajan, K., Ramalingam, G., Vaswani, K.: Generalized lattice agreement. In: PODC, pp. 125–134 (2012)
6. Fischer, M.J., Lynch, N.A., Paterson, M.S.: Impossibility of distributed consensus with one faulty process. J. ACM, **32**(2), 374–382 (1985)
7. Guerraoui, R., Kuznetsov, P., Monti, M., Pavlovic, M., Seredinschi, D.: The consensus number of a cryptocurrency. In: PODC, pp. 307–316 (2019)
8. Guerraoui, R., Kuznetsov, P., Monti, M., Pavlovic, M., Seredinschi, D.: Scalable Byzantine reliable broadcast. In: DISC (2019)
9. Herlihy, M.: Wait-free synchronization. ACM Trans. Prog. Lang. Syst. **13**(1), 123–149 (1991)
10. Lamport, L.: The part-time parliament. ACM Trans. Comput. Syst. **16**(2), 133–169 (1998)
11. Shapiro, M., Preguiça, N., Baquero, C., Zawirski, M.: Conflict-free replicated data types. In: Défago, X., Petit, F., Villain, V. (eds.) SSS 2011. LNCS, vol. 6976, pp. 386–400. Springer, Heidelberg (2011). https://doi.org/10.1007/978-3-642-24550-3_29
12. Vogels, W.: Eventually consistent. Commun. ACM **52**(1), pp. 40–44 (2009)

Amoebots and Beyond: Models and Approaches for Programmable Matter

Christian Scheideler

Department of Computer Science, Paderborn University, Paderborn, Germany
scheideler@upb.de

Abstract. "Programmable matter" is a term originally coined by Toffoli and Margolus in 1991 to refer to an ensemble of fine-grained computing elements that has the ability to change its physical properties (such as shape, density, moduli, conductivity, optical properties, etc.) based on user input or autonomous sensing. There has already been a significant amount of research on programmable matter across multiple disciplines, including physics (e.g., crystals and complex fluids), chemistry (e.g., metamaterials and shape-changing molecules), bioengineering (DNA self-assembly and cell engineering), and robotics (modular robotics and nano robotics). However, so far there does not exist any particular guideline for designing and managing programmable matter. Hence, now is the ideal time to investigate reasonable models and primitives for this matter so that it can be used effectively in applications that it is destined for. I will present the amoebot model (with some recent extensions) and show that it can be used for typical applications like shape formation and coating. However, many aspects have not been considered yet though they are crucial for programmable matter to become a reality, including energy, fault-tolerance, forces (like gravity in the 3D case), and non-local coordination and movements. Therefore, I will also discuss possible extensions of the amoebot model to address these issues.

Tutorials

Hands on Blockchains

Quentin Bramas ⓘ

ICUBE, University of Strasbourg, CNRS, France
bramas@unistra.fr

Abstract. When Bitcoin was introduced 11 years ago, it created a lot of opportunities for academics and industrial. Basically, the Bitcoin protocol is a distributed database specifically built for exchanging cryptocurrency, but it works between any number of unknown participants. The core protocol was soon after used to store other kind of data for different purposes. Each application, running on its own Blockchain, is executed by all the participants that may or may not know or trust each other.

Then, Ethereum was created as a single clockchain that can execute arbitrary applications, called smart-contracts. Applications can interact together on the Ethereum blockchain and functions can be triggered by transactions. Ethereum makes it very easy to develop new distributed applications. And looking at its code, one can easily trust an applications, even if it has been developed by someone unknown. This adds value to the data stored by an application in the blockchain as a participant really owns its data (input data comes from signed transactions).

One famous example is the cryptokitties game where anyone can buy, exchange and breed virtual cats. The game would probably not have gained much attention if the virtual cat owners were just defined by rows in a standard database. Instead, cats are associated with the public key of their owner and only the owner of the private key can sell a cat, without intermediary.

In this tutorial I will present quickly how Ethereum makes that possible. Then, I will develop a small application. To achieve this, I will present the Solidity language, how to compile, deploy, and interact with a smart-contract. I will also present several important cryptographic primitives that allow, for instance, proving something without revealing too much information.

Keywords: Blockchain · Distributed ledger technologies · Ethereum · Smart-contract.

The Theory of Blockchains

Antonella Del Pozzo

CEA LIST, PC 174, 91191, Gif-sur-Yvette, France
antonella.delpozzo@cea.fr

Abstract. The blockchain technology appeared for the first time in 2008 in the white paper of Satoshi Nakamoto, where it was designed to be the fully distributed ledger behind the first decentralized cryptocurrency: Bitcoin. Briefly, the blockchain is an append-only chain of blocks, such that in the Bitcoin case, it contains all the Bitcoin transactions that, once confirmed in the blockchain, can be reverted only with a small probability that becomes negligible with time.

Blockchain is not only Bitcoin. Indeed after 2008, different kind of blockchains have been further defined and for different purposes rather than cryptocurrencies, making clear that the Blockchain technology can serve as a more general notarization tool that can be employed when there is the need to notarize information produced by entities that do not necessarily trust each other. Indeed, with the blockchain they can trust that the information in the blockchain are immutable.

The goal of this tutorial is to understand the main mechanisms behind the Bitcoin blockchain such as the blockchain object structure itself, how the blocks are appended, and the kind of information contained in the blocks. Indeed, blocks do not contain only transactions but also non-turing complete scripts allowing flexibility in the transaction management or the creation of side chains payment channels.

The takeaway of this tutorial is the understanding of the blockchain technology itself, its potentialities and limitations, using the Bitcoin blockchain as a case study.

Contents

Invited Paper: On the Characterization of Blockchain Consensus Under Incentives

Sara Tucci-Piergiovanni[✉]

CEA LIST, PC 174, 91191 Gif-sur-Yvette, France
`sara.tucci@cea.fr`

Abstract. One of the novel aspects of blockchains is the intertwining of consensus properties and incentives. An incentive model determines participant behaviors and then the possibility to reach consensus. In this paper we propose a methodological approach to characterize an incentive model for blockchain consensus. An incentive model is defined through the characterization of an oracle, along with its failure model, and blockchain participants behaviors. The oracle assures Safety properties at the expense of Liveness, since Liveness is in the hands of participants that can behave obediently, strategically or in an adversarial way. We then apply the proposed methodology to define and analyze incentive models of popular blockchain solutions. The paper concludes on future research directions that can take advantage of the proposed characterization.

Keywords: Blockchains · Consensus · Incentives models

1 Introduction

Consensus in blockchains means to get an agreement on a unique version of a ledger where each block in the blockchain offers an updated version of the ledger state chained to a previous block. In an ideal blockchain, there is a single sequence of blocks on which participants agree. In contrast to this situation, forks might happen leading to multiple versions of the ledger.

The fairly unique aspect of blockchains is that consensus is driven by incentives. In blockchains an incentive model implements specific financial arrangements for participants influencing their behavior.

To make an example, the maintenance of the ledger in Bitcoin proceeds as follows: participants create blocks including users transactions along with the pointer to the previous block and the solution of a cryptopuzzle; diffuse each newly created block in the network; and store blocks arranged in a tree structure. In Bitcoin, the protocol embeds a rule, which is the longest chain rule, which allows selecting a chain in the tree. This way, blockchain creators chain their newly created block to the selected chain. It is important to highlight that strategic participants do not obey blindly to the longest chain rule but they act strategically selecting the longest chain only if this maximize their utility.

M. Ghaffari et al. (Eds.): SSS 2019, LNCS 11914, pp. 1–15, 2019.
https://doi.org/10.1007/978-3-030-34992-9_1

Recently, it has been shown that the longest chain rule under the Bitcoin incentive model is the best choice for strategic agents, i.e., the Bitcoin consensus mechanism along with its incentive model is incentive compatible [9][1]. The Bitcoin's incentive model provides rewards for block creators and takes into account proof-of-work expenditures. We can then state that incentive models and the possibility of reaching an agreement on a common chain are intertwined, therefore, to preserve consensus properties proper incentive models must be defined.

This paper considers the problem of characterization of blockchain consensus and related incentive models. The objective is to capture incentive models for consensus in abstract way, without entering in the detail of specific (often complex) protocols. To this end, we propose a specification of blockchain consensus that extends the abstract formal specification of blockchains proposed by Anceaume et al. in [7]. Anceaume et al. [7] model a blockchain as a concurrent abstract data type that can be concurrently accessed by an unlimited number N of participants. The data type is a rooted tree, called block tree, in which vertices represent blocks, its root corresponds to the genesis block, and each root-to-leaf path is a blockchain. The two supported operations are *append*, which adds a new vertex to the tree, and *read*, which returns a blockchain of the tree selected according to some *fitting function*.

The main feature of [7]'s formalisation is the presence of an *oracle* that helps to append blocks to the tree. Intuitively, the oracle, called by participants during the *append()* operation, controls the maximum branching factor of the block tree. The oracle allows a vertex in the tree to have up to some parameter k children. In [7] it has been shown that strong consistency, i.e., no-forks, can be achieved only assuming an oracle with branching factor $k = 1$ and that this oracle is at least as strong as Consensus. The relationship within blockchain consistency and consensus has been then formally established. The branching factor of an oracle, moreover, fully characterizes a blockchain, where an oracle with $k = 1$ characterizes blockchains resorting to consensus and $k > 1$ characterizes consensus-free blockchains. In other terms, the essence of any blockchain is captured by the assumed oracle, which is a simple abstraction.

In this paper we then characterize incentive models by augmenting oracles (not the block tree) with costs and rewards associated to its execution. In this respect we will focus on characterization of consensus-based blockchains, i.e. blockchains using an oracle with branching factor equal to 1. Let us note that since consensus cannot run among an unbounded number of processes, current solutions, e.g. [19,21], attempt to solve the problem through rotating committees of fixed size n. In this case the production of a single block needs coordination among the n committee members and rewards are distributed among members once the block is produced. The oracle then needs to feature a selection phase: for each height of the chain n committee members out of the total number

---

[1] Incentive compatibility can be established considering different solution concepts, such as dominant strategies and Nash equilibria. Incentive compatibility of Bitcoin has been shown assuming Nash equilibria as solution concept. Formal definitions of these notions will be presented later in the paper.

of participants $N \geq n$ are selected, members that need to agree on the next block. After the selection phase, a voting phase among the n members starts. We abstract the voting phase considering a function f that maps votes on a set of proposals to a joint decision. Since participants may decide to vote or not and voting has a cost, we define a quorum ν, the minimum number of votes necessary to trigger a decision. Once the voting phase completes, if a valid decision is reached, the oracle distributes rewards to participants.

The incentive model characterization is then completed by proceeding to the following steps: (1) definition of the failure model of the oracle – the oracle can be either correct or it can lose/corrupt votes, (2) definition of the type of participants, which can obedient, strategic or adversarial and (3) definition of the rewarding function, i.e., how rewards are distributed to the committee.

Different incentive models capturing current popular blockchain solutions are then analyzed. Incentive models are analyzed exploring their influence on Liveness (eventually a block is decided) only. Safety properties are indeed always guaranteed by the oracle at the expense of Liveness, driven by the whole incentive model. The paper concludes highlighting that the quest for liveness-preserving incentive models is in its early days.

The paper is organized as follows. Section 2 introduces blockchain oracles previously presented in [7]. Section 3 characterizes incentive models for consensus-based blockchains. Section 4 presents analyses of incentive models of current popular solutions and finally Sect. 5 discusses research directions.

2   Blockchain Core Abstractions

Anceaume et al. [7] present an abstract and formal characterization of blockchains that allows to capture their properties independently from protocol-specific mechanisms (Proof-of-Work, Proof-of-Stake, Consensus, etc) and system model assumptions (synchrony, type of faults, number of participants). The main feature of the characterization is the presence of so-called blockchain oracles that (i) abstract the validation process of blocks in the chain, (ii) regulate forks and (iii) abstract the communication medium.

In this section we briefly present this characterization, but for readability we omit all the formal notations giving only the essential concepts.

2.1   The Blocktree and Oracle Models

Following [7], any blockchain is characterized by a concurrent data structure representing the blockchain and the type of oracle that helps maintaining it, specified as follows.

The Blockchain Tree. A blockchain is a direct rooted tree bt that can be accessed by $read()$ and $append(b)$ operations. Each vertex of the tree is a block and any edge points backward to the root, called genesis block and denoted as b_0. The height of a block refers to its distance to the root.

The read operation $read()$ returns a branch of the tree, from the genesis to one leaf, selected through a so-called fitting function $f(bt)$, which is protocol specific. Each walk, or chain, in the tree as has an associated score that is monotonically increasing w.r.t. inclusion of blocks.

The $append(b)$ operation takes one parameter, the block to be appended. To be appended a block must be valid. The validity of a block is protocol-specific. The $append(b)$ operation returns true if the block is appended to the tree, false otherwise. If the block has been appended it is valid.

Concurrent Specification. Both append and read operation can be invoked concurrently by blockchain participants. Under concurrency, the data type satisfies Block Validity: Blocks in the blockchain returned by the read operation are valid with respect to a validity predicate; Local monotonic read: The score of the blockchain returned by subsequent reads from the same process is monotonically nondecreasing; and the Ever-growing tree property: The score of the returned blockchain eventually grows. As for consistency, two alternative consistency criteria are defined, namely: (i) *Strong consistency*: In addition to the above properties, for any two blockchains returned by a read operation, one is a prefix of the other; or alternatively (ii) *Eventual consistency*: In addition to the above properties, if a blockchain with score s is returned, then at most a finite number of read operations return blockchains that do not share the same prefix up to score s.

Strongly consistent blockchains do not allow forks, while eventually consistent blockchain admit occurrences of forks that are eventually solved.

Blockchain Oracles. Oracles are generic modules able to abstract the generation of valid blocks, the communication medium and to regulate the branching factor of the block tree. During the $append(b)$ operation, the oracle is called through a $getValidBlock(b^*, b)$ operation that takes two parameters, the proposed block b and a block of the block tree b^*, chosen as proposed parent of b. Let us note that b^* is freely chosen by the invoking process and that the block b is valid only if the $getValidBlock(b^*, b)$ operation successfully terminates. It is assumed that only the oracle is able to make the block valid, i.e. to implement the $getValidBlock(b^*, b)$ operation.

The oracle releases valid blocks depending on the merit parameter $\alpha_i \in [0, 1]$ of the invoking process i. In [7], for each α_i the oracle endows an infinite tape $tape_{\alpha_i}$ of elements in the set $\{\top, \bot\}$. Each time the getValidBlock is invoked by a process with merit α_i, an element is popped by $tape_{\alpha_i}$. If the popped element is $\top$, then the valid block is released, otherwise the operation returns *false*. In this deterministic version[2] the oracle, if invoked infinitely often by i with $\alpha_i > 0$ with the same pair of blocks (b^*, b), will eventually release a valid block guaranteeing a form of fairness. In this paper, for sake of generality and to ease the presentation we assume the presence of a higher-level function $select(\alpha_i)$ that

---

[2] Even if oracles defined in [7] are deterministic, probabilistic versions can be easily derived by associating a probability to pop $\top$ proportional to the merit.

Algorithm 1. getValidBlock operation at b^*

1: **upon** $\langle getValidBlock(b^*, b_i) \rangle$ **from** process i with α_i **do**
2:   **if** $(select(\alpha_i))$ **then**
3:     $validBlocks = validBlocks \cup \{b_i\}$
4:     $return\ b_i$
5:   **else**
6:     $return\ false$

Algorithm 2. setValidBlock operation at b^*

1: **upon** $\langle setValidBlock(b^*, b_i) \rangle$ **from** process i **do**
2:   **if** $(b^*.children.size < k) \wedge (b_i \in validBlocks) \wedge (b_i \notin b^*.children)$ **then**
3:     $b^*.children \leftarrow b^*.children \cup \{b_i\}$
4:   $return\ b^*.children$

---

accesses the tape and returns true if the popped element is $\top$. The pseudo-code of the getValidBlock operation is shown in Algorithm 1.

Once a process gets a valid block from the oracle, it can now set it through a $setValidBlock(b^*, b)$. As soon as the setValidBlock operation returns we have two cases: if b is returned then $append(b)$ returns true, otherwise $append(b)$ returns false. It is possible that more than one participant gets a valid block to append to b^*, in that case the oracle regulates forks depending on its *branching factor* k. If the oracle has a bounded branching factor, then the oracle guarantees that the returned set of any successfully executed $setValidBlock(b^*, b)$, returns a set of children of bounded size k containing successfully set blocks, then possibly not containing b. If k is unbounded, then the oracle returns a set of children containing b. The pseudo-code of the setValidBlock operation is shown in Algorithm 2. In case of unbounded branching factor, the pseudo-code is either modified by removing the first term of the *if* condition or by considering $k = \infty$.

In [7] it has been shown that (i) the oracle with $k = 1$ is equivalent to Consensus and (ii) Consensus is necessary for strongly consistent blockchains.

2.2 Oracles, a Closer Look

Oracles, as already mentioned, abstract away the consensus mechanism used in blockchains. Let us to have a closer look at the oracle characteristics. The getValidBlock operation abstracts away the validation process of blocks. For instance, if we think to the proof-of-work mechanism, we can see the getValidBlock invocation as a query to an oracle that gives to the process the solution of the proof-of-work or false otherwise.

The setValidBlock operation, has a subtler role. Let us take our weakest oracle, with $k = \infty$. The setValidBlock operation only means in this case to make the block *visible* in the blockchain. Once the setValidBlock returns, the block is accepted by the blockchain, i.e., it is appended. We can think to the set operation as a successful write of a new block in the blockchain state. It is important to stress out that actual update of a state change is far from being trivial in a open peer-to-peer network, subject to participant churn. For that reason, it is extremely useful to resort to oracles. In this perspective, all the blocks returned by setValidBlock operations can be viewed as stable vertexes in the tree: if some processes sets a given block b, than successive setValidBlock will report b in the returned set[3].

The Shared Burden Between Oracles and Blockchain Participants. Oracles, by definition, are local to one particular vertex of the tree: each participant chooses to call the getValidBlock on the parent vertex of her choice, and to call setValid-Block to try to append the block. This suggests that it is up to participants to guarantee the consistency of the blockchain and not to oracles. For instance, if there is just one participant in the blockchain that keeps calling the oracle to append a block to the genesis one, and the oracle has $k = \infty$, the ever-growing tree property will never be guaranteed. At some point the process must jump to a leaf to make the tree growing. This is also true by assuming, for the same scenario, an oracle with $k = 1$: the calling participant will always terminate her append operation by returning false, but the tree will stop growing. In case of weaker oracles with $k > 1$, if there are two processes jumping alternatively and concurrently on two parallel branches of the tree, it is up to them to eventually select and read the same chain to guarantee eventual consistency and repair forks. The separation of duties between oracles and blockchain participants described so far is particularly meaningful when strategic behavior is assumed. Blockchain participants thanks to the oracle have the possibility to update the block tree, but it is up to them to choose the chain they prefer when reading and appending; this choice is done strategically. In this respect it is interesting to highlight that stronger oracles reduce the strategic space with respect to weaker ones, but even assuming our strongest oracle with branching factor $k = 1$, strategic choices are relevant to guarantee progress.

2.3   Enriching Oracles with Incentives

In all the implementations employing the proof-of-work, a cost must be associated to block validation. For those implementations exempted by the proof-of-work, to avoid spamming – a process maliciously sending too many blocks – either a mini proof-of-work is employed or a so-called slashing mechanism allows to burn some coins of the malicious process [14,19]. In all cases we can abstract a cost associated with the generation of valid blocks. This cost can be allocated to the process either at the end of the invocation of the getValidBlock or when then process invokes the setValidBlock operation – this is to take into account a cost only when a block is sent in the system.

As for rewarding, current solutions can roughly be divided in two families: those where only one participant is accountable for the appended block and those where more than one participant is accountable for it. Rewarding is then given

---

[3] In [7] the consumeValidBlock, called in [7] consumeToken operation, has been reduced to the Generalized Lattice Agreement abstraction [12].

to accountable participants for the block. We are interested in the second family where the oracle has branching factor $k = 1$, i.e., an oracle implemented through Consensus in a deterministic setting. In current solutions based on Consensus [19, 21] a block is created coordinately by committees of n participants selected out of N blockchain participants and the reward is shared among committee members. We will define an incentive model for this class of blockchains. We will introduce all the elements needed to define a so-called Consensus Incentive Model: (i) an adaptation of the oracle with branching factor $k = 1$ to model selection and rewarding, (ii) a failure model for the oracle, (iii) participants characterization in terms of their type (obedient, adversarial, strategic) and behavior.

3   Consensus Incentive Model Definition

Oracle Characterization. The oracle is an oracle with branching factor $k = 1$ called hereafter Consensus oracle, with a cost associated to the invocation of setValidBlock. The oracle, as current solutions, e.g. [11, 14, 18, 19], selects committee members that have most merit and share the reward among committee members, even though more sophisticated models have been devised[4]. The oracle is specified as follows (Algorithms 3 and 4). The oracle when invoked at a given block b^* through $getValidBlock(b^*, *)$ grants valid blocks up to n invokers, by following a given selection criterion and puts the process i in a validator set. Once the process i gets a valid block it may decide to try to append the block in the blockchain, through $setValidBlock(b^*, b)$. The operation starts a voting phase to collect proposals. When the phase ends, only if $\nu \leq n$ proposals are collected a decision is taken through a function f that maps proposals to the block to be set. Note that the duration of a voting phase depends on the system model assumed. In synchronous systems the voting phase lasts at least the time required to gather the ν proposals, but it could last more and gather more than ν proposals. In case ν proposals are gathered, but the decision is not valid the oracle returns $false$. The reward function takes as parameter the validator set, the proposal and the decision. If all the validators are rewarded no matter if they voted or not, then we have a *reward all* rewarding scheme; alternatively, if only voters get the reward, we have a *reward only voters* scheme.

Failure models of the oracle. The oracle can be:

- *correct*: The pseudo-code (Algorithms 3 and 4) is correctly executed.
- *vote corruption*: the proposals set at line 7 can contain up to c non-valid votes, because of a failure of the valid block test.
- *vote omission*: in the proposals set at line 7 up to m votes can be lost.

The oracle is not a strategic entity, the failure model assumed abstracts away possible system failures. The *reward* function is always executed correctly.

---

[4] Some implementations [11, 14] provide a delegation mechanism in which a participant can delegate its merit to another one, in this case the reward is shared among delegators as well.

Algorithm 3. Consensus Oracle with Incentives: getValidBlock operation at b^*

```
1: upon ⟨getValidBlock(b*, bᵢ)⟩ from process i with αᵢ do
2:    if ((select(αᵢ) ∧ (n ≤ j)) then
3:       validBlocks = validBlocks ∪ {i}
4:       validatorSet ← validatorSet ∪ {bᵢ}
5:       j + +
6:       return bᵢ
7:    else
8:       return false
```

Algorithm 4. Consensus Oracle with Incentives: setValidBlock operation at b^*

```
1: upon ⟨setValidBlock(b*, bᵢ)⟩ from process i do
2:    if (b*.children.size < 1) ∧ (bᵢ ∈ validBlocks) then
3:       if voting phase not started then
4:          start voting phase
5:       b*.proposals ← b*.proposals ∪ {bᵢ}

6: upon ((end voting phase) ∧ (proposals.size = ν)) do
7:    decision ← f(proposals)
8:    reward(validatorSet, proposals, decision)
9:    if decision ∈ validBlocks then
10:      return b*.children
11:   else
12:      return false
```

Participants Characterization

Participants Action Space. Each participant has a strategy s_i from an action set. In one single voting phase,

$s_i \in \{0, 1, \perp\}$, where 1 maps to the decision of sending a vote, 0 of not sending a vote, and $\perp$ of sending an invalid vote.

Types of Participants. We define then three types of participants:

- *strategic*: a player that chooses the strategy that maximizes her payoff
- *obedient*: a player whose strategy is fixed to 1 under the condition that utility is greater than a given threshold γ. In other terms the obedient participant has limited resources, needing then some degree of fairness [5,15].
- *adversary*: a player whose strategy is disrupting system properties. In the one voting phase case, it is either 0 or $\perp$.

Later we will extend our oracle to a *multi-ballot oracle*, to include in the action set the action of sending two different votes in two different ballots in the same voting phase. This is to include the classical Byzantine misbehavior of sending different messages to different processes. This extended action set will be discussed for solutions based on BFT consensus.

Properties to Guarantee. Properties to guarantee at each height of the chain are traditional Consensus properties: Agreement (at most one decision is taken),

Validity (the decision value must be valid) and Termination (eventually a decision is taken). Agreement and Validity are safety properties while Termination is a Liveness property. In our proposed oracle construction participant behaviors can only hinder Liveness, leading to never return a valid block (line 10). This is easy to see if the oracle is assumed correct. In case of votes corruptions (modeling a vote for an invalid block), the decision function f is able to filter out corrupted proposals. If no valid proposal can be chosen, the f returns $\perp$ and the block is not returned.

The agreement property is by construction not threatened by the given action set, but easily guaranteed with a multi-ballot oracle capturing PBFT solutions as we will see later.

The main principle is to have an oracle assuring Safety at the expense of Liveness and to study how participants behavior under the assumed incentive model influences Liveness.

Utility Functions. Before being able to establish participant strategies we need to define utility functions. A utility function depends on the oracle failure model and the action set. The simplest utility function, defined assuming a correct oracle, is as follows.

$$u_i = \Sigma_{\kappa=0}^{n_i^r} R_i^\kappa - \Sigma_{\ell=0}^{n_i^v} C_i^\ell \quad (1)$$

where

- n_i^r is the number of setValidBlocks operations successfully executed and such that i belongs to $validatorSet$ (reward all) or i's proposal belongs to $proposals$ (reward only voters) at line 8.
- n_i^v is the number of setValidBlocks operations successfully executed and such that i voted, i.e., i's proposal belongs to $proposals$ at line 5.

Weaker oracles can also envisage punishments. For instance an oracle suffering from vote corruptions, can contain in the proposal set some invalid blocks. We consider that the decision function f could in this case select an invalid block but that the decision is not returned, leading to Liveness violation. The participants, if an invalid block is decided or even proposed, can be punished.

Participants Strategies. Once the utility function is in place participant strategies can be defined.

Strategies of obedient participants correspond to the expected behavior from the designer point of view. The strategy under our oracle models and action space is then naturally 1 as long as the participant can pay the inflicting costs. For obedient participants it is assumed that the participant enters the system with an endowment and that such endowment is modified solely by costs and rewards defined by her utility function. The threshold γ must be set to make sure that the participant is rewarded often enough to not exhaust her resources. In the weakest model γ is greater than zero.

Strategies for adversarial players can be determined considering strategies that undermine properties to guarantee, i.e. Termination. It assumed that the utility function is known by adversarial processes.

Strategies for strategic participants can be determined in the framework of game theory. Game theory focuses on predicting individual players' strategy and payoffs. For each combination of players and possible strategies, there is a payoff. Game theory analyzes which strategies strategic players will play in the game.

Given a game, it is interesting to look for dominant strategies. A dominant strategy for a player is a strategy leading to the best payoff, no matter how other players may play. However, dominant strategies there not always exist.

Nash Equilibrium is an equilibrium where each player's strategy is optimal given the strategies of all other players. A Nash Equilibrium exists when no player would take a different action as long as every other player remains the same. Nash Equilibria are self-enforcing; when players are at a Nash Equilibrium they have no desire to deviate, otherwise they will be worse off. Interestingly, given any finite game a Nash equilibrium there always exists.

More formally a Nash equilibrium can be defined as follows:

Nash equilibrium. Let $(\mathcal{S}, \mathcal{U})$ a game with n players where $\mathcal{S}_i$ is the strategy set for the agent, $\mathcal{S} = \mathcal{S}_1 \times \ldots \times \mathcal{S}_n$ is the set of strategy profiles and $\mathcal{U}$ is the set of utility functions $\mathcal{U} = u_1 \times \ldots \times u_n$ mapping a strategy $s_i \in \mathcal{S}_n$ to a payoff $u_i(s_i)$.

A strategy $s_i \in S_i$ for the agent i is a mapping from each state of the game to an action in the action space of the agent. A strategy tells a player what to do for every possible state of the game throughout the game[5]. When each agent chooses a strategy from its strategy set we get a strategy profile $s = (s_1, \ldots, s_n) \in \mathcal{S}$.

Let us denote with $(s|^i, s_i^*)$ the fact that i deviates from s by doing $s_i^* \in S_i$. A strategy profile s is a pure Nash Equilibrium if and only if for each i, and for all strategies $s_i^* \in S_i : u_i(s|^i, s_i^*) \leq u_i(s)$.

In the following we will analyze incentive models with strategic players under the Nash equilibrium solution concept.

Failure Models of Participants.

We assume that obedient and strategic participants are able *to correctly execute their strategy*, i.e. they are correct processes with respect to their strategy. We also assume that adversarial participants can execute their strategy in the worst case, but they can be affected by unexpected failures, i.e. they are Byzantine processes[6]. Note that our model separates failures

---

[5] A strategy can be viewed as an algorithm. The state in game theory is called the information set that is evaluated each time it is updated, to select the next action or move.

[6] Even if Byzantine failures are defined as arbitrary deviations from the prescribed behavior, an adversarial argument is assumed to prove protocols under Byzantine processes. This way the strategy of the Byzantine participant is determined.

from strategies (that could deviate or not from designer's one). This separation, inspired by [13], defines a slightly different model than the BAR model [2]. Our obedient participants are neither altruistic (they do not maximize the benefits of others in the general case) nor correct since they have limited resources. Moreover, any strategic player is assumed correct with respect to her chosen strategy (on the contrary the participant must be assumed Byzantine).

Efficiency. Any incentive model can be analyzed under the efficiency point of view. Given ν the threshold to produce a block, efficiency is the ratio between ν and the number of votes determined by chosen strategies.

4 Analysis of Current Solutions

In this section we illustrate first how to capture current solutions under an incentive model and then we analyze the incentive model itself. The section has a pedagogical purpose and let us observe how assumptions on the particular combination of oracle failure model, participants types and the reward schemes impact Liveness. We analyze two reward schemes: *reward only voters* and *reward all*.

Strategic Incentive Models for Synchronous Leader-Based Solutions in a One Shot Game. In solutions like [11,14], a single leader is elected for each height of the chain. The leader sends its block and the $n-1$ other processes send a vote for the block. We can abstract these solutions by our oracle without failures, n participants and a decision function f that selects the proposal of the leader where all the proposals are votes. We need for the block to be produced at least ν votes, otherwise liveness will not be guaranteed. Assuming a synchronous system means to collect at the end of the vote phase all the votes to potentially reward them (the oracle is assumed correct, no message losses occur). We assume n strategic participants, utility function (1) for only one height of the tree with $R > C$, assuming a one shot game. Strategies of participants are as follows [3].

Reward Only Voters. For the reward only voters scheme we have two cases: $\nu > 1$ and $\nu = 1$. The case of $\nu > 1$ has two Nash equilibria. In the first equilibrium all the participants send a vote, i.e., they call the setValidBlock operation. In the second equilibrium nobody sends a vote; violating Liveness. In the case of $\nu = 1$ we have a single Nash equilibrium where all n participants vote. The efficiency of the mechanism is $\frac{\nu}{n}$.

Reward All. For the reward all mechanism, we have two cases: $\nu > 1$ and $\nu = 1$. The case of $\nu > 1$ has multiple Nash equilibria, where either ν participants send a vote or nobody sends a vote; violating Liveness. In the case of $\nu = 1$ we have n Nash equilibria, in each equilibrium, exactly one process sends a vote. Efficiency is optimal.

Note that a good Nash equilibrium for both schemes is reached when $\nu = 1$, because the participant is pivotal: in the strategy profile where nobody send the vote, she will be better off by deviating, i.e. sending the vote. This is true for all the participants if only voters are rewarded. In the reward all scheme, this is true for ν participants.

Comparing the two mechanisms the *reward all* is more efficient, in the sense that less money is distributed to participants in good equilibria. Both of them, however, if $\nu > 1$ have bad equilibria resulting in a coordination failure and liveness violation.

Adversarial-Strategic Incentive Model for Synchronous Leader-Based Solutions in a Repeated-Game. In [3] a more complex model has been considered, where a combination of adversarial and strategic players is assumed and more leaders can be elected in a round-robin fashion to produce a block at a given height of the chain. The oracle used is an oracle accepting invalid blocks but able to filter invalid decisions out. The reward scheme assumed is a *reward only voters* with a cost inflicted to all the committee members if an invalid block is selected by the decision function f, i.e. if the leader proposed an invalid block that has been accepted by ν participants. The study assumes as well that participants have the possibility to check upfront if the leader's proposal is valid or not at some additional cost. Moreover, it is assumed that the reward is greater than the inflicted cost and the cost of accepting an invalid block is greater than the reward (see [3] for the utility function formal definition).

It is easy to see that the adversarial strategy to inflict the maximum damage to other participants is to propose an invalid block and hope to get a sufficient number of votes to have the block accepted. In this case all of them will be penalized and Liveness threatened.

As for strategic players, strategies have been found considering multiple voting rounds (for the same height of the chain) where for each round a leader is chosen in a round robin fashion. The study finds multiple Nash equilibria (we encourage the reader to look at the entire paper). Among many bad equilibria leading to coordination failure and liveness violation, the rewarding mechanism in the proposed model shows a good equilibrium if $\nu > a$ where a is the upper bound on adversarial participants. In good equilibrium invalid blocks (proposed by adversarial players) are rejected, while valid blocks (proposed by strategic players) are accepted. This implies that, if round $t = a + 1$ is reached, the players know that during all the previous a rounds the proposers were Byzantine (to draw this inference, the strategic players use their anticipation that all participants play equilibrium strategies). Consequently, at round $a + 1$, the proposer must be strategic, and all players anticipate the proposed block is valid. So, no strategic player needs to check the validity of the block but all send a vote, which brings them an expected gain equal reward minus the cost to send a vote. This is larger than their gain from deviating by not sending a message or by checking the block.

Obedient-Adversarial Incentive Models for BFT Blockchains. Blockchains as [19,21] use BFT Consensus protocols, e.g. [4,6,22], inside committees with n processes where at most $f = \lfloor \frac{n-1}{3} \rfloor$ participants are Byzantine faulty. The other $n - f$ processes are assumed correct.

For this class of applications, the model assumed is abstracted by our Consensus oracle with f adversarial participants, $n - f$ obedient participants and $\nu = n - f$. These solutions are leader-based, then multiple voting phases can be activated with leaders selected in a round-robin fashion. The eventual synchrony assumption is dealt with an adaptive timeout meant to catch the unknown message delay eventually on some voting phase. We abstract this behavior assuming that at the end of a single voting phase all sent votes are collected, but the oracle can lose at most one third of sent votes. After an finite but unknown number of voting phases, the oracle ceases to lose messages. To capture these solutions we need as well to extend our action set, to let an adversarial participant to behave like a Byzantine process sending two different votes to different processes, instead of one single vote. To this aim the oracle can be extended to a n ballot voting phase where each participant votes in n different ballots, one for each participant i. Each ballot has n entries indexed by the participant identifier and ballots are filled up to at least ν votes. This way an adversary j can vote different blocks in different ballots. The decision function f takes the first entry that have two third of common valid values. Let us to make an example. Process $p1$ is Byzantine, while $p2, p3, p4$ are correct. Values proposed on the n ballots are as follows $p1 = \{1,1,0,1\}$, $p2 = \{2,2,2,2\}$, $p3 = \{3,3,3,3\}$, $p4 = \{4,4,4,4\}$. The oracle can lose at most one vote per ballot. Values lost by the oracle are the values proposed by pi for the $(i+1)^{th}$ ballot. The ballots state, where $\{\}_i$ is the ith ballot, at the end of the vote phase is as follows: $\{1,2,3,\perp\}_1$, $\{\perp,2,3,4\}_2$, $\{0,\perp,3,4\}_3$, $\{1,2,3,\perp\}_4$. The oracle takes a decision that must be the same for obedient participants, without knowing who the Byzantines are. The oracle takes the common value contained in the first entry of two-thirds of ballots (if any), then 2 in the scenario above, and notifies all about the decision. We say that p_1, p_2 and p_4 voted for the decided value because their corresponding ballot contained the decided value. Note that "more than two-thirds correct" processes assumption neutralizes the effect of the byzantine misbehavior, i.e. in these conditions the decision value there always exists. Under these conditions, the adversarial behavior of a Byzantine will then try to hinder the participation of obedient participants to violate Liveness (to lower the threshold of correct processes). This behavior depends on the rewarding scheme. Let us recall that unlike a correct process an obedient one has limited resources, dictated by her endowment. An obedient process must be rewarded often enough to maintain a balance greater than zero. If not, the process with no endowment will not be able to cover the cost of sending her votes. Let us consider rewarding as follows. *Rewarding the Processes that Voted the Decided Value.* In the above mentioned scenario if the oracle does not lose the vote of the Byzantine, then the Byzantine gets a reward and $p3$ does not. $p3$ can be excluded from the reward·also in the case the Byzantine sends the same valid value, but the oracle loses the first entry of $p3$. If these scenarios repeat too long, $p3$ can exhaust her resources and will not be able to vote anymore. The best strategy for the adversarial participant to hinder Liveness is then to participate with valid values hoping $p3$ reaches endowment 0 and then to stay silent forever.

Rewarding All. This scheme gives a reward to all the processes, even those that do not send her vote. Note that this is the safer method here, even if less efficient, because a Byzantine has no power to make the obedient worse off.

5   Discussion and Research Directions

The proposed characterization highlighted the interlacement of participants' strategy and rewarding schemes in various incentive models based on oracles with branching factor equal to 1. Recent research [20] is now focusing on BFT consensus to see how to increase the threshold f of corruption faults. In this direction, some effort is spent in detecting corrupted processes [10] because this capability could allow to exclude them from the reward, or to punish them, hoping to tolerate the presence of more than one third of adversarial processes. As a recommendation it is important not to sacrify liveness for efficiency here and to clearly state the incentive model under study. In our characterization, for instance, an adversary is punishment-insensitive. An interesting research question would be to characterize adversarial participants sensitive to rewarding/punishment by assuming limited resources for them. Another interesting line of research is to consider preferences about transactions contained in blocks. Approaches that explored leader election [1] and consensus [8,16] under these assumptions could be analyzed under our incentive models. Finally, obedient models could be refined to consider more sophisticated fairness-related concepts and the interlacement between merit and participant endowment [5,15,17]. We hope that the provided characterization can help in exploring these research directions.

Acknowledgments. This position paper assembles ideas and results emerged through research conducted with E. Anceaume, A. del Pozzo, M. Potop-Butucaru and O. Gurcan. A special thanks goes to Y. Amoussou-Guenou working hard in this middle earth between distributed computing and economy and to Prof. Bias that literally opened us the door to economic methodologies.

References

1. Abraham, I., Dolev, D., Halpern, J.Y.: Distributed protocols for leader election: a game-theoretic perspective. ACM Trans. Econ. Comput. **7**(1), 4:1–4:26 (2019)
2. Aiyer, A.S., Alvisi, L., Clement, A., Dahlin, M., Martin, J.P., Porth, C.: Bar fault tolerance for cooperative services. In: Proceedings of the 20th ACM Symposium on Operating Systems Principles, SOSP 2005, pp. 45–58 (2005)
3. Amoussou-Guenou, Y., Biais, B., Potop-Butucaru, M., Tucci-Piergiovanni, S.: Rationals vs Byzantines in consensus-based blockchains. CoRR abs/1902.07895 (2019)
4. Amoussou-Guenou, Y., Del Pozzo, A., Potop-Butucaru, M., Tucci-Piergiovanni, S.: Dissecting tendermint. In: NETYS 2019 (2019)

5. Amoussou-Guenou, Y., Del Pozzo, A., Potop-Butucaru, M., Tucci-Piergiovanni, S.: Correctness and fairness of Tendermint-core blockchain protocols. Research report (2018). https://hal.archives-ouvertes.fr/hal-01790504
6. Amoussou-Guenou, Y., Pozzo, A.D., Potop-Butucaru, M., Tucci-Piergiovanni, S.: Correctness of tendermint-core blockchains. In: 22nd International Conference on Principles of Distributed Systems, OPODIS 2018, Hong Kong, China, 17–19 December 2018, pp. 16:1–16:16 (2018)
7. Anceaume, E., Del Pozzo, A., Ludinard, R., Potop-Butucaru, M., Tucci-Piergiovanni, S.: Blockchain abstract data type. In: SPAA 2019 (2019)
8. Bei, X., Chen, W., Zhang, J.: Distributed consensus resilient to both crash failures and strategic manipulations. CoRR abs/1203.4324 (2012)
9. Biais, B., Bisière, C., Bouvard, M., Casamatta, C.: The blockchain folk theorem. Rev. Financ. Stud. **32**, 1662–1715 (2019)
10. Civit, P., Gilbert, S., Gramoli, V.: Polygraph: accountable byzantine agreement. IACR Cryptology ePrint Archive 2019/587 (2019)
11. David, B., Gaži, P., Kiayias, A., Russell, A.: Ouroboros praos: an adaptively-secure, semi-synchronous proof-of-stake blockchain. In: Nielsen, J.B., Rijmen, V. (eds.) EUROCRYPT 2018. LNCS. Part II, vol. 10821, pp. 66–98. Springer, Cham (2018). https://doi.org/10.1007/978-3-319-78375-8_3
12. Falerio, J.M., Rajamani, S.K., Rajan, K., Ramalingam, G., Vaswani, K.: Generalized lattice agreement. In: ACM Symposium on Principles of Distributed Computing, PODC 2012, Funchal, Madeira, Portugal, 16–18 July 2012, pp. 125–134 (2012)
13. Feigenbaum, J., Shenker, S.: Distributed algorithmic mechanism design: recent results and future directions, distributed computing column. Bull. EATCS **79**, 101–121 (2003)
14. Goodman: Tezos: a self amending crypto ledger. https://tezos.com/static/white-paper-2dc8c02267a8fb86bd67a108199441bf.pdf
15. Gürcan, Ö., Del Pozzo, A., Tucci-Piergiovanni, S.: On the bitcoin limitations to deliver fairness to users. In: Panetto, H., et al. (eds.) On the Move to Meaningful Internet Systems, vol. 10573, pp. 589–606. Springer, Cham (2017). https://doi.org/10.1007/978-3-319-69462-7_37
16. Halpern, J.Y., Vilaça, X.: Rational consensus: extended abstract. In: Proceedings of the 2016 ACM Symposium on Principles of Distributed Computing, PODC 2016, Chicago, IL, USA, 25–28 July 2016, pp. 137–146 (2016)
17. Karakostas, D., Kiayias, A., Nasikas, C., Zindros, D.: Cryptocurrency egalitarianism: a quantitative approach. In: Tokenomics International Conference on Blockchain Economics, Security and Protocols 2019 (2019)
18. Kwon, J., Buchman, E.: Cosmos: A Network of Distributed Ledgers. https://cosmos.network/resources/whitepaper. Accessed 22 May 2018
19. Kwon, J., Buchman, E.: Tendermint. https://tendermint.readthedocs.io/en/master/specification.html. Accessed 22 May 2018
20. Malkhi, D., Nayak, K., Ren, L.: Flexible byzantine fault tolerance. In: ACM CCS (2019)
21. Various: The Libra Blockchain. https://developers.libra.org/docs/assets/papers/the-libra-blockchain.pdf
22. Yin, M., Malkhi, D., Reiter, M.K., Golan-Gueta, G., Abraham, I.: Hotstuff: BFT consensus with linearity and responsiveness. In: Proceedings of the 2019 ACM Symposium on Principles of Distributed Computing, PODC 2019, Toronto, ON, Canada, 29 July–2 August 2019, pp. 347–356 (2019)

Brief Announcement FORGIVE & FORGET: Self-stabilizing Swarms in Spite of Byzantine Robots

Yotam Ashkenazi[1], Shlomi Dolev[1](✉), Sayaka Kamei[2], Fukuhito Ooshita[3], and Koichi Wada[4]

[1] Department of Computer Science, Ben-Gurion University of the Negev, Be'er Sheva, Israel
{yotamash,dolev}@post.bgu.ac.il
[2] Department of Information Engineering, Graduate School of Engineering, Hiroshima University, Higashihiroshima, Japan
s-kamei@se.hiroshima-u.ac.jp
[3] Graduate School of Science and Technology, Nara Institute of Science and Technology, Ikoma, Japan
f-oosita@is.naist.jp
[4] Department of Applied Informatics, Faculty of Science and Engineering, Hosei University, Tokyo, Japan
wada@hosei.ac.jp

Abstract. In this paper, we consider the case in which a swarm of robots collaborates in a mission, where a few of the robots behave maliciously. These malicious Byzantine robots may be temporally or constantly controlled by an adversary. The scope is synchronized full information robot operations, where a robot that does not follow the program/policy of the swarm is immediately identified and can be remembered as Byzantine. As robots may be suspected of being Byzantine due to benign temporal malfunctions, it is imperative to Forgive & Forget (F&F), otherwise, a robot cannot assume collaborative actions with any other robot in the swarm. Still, remembering for a while may facilitate a policy of surrounding, isolating and freezing the movement of the misbehaving robots, by several robots, allowing the rest to perform the swarm task with no intervention.

We demonstrate the need to periodically F&F to realize swarm several tasks including patrolling/cleaning in the presence of possible Byzantine robots. The policy for achieving the task consists of blocking the movement of the Byzantine robot(s) by some of the robots, while the rest patrol/clean the plane.

This work was supported in part by JSPS KAKENHI No. 17K00019, 18K11167 and 19K11828, Frankel Center for Computer Science, Rita Altura Trust Chair in Computer Science, the Ministry of Science and Technology, Israel & JST SICORP (Grant#JPMJSC1806) and the German Research Funding Organization (DFG, Grant#8767581199).

M. Ghaffari et al. (Eds.): SSS 2019, LNCS 11914, pp. 16–21, 2019.
https://doi.org/10.1007/978-3-030-34992-9_2

1 Introduction

Swarms of robots acting towards a common task are already part of our lives, be it a swarm of autonomic cars, a swarm of the unmanned aerial vehicle (UAV), or a swarm of nano-robots.

When dealing with robots in practice, we better assume that some of the robots are Byzantine, faulty or malicious. These robots may not follow the algorithm either because of a fault or because of a malicious adversarial takeover. Such malicious takeover may imply the most disturbing behavior of the maliciously controlled robot. Obviously, when all participants are Byzantine the swarm can be regarded as malicious as well, not following actions for achieving the planned goal. Since faults and takeovers can be accumulated over time, the possibility for swarm participants to stop functioning as they should do grows with time. *Self-stabilizing* algorithms may cope with such faults, imposing automatic recovery of individuals in the swarm, and regaining collaboration among the recovered participants.

Many researches were made on how to cope with a given threshold (e.g., less than one third) of Byzantine participants. However, these researches are not aimed to cope with temporal periods in which all (or almost all) participants are Byzantine. The correctness of (non-stabilizing) algorithms is based on the consistency of the initial configuration and the preservation of the consistency as long as the threshold on the number of Byzantine is respected. This approach is too optimistic, the approach of self-stabilization is more promising, recreating the consistency thread from any arbitrary configuration whenever the minimal conditions hold (e.g., less than one-third of the participants are Byzantine).

In this paper, we consider the case of a self-stabilizing robot swarm in the presence of Byzantine robots. Typically, Byzantine robots are assumed to be Byzantine forever. Correct robots may detect and record the identity of Byzantine robots in their variables, so they can ignore or take a countermeasure to the Byzantine robots activities. In the scope of self-stabilization, such records can be set to different records for each participant, where the records of the other participants are not mutually known. For example, starting in a configuration in which each participant has a (possibly wrong) record that all other participants (but itself) were identified as Byzantine. Then all may try to take a countermeasure to all, even if none is actually Byzantine. Thus, in fact, nullifying possible swarm collaboration. Note that the unknown records of the other participants may imply that their observable moves will be regards as Byzantine by others.

Many algorithms are based on failure detectors, where each participant lists the suspected participants. Such an abstraction is useful in a fault-prone system, excluding the suspected participants' actions and concentrating on the non-suspected participants to gain progress. In our settings, where the program and identifiers of all participants, as well as all actions and all inputs of all participants, are observed by each participant, an indication on a Byzantine participant is immediate following a step that does not obey the program and inputs. Once there is an indication that a participant is Byzantine, other non-Byzantine robots may surround the Byzantine robot and block its movements, allowing the rest of the robots achieve the swarm task with no intervention.

However, in the scope of self-stabilization, the recorded indication on a participant being Byzantine may be corrupt as well, and therefore we propose the *forgive and forget* (F&F) framework. The period in which the robots remember the indications (before the robots simultaneously forgive and forget) has to be tuned to allow us to perform the task in addition to the possible need of capturing the Byzantine participants.

The F&F approach is useful in coping with unreliable indication on whether a participant was Byzantine. Even in case that the indication is reliable, such F&F approach allows the (self) recovery of a robot (say, by a periodical restart). We assume that many of the Byzantine participants can recover after a while, say by periodically rebooted, patched with new software parts, or scanned and cleaned from malware. Thus, the impression of one robot that another is Byzantine should be constantly reexamined and verified. Obviously, a robot may suspect all other participants being Byzantine, and if the suspicion is not periodically verified, global collaboration is at risk. This is why, in this research, we will assume that all the robots periodically and (to simplify arguments, use self-stabilizing Byzantine clock synchronization to impose that the robots (see [1] for a new self-stabilizing (for non-two faced) Byzantine pulse and clock synchronizations algorithms that are of independent interest) simultaneously forget their suspicions in their Byzantine suspicion list.

We will present a method to decide when should the robots forget their suspicions and reset the Byzantine list. The method reset the suspicions in a way that ensures that the robot in the swarm can repeatedly succeed in achieving a useful task in spite of the presence of several Byzantine participants. The task is repeatedly achieved even though many, or even all of the participants were Byzantine in some instances in the past.

The proposed method is based on calculating the minimum continuous time we can suspect a Byzantine robot before we consider the robot as a non-Byzantine robot again. Roughly speaking, once the suspicion lists are simultaneously reset, the Byzantines may avoid identifying themselves as such, and while doing so assist in completing the task. Alternatively, the Byzantines deviate from their program and are discovered as Byzantine by all non-Byzantine participants. Once discovered, our method suggests to surround and block the Byzantine movement by several robots, and let the other robots complete the task/mission with no (further) interventions.

We demonstrate our approach by presenting several games.[1] These games are based on periodical restart (respected by the non-Byzantine robots), where every K time units (measured in number of movement steps a robot can take) the non-Byzantine robots reset their Byzantine suspected list and show that K is big enough to ensure that if a robot exhibits Byzantine behavior before the

---

[1] Note that, this paper includes only the simplest instance of the ring cleaning game, we considered more games (e.g., fan game) and details can be found in [1].

tasks/missions are achieved, then some non-Byzantine robots can catch/block the Byzantine robot while other non-Byzantine robots complete the tasks.

Cleaning Game. In a cleaning game, Byzantine robots contaminate all the tiles of the board, and the non-Byzantine robots cooperate to clean the board infinitely often. Non-Byzantine robots can clear tiles if they visit the tiles. In addition, non-Byzantine robots can stop the contamination behavior of a Byzantine robot by capturing the Byzantine robot. If a Byzantine robot is not captured, it can contaminate all tiles instantaneously by staying at the same tile for even one step. We say the board is clean if all tiles except for tiles occupied by Byzantine robots are clean.

Definition 1 (Cleaning game task). *The cleaning game task is defined by a set of executions LE such that each E in LE consists of infinitely many configurations in which the board is clean.*

2   Cleaning a Ring Board

In this section, we demonstrate the usefulness of our F&F approach by considering a simple cleaning game for the case of a small ring.

Consider a 3×3 board where robots can move only on the outward bound of the board. That is, the board is regarded as an 8-tile ring board. The eight tiles are named as $t_1, t_2, \ldots, t_8$ clockwise from the left upper tile. Four robots, at most one of which is Byzantine, reside on this ring board.

The Ring Algorithm in a Nutshell. In the proposed algorithm, all robots move clockwise. When robots identify that another robot stops, the robot adds the Byzantine robot to the Byzantine list. Each Byzantine robot has two neighbors, r_i and r_j, their goal is to move to the Byzantine direction from both sides. The third non-Byzantine robot r_c moves to the Byzantine direction using the shortest path. If distances are equal, r_c moves clockwise. When the cleaning robot r_c cannot move and the Byzantine robot is blocked, r_c starts to clean the board. Every K steps all robots reset the Byzantine list and continue with cleaning the borders (the board can still be infected from actions taken prior to the reset).

Execution Example. Each non-Byzantine robot is represented by a green square, each detected Byzantine robot is represented by a red square, Byzantine robot that acts correctly so far (thus, sill not detected as Byzantine) is represented by a yellow square. Contaminated tiles are marked with diagonal lines and cleaned tiles marked as white squares. The next figure depicts the following situations:

(01) Initial configuration, (02) non-Byzantine robots move clockwise, Byzantine stops (yielding an immediate detection by non-Byzantine robots) and contaminates the board, (03–04) non-Byzantine robots try to block (the board is still infected), (05) non-Byzantine robots blocked the Byzantine (the board is still infected), (06–08) Two robots blocked the Byzantine, one robot cleans the board,

(09) Two non-Byzantine robots blocked the Byzantine, the board is clean, (10) Two non-Byzantine robots blocked the Byzantine until resets.

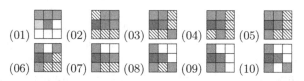

We now demonstrate that when $K = 5$ the Byzantine wins, as the non-Byzantine robots are unable to fulfill their cleaning task. Notice that the non-Byzantine robots forgive and forget, when the Byzantine changes color from red to yellow. Let k $(0 \leq k \leq K)$ be a variable to countup until the resets.

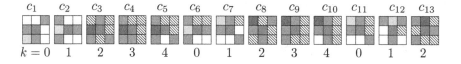

Lemma 1. *When our swarm algorithm is applied to a ring board with four robots, one of which is Byzantine, if the (non-Byzantine) robots forget when $K = 5$, then the cleaning requirement does not hold.*

We demonstrated a scenario which c_{13} and c_3 are the same with an indentation of two steps with $k = 2$. This scenario can be reproduced by Byzantine robot infinitely often, preventing the non-Byzantine robots to complete the cleaning task.

Lemma 2. *When our swarm algorithm is applied to a ring board with four robots, one of which is Byzantine, if the (non-Byzantine) robots forget when $K = 15$, then the cleaning requirement holds.*

Sketch of Proof. We first establish that at most four steps are needed to block the Byzantine robot since non-Byzantine robots detect it. Following figures represents an example of such execution (in the second configuration from left, the Byzantine is detected, and in the first configuration from right, it is blocked).

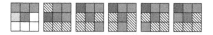

Next we show that we can clean the board after blocking, which takes, in the worst case scenario, five steps. Following figures represent an example of such execution (cleaning starts in the first configuration from left).

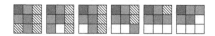

Cleaning the board with no Byzantine can take up to four steps. Assume all four robots create a row, and there are four free tiles to clean. Consider the case that the robots forgive and forget while the board is not clean. If the Byzantine moves four steps the non-Byzantine robots will be able to complete the task. The non-Byzantine robots will be able to block the Byzantine after five more steps and then cleaning the board in five more. If we choose K to be equal to fifteen the robots will be able to complete the task infinitely often.   □

Reference

1. Ashkenazi, Y., et al.: Forgive & Forget: self-stabilizing swarms in spite of Byzantine robots. Department of Computer Science BGU, Technical report #04-19 (2019). https://www.cs.bgu.ac.il/frankel/reports.html

Stationary and Deterministic Leader Election in Self-organizing Particle Systems

Rida A. Bazzi$^{(\boxtimes)}$ and Joseph L. Briones

Arizona State University, Tempe, AZ, USA
{bazzi,jbrione3}@asu.edu

Abstract. We propose the first stationary and deterministic protocol for the leader election problem for non-simply connected particle systems in the geometric Amoebot model in which particles have no unique identifiers but have common chirality. The solution does not require particle movement to break symmetry (stationary) and does not allow particles to make probabilistic choices (deterministic). We show that leader election is possible if and only if the proposed protocol succeeds in electing a unique leader. We show that if the protocol fails to elect a leader, it will always succeed in finding a finite set of $k \leq 6$ leader candidates and the system must have k-symmetry that prevents the selection of less than k candidates. The protocols runs in $O(n^2)$ steps, where n is the number of particles in the system. Other solutions to the leader election problem in the Amoebot model are either probabilistic, assume that the system is simply connected, and/or require stronger primitives to break symmetry.

1 Introduction

Leader election is a fundamental problem that has been studied under a variety of system assumptions including message passing systems [18], shared memory systems [19], radio networks [13] and swarm robotics [10,15]. It is a prototypical symmetry breaking problem [14] upon which solutions to problems that can benefit from a central coordinator can be based [17]. The goal of leader election is to select a unique node of the system as the leader. In systems with unique identifiers, the requirement for leader election is that there is a unique node identifier agreed upon by all nodes in the system as the identifier of the leader. In anonymous systems, the requirement is restated so that one unique node self-identifies as the leader and every other node agrees that a leader has been self-identified. In this paper, we are interested in solving the leader election problem in self-organizing particle systems. Specifically, we consider the well-studied Amoebot model [5]. In the Amoebot model, particles occupy cells in a hexagonal grid. Particles have finite memory, can communicate with adjacent particles, and can expand into unoccupied adjacent cells. Electing a leader in a particle system can facilitate solving problems such as shape-formation [8,16], object coating [6,7] and system compression [3].

© Springer Nature Switzerland AG 2019
M. Ghaffari et al. (Eds.): SSS 2019, LNCS 11914, pp. 22–37, 2019.
https://doi.org/10.1007/978-3-030-34992-9_3

Solutions to the leader election problem in the Amoebot model can be compared according to the following measures:

- Connectivity of system graph. The system graph is the graph whose vertices are the cells of the grid with edges between two cells if and only if both cells are occupied or both cells are unoccupied. A system is said to be *simply connected* if the unoccupied cells form one connected component (system has no holes), otherwise the system is *generally connected*. Typically, the occupied cells are assumed to form one connected component. Otherwise, there is no way to achieve coordination between different connected components without modifying the system model [11].
- Use of randomness. Local random choices by the particles can help in breaking symmetry with high probability.
- Use of movement to break symmetry. When two adjacent particles attempt to move to an empty cell adjacent to both of them, it is typically assumed that one of them will succeed which helps in breaking local symmetry. We say that a solution is *stationary* if it does not require particle movement.
- Scheduler assumption. Particles read, write and move. If the scheduler is strong, reading, writing and moving can be done together in one atomic step. For weak schedulers, reading, writing and moving are individually atomic, but cannot be combined in one atomic step. As is well known from work in shared memory systems [1], the agreement problem can be solved under the strong scheduler assumption, but not under the weak one (test&set objects have higher consensus number than shared registers (read and write memory)).
- Chirality assumption. Chirality is the local sense of rotational direction. Some solutions assume that all particles share the same chirality while other solutions have particles agree on a common chirality using a chirality agreement algorithm that relies on movement (and possibly the scheduler) to break symmetry.
- Running time. The running time is measured in rounds. In each round, each particle is activated at least once at which time, depending on the model, it can read, write, and/or move.

Table 1 summarizes the results of this and other works. We comment further on the differences between the various works. Prior to this work, solutions for the leader election problem in generally connected systems were randomized [4,8]. Further, the more efficient solution in [4] has a success probability that depends on the number of particles in the system which makes it unusable for systems with a small number of particles (for example, for 100,000 particles, the probability of failure is only guaranteed to be no more than 14%). The only existing deterministic solution for leader election prior to this work are [16] and [12], but they only work for simply connected systems and [12] relies on a stronger scheduler.

To overcome the limitation of earlier work, we come up with a novel approach to determine a small number of candidate leaders (1, 2, 3, or 6) on the unique outer border of the system. The algorithm starts with many candidate

Table 1. Comparing solutions to the leader election problem: "Holes" refers to generally connected systems. n is the number of particles. L_{max} is the length of the longest border and L is the length of the outer border. [12] also computes local identifiers and consider other lattice models. Its model of the scheduler is one that does not allow simultaneous computations for particles at distance 2 or less. It is strictly stronger that the weak scheduler of this paper but weaker than the stronger scheduler of [9]. The running time of [12] is worst-case linear. It is a function of the the radius $r(G)$ and largest height $mtree(G)$ of a subtree of the particles graph G. [9] uses the results of this paper (previously reported in a brief announcement) in its solution. Both [12] and [9] always elect a unique leader, but use a strong scheduler (see discussion above).

Paper	Deterministic	Holes	Weak scheduler	Stationary	No chirality	Time
[8]	✗	✓	✓	✓	✗	$O(L_{max})$ expected
[4]	✗	✓	✓	✓	✗	$O(L)$ w.h.p.
[16]	✓	✗	✓	✗	✓	$O(n^2)$
[12]	✓	✗	✗	✓	✗	$r(G) + mtree(G) + 1$
[9]	✓	✓	✗	✗	✓	$O(Ln^2)$
This paper	✓	✓	✓	✓	✗	$O(n^2)$

leaders each of which is associated with a *stretch*, a structure that we introduce in this paper. A stretch is a sequence of adjacent border nodes with a *head* and a finite *count* that summarizes a relevant geometric property of the nodes forming the stretch[1]. Stretches go through a stretch expansion phase in which stretches expand to merge with other stretches. Stretches operate independently and asynchronously until termination is detected at which point either one stretch remains on the outer border and its head is elected as the leader or a finite number of symmetric stretches remain on the outer border in which case their heads are the candidate leaders. After the candidate leaders are determined, the solution proceeds as in [16]. Candidate leaders grow trees that are then compared to break symmetry. If breaking symmetry is not possible, then we show that deterministic leader election is not possible. Unlike [16], in which candidate leaders are adjacent, in our setting, candidate leaders are on the outer border and coordinating their actions for tree comparison requires more care. Our solution assumes common chirality but does not require particle movement. Other solutions that do not assume common chirality use particle movement [16] or particle movement together with stronger assumptions on the scheduler [9] to agree on chirality. Emek et al. [9] showed that the results of this paper, which were previously reported in a brief announcement, can be used (as a subroutine) in developing a deterministic leader election algorithm that always succeeds in finding a leader. They use the strong scheduler model which is strictly stronger than our model. They proposed an interesting deterministic algorithm that always elects a unique leader. The solution takes advantage of particle movement and the strong scheduler.

---

[1] We should note here that the concept of a border *node* is a logical construct introduced by this paper and is different from particles.

The rest of the paper is organized as follows. Section 2 presents the system model. Section 3 gives an overview of our solution. Section 4 presents the leader election algorithm on the outer border. Section 5 presents the algorithm for electing a unique leader or determining that the system has symmetry that prevents deterministic leader election. Section 6 concludes the paper.

2   System Model

The geometric Amoebot model [5] is an amoeba-inspired model for programmable matter, in which anonymous particles with finite memory occupy cells in a hexagonal lattice. We assume that particles have the same sense of rotational direction (chirality). In the model, particles can occupy one cell (contracted) or two cells (expanded) and no cell can be occupied by more than one particle. Without loss of generality we consider a system in which each particle initially occupies one cell. Since our leader election algorithm does not involve any expansion, each particle has six ports ordered clockwise from port 0 through port 5, one port on each of the adjacent cells. The ports are used to communicate with other particles in adjoining cells and to sense if an adjacent cell is occupied by another particle or is empty. A communication edge between two particles consists of a pair of corresponding ports. For example, between cells A and B in Fig. 1, port 3 of A and port 4 of B form a communication edge between A and B. Since we are considering a system of contracted particles, there is a unique edge between any two adjacent particles. Two adjacent particles know the port numbers that form the edge between them. Particles communicate by writing to their local memory and reading the local memory of adjacent particles. These actions are atomic, but a particle cannot read the memory of adjacent particles and write to its own memory in one atomic step. Reading memory of adjacent particles and writing to local memory allows for a simple message passing between particles in which adjacent particles can have a sequence of distinct messages using an alternating bit or a finite sliding window protocol [2]. We assume that the system graph whose vertices are the particles with edges between vertices corresponding to adjacent particles is connected, but, unlike [16], we do not assume that the graph is simply connected.

In the algorithms we present, we introduce six virtual nodes for each particle, one node per port, ordered clockwise from 0 to 5. These nodes are represented in Fig. 1 by black dots at the vertices of the cells occupied by the particle. Port i is on the edge between node $i - 1 \mod 6$ and node i. It follows that for two adjacent particles, two nodes of one particle overlap with two nodes of the other particle. We treat these overlapping nodes as one and implicitly assume in the algorithm that the two adjacent particles treat them as such. For example, in Fig. 1, node 3 of A overlaps with node 5 of B and node 4 of A overlaps with node 4 of B. When running an algorithm that requires node 3 of A to send a message to the next node in clockwise order on the border, A sends a message B requesting that the node corresponding to node 3 of A (node 5 of B) sends the

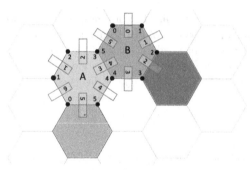

Fig. 1. Particle system surrounded by unoccupied cells

message to the next node (on the border) in the clockwise direction and B will send the message to its own node 6. In the pseudocode we present, we omit the details of this *handover* and treat each node including the overlapping nodes as independent.

Nodes execute steps when they are activated by the scheduler. Nodes can be activated in any order and a node can be activated any finite number of times before another node is activated. In a *step*, a node can read the local memory of adjacent nodes, do some internal computation or write to the memory of adjacent nodes, but cannot in one atomic step read and write to the memory of adjacent nodes. A *round* is a sequence of steps in which each particles executes at least one step. We measure the complexity of the algorithms in terms of the number of rounds that they require.

3   Leader Election: Overview

The algorithm has two main phases. In the first phase, a small number of candidate leaders (1, 2, 3, or 6) are selected on the unique outer border of the system and in the second phase further reduction of this number is attempted. If the algorithm does not elect a unique leader in the second phase, we show that the system must have symmetry that prevents deterministic leader election.

In the first phase, the deterministic leader election algorithm starts by running separate instances of a *border leader election* algorithm on all the borders of the system (a node is on a border if it is adjacent to an empty cell). An instance of a border leader election algorithm is designed to work correctly if the participants in the election consist of all nodes of a border. On the inner borders, if any, it is guaranteed that no leader is elected. On the unique outer border, 1, 2, 3, or 6 candidate leaders are selected. Each leader candidate also has what we call a *stretch*, a sequence of contiguous nodes, associated with it. If there is a unique leader, the algorithm terminates, otherwise, the outer border has symmetry that prevents this phase from electing a unique leader. This initial phase is the more involved phase of the algorithm and is done in a sequence of phases that are not strictly synchronized. It is the phase that allows us to deal with systems with holes. Unlike the relatively simple, and elegant, *erosion* phase of [16], our first phase requires subtle coordination between particles on the outer border. In the second phase, the candidate leaders attempt to further reduce their numbers. We use the tree comparison approach of [16]. Each particle with a leader node tries to recruit as many particles as it can to form a tree with the particle itself as the root of the tree. After all particles in the systems have joined a tree, each root compares its tree to the tree of the root to its *right* on the outer border according to an order relation. Every candidate leader then shares the results of these comparisons will all other candidate leaders on the border. If the results of all these comparison are equality, then there is symmetry in the system and deterministic leader election is not possible. If the result of one of the comparisons is inequality, then one or more candidate leaders are eliminated and the process is repeated a constant number of times until either there is one unique remaining leader or there are multiple leaders who are all tied in the tree comparison. If there is a unique leader, we are done, otherwise, there is symmetry that prevents deterministic leader election. We should note that the tree comparison requires more coordination in our case because the candidate leaders are not adjacent as is the case in [16].

4   Leaders on the Outer Border

This phase operates in a sequence of stages that are not strictly synchronized. The first two stages identify border nodes and execute an initial labeling of border nodes. The other two stages, stretch expansion and termination detection are more involved.

4.1   Border Nodes and Vertex Labeling

1. **Identifying border nodes.** In the initialization phase, each particle senses its surrounding to determine if one or more cells around it are unoccupied. A particle can be on more than one border, but each node can be on at most one border. In this step, the nodes of the particle that are adjacent to an unoccupied cell are identified and their successors and predecessors on the border are identified. Border nodes that are overlapping between adjacent particles are also identified (nodes 3 and 5 in Fig. 1 for example). These overlapping nodes are treated as one node in what follows.
2. **Initial vertex labeling.** Each particle labels its nodes with a unary label which is $+1$ for border nodes that belong to only one particle and -1 for border nodes that are shared between adjacent particles (Fig. 2).

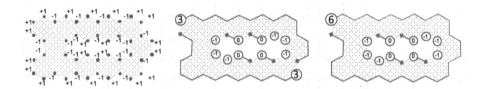

Fig. 2. Vertex labeling and initialization of stretches

Fig. 3. Intermediate step with two stretches remaining on the outer border

Fig. 4. Final configuration: termination detected on the outer border

---

Algorithm 1. Stretch Expansion

1: **function** ATTEMPTEXPANSION()
2:      ▷ s and s' are two adjacent stretches. s' is to the right of s.
3:      **if** $s.count > 0 \land s > s' \land s.count + s'.count \leq 6$ **then**
4:          MERGE(s, s')
5:      **else if** $s =_{lex} s' \land s.count = 1, 2, 3,$ or 6 **then**
6:          DETECTTERMINATION()   ▷ initiate termination detection
7: _____
8: **function** MERGE(s, s')
9:      $merge\_pending \leftarrow$ **true**
10:     send MERGE_REQUEST message to s'
11:     **if** MERGE_ACK message received **then**   ▷ ack can only be from s'
12:         $s.tail.successor \leftarrow s'.head$ ▷ Link s' to s
13:         $s'.head.predecessor \leftarrow s.tail$
14:         $s.count \leftarrow s.count + s'.count$ ▷ s' is no longer a leader candidate
15:         $s'.count \leftarrow 0$
16:         $merge\_pending \leftarrow$ **false**
17:     **else if** MERGE_NACK message received **then**   ▷ nack also from s'
18:         $merge\_pending \leftarrow$ **false**

4.2   Stretch Expansion

After labeling each node on the outer border, nodes initialize stretches and start stretch expansion. A stretch is a sequence of contiguous nodes. Initially, all nodes on the outer border are stretches of size 1 with a counter equal to their unary label. After stretch initializations, each border (inner or outer) will be completely covered by stretches. The leftmost node in a stretch is considered the leader (or head) of the stretch. The rightmost node in a stretch is called the tail of a stretch. Within a stretch, each node has a predecessor and a successor pointer that point to the successor and predecessor nodes within the stretch. The leader node's predecessor pointer is null and the tail node's successor pointer is null. Every other node in the stretch points to its predecessor and to its successor. The leader of a stretch maintains a counter which is equal to the sum of the unary labels of the nodes in the stretch. The counter value never exceeds 6.

Figures 3 and 4 illustrate stretches. The node with the integer value is the head with the stretch count and the tail is the node with a diamond. We note here that the sum of all counters of all stretches on the outer border is guaranteed to be equal at all times to $+6$. The sum is -6 on each inner border (if any).

Stretches interact with one another and are eliminated through stretch expansion. Stretches can expand by merging with other adjacent stretches to their right (clockwise). This is shown in Algorithm 1. When two stretches merge, the leader of the stretch on the left (s in the algorithm) becomes the leader of the resulting stretch and its new count is the sum of its old count and the count of the stretch being merged into. A merge is allowed only if the sum of the two counts is less than or equal to 6. In the algorithm $s > s'$ means that the count of s is larger than that of s' or the two counts are equal but s is lexicographically larger than s' (the sequence of unary labels of s viewed as a string is lexicographically larger than that of s'). To avoid deadlocks, we require that $s > s'$ for the merge to occur. For s to merge with s', s must receive an explicit ack message from s', at which point the merge is executed, or an explicit nack, at which point a new attempt at stretch expansion can be made. The code does not explicitly show the sending of ack and nack messages. The rule is very simple. Each stretch head maintains a boolean variable *merge_pending* to indicate if it has a pending merge request. If a stretch has a pending merge request, it rejects every merge request it receives, otherwise it accepts the merge request. We can show that before termination is detected (see below), there will always be one stretch that does not reject a merge request, thereby ensuring progress. If the condition for merge does not hold, and s and s' are lexicographically equal and have a count of 1, 2, 3, or 6, then termination detection is initiated. In fact, in this case, it is possible that two, or more, adjacent, lexicographically equal stretches (and hence geometrically congruent), encompass the entire outer border in which case leader election would not be possible on the outer border. A stretch that is being compared lexicographically to another stretch will not necessarily refuse the merge request but will do CLEANUP() (Algorithm 2) before accepting the request.

Lexicographic comparison is done by iteratively comparing the unary labels of the nodes of each stretch, starting with the head and finishing with the tail. This process is driven by the leader of s who repeatedly requests the next unary label of a node in its stretch and requests from the leader of s' the next unary label of a the next node in s' using the RETRIEVENEXTLABEL( ) function (Algorithm 2). If two stretches s and s' are lexicographically equal, we write $s =_{lex} s'$ After lexicographic comparison, the CLEANUP() function (Algorithm 2) removes any auxiliary state information related to the comparison.

If s and s' are lexicographically equal, it is possible that the outer border consists of identical stretches that cover it completely and that there is symmetry that prevents further stretch expansion. To determine if this is the case, termination detection is initiated. Termination detection is also needed if there is a single stretch on the outer border with a count of 6.

Algorithm 2. Lexicographic Comparison: Supporting Functions

1: **function** RETRIEVENEXTLABEL()
2:      ▷ Initially all values of $p.retrieved$ are false.
3:      $p \leftarrow s.head$
4:      **while** $p.retrieved \wedge p \neq s.tail$ **do**   ▷ Find node with non-retrieved label
5:          $p \leftarrow p.successor$
6:      **if** $p.retrieved$ **then**   ▷ If all nodes are retrieved
7:          **return** $END\_OF\_STRETCH$ to $s.head$
8:      **else**
9:          $p.retrieved \leftarrow$ **true**
10:          **send** $p.label$ to $s.head$ ▷ pass $p.label$ to $p.predecessor$
11:   ▷ until it reaches $s.head$
12: _____
13: **function** CLEANUP()
14:      $p \leftarrow s.head$
15:      **while** $\neg p.successor$ **do**
16:          $p.retreived \leftarrow$ **false**
17:          $p \leftarrow p.successor$

Algorithm 3. Termination Detection

1: **function** DETECTTERMINATION()
2:      $terminate \leftarrow$ **true**
3:      **for** $i \leftarrow 1, 6/s.count$ **do**
4:          $s' \leftarrow s$
5:          **for** $j \leftarrow 1, i - 1$ **do**
6:              $s' \leftarrow s'.left$ ▷ Rotate to the stretch left of s'
7:              $terminate \leftarrow (terminate \wedge (s'.count = s.count))$
8:          $s' \leftarrow s'.left$
9:          $terminate \leftarrow (terminate \wedge (s'.count = s.count)) \wedge s' =_{lex} s$
10:      **return** $terminate$

4.3   Termination Detection

To detect termination, a stretch attempts to establish that the whole border on which it resides is covered with $k + 1$ identical stretches that have the same positive count ($k = 1, 2, 3,$ or 6). It relies on the following theorem that can be shown to hold for the Amoebot model:

Theorem 1 (Termination detection theorem). *If there exists a sequence of $1 + 6/c$ adjacent and lexicographically equal stretches whose common count is c, $c = 1, 2, 3$ or 6, then the border is fully covered by these stretches and the last stretch in the sequence is also the first stretch in the sequence.*

Since the sum of all counters of stretches on each inner border is always -6, the condition for termination detection cannot hold on any inner border and can only hold on the unique outer border. Termination detection is done counterclockwise, opposite to the direction of merges (Algorithm 3). In the algorithm, s

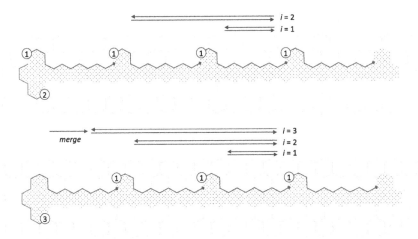

Fig. 5. Merging stretches during termination detection

is a stretch with count c ($c = 1, 2, 3$, or 6) attempting termination detection. If s is lexicographically identical to k/c stretches to its left, then it detects termination. In each iteration, the algorithm also checks (in an inner loop) that the counts of the first $i - 1$ previously checked stretches have not changed. These checks ensure that previously checked stretches indeed have not changed and, therefore, if a stretch detects termination, then the termination condition indeed holds. To implement the pseudocode by the particles at a low level, we still need to specify how to handle: (1) overlapping termination messages and responses and (2) merges happening during termination detection.

4.4   Overlapping Termination Detection Messages

In the discussion that follows, we assume that stretches do not merge during termination detection and we only concentrate on overlapping termination detection messages and responses. For such messages, we use a time to live (*TTL*) for the message and keep track of the stretches that are traversed in a *traversed* value. The *TTL* value specifies how many stretches the message needs to travel. The *traversed* value keeps track of how many stretches the message has traveled. The value of *traversed* is incremented every time they are handled by a stretch head (in either direction). A recipient knows that it is receiving a message addressed to it if the message's *TTL* and *traversed* are equal. In addition to the basic operation, we allow a message to not reach its destination because it encounters an earlier condition that necessitates short-circuiting the handling of the message. The response message has a *TTL* equal to the value of *traversed* to ensure that the sender receives the response. Since termination detection messages have a finite *TTL* (no more than 6), messages originating from different stretches will be received by their intended recipients (again we assume here that stretches are not changing). Next we discuss how to handle stretch merges that occur during termination detection.

4.5   Stretches Merging During Termination Detection

Since termination detection can be initiated while the stretches are still changing, it is possible that stretches that have been checked in loop i of the algorithm (Algorithm 3) have been merged with other stretches when loop $i+1$ is executed.

We consider the case shown in Fig. 5 as an example. In the upper part of the figure, we illustrate two rounds of the termination detection loop in which A established that $A =_{lex} A_1$ and then that $A =_{lex} A_2$. In general, it possible that a previously checked stretch is absorbed by another stretch. In Algorithm 3, we guard against this by checking that the counters of the first $i - 1$ stretches are unchanged. An unchanged counter ensures that the stretch itself is unchanged because only stretches with positive count (1, 2, 3, or 6) initiate termination detection, so previously checked stretches must have positive count and if a stretch with a positive count gets absorbed by another stretch, the count of the resulting stretch will be different from the count of the stretch that got absorbed. Checking for counter change is straightforward. When the stretch header of A_1 handles a termination detection message addressed to A_2, it checks that the *counter* value in the message data is the same as its own counter value. Otherwise, the termination detection message is short-circuited (prevented from reaching its destination) and a response to that effect is sent to the initiator. So, we conclude that if a message intended to A_2 actually reaches A_2, then all stretches between A and A_2 have not changed. Another possibility is that a merge message has already been sent by A_2 and not yet received by A_1 (lower part of Fig. 5). In this case, the termination detection message is guaranteed to cross paths with the merge message at some node The *shortcircuit* predicate will also hold and a negative response is sent back to the initiator. What makes this work is the fact that *merge* messages travel to the right and termination detection messages from an initiator travel to the left.

One last concern is whether the stretch head that initiates a termination detection message is guaranteed to still be the head of the same stretch when the response is sent back. This will not be the case if the stretch of the initiator is absorbed by another stretch to its right. We argue that this cannot happen without the termination detection message having encountered the merge message that will eliminate the initiator. There are two cases to consider: (1) the initiator already checked that its stretch is lexicographically equal to the stretch to its left, or (2) the initiator has not finished checking that it is lexicographically equal to the stretch to its left. In the first case, the stretch to the left of the initiator and that is lexicographically equal to it will not initiate a merge. The merge can only happen if the stretch to the left of the initiator is first merged with another stretch. In that case, the initiator message will either discover the changed count or will encounter the merge message as explained above. In the second case, the merge message and the initiator message will occur between the initiator and the head of the stretch to its left. In this case the *traversed* value on the termination detection message must be 0. If the termination detection messages encounters a merge message when the value of *traversed* is 0, the response message is simply not sent.

4.6 Progress

We can show that if the condition for termination detection does not hold, two stretches on the outer border will merge. Since the number of stretches is finite, eventually the condition for termination detection will hold.

5 Global Leader Election

Once termination has been detected and verified on the outer border, the stretches attempt to elect a unique leader. If there is one remaining stretch, the particle that contains the head node of the stretch is the leader (the head node of a stretch with positive count cannot be shared by two particles). If there are 2, 3 or 6 remaining leader candidates, a new global leader election phase is needed. The goal of this phase is to elect a single leader, if possible. In what follows we assume that particles that have more than one candidate leader node first reduce the total number of candidates by eliminating all but one of the candidate leader nodes that they have, so all candidate leader nodes will be on separate particles.

5.1 Trees to Break Symmetry

To elect a leader or determine that there is symmetry that prevents deterministic leader election, we adopt the tree-growing approach of [16]. The idea is to have each particle containing a leader grow a tree that includes other particles in the system with itself as the root until every particle in the system is included in one of the trees. These trees are compared using a *tree comparison operator* (implicitly defined in [16]) to break symmetry and reduce the number of candidates. We refer the reader to [16] for the details of tree building and comparison. The two crucial properties that we need and that are satisfied by the trees and the comparison on them as defined in [16] are the following:

1. **Trees and their neighborhoods are congruent (strong congruence).**
   If two trees are equal according to the comparison, then there is a mapping between nodes of the two trees such that the neighborhoods of a node and the node it is mapped to *look the same*. More formally, if two trees T and T' are equal according to the comparison then there exists a mapping $\mathcal{I}$ from the set of cells occupied by T and the cells that are adjacent to those cells (whether occupied or not) to the cells of the system such that: (1) the cell occupied by the root of T maps to the cell occupied by the root of T'; (2) cells that are occupied by particles of T maps to cells that are occupied by particles of T'; (3) unoccupied cells map to unoccupied cells; (4) if two cells

c_{p_1} and c_{p_2} are occupied respectively by two particles p_1 and p_2 of T such that p_1 is a parent/child of p_2, then $\mathcal{I}(c_{p_1})$ is a parent/child particle of $\mathcal{I}(c_{p_2})$; (5) and, for every cell c_p that is occupied by a particle p of T, there exists an ordering of the 6 cells $c_1, c_2, \ldots, c_6$ that are adjacent to c_p and the 6 cells $c'_1, c'_2, \ldots, c'_6$ that are adjacent to $\mathcal{I}(c_p)$ such that $\mathcal{I}(c_i) = c'_i$.

2. **Order relation on trees.** The tree comparison operator defines an order relation on trees.

The properties of the mapping are guaranteed through the use of *neighborhood codes* [16]. In our setting we need more involved coordination for comparing the trees because the leader candidates are not adjacent as in [16] and we can have up to 6 candidates instead of 3. Also, since leaders are not adjacent, we require a particle on a stretch to be added to the tree of its leader or the closest leader to its left and that for equal trees stretches map to each other. This ensures that if all trees are equal they can be mapped to one another by rotation.

5.2   Coordination for Breaking Symmetry

Comparing the trees and reducing the number of leader candidates might take more than one iteration (phases) to conclude and coordination between the leader candidates is needed to be able to proceed from one phase to the next.

A leader candidate can send a message to all other candidates by setting a TTL equal to the number of candidates and sending the message in one direction around the border. Once the sender receives the message back with $TTL = traversed$, it knows that every other candidate has received its message. This is not enough to go from one phase to the next. What is needed is the knowledge that all leader candidates have finished the previous phase. This can be easily established once an end-of-phase message is received from every other candidate (the number of candidates in a given phase is known). There is no concern for messages from different candidates to be mixed up because end-of-phase messages received by a particular candidate will have different $TTLs$ depending on the initiator of the message. So, if there are k leader candidates and every one of them sends an end-of-phase message with TTL equal to k, a leader candidate knows that all other candidates finished the previous phase once it receives end-of-phase messages with $TTLs$ 1 through k.

After growing the trees, each leader compares its tree to that of the leader to its right. The result of the comparison can be $<$, $=$, or $>$. Every leader sends the results of its comparison to all other leaders. Once all leaders have received results of comparisons from all other leaders, they send a message to every other leader to that effect. If all comparisons results are $=$, we show below that the system's symmetry prevents deterministic leader election. If one comparison result is not $=$, then we would like all leaders to agree on some leaders to be eliminated. The leader candidates to be eliminated are determined as follows. Each leader calculates maximal non-descending sequences of leaders ordered in clockwise direction by the results of tree comparisons. The candidates that are eliminated are those candidates that are the leftmost elements of those sequences.

For example, if the results of the comparison starting from a given leader l_1 are: $l_1 = l_2$, $l_2 = l_3$, $l_3 < l_4$, $l_4 > l_5$, $l_5 = l_6$, and $l_6 < l_1$, then l_5 should be eliminated because it is the leftmost element of the maximal non-descending sequence $l_5 = l_6 < l_1 = l_2 = l_3 < l_4$. All leader candidates calculate the same non descending sequences and choose the same candidates to be eliminated (defined by relative TTL distance). So, at the end of this step we know that either the number of leaders will stay the same and there is symmetry or that the number of leaders will be reduced. We do not have to worry about all leaders being eliminated because the tree comparison operator induces a total order and the largest tree will not be eliminated. Once a leader elimination phase ends, the remaining leaders resume from the tree-growing phase and after the trees are grown, they proceed to the tree comparison phase. This process continues until no further candidates can be eliminated at which time either there is a unique leader in the system or there are multiple leaders left whose trees are strongly congruent.

5.3   Impossibility of Leader Election with k-symmetry $(k > 1)$

If the algorithm terminates with $k > 1$ leaders whose trees are strongly congruent, we show that deterministic and stationary election of less than k candidate leaders is not possible in the system. The idea of the proof is simple. We divide the particles into k groups of equal size, say l, corresponding to the particles in each tree. We choose one of the trees, say T_1, and we denote the j'th particle in the depth order traversal of T_1 with p_{1j}, $1 \leq j \leq l$. For the other trees, we number the particles so that the j'th particle of a tree is the image of the j's particle of T_1 under the strong congruence mapping. We give a schedule of activations of the particles so that after any number of rounds, particles p_{ij} has the same state as particle $p_{i'j}$, $1 \leq i, i' \leq k$ and $1 \leq j \leq l$. It follows that if any particle in one group is a leader, $k - 1$ other particles in the other groups are also leaders. The schedule is the following: all particles read the memory of their neighbors, then all particles write to their own memory and the cycle repeats any number of times. It is straightforward to show that if the state of p_{ij} and $p_{i'j}$, $1 \leq i, i' \leq k$ and $1 \leq j \leq l$, are the same at the beginning of a cycle of activations, their states will be the same at the end of the cycle.

6   Conclusion

In this paper we established the conditions under which stationary and deterministic leader election is possible in generally-connected particle systems with a weak scheduler and in which particles have common chirality: (1) leader election is possible if and only if the system does not have k-symmetry ($k = 2, 3,$ or 6) (the only possible symmetries greater than one in the hexagonal grid), and (2) if the system has k-symmetry ($k = 2, 3,$ or 6), then the election of k candidate leaders is possible and the election of less than k leaders is not possible. The paper leaves open a number of interesting problems. If we allow a strong scheduler, then leader election can be possible in the presence of k-symmetry $(k > 1)$ even without movement. For example, in a system of two adjacent particles (a system with 2-symmetry), each particle can have a flag, initially false, to indicate if it is the leader. The first particle to be activated can atomically read the flag of the other particle and declare itself a leader if the other particle is not already a leader. This is clearly more than what can be achieved in the presence of a weak scheduler. Characterizing what is possible to achieve in the presence of a strong scheduler but without movement is an open problem. Another interesting open problem is deterministic and stationary leader election for systems with a weak scheduler and no common chirality; we conjecture that the election of a finite number of leaders is possible by building on the results of this paper.

Acknowledgements. We would like to thank Shay Kutten for helpful discussions about the topic of this paper and for suggesting the example with two particles that we used in the conclusion section.

References

1. Abu-Amara, H.H.: Fault-tolerant distributed algorithms for agreement and election. Ph.D. thesis, University of Illinois, Champaign, IL, USA (1988)
2. Afek, Y., Brown, G.M.: Self-stabilization of the alternating-bit protocol. In: SRDS 1989, pp. 80–83. IEEE (1989)
3. Cannon, S., Daymude, J.J., Randall, D., Richa, A.W.: A Markov chain algorithm for compression in self-organizing particle systems. In: PODC, pp. 279–288. ACM (2016)
4. Daymude, J.J., Gmyr, R., Richa, A.W., Scheideler, C., Strothmann, T.: Improved leader election for self-organizing programmable matter. In: Fernández Anta, A., Jurdzinski, T., Mosteiro, M.A., Zhang, Y. (eds.) ALGOSENSORS 2017. LNCS, vol. 10718, pp. 127–140. Springer, Cham (2017). https://doi.org/10.1007/978-3-319-72751-6_10
5. Derakhshandeh, Z., Dolev, S., Gmyr, R., Richa, A.W., Scheideler, C., Strothmann, T.: Brief announcement: amoebot-a new model for programmable matter. In: SPAA, pp. 220–222. ACM (2014)
6. Derakhshandeh, Z., Gmyr, R., Porter, A., Richa, A.W., Scheideler, C., Strothmann, T.: On the runtime of universal coating for programmable matter. In: Rondelez, Y., Woods, D. (eds.) DNA 2016. LNCS, vol. 9818, pp. 148–164. Springer, Cham (2016). https://doi.org/10.1007/978-3-319-43994-5_10
7. Derakhshandeh, Z., Gmyr, R., Richa, A.W., Scheideler, C., Strothmann, T.: Universal coating for programmable matter. Theoret. Comput. Sci. **671**, 56–68 (2017)
8. Derakhshandeh, Z., Gmyr, R., Strothmann, T., Bazzi, R.A., Richa, A.W., Scheideler, C.: Leader election and shape formation with self-organizing programmable matter. Comput. Mol. Program. DNA **21**, 117–132 (2015)
9. Emek, Y., Kutten, S., Lavi, R., Moses Jr., W.K.: Deterministic leader election in programmable matter. In: ICALP, 9–12 July, pp. 140:1–140:14 (2019)
10. Flocchini, P., Prencipe, G., Santoro, N.: Distributed computing by oblivious mobile robots. Synth. Lect. Distrib. Comput. Theory **3**(2), 1–185 (2012)
11. Flocchini, P., Prencipe, G., Santoro, N., Widmayer, P.: Arbitrary pattern formation by asynchronous, anonymous, oblivious robots. Theoret. Comput. Sci. **407**(1–3), 412–447 (2008)

12. Gastineau, N., Abdou, W., Mbarek, N., Togni, O.: Distributed leader election and computation of local identifiers for programmable matter. In: Gilbert, S., Hughes, D., Krishnamachari, B. (eds.) ALGOSENSORS 2018. LNCS, vol. 11410, pp. 159–179. Springer, Cham (2019). https://doi.org/10.1007/978-3-030-14094-6_11
13. Ghaffari, M., Haeupler, B.: Near optimal leader election in multi-hop radio networks. In: SODA, pp. 748–766 (2013)
14. Itai, A., Rodeh, M.: Symmetry breaking in distributed networks. Inf. Comput. **88**(1), 60–87 (1990)
15. Karpov, V., Karpova, I.: Leader election algorithms for static swarms. Biol. Inspired Cogn. Archit. **12**, 54–64 (2015)
16. Luna, G.A.D., Flocchini, P., Santoro, N., Viglietta, G., Yamauchi, Y.: Shape formation by programmable particles. In: 21st International Conference on Principles of Distributed Systems, OPODIS 2017, pp. 31:1–31:16 (2017)
17. Lynch, N.A.: Distributed Algorithms. Elsevier, Amsterdam (1996)
18. Peleg, D.: Time-optimal leader election in general networks. J. Parallel Distrib. Comput. **8**(1), 96–99 (1990)
19. Styer, E.F.: Symmetry Breaking on networks of processes. Ph.D. thesis, Georgia Institute of Technology, Atlanta, GA, USA (1989)

Robust Privacy-Preserving Gossip Averaging

Amaury Bouchra Pilet[1,2(✉)], Davide Frey[1], and Francois Taiani[1]

[1] Univ Rennes, Inria, CNRS, IRISA, Rennes, France
Amaury.Bouchra-Pilet@IRISA.fr
[2] École Normale Supérieure «Ulm», Paris, France

Abstract. Decentralized solutions are emerging as promising candidates to overcome the privacy risks associated with centralized data services. Such solutions suffer however from their own range of privacy vulnerabilities, arising from untrusted and malicious peers. In this paper, we consider the emblematic problem of privacy-preserving decentralized averaging, and propose a novel gossip protocol that exchanges noise for several rounds before starting to exchange actual data. This makes it hard for an honest but curious attacker to know whether a user is transmitting noise or actual data. Our protocol and analysis do not assume a lock-step execution, and demonstrate improved resilience to colluding attackers. We prove the correctness of this protocol as well as several privacy results. Finally, we provide simulation results about the efficiency of our averaging protocol.

1 Introduction

The recent evolution of applications like the Internet of Things has fostered interest in protocols that enable large networks of devices to perform collaborative computations on their own data. For these protocols, privacy protection acquires paramount importance, since the data being processed may be personal or otherwise confidential, e.g. location or medical data. While it is possible to simply centralize data processing, centralization raises privacy and durability issues. The provider of a centralized service has access to the personal data of all users and may cut off the service at any time. To prevent these issues, several authors have proposed decentralized solutions to the problem of data aggregation [8,20], including peer-to-peer protocols that compute the average of values initially held by individual peers, an important topic in a range of statistical and machine-learning applications [7].

These peer-to-peer solutions remove the need for a central server, and the risk of being spied by the server's operator or by third parties that may obtain access to server-side data. While this goes in the direction of privacy protection, these algorithms also have serious disadvantages for privacy. They require users to send information to unknown peers, which may allow not only big companies or governmental agencies, but also criminal organizations or "curious" people to

M. Ghaffari et al. (Eds.): SSS 2019, LNCS 11914, pp. 38–52, 2019.
https://doi.org/10.1007/978-3-030-34992-9_4

access personal data rather easily, without having to compromise heavily secured servers and communications.

In order to overcome this central weakness of peer-to-peer averaging protocols, several algorithms have been proposed that allow gossip averaging while protecting their users' privacy [1,2,8,26]. These algorithms unfortunately suffer from a number of limitations that expose them to some eavesdropping attacks [1], constrain how peers must coordinate their exchanges [8,23,26], or require the use of costly cryptographic primitives such as homomorphic encryption [19].

In this paper, we propose to overcome these limitations with a novel peer-to-peer protocol for decentralized averaging. Following earlier solutions [1,26,28], our approach injects randomized values into the averaging process while ensuring deterministic convergence to the exact wanted value. In contrast to earlier attempts, the successive values exposed by an individual node in our protocol's early stage are independent of this node's initial value. Contrary to [26], we also do not assume a fully lock-step execution model to implement our protocol and perform our analysis.

Our design relies on a random peer sampling (RPS) service [21]. For our algorithm to attain its goals in a context where the RPS service itself may be attacked, this RPS service need to be resilient to attacks, especially attacks trying to isolate a specific peer, surrounding it by malicious peers. One example of such an attack-resilient RPS protocol is Brahms [4].

Overall, the random injection we present combined with random peer sampling removes the need for peers to tightly coordinate their actions, allows peers to protect their privacy without having to make explicit privacy protection requests to others, and improves resilience to attacks involving the use of numerous malicious peers controlled by an attacker. We evaluate our protocol by first proving its correctness as well as several privacy properties. We present simulation results in combination with two different peer sampling protocols.

2   System Model and Problem

System Model. We consider an asynchronous decentralized system consisting of a large number of peers $\{p_1, \ldots, p_i, \ldots, p_N\}$ that can communicate through message passing. We use the terms "node" and "peer" interchangeably throughout this paper. We assume the network is reliable (messages do not get lost), but we do not make any assumption regarding the synchrony of the network or of the execution at different peers. Messages may take an arbitrary (albeit finite) time to arrive, while the protocol's execution evolves independently at each peer. Peers synchronize only in pairs for the duration of a message exchange. A node involved in a message exchange with another node simply waits for a response or for a failure-detection timeout before engaging in other exchanges.

We assume peers have access to a random peer sampling service (RPS for short) [21] that provides each individual peer with a sample stream of other peers present in the network. For our analysis, we assume this RPS protocol is resilient to attempts to bias its results by individual peers [4].

```
 1  Function gAvg(val):
 2  │   while true do
 3  │   │   peer ← randomPeer()
 4  │   │   sendTo(peer, val)
 5  │   │   rcv ← recvFrom(peer)
 6  │   │   val ← val+rcv/2
 7  Function answer():
 8  │   rcv ← recvFrom(peer)
 9  │   sendTo(peer, val)
10  │   val ← val+rcv/2
```

Algorithm 1. Non-private Gossip Averaging

The Private Averaging Problem. Each peer p_i possesses a local initial value val_i and wants to compute the average value of all the peers in the system $\frac{1}{N}\sum_i val_i$, while giving other peers as little information as possible regarding its own local value. We consider honest but curious attackers in the sense that they observe exchanged values, but they do not inject fake values or try to prevent the algorithm from computing a correct result.

We consider the two kinds of attackers defined in [1], possibly coexisting in a single entity. *Edge attackers* eavesdrop on data exchanged by other peers; they may, for example, try to obtain a user's value by retrieving all the values s/he exchanges with other users. *Node attackers* use a set of peers under their control to get information from other peers.

Algorithm 1 [20] describes the classical approach to decentralized averaging (which does not protect privacy). Each peer starts with an initial value. Pairs of peers regularly exchange their local values (lines 4–5) and replace them with their average (line 6)—the function ANSWER is invoked when receiving a message sent using SENDTO. While this method ensures eventual convergence of all values to their average, provided that the network's graph is connected, peers have to expose their local values to potential strangers, thus raising critical privacy issues if this value is sensitive and other peers cannot be fully trusted. Our private-averaging problem consists in computing the same average while hiding initial values from edge and node attackers.

3   Privacy-Preserving Averaging

We address private averaging with a protocol that uses random values to protect the initial values of peers. Unlike some existing work [26], our protocol does not require synchronous lock-step rounds. Moreover, a peer can protect its value without any of its neighbors doing so and without explicitly informing anyone. In particular, we can have a mix of peers using privacy protection and peers not using it.

```
1  Function privGAvg(val, privLvl):
2  |    err ← 0
3  |    for i = 0 to i = privLvl − 1 do
4  |    |    peer ← randomPeer()
5  |    |    fakeVal ← rand()
6  |    |    err+ = val − fakeVal
7  |    |    sendTo(peer, fakeVal)
8  |    |    rcv ← recvFrom(peer)
9  |    |    val ← (fakeVal+rcv)/2
10 |    val+ = err
11 |    while true do
12 |    |    peer ← randomPeer()
13 |    |    sendTo(peer, val)
14 |    |    rcv ← recvFrom(peer)
15 |    |    val ← (val+rcv)/2
16 Function answer():
17 |    rcv ← recvFrom(peer)
18 |    if i < privLvl then
19 |    |    fakeVal ← rand()
20 |    |    err+ = val − fakeVal
21 |    |    sendTo(peer, fakeVal)
22 |    |    val ← (fakeVal+rcv)/2
23 |    sendTo(peer, val)
24 |    val ← (val+rcv)/2
```

Algorithm 2. Private Gossip Averaging

3.1 The Algorithm

The solution we proposed is described in Algorithm 2. (As in Algorithm 1 the function ANSWER is called upon receiving a message.)

Our proposal consists in adding a simple *privacy-generation phase* to the averaging algorithm. The privacy-generation phase (lines 2–10) behaves exactly in the same way as the averaging protocol. However, a peer performing this phase sends a random value *fakeVal* (line 7) instead of its current value to the selected peer. The difference between its current value and the sent random value being accumulated in a local error variable *err* (line 6).

After having performed the desired number of random exchanges (determined by the protocol parameter *privLvl* at line 3), the peer sums its accumulated error variable and its original value (line 10), and then it continues with the original averaging protocol by using this sum as its value (lines 11–15). Furthermore, all communication between peers is encrypted using standard techniques. So we assume that an eavesdropper may only know the time, the sender and the receiver of any communication, not its content.

Different termination conditions can be used with this algorithm: time based, communication-round based, value-change based, etc. The best termination condition being dependent on the considered use-case, and this paper being about private gossip averaging in general, we do not suggest any specific condition for our algorithm.

An example of the execution of the algorithm is shown in Fig. 1 on a toy example of four peers with $privLvl = 1$, assuming a static synchronous network for simplicity's sake. Each peer is represented by a circle with two numbers. The top number is the current value val held by the peer: its initial value at the start of the protocol (Step 1, top-left corner), which should converge arbitrarily close to the network's average (here 6) as the protocol progresses. (In this particular example, peers do converge to the exact average value at Step 6, bottom-right corner.) The bottom number if the error err progressively accumulated by each peer.

During the first two rounds (shown at Steps 2 and 3), the peers execute the privacy-generation phase of Algorithm 2. During this phase, peers exchange fully random values, while accumulating the resulting error in their local err variable. At Step 4, each peer executes line 10 of Algorithm 2 and corrects its initial value val with the error accumulated so far. Note how at each step the overall average value of the network remains 6. Finally, at Steps 5 and 6, our protocol executes a standard decentralized averaging protocol with the corrected values.

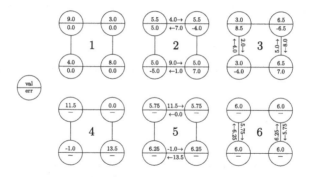

Fig. 1. Four peers executing our algorithm

3.2 Peer-Sampling Adjustments

As mentioned above, we assume a Byzantine-resilient peer sampling protocol [4]. This ensures that peers can select their communication partner from a uniformly random sample of the network. However, a malicious peer could still try to contact and exchange information with a target peer by bypassing the peer sampling mechanism. For this reason, we introduce an adjustment to connection establishment. When a peer receives a connection request from another one, A, it never answers directly. Rather, it waits for another peer B's contact and forwards to it the information received from A. Then it replies to A with the response received from B. This simple mechanism makes it difficult for an attacker to target a peer in order to monitor its exchanges.

4   Evaluation

We evaluate the performance of our averaging algorithm by means of a theoretical analysis and an experimental evaluation of its performances in conjunction with different peer sampling protocols.

4.1   Averaging Correctness

We start by proving that, in a classical gossip averaging algorithm as proposed in [20], if some of the values exchanged by peers are replaced by random values and, if later, an appropriate correction is done on the value of peers, then, the algorithm will converge the same way as without these operations. Let us consider a classical theoretical continuous-time model of gossip averaging where each peer is a vertex of a complete graph $K(V)$ (which we use to capture the RPS protocol we rely on). For conciseness, we note in the following $x_i(t)$ the value of the variable val_i of peer p_i at time t, and $y_{i,k}$ is the kth value *fakeVal* used by peer p_i.

Theorem 1 (Correctness Theorem). *Let $p_i \in \{p_0, \ldots, p_k\} \subseteq V$ be a peer, and let us replace its value, $x_i(t)$, by the values $y_{i,0}; \ldots; y_{i,r_i}$ at times $t_{i,0}; \ldots; t_{i,r_i}$ (i.e. $\forall_{0 \leqslant i \leqslant k} x_i(t_{i,j} + \varepsilon) = y_{i,j}$). If we later add (at time t_*) the value $\sum_{j=0}^{r_i}(x_i(t_{i,j}) - y_{i,j})$ to $x_i(t_*)$ (i.e. $\forall_{0 \leqslant i \leqslant k} x_i(t_* + \varepsilon) = x_i(t_*) + \sum_{j=0}^{r_i}(x_i(t_{i,j}) - y_{i,j}))$, then, executing a classical gossip averaging algorithm on this graph will ultimately make every vertex's value converge to the average of all initial values.*

Proof. Let us consider a single peer p_i and a single replacement occurring at time $t_{i,j}$, $j \in [0, r_i]$. In this transformation, we remove $x_i(t) - y_{i,j}$ from the value of x_i. This modifies the average of all values into $avg(t + \varepsilon) = avg(t) - \frac{x_i(t) - y_{i,j}}{|V|}$. Since averaging operations have no effect on the average value, if we later, at time t', add $x_i(t) - y_{i,j}$ to the value of any peer, we restore the original value of avg. Iterating this reasoning over all values of i and j proves the theorem.   □

Now we know that we can replace the value of a peer by a random value an arbitrary number of times. The averaging algorithm will still work, provided that we later add to the value of our peer the sum of all the differences between the value our peer had at the time of replacement and the random value. The correctness of our algorithm follows trivially.

Property 1 (Correctness property). *If the Gossip Privacy Protector algorithm is applied by any number of peers at the start of a classical gossip averaging algorithm, this does not change the value to which all peers will converge.*

The proof is a simple application of our Correctness Theorem.

4.2   Attack Resilience

We now show that our protocol has good privacy properties. We start by observing that by definition, all the random values sent by peers are independent of each other and of the original value of the peer. We then distinguish two types of attack: direct and indirect. Throughout the analysis, we let l (instead of $privLvl$) denote the privacy level chosen by the considered peer. In addition, we let $\tau = \frac{\#\text{corrupted peers}}{\#\text{peers}}$ represent the fraction of corrupted peers. Other than that, we use the same notation as in the algorithm (with time indexes added).

Direct Attacks. In direct attacks, the attacker tries to learn the value of a peer by communicating directly with it. We prove two properties in this context.

Property 2 (Deterministic Privacy Property)
An attacker needs to get all the random values sent to compute the exact original value of the peer.

Proof. We will use the same notation as in the algorithm (with time indexes added, and l instead of $privLvl$). The first non-random value is exactly the value exchanged at the l-th round, val_l.

$$val_l = val_{l-1} + err_{l-1} = \frac{fakeVal_{l-1} + rcv_{l-1}}{2} + \sum_{i=0}^{l-1}(val_i - fakeVal_i)$$

$$val_l = \frac{fakeVal_{l-1} + rcv_{l-1}}{2} + \sum_{i=1}^{l-1}\left(\frac{fakeVal_{i-1} + rcv_{i-1}}{2} - fakeVal_i\right) + val_0 - fakeVal_0$$

$$val_l = \sum_{i=0}^{l-1}\left(\frac{rcv_i - fakeVal_i}{2}\right) + val_0 \implies val_0 = val_l - \sum_{i=0}^{l-1}\left(\frac{rcv_i - fakeVal_i}{2}\right)$$

We see that computing the original value from the first non-random value requires knowledge of all the random values (and of the answers from contacted peers). All later non-random values having even more noise from other peers and no non-random values being transmitted before, this proves the property.   □

Property 3 (Probabilistic Privacy Property)
If the attacker lacks k random values (but not necessarily the associated answers), the best s/he can do is take their expected values (if known), with a level of uncertainty at least as large as that of guessing the value of $\frac{\sum_{i=0}^{k} fakeVal_i}{2}$.

Proof. Since peers generate independent random values, it is impossible to guess anything from them, except by using their relation to val_l.
Since $val_0 = val_l - \sum_{i=0}^{l-1}\left(\frac{rcv_i - fakeVal_i}{2}\right)$, if the attacker lacks k values, s/he has to guess them without any hint. Therefore, the best s/he can do consists in taking their expected values and nothing will give her a lower level of uncertainty than the natural level of uncertainty of these random variables.   □

From the Deterministic Privacy Property, we can derive that, if the peer sampling is effectively random, the chance that an attacker will get the exact original value of a peer is $\leqslant \tau^l$. See Fig. 2a for a plot.

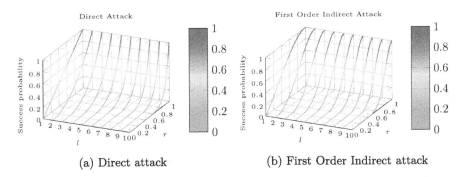

(a) Direct attack (b) First Order Indirect attack

Fig. 2. Success probabilities for direct and indirect attacks (upper bounds). l: privacy level, τ: fraction of corrupted peers

Indirect Attacks. In indirect attacks, the attacker exploits information exchanged between the target and other peers. For example, the attacker may guess the value of a peer p by observing that p communicates with another peer q whose value changes from val'_q to val''_q as a result of the exchange. This is only possible if the attacker is both a node and an edge attacker. The attacker requires fake peers to obtain the values of nodes, and eavesdroppers to know which values s/he needs. Also, note that the attacker needs to get values older than the ones s/he wants to obtain. Moreover, the attack will only work if all involved peers, but the target, have ended their privacy-generation phases since values sent by a peer generating privacy are all independent of each other. This implies that, if all peers start the process at the same time, it is unlikely that these attacks will be possible. Except for the very simple case of spying the target peer's neighbors, this kind of attack will probably be very impractical for most, if not all, cases on a real network.

Indirect attacks rely on the following principle: if peer i exchanges values with peer j at time t, it is possible to compute the exchanged values by knowing $v_j(t)$ (the value sent by j) and $v_j(t+1) = \frac{v_i(t)+v_j(t)}{2}$. This attack can be iterated: if the attacker lacks one, or both, of $v_j(t)$ and $v_j(t+1)$, s/he may get it using the same attack. We call this a Higher Order Indirect Attack, the base case being a First Order Indirect Attack.

First order indirect attacks give the attacker a second chance to get a value with respect to a direct attack, but require two contacts instead of one. This increases the upper bound of the probability of an exact evaluation from $\leqslant \tau^l$ to $\leqslant (\tau + (1-\tau)\tau^2))^l = (\tau + \tau^2 - \tau^3))^l$. See Fig. 2b for a plot.

Higher-order indirect attacks are very unlikely in practice. But from a theoretical point of view, they can be very powerful. We will prove two results showing the limitations of these attacks. We consider a node-edge attacker which is a universal eavesdropper (s/he can see all messages). The attacker may take an arbitrarily long (finite) time to get the desired information and may have to use arbitrarily old information. We state our first theorem under Assumption 1, which holds with high probability in the presence of a uniform random

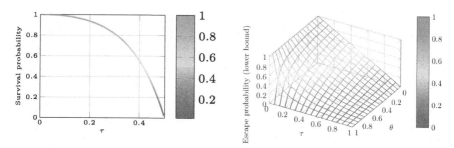

(a) Survival probability (lower bound) given by our HO Survival Theorem

(b) Escape probability (lower bound) given by our Higher Order Escape Theorem

Fig. 3. Lower bounds on survival and escape probabilities. τ: fraction of corrupted peers, θ fraction or unsafe edges

peer sampling over an infinite set of peers. The theorem states that as long as the assumption holds, the target can survive a higher-order indirect attack from a universal eavesdropper. We therefore refer to it as *Higher Order Survival Theorem*. We plot the theorem result in Fig. 3a.

Assumption 1. *There is neither a pair of consecutive exchanges between the same two peers, nor a pair of exchanges with one (or both) peer(s) in common separated by only one exchange.*

Theorem 2. (Higher Order Survival Theorem)
As long as Assumption 1 holds, and $\tau < 1/2$, a universal attacker will never obtain the value of the target peer with probability $\geqslant 1 - \frac{1-2\tau(1-\tau)-\sqrt{1-4\tau(1-\tau)}}{2(1-\tau)^2}$.

Proof. We model the tree of exchanges induced by the higher order attack by a Galton–Watson process. The number of descendants of an individual is the number of exchanges made just before and after the exchange corresponding to the individual which are not made with a corrupted peer. This gives the following probabilities: $p_0 = \tau^2$, $p_1 = 2\tau(1-\tau)$, $p_2 = (1-\tau)^2$. The average number of descendants is $m = p_1 + 2p_2 = 2\tau(1-\tau) + 2(1-\tau)^2$. Let us analyze the branching process' behavior as a function of τ.

$$m = 2\tau(1-\tau) + 2(1-\tau)^2 = 2\tau - 2\tau^2 + 2 - 4\tau + 2\tau^2 = 2 - 2\tau$$

So $m = 1 \Leftrightarrow 2\tau = 1 \Leftrightarrow \tau = 1/2$. Since $p_1 = 2\tau(1-\tau) = 2^{1/2^2} = 1/2 < 1$, the probability of extinction in the critical case is 1.

For the super-critical case, $m > 1$, let us analyze the generating function.

$$\varphi(s) = \sum_{n \geqslant 0} p_n s^n = p_2 s^2 + p_1 s + p_0 = (1-\tau)^2 s^2 + 2\tau(1-\tau)s + \tau^2$$

To solve $\varphi(s) = s$, we search for the roots r of the polynomial $\varphi(s) - s$.

$$\varphi(s) - s = (1-\tau)^2 s^2 + (2\tau(1-\tau) - 1)s + \tau^2$$

$$r = \frac{1 - 2\tau(1-\tau) \pm \sqrt{1-4\tau(1-\tau)}}{2(1-\tau)^2}, (0 \leqslant \tau < 1/2, 1 - 4\tau(1-\tau) > 0)$$

Of the two solutions, the lower one living in [1] is always $r = \frac{1-2\tau(1-\tau)-\sqrt{1-4\tau(1-\tau)}}{2(1-\tau)^2}$ (the other being always 1 for $0 \leqslant \tau < 1/2$). The probability of extinction (the attacker obtains the information) in the super-critical case is $\frac{1-2\tau(1-\tau)-\sqrt{1-4\tau(1-\tau)}}{2(1-\tau)^2}$, so, the probability of survival (the attacker does not obtain the information) is: $1 - \frac{1-2\tau(1-\tau)-\sqrt{1-4\tau(1-\tau)}}{2(1-\tau)^2}$.   □

The above theorem relies on a major assumption (Assumption 1) and does not provide anything for $\tau \geqslant 1/2$. We therefore introduce a second theorem that does not have these limitations but that is not applicable to universal eavesdroppers. Let $\theta = \frac{\#\text{unsafe edges}}{\#\text{edges}}$ in the sub-graph from which all corrupted peers have been removed. We assume $0 < \tau < 1$ (if not, there is either no attacker or the attacker controls the whole network) and $0 < \theta < 1$ (if not, we have either a universal eavesdropper or no eavesdropper at all). In this case we have no assumption that needs to remain valid for an extended period. Rather, the theorem states the probability that the target will escape from the attack. We therefore refer to this theorem as *Higher Order Escape Theorem* as opposed to survival. We plot the result in Fig. 3b.

Theorem 3 (Higher Order Escape Theorem)
There is a $\geqslant 1 - \frac{\tau}{1-\theta(1-\tau)}$ probability that the attacker will never get the value of an exchange.

Proof. This theorem relies on the fact that, if, at some time, a communication is made via a safe edge (which the eavesdropper cannot spy), then, the attacker will never know which peer to spy for the next step of the indirect attack. Since we do not assume that the same peer will not be randomly selected several times, and we only want a lower bound on the probability of success of an attack, we will consider a worst case, where only one new exchange needs to be captured at each step.

For the first step, the probability of the exchange not being captured is $1 - \tau$ and then, its probability of happening on a safe edge is $1 - \theta$, which gives a $(1 - \tau)(1 - \theta)$ probability that the attacker will never learn the value of the exchange. If the communication happens on an unsafe edge, then, at each step, the chance of reaching a safe edge is multiplied by $\theta(1 - \tau)$ (the previous edge was not safe and the next peer is not corrupted). So, at step k (assuming that the first step is 0), the probability of not having reached a corrupted peer and reaching a safe edge for the first time at this step is $\theta^k(1 - \tau)^{k+1}(1 - \theta)$. This is a geometric series, so we can compute its sum.

$$E_n = \sum_{k=0}^{n} \theta^k(1-\tau)^{k+1}(1-\theta) = (1-\tau)(1-\theta)\frac{1 - \theta^{k+1}(1-\tau)^{k+1}}{1 - \theta(1-\tau)}$$

Since $\theta(1-\tau) < 1$, we have $E_{+\infty} = (1-\tau)(1-\theta)\dfrac{1}{1-\theta(1-\tau)} = \dfrac{(1-\tau)(1-\theta)}{1-\theta(1-\tau)}$

$$E_{+\infty} = \frac{1-\theta-\tau+\theta\tau}{1-\theta+\theta\tau} = 1 - \frac{\tau}{1-\theta+\theta\tau} = 1 - \frac{\tau}{1-\theta(1-\tau)}$$

$\square$

4.3   Averaging Performance

We now report on our simulations[1] of a system consisting of 1000 peers with uniformly distributed values between -100 and 100. We first evaluate the convergence speed of the averaging protocol when running with a perfect peer sampling (complete graph, Fig. 4a) and a Req-Pull peer sampling [3] (Fig. 4b) in a simulator. To this end, we plot the values of 40 peers for each case omitting the privacy-generation phase. We first let the peer sampling construct local views before starting the averaging process. The time unit is a tick of c++ std::chrono::steady_clock (10^{-9} s on our test system).

Figure 5 shows instead the convergence time (no peers with $>1\%$ error) of the averaging process depending on the required privacy level, with different peer sampling protocols and network sizes. We observe that neither the peer sampling used nor the number of nodes has great influence on the convergence time. We see that the convergence time grows linearly with the privacy level. It is worth noting that, in our theoretical analysis, we proved that the probability of success of a direct attack decreases exponentially with the privacy level, compared to the linear growth of the convergence time.

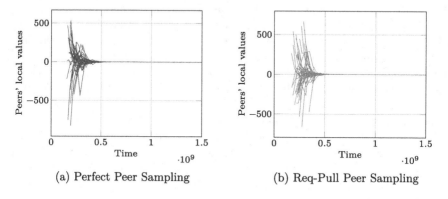

(a) Perfect Peer Sampling   (b) Req-Pull Peer Sampling

Fig. 4. Value convergence with different peer sampling protocols.

---

[1] The code used for these experiments is available at https://github.com/ALRBP/Private_Gossip_Average.

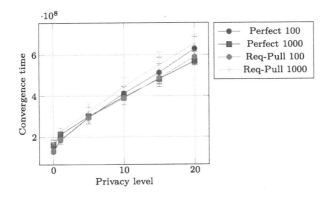

Fig. 5. Convergence time versus privacy level.

5 Related Work

Several authors have tackled the problem of privacy-preserving averaging in a decentralized setting, mostly within the automatic control community, where this problem is known as average consensus [5,13,16,18,19,31]. But most of these algorithms either rely on central components or assume static networks that operate in synchronous rounds with a lock-step execution model.

For example, [13] requires a central authority that distributes concealing factors before the beginning of the computation. Nodes then use this information to perturb the values sent during the averaging algorithm. Similarly, [27] assumes synchronous lock-step rounds and proposes a solution in which nodes exchange values together with a decaying noise. Moreover, due to the nature of this noise, the algorithm does not converge to the average in an exact sense but only in a mean-square sense. The work in [16] considers an ad hoc network setting, and thus also relies on an update rule that reaches all neighbors at the same time. But unlike [27] it provides deterministic convergence to the average value.

The work in [26] presents close similarities to our approach. Each node adds a zero-average noise to message exchanges for an independently chosen number of steps. The paper establishes topological conditions for the effectiveness of its privacy preservation, but this relies on the assumption that nodes communicate with all their neighbors at each communication step. Moreover, like in most other protocols, nodes start by exchanging their value plus some noise, whereas our protocol only exchanges noise for several rounds and adds the actual value only later in the process.

The work in [12] adopts an approach similar to [26] in the context of a push-sum averaging protocol [22] but still assumes that the network evolves in synchronous lock-step rounds. The authors of [38] further show that [26] is vulnerable to attacks on some topologies (for example on a ring). Our approach does not suffer from the same vulnerability thanks to its dynamic topology and the use of encrypted communication for all exchanges.

Some authors [14,24,37] have combined noise addition with homomorphic cryptography [11]. For example, [24] improves [27] by using homomorphic encryption in order to establish a confidential interaction protocol between nodes, while [30] proposes an averaging protocol that exploits partial homomorphic encryption in pairwise interactions. The work in [8] applies homomorphic encryption in an asynchronous setting, but it requires random noise to be generated by pairs of peers, which implies that a peer cannot protect its value without collaborating with another peer that is also protecting its value. Overall, even if partially homomorphic cryptographic starts to be viable, these approaches still incur a computational cost that appears unsuitable for small, battery-operated, devices. Moreover, in the case of our protocol, simple public-key encryption is enough to protect communication against eavesdroppers.

Other authors have combined the idea of adding noise with that of state decomposition. [35] divides each node into two virtual nodes, one that talks to the original node's neighbors and the other that only talks to the first virtual node. The approach has the advantage that the attacker cannot estimate the value with any guaranteed accuracy, but only if the target has at least one neighbor which is not under the influence of or observable by the attacker. [1] creates instead virtual nodes to divide the node's original value into random shares. But this allows attackers that control a sufficiently large sample of the network to guess the actual value of the peer with a low level of uncertainty. Finally, [32] considers a random-share protocol incorporating wiretap codes [36] to reduce message overhead. This yields low message complexity but only thanks to the use of a broadcast channel that is not available in large-scale networks.

Another recent contribution [34] proposes a hybrid architecture where a set of servers collect data from a set of nodes and compute the average. The paper focuses on providing nodes with heterogeneous privacy guarantees with respect to different privacy violators. It uses noise addition and employs KL divergence to measure a privacy preserving degree (PPD). However, it cannot compute the precise average but just an approximation.

This is similar to what happens with differential privacy [9,10]. Converging to a perturbed value guarantees that attacker cannot guess information about the original data distribution from the value of the final average. But this comes at the cost of being unable to converge to the exact average value, as proven in [29] and in [15]. As a further example, [33] proposes a differentially private averaging protocol with an optimization that groups node interactions to minimize network usage, while [2] proposes a differentially private protocol that uses homomorphic encryption in the context of k-means computation. Finally, some authors have started to tackle the problem of dealing with dishonest nodes that lie on their values [17]. This could be an interesting improvement for our work.

6 Conclusion

We proposed a novel protocol for privacy-preserving gossip averaging that addresses several limitations of previous approaches. While our protocol cannot be completely immune to attacks, we characterized its guarantees by formally proving several privacy properties. Now that we have a working averaging protocol, we plan to apply it in the context of higher-level applications such as decentralized machine-learning. Another direction consists in exploring its relations with recent work in the context of multi-party computation [6, 25].

References

1. Allard, T., Frey, D., Giakkoupis, G., Lepiller, J.: Lightweight privacy-preserving averaging for the Internet of Things (2016)
2. Allard, T., Hébrail, G., Masseglia, F., Pacitti, E.: Chiaroscuro: transparency and privacy for massive personal time-series clustering. In: ACM SIGMOD 2015, pp. 779–794 (2015)
3. Allavena, A., Demers, A., Hopcroft, J.E.: Correctness of a gossip based membership protocol. In: Proceedings of the Twenty-fourth Annual ACM Symposium on Principles of Distributed Computing, pp. 292–301. ACM (2005)
4. Bortnikov, E., Gurevich, M., Keidar, I., Kliot, G., Shraer, A.: Brahms: Byzantine resilient random membership sampling. Comput. Netw. **53**(13), 2340–2359 (2009)
5. Clifton, C., Kantarcioglu, M., Vaidya, J., Lin, X., Zhu, M.Y.: Tools for privacy preserving distributed data mining. ACM SIGKDD Explor. Newsl. **4**, 28–34 (2002)
6. Coretti, S., Garay, J., Hirt, M., Zikas, V.: Constant-round asynchronous multi-party computation based on one-way functions. In: Cheon, J.H., Takagi, T. (eds.) ASIACRYPT 2016. LNCS, vol. 10032, pp. 998–1021. Springer, Heidelberg (2016). https://doi.org/10.1007/978-3-662-53890-6_33
7. Danner, G., Jelasity, M.: Fully distributed privacy preserving mini-batch gradient descent learning. In: Bessani, A., Bouchenak, S. (eds.) DAIS 2015. LNCS, vol. 9038, pp. 30–44. Springer, Cham (2015). https://doi.org/10.1007/978-3-319-19129-4_3
8. Dellenbach, P., Bellet, A., Ramon, J.: Hiding in the crowd: a massively distributed algorithm for private averaging with malicious adversaries. CoRR (2018)
9. Dwork, C.: Differential privacy: a survey of results. In: Agrawal, M., Du, D., Duan, Z., Li, A. (eds.) TAMC 2008. LNCS, vol. 4978, pp. 1–19. Springer, Heidelberg (2008). https://doi.org/10.1007/978-3-540-79228-4_1
10. Dwork, C., McSherry, F., Nissim, K., Smith, A.: Calibrating noise to sensitivity in private data analysis. In: Halevi, S., Rabin, T. (eds.) TCC 2006. LNCS, vol. 3876, pp. 265–284. Springer, Heidelberg (2006). https://doi.org/10.1007/11681878_14
11. Frikken, K.B.: Secure multiparty computation. In: Atallah, M.J., Blanton, M. (eds.) Algorithms and Theory of Computation Handbook, pp. 14.1–14.16. Chapman & Hall/CRC (2010)
12. Gao, H., Zhang, C., Ahmad, M., Wang, Y.: Privacy-preserving average consensus on directed graphs using push-sum. In: 2018 IEEE Conference on Communications and Network Security (CNS), pp. 1–9. IEEE (2018)
13. Gupta, N., Chopra, N.: Confidentiality in distributed average information consensus. In: 2016 IEEE 55th Conference on Decision and Control (CDC), pp. 6709–6714. IEEE (2016)
14. Hadjicostis, C.N.: Privacy preserving distributed average consensus via homomorphic encryption. In: 2018 IEEE Conference on Decision and Control (CDC), pp. 1258–1263. IEEE (2018)
15. He, J., Cai, L.: Differential private noise adding mechanism: basic conditions and its application. In: 2017 American Control Conference (ACC), pp. 1673–1678. IEEE (2017)
16. He, J., Cai, L., Cheng, P., Pan, J., Shi, L.: Consensus-based privacy-preserving data aggregation. IEEE Trans. Autom. Control (2016)

17. He, J., Cai, L., Cheng, P., Pan, J., Shi, L.: Distributed privacy-preserving data aggregation against dishonest nodes in network systems. IEEE Internet Things J. **6**(2), 1462–1470 (2019)
18. He, J., Cai, L., Zhao, C., Cheng, P., Guan, X.: Privacy-preserving average consensus: privacy analysis and optimal algorithm design. IEEE Trans. Signal Inf. Process. Netw. **5**(1), 127–138 (2019)
19. Huang, Z., Mitra, S., Dullerud, G.: Differentially private iterative synchronous consensus. In: Proceedings of the 2012 ACM Workshop on Privacy in the Electronic Society, WPES 2012, pp. 81–90. ACM (2012)
20. Jelasity, M., Montresor, A., Babaoglu, O.: Gossip-based aggregation in large dynamic networks. ACM Trans. Comput. Syst. **23**(3), 219–252 (2005)
21. Jelasity, M., Voulgaris, S., Guerraoui, R., Kermarrec, A.-M., van Steen, M.: Gossip-based peer sampling. ACM ToCS **25**(3), 8 (2007)
22. Kempe, D., Dobra, A., Gehrke, J.E.: Gossip-based computation of aggregate information. In: Proceedings of the Twenty-Fourth Annual ACM Symposium on Principles of Distributed Computing, pp. 482–491 (2003)
23. Lepiller, J.: Private decentralized aggregation (2016)
24. Liu, Q., Ren, X., Mo, Y.: Secure and privacy preserving average consensus. In: 2017 11th Asian Control Conference (ASCC), pp. 274–279. IEEE (2017)
25. Liu-Zhang, C.-D., Loss, J., Maurer, U., Moran, T., Tschudi, D.: Robust MPC: asynchronous responsiveness yet synchronous security. In: Theory and Practice of Multi-Party Computation Workshops (2019)
26. Manitara, N.E., Hadjicostis, C.N.: Privacy-preserving asymptotic average consensus. In: 2013 European Control Conference (ECC), pp. 760–765. IEEE (2013)
27. Mo, Y., Murray, R.M.: Privacy preserving average consensus. In: 53rd IEEE Conference on Decision and Control, pp. 2154–2159. IEEE (2014)
28. Nédelec, B., Tanke, J., Molli, P., Mostéfaoui, A., Frey, D.: An adaptive peer-sampling protocol for building networks of browsers. World Wide Web **21**, 629–661 (2017)
29. Nozari, E., Tallapragada, P., Cortés, J.: Differentially private average consensus: obstructions, trade-offs, and optimal algorithm design. Automatica **81**, 221–231 (2015)
30. Ruan, M., Gao, H., Wang, Y.: Secure and privacy-preserving consensus. IEEE Trans. Autom. Control (2019)
31. Sheikh, R., Kumar, B., Mishra, D.K.: A distributed k-secure sum protocol for secure multi-party computations. J. Comput. **2**, 68–72 (2010)
32. Thobaben, R., Dán, G., Sandberg, H.: Wiretap codes for secure multi-party computation. In: 2014 IEEE Globecom Workshops (GC Wkshps), pp. 1349–1354. IEEE (2014)
33. Wang, A., Liao, X., He, H.: Event-triggered differentially private average consensus for multi-agent network. IEEE/CAA J. Automatica Sinica **6**(1), 75–83 (2019)
34. Wang, X., He, J., Cheng, P., Chen, J.: Privacy preserving collaborative computing: heterogeneous privacy guarantee and efficient incentive mechanism. IEEE Trans. Signal Process. **67**(1), 221–233 (2018)
35. Wang, Y.: Privacy-preserving average consensus via state decomposition. IEEE Trans. Autom. Control (2019)
36. Wyner, A.D.: The wire-tap channel. Bell Syst. Tech. J. **54**(8), 1355–1387 (1975)
37. Yin, T., Lv, Y., Yu, W.: Accurate privacy preserving average consensus. IEEE Trans. Circuits Syst. II: Express Briefs (2019)
38. Zhou, H., Yang, W., Yang, C.: Privacy preserving consensus under interception attacks. In: 2017 36th Chinese Control Conference (CCC), pp. 8485–8490. IEEE (2017)

Synchronous *t*-Resilient Consensus in Arbitrary Graphs

Armando Castañeda[1(✉)], Pierre Fraigniaud[2], Ami Paz[2], Sergio Rajsbaum[1], Matthieu Roy[3,4], and Corentin Travers[5]

[1] UNAM, Mexico City, Mexico
{armando.castaneda,rajsbaum}@im.unam.mx
[2] CNRS and Université de Paris, Paris, France
{pierref,amipaz}@irif.fr
[3] Laboratorio Solomon Lefschetz - UMI LaSoL, CNRS, CONACYT, UNAM, Cuernavaca, Mexico
[4] LAAS-CNRS, CNRS, Université de Toulouse, Toulouse, France
roy@laas.fr
[5] CNRS and University of Bordeaux, Bordeaux, France
travers@labri.fr

Abstract. We study the number of rounds needed to solve consensus in a synchronous network G where at most t nodes may fail by crashing. This problem has been thoroughly studied when G is a complete graph, but very little is known when G is arbitrary. We define a notion of radius that considers all ways in which t nodes may crash, and present an algorithm that solves consensus in radius rounds. Then we derive a lower bound showing that our algorithm is optimal for vertex-transitive graphs, among oblivious algorithms.

Keywords: Crash failures · Consensus · Combinatorial topology · Distributed graph algorithms

1 Introduction

The Problem. We consider a synchronous message-passing distributed system, where at most t out of n nodes may fail by crashing. The nodes communicate by sending messages to each other over the edges of an undirected graph G. In the *consensus* problem each node is given an input, and after some number of rounds produces an output, such that all outputs are the same and must be equal to one of the inputs.

One of the earliest and most well-know facts in distributed computing is that the number of rounds needed to solve consensus when G is the complete graph, K_n, is $t + 1$. Namely, consensus can be solved in $t + 1$ rounds, for $t < n$, and any algorithm requires this number of rounds in the worst case. The round

Supported by ANR Project DESCARTES, INRIA Project GANG, UNAM-PAPIIT IA102417 and IN109917, and Fondation des Sciences Mathématiques de Paris.

© Springer Nature Switzerland AG 2019
M. Ghaffari et al. (Eds.): SSS 2019, LNCS 11914, pp. 53–68, 2019.
https://doi.org/10.1007/978-3-030-34992-9_5

complexity to solve consensus in K_n has been thoroughly studied, but not for graphs other than the complete graph.

1.1 Results

This paper studies the number of rounds needed to solve consensus, as a function of G and t. It presents two main contributions.

First, it shows that for any given $(t + 1)$-vertex-connected graph G, it is possible to solve consensus tolerating t failures, in $\mathsf{radius}(G, t)$ rounds. Roughly, the *eccentricity of v against t failures*, $\mathsf{ecc}(v, t)$, is the smallest number of rounds needed for a node v to broadcast its input value, independently of the failure pattern (when and how nodes crash). Then, $\mathsf{radius}(G, t)$ is equal to the smallest $\mathsf{ecc}(v, t)$, over all nodes v. For example, $\mathsf{radius}(K_n, t) = t + 1$ for the complete graph and $\mathsf{radius}(C_n, 1) = n-1$ for the cycle. For the wheel, $\mathsf{radius}(W_n, 2) = n-1$ and $\mathsf{radius}(W_n, 1) = 1 + \lfloor (n - 1)/2 \rfloor$.

Second, we present a corresponding lower bound, showing that our algorithm is optimal among oblivious algorithms, in any graph that is vertex-transitive. In an *oblivious* algorithm, the decision value of a node is based on the set of input values it has seen so far, and not on the particular failure pattern. Roughly speaking, a graph is *vertex-transitive* if it is highly symmetric. This is a large and well studied class of graphs (see, e.g., [18]).

The question of achieving consensus in a network prone to failures was intensively studied when the communication pattern is the *complete graph*. However, it seems difficult to obtain direct generalizations of these classical upper and lower bound techniques from a complete graph to a *general graph*. Instead, both our upper and lower bounds use novel ideas, that we discuss next.

Our Upper Bound Techniques. In a classic algorithm to solve consensus on a complete graph, e.g. [29], nodes repeatedly send all the inputs they know, and at the end of round $t + 1$, each node that has not crashed, decides the smallest input value among the values it has seen. The usual agreement argument is that among the $t + 1$ rounds there must be at least one in which no node crashes. All nodes that are alive at the end of such a round have seen the same set of inputs, i.e., there is *common knowledge* [14] on a set of inputs. This argument holds only under the assumption that the graph is complete. We use a similar algorithm on an arbitrary graph, but apply a more fine-grained argument, of *information flow*, to prove its correctness and running time.

Given a node v and its $\mathsf{ecc}(v, t)$, we show that at the end of round $\mathsf{ecc}(v, t)$, either all alive nodes have received v's input, or none has. For the complete graph, $\mathsf{ecc}(v, t) = t + 1$ for all nodes v, and indeed, for any node v, either all nodes have received the input of v by round $t+1$, or no node will ever receive it. This implies the correctness of the algorithm for the complete graph described above. Notice that the eccentricity is not less than $t + 1$, because the adversary may create a *hidden path*, $v_1, \ldots, v_t$ such that $v_1 = v$ and each v_i, $1 \le i \le t - 1$, fails in round i and sends a message to only v_{i+1} before failing.

We use this information flow perspective to derive simple consensus algorithms for arbitrary graphs. Each node repeatedly forwards all the pairs (v, in_v) it knows about, where in_v is the input value of node v. Then, an algorithm is specified by two functions: $R(G, t)$ which returns the the number of rounds to execute, and $D(G, t)$ which tells a node which value to decide, among the input values it has seen. After $R(G, t)$ rounds, the active nodes have the same view of the inputs of a carefully chosen *subset* of $t + 1$ nodes, thus, after $R(G, t)$ rounds, $D(G, t)$ can pick deterministically the input of one of these nodes. Remarkably, our lower bound shows that this is not necessarily the case after less rounds.

Our Lower Bound Techniques. There are several lower bound proofs for the number of rounds to solve consensus under crash failures for the case when G is a complete graph. The classic $t + 1$ lower bound proof style proceeds by a rather complex backward induction (a detailed description appears in [25]). Later on, simpler forward induction proofs were discovered [1,26], following the classical bivalency arguments that were originally developed for proving the impossibility of solving consensus in asynchronous systems [17].

The aforementioned proofs hold for general graphs as well, namely, $t + 1$ rounds is a lower bound for solving consensus on any graph G. However, in general graphs this bound is very weak, as it does not take into consideration the graph's structure. An obvious example is a cycle with $t = 1$: our lower bound is $n - 1$, while the standard approaches give a lower bound of 2 rounds.

Our lower bound technique is different from both the backward and the forward arguments. It is inspired by the topological techniques for distributed computing [20], though we do not use topology explicitly in the current paper. Our lower bound technique is similar to the connectivity analysis of the *protocol complex*, the structure of states at the end of executions of an algorithm after a certain number of rounds. However, instead of working with the protocol complex, we consider an *information flow* directed graph version based on failure patterns, without including input values. We prove that consensus is solvable by an oblivious algorithm if and only if all connected components of the information flow graph have a *dominating* vertex, namely, a vertex with an edge from it to any other vertex in its connected component. In [6], we study these information flow techniques and their relation to *set agreement* and *approximate agreement*.

The seminal paper [14] shows that, as soon as there is common knowledge of a *clean* round (where a node that crashes does not send any messages), it is also common knowledge that nodes have identical views of the initial configuration. As a consequence, any action that depends on the system's initial configuration can be carried out simultaneously in a consistent way by the set of active nodes at any round $k \geq t + 1$, if it can be carried out at all. Our lower bound is larger than $t + 1$ on general graphs, and hence shows how the round in which nodes have common knowledge of a subset of the input configuration is affected also by the structure of the graph.

1.2 Related Work

Consensus in the failure-prone synchronous model has been thoroughly studied since the beginning of the distributed computing field in the late 1970's [34]. A variety of aspects have been considered, including the number of rounds (in great detail, including worst case, early deciding, simultaneous, unbeatability, etc.), number and size of messages, variants of consensus, in static and dynamic networks, and under various failure models. We only mention some of the most relevant papers, among a vast literature, which even surveys e.g. [8,29] and textbooks on the field cover only partially, e.g., [4,25,30].

For general graphs, since early on there has been an interest in characterizing the graphs where consensus is solvable, initially for Byzantine failures [11,12,16]. It was observed early on [24] that $t+1$ connectivity is necessary and an exponential algorithm was described. The algorithms for Byzantine settings also work in our model. However, they have not been optimized for the number of rounds, and furthermore, our setting requires only $t + 1$ vertex-connectivity, while an algorithm tolerating Byzantine failures requires $n \geq 3t + 1$, and vertex-connectivity at least $2t + 1$ [12]. Very recently, consensus algorithms for general graphs were designed, for *local broadcast* Byzantine failures [22]. One algorithm works in the local broadcast model on a graph under the weakest requirements—minimum degree $2t$, and $(\lfloor 3t/2+1 \rfloor)$ vertex-connected; however, it has an exponential time complexity. A different consensus algorithm terminates in $3n$ rounds, but only assuming the graph is $2t$-connected. There has also been work in characterizing the *directed* graphs for which fault tolerant synchronous consensus is solvable, both under crash and under Byzantine failures [32,33].

We are not aware of any previous lower bounds techniques for arbitrary graphs. The $t + 1$ lower bound on the number of rounds to solve consensus in K_n was originally proved in [15] for Byzantine failures, and was later extended to the case were digital signatures can be used [11], and finally to crash failures (see, e.g., [19]).

Our lower bound technique is mainly inspired by the topological techniques for distributed computing [20], and more specifically by the topological structure of the executions of a synchronous algorithm after a certain number of rounds [21]. Indeed, the technique used for deriving our second algorithm is reminiscent of topological existential upper bounds proofs used in the past [3]. Hidden paths have played an important role in the design of *early-deciding* consensus algorithms in the complete graph [7].

Research on *dynamic networks* also characterizes families of networks for which consensus (or a variant of it) is solvable [9,10,27,31,35]. Interestingly, dynamic networks research and works on synchronous fault-tolerant consensus [32,33] share the idea of picking a node as a source, and having all nodes deciding on the input of this source. In Theorem 3 we present an information flow characterization for consensus, in terms of such a source. Our notion of a core set (see Sect. 3.2) can be seen as a refinement of such notions, defined in order to optimize the number of rounds. Interestingly, [27] presents a topological solvability characterization of consensus using the *point set topology* techniques introduced in [2].

The line of work on *almost everywhere agreement* initiated in [5,13], was motivated by the impossibility of tolerating t crashes when the network is not $t + 1$ connected (these works also consider Byzantine failures). They present algorithms for networks where consensus is actually unsolvable due to weak connectivity.

2 Preliminaries

Model of Computation. We consider the standard synchronous message-passing model of computation where at most t nodes may fail by crashing. A set of $n \geq 2$ nodes V communicate through bidirectional channels E defining a graph $G = (V, E)$. In the remainder of the paper, we fix G and t, and assume $t < \kappa(G)$, the *vertex connectivity* of G, i.e., the minimum number of nodes whose deletion disconnects G.

An *execution* proceeds in a infinite sequence of synchronous rounds, starting in round 1. In every round, each node v first performs some local computation, then sends a message to each of its neighbors in G, denoted $N(v)$, and then receives the messages sent to it from $N(v)$ in that round. When a node crashes in round r, it fails to send its message to some of its neighbors in round r, and sends no message in subsequent rounds.

A *failure pattern* φ for G, t specifies, for each node that fails, in which round number it fails, and which messages it fails to send. It is a set of triples of the form (v, F_v, f_v), indicating that v crashes in round f_v, in which it does not send the messages to $\emptyset \neq F_v \subseteq N(v)$. Since at most t nodes can fail, $|\varphi| \leq t$, and since nodes do not recover from a failure, if $(v, F_v, f_v), (u, F_u, f_u) \in \varphi$ then $v \neq u$.

For an execution with failure pattern φ, the *faulty* nodes are those that appear in a triplet in φ; the others are the *correct* nodes. A node is *active* in round r in φ if it is correct, or if it fails in a round later than r. A node that crashes with $F_v = N(v)$ is said to crash *cleanly* in φ.

Consider any input assignment to the nodes, and a failure pattern φ. Our algorithms are of the following form. Initially, for each node v with input in_v, its *view* is $\{(v, in_v)\}$. In each round, each node v sends its *view* to $N(v)$, and at the end of the round it updates its *view* with the new input value-pairs it receives.

We say that u *hears from* v in φ, if in some round u receives a message containing the input of v. Similarly, we way that u hears from v *by round* r in φ if u receives a message with v's input in round r, or before. In other words, there is a *causal path* from u to v [23] in an infinite execution with failure pattern φ. Clearly, the existence of such a path depends on φ, but not on the input assignment. Thus, to analyze the structure of all possible failure patterns, we ignore the input values. This is what we do next, where we may identify φ with the infinite execution with that failure pattern.

Eccentricity and Radius in Failure Patterns. Let $\mathsf{dist}_G(u,v)$ denote the distance between nodes u and v in $G = (V, E)$. The *eccentricity* of a node $v \in V$ is defined as $\mathsf{ecc}_G(v) = \max_{u \in V} \mathsf{dist}_G(u, v)$. The *diameter* of a graph is defined as $\max_{v \in V} \mathsf{ecc}_G(v)$, and its *radius* as $\min_{v \in V} \mathsf{ecc}_G(v)$. We generalize the notions of eccentricity and radius to the synchronous t-resilient model.

In the following, failure patterns are denoted by lower case Greek letters $\varphi, \psi, \ldots$, and sets of failure patterns are denoted by upper case Greek letters $\Phi, \Psi, \ldots$. We denote by $\Phi_{\mathrm{all}}^{(t)}$ the set of *all* failure patterns for G and t. The failure pattern in which no nodes crash is $\varphi_\varnothing$, and hence $\Phi_{\mathrm{all}}^{(0)} = \{\varphi_\varnothing\}$.

Definition 1. *Given a node $v \in V$ and a failure pattern $\varphi \in \Phi_{\mathrm{all}}^{(t)}$, the eccentricity $\mathsf{ecc}_G(v, \varphi) \in \mathbb{N} \cup \{\infty\}$ of v in φ is the minimum number of rounds required for all correct nodes to hear from v (i.e., there is causal path from v to every correct node), or ∞ if not all correct nodes hear from v. If $\mathsf{ecc}_G(v, \varphi) \in \mathbb{N}$, we say that v* floods *to the correct nodes in φ.*

Consider any φ. Notice that since G is at least $(t+1)$-connected, and at most t nodes crash, if a correct node u hears from v, then every correct node receives a message from v (because it can get from u to every correct node). We thus have the following claim.

Fact 1. *For every $v \in V$, and every $\varphi \in \Phi_{\mathrm{all}}^{(t)}$, if $\mathsf{ecc}_G(v, \varphi) = \infty$ then no correct node hears from v in φ.*

Definition 2. *For $v \in V$ and $\Phi \subseteq \Phi_{\mathrm{all}}^{(t)}$, such that there is at least one $\varphi \in \Phi$ with $\mathsf{ecc}_G(v, \varphi) \in \mathbb{N}$, let*

$$\mathsf{ecc}_G(v, \Phi) = \max\{\mathsf{ecc}_G(v, \varphi) \ : \ \varphi \in \Phi, \mathsf{ecc}_G(v, \varphi) \in \mathbb{N}\}.$$

Notice that there is at least one $\varphi \in \Phi$ with $\mathsf{ecc}_G(v, \varphi) \in \mathbb{N}$, for any Φ containing failure patterns where v is correct.

Lemma 1. *For $v \in V$ and $\varphi \in \Phi_{\mathrm{all}}^{(t)}$, let A be the set of all active nodes in round $\mathsf{ecc}_G(v, \Phi_{\mathrm{all}}^{(t)})$ under φ. Either all nodes in A hear from v by round $\mathsf{ecc}_G(v, \Phi_{\mathrm{all}}^{(t)})$, or no node in A hears from v by round $\mathsf{ecc}_G(v, \Phi_{\mathrm{all}}^{(t)})$ in φ.*

Proof. Let $\varphi' \in \Phi_{\mathrm{all}}^{(t)}$ be the failure pattern identical to φ in the first $\mathsf{ecc}_G(v, \Phi_{\mathrm{all}}^{(t)})$ rounds, but with all the nodes of A correct in φ'. Then, the nodes in A have the same view in both φ and φ' in round $\mathsf{ecc}_G(v, \Phi_{\mathrm{all}}^{(t)})$.

If $\mathsf{ecc}_G(v, \varphi') \in \mathbb{N}$, by Definition 1, all nodes in A hear from v by time $\mathsf{ecc}_G(v, \varphi')$, which is at most $\mathsf{ecc}_G(v, \Phi_{\mathrm{all}}^{(t)})$, by Definition 2. The same is true for φ, as φ and φ' are identical in the first $\mathsf{ecc}_G(v, \Phi_{\mathrm{all}}^{(t)})$ rounds.

If $\mathsf{ecc}_G(v, \varphi') = \infty$, no node in A hears from v in φ', by Fact 1, and then no node in A hears from v by round $\mathsf{ecc}_G(v, \Phi_{\mathrm{all}}^{(t)})$ in φ because φ and φ' are identical in the first $\mathsf{ecc}_G(v, \Phi_{\mathrm{all}}^{(t)})$ rounds. $\qquad\square$

Definition 3. *Let $\Phi \subseteq \Phi_{\text{all}}^{(t)}$ such that for every $v \in V$ there is at least one $\varphi \in \Phi$ with $\text{ecc}_G(v, \varphi) \in \mathbb{N}$. The radius of G with respect to Φ is defined as $\text{radius}(G, \Phi) = \min_{v \in V} \text{ecc}_G(v, \Phi)$.*

For $t = 0$, our notion of eccentricity and radius coincides with the classical graph-theoretic definition, i.e., $\text{ecc}_G(v, \Phi_{\text{all}}^{(0)}) = \text{ecc}_G(v)$ and $\text{radius}(G, \Phi_{\text{all}}^{(0)}) = \text{radius}(G)$. Moreover, in the complete graph K_n, we have $\text{radius}(K_n, \Phi_{\text{all}}^{(t)}) = t+1$, which together with Lemma 1 implies the correctness of the simple algorithm discussed in the Introduction.

3   Consensus Algorithms in Arbitrary Graphs

We consider the usual *consensus* problem in which each node starts with an input value, defined by the following properties. **Termination**: Every correct node decides a value; **Validity**: The decision of a node is equal to the input of some node; **Agreement**: The decisions of any pair of nodes are the same.

Oblivious Algorithms. Recall that in our algorithms, a node resends to its neighbors the set of input values it has received, each one together with the name of the node that has the corresponding input value. Thus, to specify a consensus algorithm, we define a function $\text{R}(G, t)$ that returns a round number, stating that all correct nodes decide in round $\text{R}(G, t)$. Also, we define a *decision function* $\text{D}(G, t)$ used by a node to select a consensus value from its view (possibly taking in consideration the names of the nodes that proposed this inputs). Namely, in a t-fault tolerant oblivious consensus algorithm for G, after $\text{R}(G, t)$ rounds of communication (independently of the failure pattern or the input assignment), each node selects a value from its view, as specified by the function $\text{D}(G, t)$. We stress that $\text{R}(G, t)$ and $\text{D}(G, t)$ are not computed by the nodes, they are given as part of the algorithm (alternatively, if the nodes "know" G and t, then they can compute these functions locally).

3.1   A Naive Algorithm

We describe algorithm $\text{P}_{\text{ecc}}^{G,t} = (\text{R}_{\text{ecc}}(G, t), \text{D}_{\text{ecc}}(G, t))$, based on a simple idea. Let us order the n vertices of G as $v_1, \ldots, v_n$, with

$$\text{ecc}_G(v_i, \Phi_{\text{all}}^{(t)}) \leq \text{ecc}_G(v_{i+1}, \Phi_{\text{all}}^{(t)}) \tag{1}$$

for $1 \leq i < n$. In particular, we have $\text{radius}(G, \Phi_{\text{all}}^{(t)}) = \text{ecc}_G(v_1, \Phi_{\text{all}}^{(t)})$.

Let $\text{R}_{\text{ecc}}(G, t) = \text{ecc}_G(v_{t+1}, \Phi_{\text{all}}^{(t)})$, and $\text{D}_{\text{ecc}}(G, t)$ be the function that returns the input of the smallest[1] node among the nodes in $\{v_1, \ldots, v_{t+1}\}$.

Theorem 1. *Algorithm $\text{P}_{\text{ecc}}^{G,t}$ solves consensus in $\text{ecc}_G(v_{t+1}, \Phi_{\text{all}}^{(t)})$ rounds.*

---

[1] Assuming V is a totally ordered set.

Proof. The algorithm satisfies termination as all correct nodes run $\mathsf{R}_{\mathrm{ecc}}(G,t) = \mathrm{ecc}_G(v_{t+1}, \Phi_{\mathrm{all}}^{(t)})$ rounds. For validity, the definition of $\mathrm{ecc}_G(v_{t+1}, \Phi_{\mathrm{all}}^{(t)})$ and Eq. 1 imply that all nodes receive at least one input of a node in $\{v_1, \ldots, v_{t+1}\}$ by round $\mathrm{ecc}_G(v_{t+1}, \Phi_{\mathrm{all}}^{(t)})$, in every $\varphi \in \Phi_{\mathrm{all}}^{(t)}$. For agreement, consider any $\varphi \in \Phi_{\mathrm{all}}^{(t)}$ and the set A of all nodes that are active in round $\mathrm{ecc}_G(v_{t+1}, \Phi_{\mathrm{all}}^{(t)})$ in φ. Lemma 1 and Eq. 1 imply that either all nodes in A have received v_i's input, $1 \le i \le t+1$, in round $\mathrm{ecc}_G(v_{t+1}, \Phi_{\mathrm{all}}^{(t)})$ in φ, or none of them has received it in that round. Therefore, all nodes in A have the same view of the inputs of the nodes $v_1, \ldots, v_{t+1}$, hence $\mathsf{D}_{\mathrm{ecc}}(G,t)$ returns the same value to all of them. □

It is easy to come up with graphs for which this solution is not optimal, in terms of number of rounds.

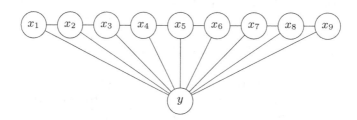

Fig. 1. A graph for which $\mathsf{P}_{\mathrm{ecc}}^{G,t}$ is not time optimal.

Lemma 2. *There is a graph G for which $\mathsf{P}_{\mathrm{ecc}}^{G,t}$ is not time optimal, with $t = 1$.*

Algorithm $\mathsf{P}_{\mathrm{ecc}}^{G,t}$ is not optimal in the graph in Fig. 1 because $v_2 = x_4$ needs many rounds in order to broadcast its input, even when $v_1 = x_5$ crashes. Instead, y broadcasts very quickly when $v_1 = x_5$ crashes. As a consequence, y is a better choice for replacing x_5 whenever this latter node crashes. More generally, the sequence $v_1, \ldots, v_n$ defined in Eq. (1) is not adaptive. In the next subsection, we define an adaptive sequence, in which the performances of v_i are measured only for failure patterns in which $v_1, \ldots, v_{i-1}$ are prevented from flooding.

3.2   An Adaptive-Eccentricity Based Algorithm

The algorithm $\mathsf{P}_{\mathrm{ecc}}^{G,t}$ is based on a *core* set of nodes $\{v_1, \ldots, v_{t+1}\}$, consisting of the first $t + 1$ nodes in order of ascending eccentricity. We show here that there is a more clever way of selecting a core set of $t + 1$ nodes. The corresponding algorithm, $\mathsf{P}_{\mathrm{adapt}}^{G,t} = (\mathsf{R}_{\mathrm{adapt}}(G,t), \mathsf{D}_{\mathrm{adapt}}(G,t))$, is similar, except that, $\mathsf{R}_{\mathrm{adapt}}(G,t) = \mathrm{radius}(G, \Phi_{\mathrm{all}}^{(t)})$. As before, $\mathsf{D}_{\mathrm{adapt}}(G,t)$ returns the input of the smallest node among the core set, but now the core set is $\{s_1, \ldots, s_{t+1}\}$, as defined next.

The first node s_1 is the same v_1 as in $\mathsf{P}_{\mathrm{ecc}}^{G,t}$. To choose the i-th node, we consider all the un-chosen nodes, and their eccentricity *only among the failure patterns where the previously selected nodes have* ∞ *eccentricity*, and take the node that minimizes this quantity.

Formally, to define the core set of $t + 1$ nodes, we construct a sequence of pairs (s_i, Φ_i), with $s_i \in V$, and $\Phi_i \subseteq \Phi_{all}^{(t)}$, for $i = 1, \ldots, t+1$, inductively, as follows. For every node $v \in V$, let $\Phi_v^\infty = \{\varphi \in \Phi_{all}^{(t)} : ecc_G(v, \varphi) = \infty\}$ and $\Phi_v^{\mathbb{N}} = \{\varphi \in \Phi_{all}^{(t)} : ecc_G(v, \varphi) \in \mathbb{N}\}$. Let $\Phi_0 = \Phi_{all}^{(t)}$, and, for $i = 1, \ldots, t+1$, let

$$
\begin{cases}
s_i = \arg\min_{v \in V \setminus \{s_1, \ldots, s_{i-1}\}} \ ecc_G(v, \Phi_v^{\mathbb{N}} \cap \Phi_{i-1}), \\
\Phi_i = \Phi_{s_i}^\infty \cap \Phi_{i-1},
\end{cases}
\tag{2}
$$

where, for $i = 1$, we interpret $\{s_1, \ldots, s_{i-1}\}$ as the empty set. In other words, $\Phi_i = \Phi_{s_1}^\infty \cap \cdots \cap \Phi_{s_i}^\infty$, and also $\Phi_i = \Phi_{i-1} \setminus \Phi_{s_i}^{\mathbb{N}}$. Observe that, for every $i = 1, \ldots, t+1$, and every $v \in V \setminus \{s_1, \ldots, s_{i-1}\}$, $\Phi_v^{\mathbb{N}} \cap \Phi_{i-1}$ is not empty as it contains the failure pattern in which all nodes $s_1, \ldots, s_{i-1}$ crash cleanly at the first round, and no other node crashes. Also note that $ecc_G(s_1, \Phi_{s_1}^{\mathbb{N}}) = radius(G, \Phi_{all}^{(t)})$.

For example, in K_n, we have $ecc_{K_n}(s_i, \Phi_{s_i}^{\mathbb{N}}) = t - i + 2$ for $i = 1, \ldots, t+1$ whenever $t < n - 1$. For $t = n - 1$, we have $ecc_{K_n}(s_i, \Phi_{s_i}^{\mathbb{N}}) = n - i$ for $i = 1, \ldots, n$. In the cycle C_n with $t = 1$, we have $ecc_{C_n}(s_1, \Phi_{s_1}^{\mathbb{N}}) = n - 1$ and $ecc_{C_n}(s_2, \Phi_{s_2}^{\mathbb{N}}) = \lfloor \frac{n-1}{2} \rfloor$. For the graph G in Fig. 1, $s_1 = x_5$ and $s_2 = y$, $ecc_G(s_1, \Phi_{s_1}^{\mathbb{N}}) = radius(G, \Phi_{all}^{(1)}) = 4$, and $ecc_G(s_2, \Phi_{s_2}^{\mathbb{N}}) = 1$.

The *core set* for G, t is $\{s_1, \ldots, s_{t+1}\}$, and the *core sequence* for G is the ordered sequence $(s_1, \ldots, s_{t+1})$. A crucial property of this sequence is that, while the sequence $(ecc_G(v_i, \Phi_{v_i}^{\mathbb{N}}))_{1 \leq i \leq t+1}$ defined in Eq. (1) is non decreasing, and may even be increasing, the sequence $(ecc_G(s_i, \Phi_{s_i}^{\mathbb{N}} \cap \Phi_{i-1}))_{1 \leq i \leq t+1}$ defined in Eq. (2) is non increasing, and is actually always decreasing. Intuitively, this is because the maximization in the computation of $ecc_G(v, \Phi_v^{\mathbb{N}} \cap \Phi_i)$ for determining s_{i+1} is taken over the set $\Phi_v^{\mathbb{N}} \cap \Phi_i$ which is smaller than the set $\Phi_v^{\mathbb{N}} \cap \Phi_{i-1}$ used for the computation of s_i.

Lemma 3. *Consider the core sequence* $(s_1, \ldots, s_{t+1})$ *and the pairs defined in Eq.* (2). *Then,* $ecc_G(s_i, \Phi_{s_i}^{\mathbb{N}} \cap \Phi_{i-1})) > ecc_G(s_{i+1}, \Phi_{s_{i+1}}^{\mathbb{N}} \cap \Phi_i))$, *for* $i \in \{1, \ldots, t\}$.

The correctness proof of $P_{adapt}^{G,t}$ is very similar to that of $P_{ecc}^{G,t}$.

Theorem 2. *Algorithm* $P_{adapt}^{G,t}$ *solves consensus in* $radius(G, \Phi_{all}^{(t)})$ *rounds.*

Finally, observe that $P_{ecc}^{G,t}$ performs in $ecc_G(v_{t+1}, \Phi_{all}^{(t)})$ rounds according to the notations of Eq (1), while $P_{adapt}^{G,t}$ performs in $radius(G, \Phi_{all}^{(t)}) = ecc_G(v_1, \Phi_{all}^{(t)})$ rounds according to the same notations.

4   The Lower Bound

In this section we show that $P_{adapt}^{G,t}$ is time optimal for vertex-transitive graphs, among oblivious algorithms. Recall that in an oblivious algorithm, the decision value of a node is based on the set of input values it has seen so far, and not on the particular failure pattern. Our algorithms $P_{ecc}^{G,t}$ and $P_{adapt}^{G,t}$ are oblivious.

4.1   Information Flow Graph

Recall that the view of a node u in a given round r is the set of all pairs (v, in_v) such that u hears from v by round r. The vertices of the *information flow graph* have the form (v, view_v), meaning that node v has view view_v in round r, and there is a *directed* edge from (v, view_v) to (u, view_u) if and only if $(v, in_v) \in \text{view}_u$, i.e., u hears from v by round r. Of course, these properties are conditioned by the actual failure pattern.

Consider a set of failure patterns $\Phi \subseteq \Phi_{\text{all}}^{(t)}$. Let u be a node that is active in round r in φ, for some $r \geq 1$. Let $\text{view}_G(u, \varphi, r)$ denote the view of u in round r in φ.

Definition 4. *The* information flow *graph in round r with respect to Φ is the directed graph $\mathbb{IF}_{G,\Phi,r}$:*

- $V(\mathbb{IF}_{G,\Phi,r}) = \{(u, \text{view}_G(u, \varphi, r)) : u \in V$ *is active in round r in $\varphi \in \Phi\}$;*
- $E(\mathbb{IF}_{G,\Phi,r}) = \{((u, \text{view}_G(u, \varphi, r)), (v, \text{view}_G(v, \varphi, r))) : u \in \text{view}_G(v, \varphi, r)\}.$

Note that a node u may have the same view in two distinct $\varphi, \psi \in \Phi$ in round r, i.e., $\text{view}_G(u, \varphi, r) = \text{view}_G(u, \psi, r)$, in which case $(u, \text{view}_G(u, \varphi, r))$ and $(u, \text{view}_G(u, \psi, r))$ correspond to the same vertex of $\mathbb{IF}_{G,\Phi,r}$. Moreover, for any two distinct nodes u, v, we have $(u, \text{view}_G(u, \varphi, r)) \neq (v, \text{view}_G(v, \varphi, r))$, even if $\text{view}_G(u, \varphi, r) = \text{view}_G(v, \varphi, r)$.

The set $\text{config}_G(\varphi, r) = \{(v, \text{view}_G(v, \varphi, r)) : v \in V$ is active in round r in $\varphi\}$ is called the *r-round configuration* for failure pattern φ. See Fig. 2 for the information flow graph of the triangle K_3, with one failure, and one communication round.

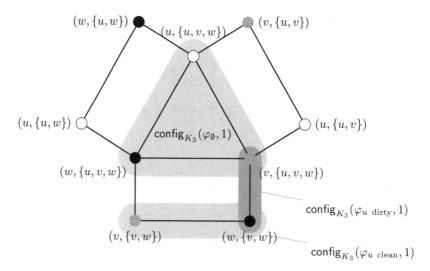

Fig. 2. $\mathbb{IF}_{K_3, \Phi_{\text{all}}^{(1)}, 1}$, with the $\text{config}_{K_3}(\varphi, 1)$ sets marked, for some $\varphi \in \Phi_{\text{all}}^{(1)}$; $\varphi_\emptyset$ denotes the failure pattern without failures, $\varphi_{u \text{ clean}}$ the failure patter where u fails cleanly in round 1 and $\varphi_{u \text{ dirty}}$ the failure patter where u fails in round 1 and sends a message only to v.

Lemma 4. *For every failure pattern $\varphi \in \Phi$, and every $r \geq 1$, the set $\mathsf{config}_G(\varphi, r)$ induces a connected subgraph of $\mathbb{IF}_{G,\Phi,r}$.*

Note that there is an edge from $(u, \mathsf{view}_G(u, \varphi, r))$ to $(v, \mathsf{view}_G(v, \psi, r))$ in $\mathbb{IF}_{G,\Phi,r}$ if and only if there exists $\varrho \in \Phi$ such that u and v are active in round r in ϱ, and $\mathsf{view}_G(u, \varphi, r) = \mathsf{view}_G(u, \varrho, r)$, $\mathsf{view}_G(v, \psi, r) = \mathsf{view}_G(v, \varrho, r)$ and $u \in \mathsf{view}_G(v, \varrho, r)$. Furthermore, if there are two failure patterns φ and ψ yielding the same view for a node v but two different views for a node u, then either the edges from the two views of u to the view of v both exist, or neither exists. This is specified in the following lemma.

Lemma 5. *Let $\varphi, \psi \in \Phi$ and $u, v \in V$ such that u and v are active in round r in both φ and ψ. If $\big((u, \mathsf{view}_G(u, \varphi, r)), (v, \mathsf{view}_G(v, \varphi, r))\big) \in E(\mathbb{IF}_{G,\Phi,r})$ and $\mathsf{view}_G(v, \varphi, r) = \mathsf{view}_G(v, \psi, r)$, then $\big((u, \mathsf{view}_G(u, \psi, r)), (v, \mathsf{view}_G(v, \psi, r))\big) \in E(\mathbb{IF}_{G,\Phi,r})$.*

4.2 The Solvability Characterization

The next result provides a solvability characterization for consensus by oblivious algorithms. In essence, it states that the number r of rounds should be large enough so that every connected component of $\mathbb{IF}_{G,\Phi,r}$ has a *dominating* node. A connected component of $\mathbb{IF}_{G,\Phi,r}$ is a connected component of the underlying, undirected graph of $\mathbb{IF}_{G,\Phi,r}$. We say that a node $v \in V$ of the graph G *dominates* a connected component C of $\mathbb{IF}_{G,\Phi,r}$, if the set $\{(v, \mathsf{view}_G(v, \varphi, r)) : \varphi \in \Phi\}$ dominates C. That is, for every $(w, \mathsf{view}_G(w, \varphi, r))$ in C, there is an arc from the vertex $(v, \mathsf{view}_G(v, \varphi, r))$ to $(w, \mathsf{view}_G(w, \varphi, r))$.

Theorem 3. *There is an oblivious algorithm solving consensus in r rounds under the set of failure patterns $\Phi \subseteq \Phi_{\mathrm{all}}^{(t)}$ if and only if every connected component C of $\mathbb{IF}_{G,\Phi,r}$ has a dominating node in V.*

The two directions of the theorem are proved by the next two lemmas.

Lemma 6. *For any $\Phi \subseteq \Phi_{\mathrm{all}}^{(t)}$, if every connected component C of $\mathbb{IF}_{G,\Phi,r}$ has a dominating node in V, then there is an oblivious algorithm solving consensus in r rounds under the set of failure patterns Φ.*

Proof. To solve consensus we only need to specify the decision function after r rounds of communication. For every connected component C of $\mathbb{IF}_{G,\Phi,r}$, pick a dominating node $v \in V$ of C. Let w be a node. The view view_w of w determines to which connected component C the vertex (w, view_w) belongs. The decision of w is the input value of the node v that dominates C.

Clearly, the algorithm satisfies termination and validity. For agreement, consider any $\varphi \in \Phi$. Let w and w' be two nodes that are active in round r in φ. By Lemma 4, the subgraph of $\mathbb{IF}_{G,\Phi,r}$ induced by $\mathsf{config}_G(\varphi, r)$ is connected. Therefore, $(w, \mathsf{view}(w, \varphi, r))$ and $(w', \mathsf{view}(w', \varphi, r))$ belongs to the same connected component C of $\mathbb{IF}_{G,\Phi,r}$, thus w and w' decide the input of the same node. □

Lemma 7. *For any* $\Phi \subseteq \Phi_{\text{all}}^{(t)}$, *if there is an oblivious algorithm solving consensus in* r *rounds under the set of failure patterns* Φ, *then every connected component* C *of* $\mathbb{IF}_{G,\Phi,r}$ *has a dominating node in* V.

Proof (Sketch of proof). We prove the contrapositive: if there is a connected component C of $\mathbb{IF}_{G,\Phi,r}$ with no dominating node in V, then there is no oblivious algorithm solving consensus in r rounds under Φ.

In the proof, we consider a standard connectivity argument a chain of failure patterns (executions). More specifically, we exhibit a sequence of failure patterns $\varphi_1, \ldots, \varphi_n$ such that (1) all nodes start with 0 in φ_1, (2) all nodes start with 1 in φ_n, and (3) there is a node v_i that has the same view in round r in both φ_i and φ_{i+1}. For proving (3), we exploit the fact that there is no node in V that dominates C, and thus it is possible to find a node that has the same view in both failure patterns, in round r. An algorithm cannot exist because the decision in φ_1 has to be 0, while the decision in φ_n has to be 1 and, then there are φ_i and φ_{i+1} with distinct decisions, which is a contradiction. $\square$

4.3   Optimality of $\mathsf{P}_{\text{adapt}}^{G,t}$ for Symmetric Graphs

To conclude, we use the characterization in Theorem 3 to show that $\mathsf{P}_{\text{adapt}}^{G,t}$ is time optimal for vertex-transitive graphs, among oblivious algorithms.

An *automorphism* of G is a bijection $\pi : V \to V$ such that, for every two nodes u and v, $\{u,v\} \in E \iff \{\pi(u), \pi(v)\} \in E$. A graph $G = (V, E)$ is *vertex-transitive* if, for every two nodes u and v, there exists an automorphism π of G such that $\pi(u) = v$. For instance, the complete graphs K_n, the cycles C_n, the d-dimensional hypercubes Q_d, the d-dimensional toruses $C_{n_1} \times \cdots \times C_{n_d}$, the Kneser graphs $KG_{n,k}$, the Cayley graphs, etc., are all vertex-transitive. The wheel, composed of a cycle and a central node, is not vertex-transitive, since the center node has degree $n - 1$ while the cycle nodes have degree 3.

Theorem 4. *If* G *is vertex-transitive, then there is no oblivious algorithm that solves consensus in fewer than* $\mathsf{radius}(G, \Phi_{\text{all}}^{(t)})$ *rounds.*

Proof (Sketch of proof). Clearly, the result holds if $\mathsf{radius}(G, \Phi_{\text{all}}^{(t)}) = 1$, as consensus is trivially not solvable in zero rounds in any graph with at least 2 nodes, even with no failures. So we assume now that $\mathsf{radius}(G, \Phi_{\text{all}}^{(t)}) \geq 2$.

In a vertex-transitive graph G, we have that for every $s \in V$, $\mathsf{radius}(G, \Phi_{\text{all}}^{(t)}) = \mathsf{ecc}_G(s, \Phi_{\text{all}}^{(t)})$. Therefore, for every $s \in V$, we can assign a failure pattern $\varphi_s \in \Phi_{\text{all}}^{(t)}$ such that $\mathsf{radius}(G, \Phi_{\text{all}}^{(t)}) = \mathsf{ecc}_G(s, \varphi_s)$. Let $\Phi = \{\varphi_s : s \in V\} \cup \{\varphi_\varnothing\}$. These execution sets $\mathsf{config}_G(\varphi, t)$ for $\varphi \in \Phi$ are depicted in Fig. 3 for the case K_3 and $t = 1$. We will show a result stronger than the one expressed in the statement of the theorem. Namely, we show that no oblivious algorithms can solve consensus in a vertex-transitive graph G under Φ in less than $\mathsf{radius}(G, \Phi_{\text{all}}^{(t)})$ rounds. That is, even if the algorithm has only to deal with the $n + 1$ failure patterns in

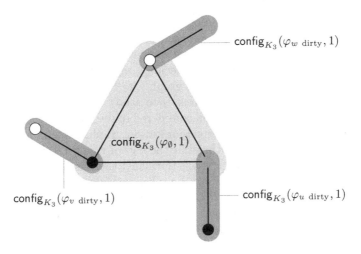

Fig. 3. The information flow graph $\mathbb{F}_{K_3,\Phi,1}$ appearing in the proof of Theorem 4, for K_3 and the failure pattern Φ defined there. $\varphi_\emptyset$ denotes the failure patter without failures, while $\varphi_{x\ \mathrm{dirty}}$ denotes the failure patter where x fails in round 1, sending a message to only one node.

$\Phi \subseteq \Phi_{\mathrm{all}}^{(t)}$, still consensus is not solvable in fewer than $\mathsf{radius}(G, \Phi_{\mathrm{all}}^{(t)})$ rounds. To establish this result, let $R = \mathsf{radius}(G, \Phi_{\mathrm{all}}^{(t)})$. Using Theorem 3, it is sufficient to prove that the following lemma:

Lemma 8. *The underlying graph of the information flow graph* $\mathbb{F}_{G,\Phi,R-1}$ *is connected and has no dominating vertex.*

The theorem directly follows from the previous lemma and the characterization in Theorem 3.   □

Theorem 5. *If* G *is vertex-transitive,* $\mathsf{P}_{\mathrm{adapt}}^{G,t}$ *is time optimal among oblivious algorithms.*

We conjecture that $\mathsf{P}_{\mathrm{adapt}}^{G,t}$ is time optimal for all graphs, among oblivious algorithms. This conjecture is grounded on the fact that Lemma 3 holds for all graphs, and not only for those that are vertex-transitive.

5   Conclusions

We have studied for the first time the number of rounds needed to solve fault-tolerant consensus in a crash prone synchronous network with arbitrary structure. We have defined a notion of *dynamic radius* of a graph G when t nodes may crash, which precisely determines the worst case number of rounds needed to solve oblivious consensus for vertex-transitive networks. The optimality of our algorithm was shown through a novel consensus solvability characterization in arbitrary networks, using the notion of *information flow*. A second consequence of the characterization is an abstract consensus algorithm that is optimal for all graphs. Our focus has been in the worst-case number of rounds. An interesting challenge would be to design early deciding algorithms (a problem that is well-studied in the case of the complete graph e.g. [8]).

An interesting future line of research is to study the case of non-oblivious algorithms (such algorithms have been considered in the past, e.g. [30]). Remarkably, for the case of the complete communication graph, there is no difference between these two types of algorithms: at the end of round $t + 1$, every pair of nodes have the same set of pairs (v, in_v) (formally, there is common knowledge on a set of inputs), hence decisions can be taken considering only this set.

Recall that, in our algorithms, $R(G, t)$ and $D(G, t)$ are hard-coded for a given G and t. It is worth exploring if our techniques are useful for the case where the graph G is not known to the nodes. Indeed, it is a challenge to combine fault-tolerant arguments with techniques of (failure-free) network computing [28]. Our results for $t = 0$ correspond to network computing. Yet, the case of $t > 0$ for arbitrary or evolving networks is an intriguing and complex research question.

References

1. Aguilera, M.K., Toueg, S.: A simple bivalency proof that t-resilient consensus requires t+1 rounds. Inf. Process. Lett. **71**(3), 155–158 (1999)
2. Alpern, B., Schneider, F.B.: Defining liveness. Inf. Process. Lett. **21**(4), 181–185 (1985)
3. Attiya, H., Castañeda, A., Herlihy, M., Paz, A.: Bounds on the step and namespace complexity of renaming. SIAM J. Comput. **48**(1), 1–32 (2019)
4. Attiya, H., Welch, J.: Distributed Computing: Fundamentals, Simulations, and Advanced Topics. Wiley Series on Parallel and Distributed Computing. Wiley, Hoboken (2004)
5. Berman, P., Garay, J.A.: Fast consensus in networks of bounded degree. Distrib. Comput. **7**(2), 67–73 (1993)
6. Castañeda, A., Fraigniaud, P., Paz, A., Rajsbaum, S., Roy, M., Travers, C.: A topological perspective on distributed network algorithms. In: Censor-Hillel, K., Flammini, M. (eds.) SIROCCO 2019. LNCS, vol. 11639, pp. 3–18. Springer, Cham (2019). https://doi.org/10.1007/978-3-030-24922-9_1
7. Castañeda, A., Gonczarowski, Y.A., Moses, Y.: Unbeatable consensus. In: Kuhn, F. (ed.) DISC 2014. LNCS, vol. 8784, pp. 91–106. Springer, Heidelberg (2014). https://doi.org/10.1007/978-3-662-45174-8_7
8. Castañeda, A., Moses, Y., Raynal, M., Roy, M.: Early decision and stopping in synchronous consensus: a predicate-based guided tour. In: El Abbadi, A., Garbinato, B. (eds.) NETYS 2017. LNCS, vol. 10299, pp. 206–221. Springer, Cham (2017). https://doi.org/10.1007/978-3-319-59647-1_16
9. Charron-Bost, B., Moran, S.: Minmax algorithms for stabilizing consensus. CoRR, abs/1906.09073 (2019)
10. Coulouma, É., Godard, E., Peters, J.G.: A characterization of oblivious message adversaries for which consensus is solvable. Theor. Comput. Sci. **584**, 80–90 (2015)
11. Dolev, D., Strong, H.: Authenticated algorithms for Byzantine agreement. SIAM J. Comput. **12**(4), 656–666 (1983)

12. Dolev, D.: The Byzantine generals strike again. J. Algorithms **3**(1), 14–30 (1982)
13. Dwork, C., Peleg, D., Pippenger, N., Upfal, E.: Fault tolerance in networks of bounded degree. In: Proceedings of the Eighteenth Annual ACM Symposium on Theory of Computing, STOC 1986, pp. 370–379. ACM (1986)
14. Dwork, C., Moses, Y.: Knowledge and common knowledge in a Byzantine environment: crash failures. Inf. Comput. **88**(2), 156–186 (1990)
15. Fischer, M.J., Lynch, N.A.: A lower bound for the time to assure interactive consistency. Inf. Process. Lett. **14**(4), 183–186 (1982)
16. Fischer, M.J., Lynch, N.A., Merritt, M.: Easy impossibility proofs for distributed consensus problems. Distrib. Comput. **1**(1), 26–39 (1986)
17. Fischer, M.J., Lynch, N.A., Paterson, M.: Impossibility of distributed consensus with one faulty process. J. ACM **32**(2), 374–382 (1985)
18. Godsil, C., Royle, G.: Algebraic Graph Theory. Graduate Texts in Mathematics, vol. 207. Springer, New York (2001). https://doi.org/10.1007/978-1-4613-0163-9
19. Hadzilacos, V.: A lower bound for Byzantine agreement with fail-stop processors. Technical report 21–83, Department of Computer Science, Harvard University, Cambridge, MA, July 1983
20. Herlihy, M., Kozlov, D., Rajsbaum, S.: Distributed Computing Through Combinatorial Topology. Morgan Kaufmann, Burlington (2013)
21. Herlihy, M., Rajsbaum, S., Tuttle, M.R.: An axiomatic approach to computing the connectivity of synchronous and asynchronous systems. Electr. Notes Theor. Comput. Sci. **230**, 79–102 (2009)
22. Khan, M.S., Naqvi, S.S., Vaidya, N.H.: Exact Byzantine consensus on undirected graphs under local broadcast model. In: Proceedings of the 2019 ACM Symposium on Principles of Distributed Computing, PODC, pp. 327–336 (2019)
23. Kuhn, F., Oshman, R.: Dynamic networks: models and algorithms. SIGACT News **42**(1), 82–96 (2011)
24. Lamport, L., Shostak, R., Pease, M.: The Byzantine generals problem. ACM Trans. Program. Lang. Syst. **4**(3), 382–401 (1982)
25. Lynch, N.A.: Distributed Algorithms. Morgan Kaufmann Publishers Inc., San Francisco (1996)
26. Moses, Y., Rajsbaum, S.: A layered analysis of consensus. SIAM J. Comput. **31**(4), 989–1021 (2002)
27. Nowak, T., Schmid, U., Winkler, K.: Topological characterization of consensus under general message adversaries. In: Proceedings of the 2019 ACM Symposium on Principles of Distributed Computing, PODC, pp. 218–227 (2019)
28. Peleg, D.: Distributed Computing: A Locality-Sensitive Approach. SIAM, Philadelphia (2000)
29. Raynal, M.: Consensus in synchronous systems: a concise guided tour. In: 9th Pacific Rim International Symposium on Dependable Computing (PRDC), pp. 221–228 (2002)
30. Raynal, M.: Fault-Tolerant Message-Passing Distributed Systems - An Algorithmic Approach. Springer, Cham (2018)
31. Santoro, N., Widmayer, P.: Agreement in synchronous networks with ubiquitous faults. Theor. Comput. Sci. **384**(2–3), 232–249 (2007)
32. Tseng, L., Vaidya, N.H.: Fault-tolerant consensus in directed graphs. In: Proceedings of the 2015 ACM Symposium on Principles of Distributed Computing, PODC, pp. 451–460. ACM (2015)

33. Tseng, L., Vaidya, N.H.: A note on fault-tolerant consensus in directed networks. SIGACT News **47**(3), 70–91 (2016)
34. Wensley, J.H., et al.: Sift: design and analysis of a fault-tolerant computer for aircraft control. Proc. IEEE **66**, 1240–1255 (1978)
35. Winkler, K., Schmid, U.: An overview of recent results for consensus in directed dynamic networks. Bull. Eur. Assoc. Theor. Comput. Sci. (EATCS) **128**, 41–72 (2019)

Tasks in Modular Proofs of Concurrent Algorithms

Armando Castañeda[1], Aurélie Hurault[2], Philippe Quéinnec[2(✉)],
and Matthieu Roy[3,4]

[1] Instituto de Matemáticas, UNAM, Mexico City, Mexico
`armando.castaneda@im.unam.mx`
[2] IRIT – Université de Toulouse, Toulouse, France
`{hurault,queinnec}@enseeiht.fr`
[3] Laboratorio Solomon Lefschetz - UMI LaSoL, CNRS, CONACYT, UNAM,
Cuernavaca, Mexico
[4] LAAS-CNRS, CNRS, Université de Toulouse, Toulouse, France
`roy@laas.fr`

Abstract. Proving correctness of distributed or concurrent algorithms is a mind-challenging and complex process. Slight errors in the reasoning are difficult to find, calling for computer-checked proof systems. In order to build computer-checked proofs with usual tools, such as Coq or TLA$^+$, having sequential specifications of all base objects that are used as building blocks in a given algorithm is a requisite to provide a modular proof built by composition. Alas, many concurrent objects do not have a sequential specification.

This article describes a systematic method to transform any *task*, a specification method that captures concurrent one-shot distributed problems, into a sequential specification involving two calls, set and get. This transformation allows system designers to *compose* proofs, thus providing a framework for modular computer-checked proofs of algorithms designed using tasks and sequential objects as building blocks. The Moir&Anderson implementation of *renaming* using *splitters* is an iconic example of such algorithms designed by composition.

Keywords: Formal methods · Verification · Concurrent algorithms · Renaming

1 Introduction

Fault-tolerant distributed and concurrent algorithms are extensively used in critical systems that require strict guarantees of correctness [23]; consequently, verifying such algorithms is becoming more important nowadays. Yet, proving distributed and concurrent algorithms is a difficult and error-prone task, due to the complex interleavings that may occur in an execution. Therefore, it is crucial to develop frameworks that help assessing the correctness of such systems.

A major breakthrough in the direction of systematic proofs of concurrent algorithms is the notion of *atomic* or *linearizable* objects [20]: a linearizable

© Springer Nature Switzerland AG 2019
M. Ghaffari et al. (Eds.): SSS 2019, LNCS 11914, pp. 69–83, 2019.
https://doi.org/10.1007/978-3-030-34992-9_6

object behaves as if it is accessed sequentially, even in presence of concurrent invocations, the canonical example being the atomic register. Atomicity lets us model a concurrent algorithm as a transition system in which each transition corresponds to an atomic step performed by a process on a base object. Human beings naturally reason on sequences of events happening one after the other; concurrency and interleavings seem to be more difficult to deal with.

However, it is well understood now that several natural one-shot base objects used in concurrent algorithms cannot be expressed as sequential objects [9,16,33] providing a single operation.

An iconic example is the *splitter* abstraction [31], which is the basis of the classical Moir&Anderson renaming algorithm [31]. Intuitively, a splitter is a concurrent one-shot problem that splits calling processes as follows: whenever p processes access a splitter, at most one process obtains stop, at most $p - 1$ obtain right and at most $p - 1$ obtain down. Moir&Anderson renaming algorithm uses splitters arranged in a half grid to scatter processes and provide new names to processes. It is worth to mention that, since its introduction almost thirty years ago, the *renaming* problem [4] has become a paradigm for studying symmetry-breaking in concurrent systems (see, for example, [1,8]).

A second example is the *exchanger* object provided in Java, which has been used for implementing efficient linearizable elimination stacks [16,24,36]. Roughly speaking, an exchanger is a meeting point where pairs of processes can exchange values, with the constraint that an exchange can happen only if the two processes run concurrently.

Splitters and exchangers are instances of one-shot concurrent objects known in the literature as *tasks*. Tasks have played a fundamental role in understanding the computability power of several models, providing a topological view of concurrent and distributed computing [18]. Intuitively, a task is an object providing a single one-shot operation, formally specified through an input domain, an output domain and an input/output relation describing the valid output configurations when a set of processes run concurrently, starting from a given input configuration. Tasks can be equivalently specified by mappings between topological objects: an input simplicial complex (i.e., a discretization of a continuous topological space) modeling all possible input assignments, an output simplicial complex modeling all possible output assignments, and a carrier map relating inputs and outputs.

Contributions. Our main contribution is a generic transformation of any task T (with a single operation) into a sequential object S providing two operations, set and get. The behavior of S "mimics" the one of T by splitting each invocation of a process to T into two invocations to S, first set and then get. Intuitively, the set operation records the processes that are participating to the execution of the task. A process actually calls the task and obtains a return value by invoking get. Each of the operations is atomic; however, set and get invocations of a given process may be interleaved with similar invocations from other processes.

We show that these two operations are sufficient for any task, no matter how complicated it may be; since a task is a mapping between simplicial complexes, it can specify very complex concurrent behaviors, sometimes with obscure associated operational semantics.

A main benefit of our transformation is that one can replace an object solving a task T by its associated sequential object S, and reason as if all steps happen sequentially. This allows us to obtain simpler models of concurrent algorithms using solutions to tasks and sequential objects as building blocks, leading to modular correctness proofs. Concretely, we can obtain a simple transition system of Moir&Anderson renaming algorithm, which helps to reason about it. In a companion paper [22], our model is used to derive a full and modular TLA$^+$ proof of the algorithm, the first available TLA$^+$ proof of it.

In Sect. 2, we explain the ideas in Moir&Anderson renaming algorithm that motivated our general transformation, which is presented in Sect. 3. Due to lack of space, some basic definitions, proofs and detailed constructions are omitted. They can be found in the extended version [7].

2   Verifying Moir&Anderson Renaming

We consider a concurrent system with n asynchronous processes, meaning that each process can experience arbitrarily long delays during an execution. Moreover, processes may crash at any time, i.e., permanently stopping taking steps. Each process is associated with a unique ID $\in \mathbb{N}$. The processes can access *base* objects like simple atomic read/write registers or more complex objects.

The original Moir&Anderson renaming algorithm [31] is designed and explained with splitters. Their seminal work first introduces the splitter algorithm based on atomic read/write registers and discusses its properties. Then, they describe a renaming algorithm that uses a grid of splitters. The actual implementation inlines splitters into the code of the renaming algorithm, and their proof is performed on the resulting program that uses solely read/write registers as base objects.

The Splitter Abstraction. A *splitter* [31] is a one-shot concurrent task in which each process starts with its unique ID $\in \mathbb{N}$ and has to return a value satisfying the following properties: (1) Validity. The returned value is right, down nor stop. (2) Splitting. If $p \geq 1$ processes participate in an execution of the splitter, then at most $p - 1$ processes obtain the value right, at most $p - 1$ processes obtain the value down, at most one process obtains the value stop. (3) Termination. Every correct process (which doesn't crash) returns a value.

Notice that if a process runs solo, i.e., $p = 1$, it must obtain stop, since the splitting property holds for any $p \geq 1$.

Figure 1 contains the simple and elegant splitter implementation based on atomic read/write registers from [31] (register names have been changed for clarity). After carefully analyzing the code, the reader can convince herself that the algorithm described in Fig. 1 implements the splitter specification. The fact that the implementation is based on *atomic* registers allows us to obtain a transition system of it in which each transition corresponds to an atomic operation

```
initially CLOSED = false
operation splitter():
(01)   LAST ← my_ID;
(02)   if (CLOSED)
(03)       then return(right)
(04)       else  CLOSED ← true;
(05)           if (LAST = my_ID)
(06)           then return(stop)
(07)           else  return(down)
(08)           end if
(09)   end if.
```

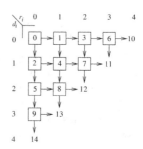

Fig. 1. Implementation of a Splitter [31].   **Fig. 2.** Renaming using Splitters.

on an object. The benefit of this modelization is that every execution of the implementation is simply described as a sequence of steps, as concurrent and distributed systems are usually modeled (see, for example, [19,35]). Although the splitter implementation is very short and simple, its TLA$^+$ proof is long and rather complex —particularly when considering that it uses a boolean register and a plain register only—(see [22] for details).

The Renaming Problem. In the M-renaming task [4], each process starts with its unique ID $\in \mathbb{N}$, and processes are required to return an output name satisfying the following properties: (1) Validity. The output name of a process belongs to $[1,\ldots,M]$. (2) Uniqueness. No two processes obtain the same output name. (3) Termination. Every correct process returns an output name.

Let p be the number of processes that participate in a given renaming instance. A renaming implementation is *adaptive* if the size M of the new name space depends only on p, the number of participating processes. We have then $M = f(p)$ where $f(p)$ is a function on p such that $f(1) = 1$ and, for $2 \le p \le n$, $p - 1 \le f(p-1) \le f(p)$.

Moir&Anderson Splitter-Based Renaming Algorithm. Moir and Anderson propose in [31] a read/write renaming algorithm designed using the splitter abstraction. The algorithm is conceptually simple: for up to n processes, a set of $n(n+1)/2$ splitters are placed in a half-grid, each with a unique name, as shown in Fig. 2 for $n = 5$. Each process starts invoking the splitter at the top-left corner, following the directions obtained at each splitter. When a splitter invocation returns stop, the process returns the name associated with the splitter. We use here an adaptive version of their algorithm that allows p participating processes to rename in at most $p(p+1)/2$ names; the original solution in [31] is non-adaptive and the only difference is the labelling of the splitters in the grid.

Splitters as Sequential Objects? Although Moir&Anderson renaming algorithm is easily described in a modular way, the actual program is not modular as each splitter in the conceptual grid is replaced by an independent copy of the splitter implementation of Fig. 1. Thus, the correctness proof in [31] deals with the

State: Sets $Participants, Stop, Down, Right$
all sets are initialized to $\emptyset$

Function set(id)
   Pre-condition: id $\notin Participants$
   Post-condition: $Participants' \leftarrow Participants \cup \{id\}$
   Output: void
endFunction

Function get(id)
   Pre-condition: id $\in Participants \wedge$ id $\notin Stop, Down, Right$
   Post-condition:
      $D \leftarrow \emptyset$
      if $|Stop| = 0$ **then** $D \leftarrow D \cup \{$stop$\}$
      if $|Down| < |Participants| - 1$ **then** $D \leftarrow D \cup \{$down$\}$
      if $|Right| < |Participants| - 1$ **then** $D \leftarrow D \cup \{$right$\}$
      Let dec be any value in D
      if $dec =$ stop **then** $Stop \leftarrow Stop \cup \{id\}$
      if $dec =$ down **then** $Down \leftarrow Down \cup \{id\}$
      if $dec =$ right **then** $Right \leftarrow Right \cup \{id\}$
   Output: dec
endFunction

Fig. 3. An *ad hoc* specification of the Splitter.

possible interleavings that can occur, considering all read/write splitter implementations in the grid.

In the light of the simple splitter based conceptual description, we would like to have a transition system describing the algorithm based on splitters as building blocks, in which each step corresponds to a splitter invocation. Such a description would be very beneficial as it would allow us to obtain a modular correctness proof showing that the algorithm is correct as long as the building blocks are splitters, hence the correctness is independent of any particular splitter implementation.

As it is formally proved in Sect. 3, it is impossible to obtain such a transition system. The obstacle is that a splitter is inherently concurrent and cannot be specified as a sequential object with a single operation. The intuition of the impossibility is the following. By contradiction, suppose that there is a sequential object corresponding to a splitter. Since the object is sequential, in every execution, the object behaves as if it is accessed sequentially (even in presence of concurrent invocation). Then, there is always a process that invokes the splitter object first, which, as noted above, must obtain stop. The rest of the processes can obtain either down or right, without any restriction (the value obtained by the first process precludes that all obtain right or all down). However, such an object is allowing strictly fewer behaviors: in the original splitter definition it is perfectly possible that all processes run concurrently and half of them obtain right and the other half obtain down, while none obtains stop.

The Splitter Task as a Sequential Object. One can circumvent the impossibility described above by splitting the single method provided by a splitter into two (atomic) operations of a sequential object. Figure 3 presents a sequential specification of a splitter with two operations, set and get, using a standard pre/post-condition specification style. Each process invoking the splitter, first invokes set and then get (always in that order). The idea is that the set operation first records in the state of the object the processes that are participating in the splitter, so far, and then the get operation nondeterministically produces an output to a process, considering the rules of the splitter. In Sect. 3, we formally prove that this sequential object indeed models the splitter defined above.

Proving Moir&Anderson Renaming with Splitters as Base Sequential Objects. Using the sequential specification of a splitter in Fig. 3, we can easily obtain a *generic* description of the original Moir&Anderson splitter-based algorithm: each renaming object is replaced with an equivalent sequential version of it, and every process accessing a renaming object asynchronously invokes first set and then get, which returns a direction to the process. The resulting algorithm does not rely on any particular splitter implementation, and uses only atomic objects, which allows us to obtain a transition system of it. This is the algorithm that is verified in TLA$^+$ in [22]. The equivalence between the concurrent renaming specification and the sequential set/get specification imply that the proof in [22] also proves for the original Moir&Anderson splitter-based algorithm.

3   Dealing with Tasks Without Sequential Specification

In this section, we show that the transformation in Sect. 2 of the splitter task into a sequential object with two operations, get and set, is not a trick but rather a general methodology to deal with tasks without a sequential specification. Our get/set solution proposed here is reminiscent to the *request-follow-up* transformation in [25] that allows to transform a *partial* method of a sequential object (e.g. a queue with a blocking dequeue method when the queue is empty) into two *total* methods: a total request method registering that a process wants to obtain an output, and a total follow-up method obtaining the output value, or *false* if the conditions for obtaining a value are not yet satisfied (the process invokes the follow-up method until it gets an output). We stress that the *request-follow-up* transformation [25] considers only objects with a sequential specification and is not shown to be general as it is only used for queues and stacks.

Model of Computation in Detail. We consider a standard concurrent system with n *asynchronous* processes, $p_1, \ldots, p_n$, which may *crash* at any time during an execution of the system, i.e., stopping taking steps (for more detail see for example [19,35]). Processes communicate with each other by invoking operations on shared, concurrent *base objects*. A base object can provide

Read/Write operations (also called *register*), more powerful operations, such as Test&Set, Fetch&Add, Swap or Compare&Swap, or solve a concurrent distributed problem, for example, Splitter, Renaming or Set_Agreement.

Each process follows a local state machines $A_1, \ldots, A_n$, where A_i specifies which operations on base objects p_i executes in order to return a response when it invokes a high-level operation (e.g. push or pop operations). Each of these base-objects operation invocations is a *step*. An *execution* is a possibly infinite sequence of steps and invocations and responses of high-level operations, with the following properties:

1. Each process first invokes a high-level operation, and only when it has a corresponding response, it can invoke another high-level operation, i.e., executions are *well-formed*.
2. For any invocation $inv(\langle \mathsf{opType}, p_i, input \rangle)$ of a process p_i, the steps of p_i between that invocation and its corresponding response (if there is one), are steps that are specified by A_i when p_i invokes the high-level operation $\langle \mathsf{opType}, p_i, input \rangle$.

A high-level operation in an execution is *complete* if both its invocation and response appear in the execution. An operation is *pending* if only its invocation appears in the execution. A process is *correct* in an execution if it takes infinitely many steps.

Sequential Specifications. A central paradigm for specifying distributed problems is that of a shared object X that processes may access concurrently [19,35], but the object is defined in terms of a *sequential specification*, i.e., an automaton describing the outputs the object produces when it is accessed sequentially. Alternatively, the specification can be described as (possibly infinite) prefix-closed set, $SSpec(X)$, with all sequential executions allowed by X.

Once we have a sequential specification, there are various ways of defining what it means for an execution to *satisfy* an object, namely, that it respects the sequential specification. *Linearizability* [20] is the standard notion used to identify correct executions of implementations of sequential objects. Intuitively, an execution is linearizable if its operations can be ordered sequentially, without reordering non-overlapping operations, so that their responses satisfy the specification of the implemented object. To formalize this notion we define a partial order on the completed operations of an execution E: op $<_E$ op' if and only if the response of op precedes the invocation of op' in E. Two operations are *concurrent* if they are incomparable by $<_E$. The execution is *sequential* if $<_E$ is a total order.

An execution E is *linearizable* with respect to X if there is a sequential execution S of X (i.e., $S \in SSpec(X)$) such that: (1) S contains every completed operation of E and might contain some pending operations. Inputs and outputs of invocations and responses in S agree with inputs and outputs in E, and (2) for every two completed operations op and op' in E, if op $<_E$ op', then op appears before op' in S.

Using the linearizability correctness criteria for sequential objects, we can define the set of *valid* executions for X, denoted $VE(X)$, as the set containing every execution E that consists of invocations and responses and is linearizable w.r.t. X. $VE(X)$ contains the behavior one might expect from any *building-block* implementation of X, e.g., any algorithm that implements X.

Tasks. A task is the basic distributed equivalent of a function in sequential computing, defined by a set of inputs to the processes and for each (distributed) input to the processes, a set of legal (distributed) outputs of the processes, e.g., [18].

In an algorithm designed to solve a task, each process starts with a private input value and has to eventually decide irrevocably on an output value. A process p_i is initially not aware of the inputs of other processes. Consider an execution where only a subset of $k \leq n$ processes participate; the others crash without taking any steps. A set of pairs $\sigma = \{(\mathrm{id}_1, x_1), \ldots, (\mathrm{id}_k, x_k)\}$ is used to denote the input values, or output values, in the execution, where x_i denotes the value of the process with identity id_i, either an input value or an output value. A set σ as above is called a *simplex*, and if the values are input values, it is an *input simplex*, if they are output values, it is an *output simplex*. The elements of σ are called *vertices*, and any subset of σ is a *face* of it. An *input vertex* $v = (\mathrm{id}_i, x_i)$ represents the initial state of process id_i, while an *output vertex* represents its decision. The *dimension* of a simplex σ, denoted $dim(\sigma)$, is $|\sigma| - 1$, and it is *full* if it contains n vertices, one for each process. A *complex* $\mathcal{K}$ is a set of simplexes (i.e. a set of sets) closed under containment. The dimension of $\mathcal{K}$ is the largest dimension of its simplexes, and $\mathcal{K}$ is *pure* of dimension k if each of its simplexes is a *face* of a k-dimensional simplex. In distributed computing, the simplexes (and complexes) are often *chromatic*: vertices of a simplex are labeled with a distinct process identities. The set of processes identities in an input or output simplex σ is denoted $\mathrm{ID}(\sigma)$.

A *task* T for n processes is a triple $(\mathcal{I}, \mathcal{O}, \Delta)$ where $\mathcal{I}$ and $\mathcal{O}$ are pure chromatic $(n-1)$-dimensional complexes, and Δ maps each simplex σ from $\mathcal{I}$ to a subcomplex $\Delta(\sigma)$ of $\mathcal{O}$, satisfying: (1) $\Delta(\sigma)$ is pure of dimension $dim(\sigma)$, (2) for every τ in $\Delta(\sigma)$ of dimension $dim(\sigma)$, $\mathrm{ID}(\tau) = \mathrm{ID}(\sigma)$, and (3) if σ, σ' are two simplexes in $\mathcal{I}$ with $\sigma' \subset \sigma$ then $\Delta(\sigma') \subset \Delta(\sigma)$. A task is a very compact way of specifying a distributed problem, and indeed typically it is hard to understand what exactly is the problem being specified. Intuitively, Δ specifies, for every simplex $\sigma \in \mathcal{I}$, the valid outputs $\Delta(\sigma)$ for the processes in $\mathrm{ID}(\sigma)$ assuming they run to completion, and the other processes crash initially, and do not take any steps.

As an example consider the *splitter* task [31]. Figure 4 shows a graphic description of the splitter task for three processes with IDs 1, 2 and 3. The input complex, shown at the left, consists of a triangle and all its faces. The output complex, at the right, contains all possible valid output simplexes (the triangle with all right outputs is not in the complex). The Δ function maps the input vertex with ID 1 to the output vertex $(1, \mathrm{stop})$, the input edge with IDs 1 and 2 to the complex with the bold edges in the output complex, and the

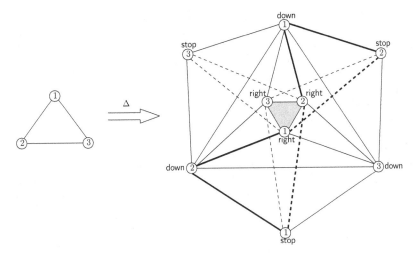

Fig. 4. The Splitter Task for Three Processes.

input triangle is mapped to the whole output complex. The rest of Δ is defined symmetrically.

Let E be an execution where each process invokes a task $\langle \mathcal{I}, \mathcal{O}, \Delta \rangle$ once. Then, σ_E is the input simplex defined as follows: (id_i, x_i) is in σ_E iff in E there is an invocation of $\mathsf{task}(x_i)$ by process id_i. The output simplex τ_E is defined similarly: (id_i, y_i) is in τ_E iff there is a response y_i to a process id_i in E. We say that E *satisfies* $(\mathcal{I}, \mathcal{O}, \Delta)$ if for every prefix E' of E, it holds that $\tau_{E'} \in \Delta(\sigma_{E'})$.

Using the satisfiability notion of tasks we can now consider the set of valid executions, $VE(T)$, for a given task $T = (\mathcal{I}, \mathcal{O}, \Delta)$: the set containing every execution E that has only invocations and responses and satisfies T. Arguably, the set $VE(T)$ contains the behavior one might expect from a *building-block* (e.g. an algorithm) that implements T.

Modeling Tasks as Sequential Objects. Intuitively, tasks and sequential specifications are inherently different paradigms for specifying distributed problems: while a task specifies what a set of processes might output when running concurrently, a sequential specification specifies the behavior of a concurrent object when accessed sequentially (and linearizability tells when a concurrent execution "behaves" like a sequential execution of the object). A natural question is if any task can be modeled as a sequential object with a single operation, namely, the object defines the same set of valid executions. A well-known example for which this is possible is the consensus distributed coordination problem that can be equivalently defined as a task or as a sequential object (see for example [19] where it is defined as an object[1] and [18] where it is defined as a task).

---

[1] Sometimes, for clarity or efficiency, the object is defined with two operations (in the style of the Theorem 1); however, consensus can be equivalently defined with one operation.

State: a pair (σ, τ) of input/output simplexes, initialized to $(\emptyset, \emptyset)$

Function set(id_i, x_i)
   Pre-condition: $\mathrm{id}_i \in \mathrm{ID} \wedge \mathrm{id}_i \notin \mathrm{ID}(\sigma)$
   Post-condition: $\sigma' \leftarrow \sigma \cup \{(\mathrm{id}_i, x_i)\}$
   Output: void
endFunction

Function get(id_i)
   Pre-condition: $\mathrm{id}_i \in \mathrm{ID} \wedge \mathrm{id}_i \notin \mathrm{ID}(\tau)$
   Post-condition: Let y_i be any output value such that $\tau \cup \{(\mathrm{id}_i, y_i)\} \in \Delta(\sigma)$.
               Then, $\tau' \leftarrow \tau \cup \{(\mathrm{id}_i, y_i)\}$
   Output: y_i
endFunction

Fig. 5. A Generic Sequential Specification of a Task $T = (\mathcal{I}, \mathcal{O}, \Delta)$.

Lemma 1. *Consider the splitter task $T_{\mathsf{spl}} = (\mathcal{I}_{\mathsf{spl}}, \mathcal{O}_{\mathsf{spl}}, \Delta_{\mathsf{spl}})$. There is no sequential object X_{spl} with a single operation satisfying $VE(T_{\mathsf{spl}}) = VE(X_{\mathsf{spl}})$.*

In a very similar way, one can prove that the following known tasks cannot be specified as sequential objects with a single operation: *exchanger* [17,36], *adaptive renaming* [4], *set agreement* [10], *immediate snapshot* [5], *adopt-commit* [6,13] and *conflict detection* [3].[2]

To circumvent the impossibility result in Lemma 1, we model any given task T through a sequential object S with two operations, set and get, that each process access in a specific way: it first invokes set with its input to the task T (receiving no output) and later invokes get in order to get an output value from T. Intuitively, decoupling the single operation of T into two (atomic) operations allows us to model concurrent behaviors that a single (atomic) operation cannot specify. In what follows, let $SSpec(S)$ be the set with all sequential executions of S in which each process invokes at most two operations, first set and then get, in that order.

Theorem 1. *For every task $T = (\mathcal{I}, \mathcal{O}, \Delta)$ there is a sequential object S with two operations, $\mathsf{set}(id_i, x_i)$ and $\mathsf{get}(id_i) : y_i$, such that there is a bijection α between $VE(T)$ and $SSpec(S)$ satisfying that: (1) each invocation or response of process id_i is mapped to an operation of process id_i, and (2) each invocation inv (response resp) with input (output) x is mapped to a completed set (get) operation with input (output) x.*

An implication of Theorem 1 is that if one is analyzing an algorithm that uses a building-block (subroutine, algorithm, etc.) B that solves a task T, one

---

[2] There are non-deterministic sequential specifications of these tasks with *unavoidable* and *pathological* executions in which some operations *guess* the inputs of future operations. See [9, Section 2] for a detailed discussion.

can safely replace B with the sequential object S related to T described in the theorem (each invocation to the operation of B is replaced with an (atomic) invocation to set and then an (atomic) invocation to get), and then analyze the algorithm considering the atomic operations of S. The advantage of this transformation is that (1) if all operations in an algorithm are atomic, we can think that each process takes a step at a time in an execution, hence obtaining a a transition system with atomic events, (2) at all times we have a concrete state of S in an execution (which is not clear in a task specification) and (3) given a state of S, an output for a get operation can be easily computed using the sequential object S (something that is typically complicated for B as it might be accessed concurrently).

The construction used (for simplicity) in the proof of Theorem 1 (in the full version of the paper) might be too coarse to be helpful for analyzing an algorithm. We would like to have a construction producing an equivalent sequential automaton modeling the task in a simpler way. Consider the simple sequential object in Fig. 5 obtained from any given task $T = (\mathcal{I}, \mathcal{O}, \Delta)$, which is described in a classic pre/post-condition form. Intuitively, the meaning of a state (σ, τ) is the following: σ contains the processes that have invoked the task so far (this represents the *participating set* of the current execution) while τ contains the outputs that have been produced so far. The main invariant of the specification is that $\tau \in \Delta(\sigma)$. It directly follows from the properties of the task: when a process invokes set(id_i, x_i), we have that $\tau \in \Delta(\sigma \cup \{(\mathrm{id}_i, x_i)\})$ because $\Delta(\sigma) \subset \Delta(\sigma \cup \{(\mathrm{id}_i, x_i)\})$, and when a process invokes get(id_i), it holds that $\tau \cup \{(\mathrm{id}_i, y_i)\} \in \Delta(\sigma)$ because $\Delta(\sigma)$ is a pure complex of dimension $dim(\sigma)$ and thus there must exist a simplex in $\Delta(\sigma)$ (properly) containing τ and with an output for id_i. One can formally prove that this sequential object and the one in the proof Theorem 1 define the same set of sequential executions.

Finally, one can obtain ad-hoc and equivalent specifications for specific tasks, like the one for splitters in Fig. 3 in Sect. 2.

4   Related Work

Linearizability Criteria. Neiger observed for the first time that some fundamental tasks, like *set agreement* [10] and *immediate snapshot* [5], cannot be modeled as sequential objects [33] (with a single operation). Motivated by the need of a unified framework for tasks and objects, he proposed *set-linearizability* [33]. Roughly speaking, a set sequential object is generalization of a sequential object in which transitions between states involve more than one operation (formally, a set of operations), meaning that these operations are allowed to occur concurrently, and their results can be *concurrency-dependent*. Set linearizability is the consistency condition for set-sequential objects, where one needs to find linearizability points (same as in linearizability) and several operations can be linearized at the same point (different from linearizability).

Later on, it was again observed that for some concurrent objects it is impossible to provide a sequential specification, and *concurrency-aware* linearizability was defined [16]. Set linearizability and concurrency-aware linearizability are very closely related, both based on the same principle: sets of operations can occur concurrently. Also, a non-automatic verification technique for reasoning about concurrency-aware objects is presented in [16].

Recently it was observed in [9] that some natural tasks specify concurrency dependencies that are beyond the set-linearizability and concurrency-aware formalisms, hence that paper proposed *interval linearizability*. In an interval-sequential object not only sets of operations can occur concurrently but some of these operations might be pending and then overlap operations in the next transition; thus each operation corresponds to an interval instead of a single point. Interval linearizability is the related consistency condition in which, for each operation, one needs to find an interval in which the operation happens. It is shown in [9] that interval-linearizability is *complete* for tasks in the sense that it is possible to specify *any* task as an interval-sequential object (with a single operation).

Although interval-sequential specifications can model any task, this approach does not seem to be useful when one is searching for machine-checked proofs of concurrent algorithms. The main reason is that by replacing a task with its equivalent interval-sequential object, we obtain a transition system in which one still needs to think in concurrent behaviors, which is usually hard to deal with. In contrast, our proposed get-set transformation allows to "decouple" the inherent concurrency in tasks in a way that in the resulting transition system all events are atomic, namely, they happen one after the other.

Mechanized Verification of Distributed Algorithms. Mechanized (or machine-assisted) verification of distributed and concurrent algorithms is usually done with model checking or theorem proving or a combination of both. Enumerative model-checking is the oldest fully automatic method with tools like Spin [21] or TLC, the TLA+ model checker [27]. To avoid the well-known problem of state explosion, various optimisations such as symmetry or reduction have been introduced, and recent work is on going on parameterized model checking, for instance with MCMT (Model Checking Modulo Theory) [14], Cubicle [11] or ByMC [26]. Nevertheless, automatic verification of a distributed/concurrent algorithm is still restricted to small finite instances of the algorithm or imposes significant constraints on its description, due to the limited expressiveness of the specification language.

Fully automatic theorem proving is based on a proof decision procedure. For useful logics, it is often semi-decidable at best and heavily depends on heuristics to achieve good performance. Recent work on SMT has made a substantial leap forward checking complex formulae combining first-order reasoning with decision procedures for theory such as arithmetic, equality, arrays. Nonetheless, the overall proof of a distributed algorithm is still largely manual and, when seeking confidence in this proof, an interactive proof assistant is the current approach. Several examples of verification of complex distributed algorithms exist: Chord with Alloy [38], Pastry with TLA$^+$ [29,30], Paxos also with TLA$^+$ [28], snapshot algorithms in Event-B [2], just to cite a few.

Several wait-free implementations of tasks have been mechanically proven (e.g. [12,34,37]). However, to the best of our knowledge, no non-trivial algorithm built upon concurrent tasks have been mechanically proved. Our intuition for this situation is that proofs cannot be made modular and compositional when using bricks which are inherently concurrent if their internal structure must be visible to take into account this concurrency. Several complex and original algorithms can be found in the literature such as Moir and Anderson renaming algorithm [31] that we have considered in this paper, stacks implemented with elimination trees [36], lock-free queues with elimination [32]. In these papers, the correctness proofs are intricate as they must consider the algorithm as a whole, including the tricky part involving wait-free objects, and they have not been mechanically checked. Our approach which exposes a more simple and sequential specification (instead of a complex concurrent implementation) seeks to alleviate this limitation.

5   Final Remarks and Future Work

In this paper, we showed a technique to circumvent the known impossibility of specifying a task as a sequential object. Our technique consists in modeling the single operation of the task with two atomic operations, set and get. This transformation leads to a framework for developing transitional models of concurrent algorithms using tasks and sequential objects as building blocks. As a proof of concept, we developed in a companion paper [22] a full and modular TLA$^+$ proof of the Moir&Anderson renaming algorithm [31].

A natural extension of our work is to apply the framework to other concurrent algorithms. Another direction is to extend our techniques to the case of *refined tasks* and *interval-sequential* objects, recently defined in [9]. These two formalisms are generalization of the task and sequential object formalism with strictly more expressiveness; particularly, contrary to the task formalism, refined task are *multi-shot*, namely, each process may perform several invocations, possibly infinitely many.

A third direction is to study if the duality between the epistemic logic approach and the topological approach shown in [15] might be useful in verifying concurrent algorithms. Generally speaking, it is shown in [15] that a task can be represented as a *Kripke model* with an *action model*, specifying the knowledge obtained by processes when solving the task. It could be interesting to explore how this knowledge could be reflected in our set/get construction and if it could be useful in proving correctness.

Acknowledgements. Armando Castañeda was supported by PAPIIT project IA102417.

References

1. Alistarh, D.: The renaming problem: recent developments and open questions. Bull. EATCS **3**, 117 (2015)
2. Andriamiarina, M.B., Méry, D., Singh, N.K.: Revisiting snapshot algorithms by refinement-based techniques. Comput. Sci. Inf. Syst. **11**(1), 251–270 (2014)
3. Aspnes, J., Ellen, F.: Tight bounds for adopt-commit objects. Theory Comput. Syst. **55**(3), 451–474 (2014)
4. Attiya, H., Bar-Noy, A., Dolev, D., Peleg, D., Reischuk, R.: Renaming in an asynchronous environment. J. ACM **37**(3), 524–548 (1990)
5. Borowsky, E., Gafni, E.: Generalized FLP impossibility result for t-resilient asynchronous computations. In: STOC 1993: Proceedings of the ACM Symposium on Theory of computing, pp. 91–100. ACM, New York (1993)
6. Borowsky, E., Gafni, E., Lynch, N.A., Rajsbaum, S.: The BG distributed simulation algorithm. Distrib. Comput. **14**(3), 127–146 (2001)
7. Castañeda, A., Hurault, A., Quéinnec, P., Roy, M.: Tasks in modular proofs of concurrent algorithms. CoRR, arXiv:1909.05537 [cs.DC] (2019)
8. Castañeda, A., Rajsbaum, S., Raynal, M.: The renaming problem in shared memory systems: an introduction. Comput. Sci. Rev. **5**(3), 229–251 (2011)
9. Castañeda, A., Rajsbaum, S., Raynal, M.: Unifying concurrent objects and distributed tasks: interval-linearizability. J. ACM **65**(6), 45 (2018)
10. Chaudhuri, S.: More choices allow more faults: set consensus problems in totally asynchronous systems. Inf. Comput. **105**(1), 132–158 (1993)
11. Conchon, S., Goel, A., Krstić, S., Mebsout, A., Zaïdi, F.: Cubicle: a parallel SMT-based model checker for parameterized systems. In: Madhusudan, P., Seshia, S.A. (eds.) CAV 2012. LNCS, vol. 7358, pp. 718–724. Springer, Heidelberg (2012). https://doi.org/10.1007/978-3-642-31424-7_55
12. Drăgoi, C., Gupta, A., Henzinger, T.A.: Automatic linearizability proofs of concurrent objects with cooperating updates. In: Sharygina, N., Veith, H. (eds.) CAV 2013. LNCS, vol. 8044, pp. 174–190. Springer, Heidelberg (2013). https://doi.org/10.1007/978-3-642-39799-8_11
13. Gafni, E.: Round-by-round fault detectors: unifying synchrony and asynchrony (extended abstract). In: Proceedings of the Seventeenth Annual ACM Symposium on Principles of Distributed Computing, PODC 1998, pp. 143–152 (1998)
14. Ghilardi, S., Ranise, S.: MCMT: a model checker modulo theories. In: Giesl, J., Hähnle, R. (eds.) IJCAR 2010. LNCS (LNAI), vol. 6173, pp. 22–29. Springer, Heidelberg (2010). https://doi.org/10.1007/978-3-642-14203-1_3
15. Goubault, É., Ledent, J., Rajsbaum, S.: A simplicial complex model for dynamic epistemic logic to study distributed task computability. In: Ninth International Symposium on Games, Automata, Logics, and Formal Verification, GandALF 2018, pp. 73–87 (2018)
16. Hemed, N., Rinetzky, N., Vafeiadis, V.: Modular verification of concurrency-aware linearizability. In: Moses, Y. (ed.) DISC 2015. LNCS, vol. 9363, pp. 371–387. Springer, Heidelberg (2015). https://doi.org/10.1007/978-3-662-48653-5_25
17. Hendler, D., Shavit, N., Yerushalmi, L.: A scalable lock-free stack algorithm. J. Parallel Distrib. Comput. **70**(1), 1–12 (2010)
18. Herlihy, M., Kozlov, D.N., Rajsbaum, S.: Distributed Computing Through Combinatorial Topology. Morgan Kaufmann, Burlington (2013)
19. Herlihy, M., Shavit, N.: The Art of Multiprocessor Programming. Morgan Kaufmann, Burlington (2008)

20. Herlihy, M., Wing, J.M.: Linearizability: a correctness condition for concurrent objects. ACM Trans. Program. Lang. Syst. **12**(3), 463–492 (1990)
21. Holzmann, G.J.: The SPIN Model Checker - Primer and Reference Manual. Addison-Wesley, Boston (2004)
22. Hurault, A., Quéinnec, P.: Proving a non-blocking algorithm for process renaming with TLA$^+$. In: Beyer, D., Keller, C. (eds.) TAP 2019. LNCS, vol. 11823, pp. 147–166. Springer, Cham (2019). https://doi.org/10.1007/978-3-030-31157-5_10
23. IEC: IEC-61508: Functional safety. https://www.iec.ch/functionalsafety/
24. Scherer III, W.N., Lea, D., Scott, M.L.: Scalable synchronous queues. Commun. ACM **52**(5), 100–111 (2009)
25. Scherer, W.N., Scott, M.L.: Nonblocking concurrent data structures with condition synchronization. In: Guerraoui, R. (ed.) DISC 2004. LNCS, vol. 3274, pp. 174–187. Springer, Heidelberg (2004). https://doi.org/10.1007/978-3-540-30186-8_13
26. John, A., Konnov, I., Schmid, U., Veith, H., Widder, J.: Parameterized model checking of fault-tolerant distributed algorithms by abstraction. In: Formal Methods in Computer-Aided Design, FMCAD 2013, pp. 201–209. IEEE, October 2013
27. Lamport, L.: Specifying Systems. Addison Wesley, Boston (2002)
28. Lamport, L.: Byzantizing paxos by refinement. In: Peleg, D. (ed.) DISC 2011. LNCS, vol. 6950, pp. 211–224. Springer, Heidelberg (2011). https://doi.org/10.1007/978-3-642-24100-0_22
29. Lu, T.: Formal verification of the pastry protocol. Ph.D. thesis, Université de Lorraine - Universität des Saarlandes, July 2013
30. Lu, T., Merz, S., Weidenbach, C.: Towards verification of the pastry protocol using TLA$^+$. In: Bruni, R., Dingel, J. (eds.) FMOODS/FORTE -2011. LNCS, vol. 6722, pp. 244–258. Springer, Heidelberg (2011). https://doi.org/10.1007/978-3-642-21461-5_16
31. Moir, M., Anderson, J.H.: Wait-free algorithms for fast, long-lived renaming. Sci. Comput. Program. **25**(1), 1–39 (1995)
32. Moir, M., Nussbaum, D., Shalev, O., Shavit, N.: Using elimination to implement scalable and lock-free FIFO queues. In: 17th ACM Symposium on Parallelism in Algorithms and Architectures, SPAA 2005, pp. 253–262. ACM (2005)
33. Neiger, G.: Set-linearizability. In: Proceedings of the Thirteenth Annual ACM Symposium on Principles of Distributed Computing, Los Angeles, California, USA, 14–17 August 1994, p. 396 (1994)
34. O'Hearn, P.W., Rinetzky, N., Vechev, M.T., Yahav, E., Yorsh, G.: Verifying linearizability with hindsight. In: 29th Annual ACM Symposium on Principles of Distributed Computing, PODC 2010, pp. 85–94. ACM (2010)
35. Raynal, M.: Concurrent Programming - Algorithms, Principles, and Foundations. Springer, Heidelberg (2013). https://doi.org/10.1007/978-3-642-32027-9
36. Shavit, N., Touitou, D.: Elimination trees and the construction of pools and stacks. Theory Comput. Syst. **30**(6), 645–670 (1997)
37. Tofan, B., Schellhorn, G., Reif, W.: A compositional proof method for linearizability applied to a wait-free multiset. In: Albert, E., Sekerinski, E. (eds.) IFM 2014. LNCS, vol. 8739, pp. 357–372. Springer, Cham (2014). https://doi.org/10.1007/978-3-319-10181-1_22
38. Zave, P.: Using lightweight modeling to understand Chord. SIGCOMM Comput. Commun. Rev. **42**(2), 49–57 (2012)

On Gathering of Semi-synchronous Robots in Graphs

Serafino Cicerone[1], Gabriele Di Stefano[1], and Alfredo Navarra[2(✉)]

[1] Dipartimento di Ingegneria e Scienze dell'Informazione e Matematica, Università degli Studi dell'Aquila, 67100 L'Aquila, Italy
{serafino.cicerone,gabriele.distefano}@univaq.it
[2] Dipartimento di Matematica e Informatica, Università degli Studi di Perugia, 06123 Perugia, Italy
alfredo.navarra@unipg.it

Abstract. We consider the *Gathering* problem where a swarm of weak robots disposed on the vertices of an anonymous graph are required to meet at one vertex from where they do not move anymore. In our recent work [*Cicerone et al.*, SIROCCO'19], we have shown how synchronicity heavily affects the design of resolution algorithms within the standard Look-Compute-Move (LCM) model. In particular, we have investigated two dense and highly symmetric topologies: complete graphs and complete bipartite graphs. We characterized all solvable configurations for synchronous robots, whereas it is known that in complete graphs asynchronous robots cannot solve the problem, ever. Instead of approaching directly the asynchronous case in complete bipartite graphs, we asked what happens in the so-called *semi-synchronous* model, that is robots are synchronized but they are not necessarily all active within all LCM cycles. It turns out that still the gathering can never be accomplished on complete graphs, whereas challenging cases arise in complete bipartite graphs. We provide a distributed algorithm solving the problem for a wide set of possible configurations. For most of the remaining ones instead we provide impossibility results and a few of ad hoc resolution algorithms studied for very specific cases. Over all, still a full characterization is missing but our study points out how difficult might be to derive a general argument that catches all peculiarities. Moreover, some of our approaches reveal new insights that might be very useful for the resolution of other tasks.

1 Introduction

The *Gathering* problem is one of the primitives widely investigated in the context of computing with mobile entities. The task aims to move a swarm of very

The work has been supported in part by the European project "Geospatial based Environment for Optimisation Systems Addressing Fire Emergencies" (GEO-SAFE), contract no. H2020-691161, and by the Italian National Group for Scientific Computation (GNCS-INdAM).

© Springer Nature Switzerland AG 2019
M. Ghaffari et al. (Eds.): SSS 2019, LNCS 11914, pp. 84–98, 2019.
https://doi.org/10.1007/978-3-030-34992-9_7

weak robots initially disposed on different vertices of a graph toward a common vertex, from where they do not move anymore. Robots are assumed to be: *Anonymous*: no unique identifiers; *Autonomous*: no centralized control; *Oblivious*: no memory of past events; *Homogeneous*: they all execute the same deterministic algorithm; *Silent*: no means of communication; *Disoriented*: no common orientation. Robots operate in standard *Look-Compute-Move* (LCM) cycles. In one cycle a robot takes a snapshot of the current global configuration (Look) in terms of robots' locations. Successively, in the Compute phase it decides whether to move toward a neighboring vertex or not, and in the positive case it moves (Move).

Cycles might be performed with respect to different levels of synchronicity:

- *Fully-Synchronous* (FSYNC): The *activation* phase (i.e. the execution of a LCM-cycle) of all robots can be logically divided into global rounds. In each round all the robots are activated, obtain the same snapshot of the environment, compute and perform their move.
- *Semi-Synchronous* (SSYNC): It coincides with the FSYNC model, with the only difference that not all robots are necessarily activated in each round.
- *Asynchronous* (ASYNC): The robots are activated independently, and the duration of each phase is finite but unpredictable.

It is worth to remark that when dealing with SSYNC (as well as for ASYNC ones) any gathering algorithm cannot rely on the concurrent movement of two or more robots as the adversary can always linearize their activation. Whereas FSYNC robots can be forced to move concurrently as they are always active. The amount of time between two LCM-cycles performed by a robot is assumed to be finite but unpredictable. In particular, in both the SSYNC and ASYNC cases it is assumed that the *adversary* determines such a time. The timing is assumed to be *fair*, that is, each robot performs its LCM-cycle within finite time and infinitely often.

Due to impossibility results within the LCM-model, robots are endowed with the so-called *multiplicity detection* capability (see e.g. [7,18]). Basically, when more than one robot resides on the same vertex x, then x is said to be occupied by a *multiplicity*. A robot is said to have the (global strong) multiplicity detection ability when it can detect the exact number of robots composing a multiplicity at any given vertex. Other weaker forms of multiplicity detection could be defined but they would lead to wider impossibility results.

While the gathering problem has been deeply investigated and fully characterized for robots moving on the Euclidean plane [7] (also with respect to given fixed points [3,4]), not much is known for the graph environment.

Recently, in [6] we have provided some general properties for the gatherability of FSYNC robots on graphs. However, a full characterization is still missing, apart for some specific topologies like complete graphs and complete bipartite graphs.

For ASYNC robots [5], the considered topologies so far are trees [8,15], rings [9–13,15,17,18], regular bipartite graphs [16], finite [8] or infinite [14] grids and hypercubes [1], also from an optimization perspective [2,3].

Most of the considered topologies are very symmetric when dealing with anonymous graphs, that is all vertices look equivalent. This choice has been done so that robots cannot exploit much topological properties. For instance if a tree or a finite grid admit only one center, then all robots can detect it and move there, even asynchronously. Contrary, in complete graphs, rings, or hypercubes, all vertices are equivalent and the synchronicity may heavily impact on feasibility.

A main observation coming out from the literature about gathering in graphs is that feasibility is very constrained with respect to synchronicity. On the one hand, dealing with ASYNC robots is much harder than considering SSYNC or FSYNC ones. On the other hand, it may happen that ASYNC robots simply cannot solve some instances, hence sensibly reducing the scope of research for resolution algorithms. In other words, the graph context seems requiring deep investigation on the different results one may achieve when switching from the ASYNC to the SSYNC or FSYNC cases.

Our Results. First, we study general properties of graphs that can be exploited in order to establish the unsolvability of the gathering task in the FSYNC (and hence also in the SSYNC or ASYNC) setting. We extend the knowledge of partitive configurations, introduced in [15] to catch more ungatherable cases. We also define what we call weak-partitive configurations that concern a wider set of configurations with respect to the partitive ones and that can be exploited in the SSYNC setting to further extend the ungatherability result. We then consider SSYNC robots on dense and symmetric graphs like complete and complete bipartite graphs where ASYNC robots cannot solve much. For complete graphs we prove that still the problem remains unsolvable. For complete bipartite graphs, instead we are able to provide some specific impossibility results and some resolution algorithms. Still a full characterization is missing but our study points out how difficult might be to derive a general argument that catches all peculiarities. Moreover, some of our approaches reveal new insights that might be very useful for the resolution of other tasks.

2   Problem Definition and Impossibility Results

The topology where robots are placed on is represented by a simple and connected graph $G = (V, E)$, with vertex set V and edge set E. The cardinality of V is represented as $|V|$ or $|G|$. A function $\lambda : V \to \mathbb{N}$ represents the number of robots on each vertex of G, and we call $C = (G, \lambda)$ a *configuration* whenever $\sum_{v \in V} \lambda(v)$ is bounded and greater than zero. A vertex $v \in V$ such that $\lambda(v) > 0$ is said *occupied*, *unoccupied* otherwise. A subset $V' \subseteq V$ is said *occupied* if at least one of its elements is occupied, *unoccupied* otherwise. A configuration is *initial* if each robot is placed on a different vertex (i.e., $\lambda(v) \leq 1$ for each $v \in V$). A configuration is *final* if all the robots are on a single vertex (i.e., $\exists u \in V : \lambda(u) > 0$ and $\lambda(v) = 0, \ \forall v \in V \setminus \{u\}$). The Gathering problem can be informally defined as the problem of transforming an initial configuration into a final one. Throughout the paper we assume that each initial configuration is composed of at least two robots (otherwise the problem is trivially solved). A *gathering algorithm* for this problem is a deterministic distributed algorithm that brings the robots in the system to a final configuration in a finite number of

LCM-cycles from any given initial configuration, regardless of the adversary. Formally, an algorithm $\mathcal{A}$ solves the Gathering problem for an initial configuration C if, for any execution $\mathbb{E} : C = C(0), C(1), \ldots$ of $\mathcal{A}$, there exists a time instant $i > 0$ such that $C(i)$ is final and no robots move after i, i.e., $C(t) = C(i)$ holds for all $t \geq i$. We say that an initial configuration $C = (G, \lambda)$ is *gatherable* if there exists a gathering algorithm for C, otherwise we say that C is *ungatherable*. For FSYNC/SSYNC robots, the *time complexity* of a gathering algorithm $\mathcal{A}$ is the maximum amount of time units (that is the number of LCM-cycles) required by $\mathcal{A}$ for processing any gatherable initial configuration.

During an execution, $\Lambda(t)$ denotes the number of occupied vertices at time t; formally, $\Lambda(t) = |\{u \in V : \lambda(v) > 0\}|$. Given a subset $V' \subseteq V$, $Rob(V')$ denotes the set containing all robots placed on vertices in V'.

Configuration Automorphisms and Symmetries. Two undirected graphs $G = (V, E)$ and $G' = (V', E')$ are *isomorphic* if there is a bijection φ from V to V' such that $\{u, v\} \in E$ if and only if $\{\varphi(u), \varphi(v)\} \in E'$. An *automorphism* on a graph G is an isomorphism from G to itself, that is a permutation of the vertices of G that maps edges to edges and non-edges to non-edges. The set of all automorphisms of G, under the composition operation, forms a group called *automorphism group* of G and denoted by $Aut(G)$. If $|Aut(G)| = 1$, that is G admits only the identity automorphism, then G is said *asymmetric*, otherwise it is said *symmetric*.

Definition 1. *Given a graph $G = (V, E)$, two vertices $u, v \in V$ are equivalent in G (or G-equivalent) if there exists an automorphism $\varphi \in Aut(G)$ s.t. $\varphi(u) = v$.*

The concept of graph automorphism can be extended to configurations in a natural way: (1) two configurations $C = (G, \lambda)$ and $C' = (G', \lambda')$ are isomorphic if G and G' are isomorphic via a bijection φ and $\lambda(v) = \lambda'(\varphi(v))$ for each vertex v in G; (2) an automorphism on a configuration $C = (G, \lambda)$ is an isomorphism from C to itself and the set of all automorphisms of C forms a group that we call automorphism group of C, denoted by $Aut(C)$. Analogously to the case of graphs, if $|Aut(C)| = 1$, we say that the configuration C is asymmetric, otherwise it is symmetric.

Definition 2. *Given any configuration $C = (G, \lambda)$, two vertices u and v are equivalent in C (or C-equivalent) if there exists an automorphism $\varphi \in Aut(C)$ s.t. $\varphi(u) = v$.*

For sake of simplicity, in the remainder we extend the same definition to robots: two distinct robots r and r' are *equivalent* if they reside on C-equivalent vertices u and v (not necessarily distinct). An important consequence of the above definition is that when two equivalent robots r and r' reside on vertices u and v, then r cannot distinguish its position at vertex u from its equivalent robot r' located at vertex v. As a consequence, no algorithm can distinguish between two equivalent robots.

Definition 3. *Let $C = (G, \lambda)$ be a configuration and $V' \subseteq V$. V' is a G-batch (C-batch, resp.) if its elements are pairwise G-equivalent (C-equivalent, resp.).*

Notice that any C-batch is also a G-batch. We simply use the term *batch* when we are not interested in distinguishing between G- or C-batches or it is clear from the context. Notice that given a batch B, any non-empty subset of B is a batch as well. A batch B is *maximal* if there does not exists a batch B' such that $B \subset B'$. If there exists at least one edge connecting vertices belonging to two distinct batches, then such batches are *adjacent*.

Lemma 1. *Let $G = (V, E)$ be a graph and $C = (G, \lambda)$ be a configuration. The partition of V into maximal C-batches is unique. The same holds for the partition of V into G-batches.*

Given any batch B, we use the following additional notions: (1) the *order* of B is denoted as $\lambda(B)$ and corresponds to the multiplicity in any vertex in B (note that, by definition, each vertex in B contains the same number of robots); (2) the *size* of B is simply $|B|$, that is the number of vertices it contains; (3) the *occupancy* of B is $|Rob(B)|$, that is the total number of robots residing in B - notice $|Rob(B)| = |B| \cdot \lambda(B)$.

In the remainder we make use of the following additional notation. B_{min} and B_{max} will denote occupied batches of minimum and maximum order, respectively. Given some integers $k_1, k_2, \ldots, k_t$, with $t \geq 2$ and $k_i > 0$ for each $1 \leq i \leq t$, we use $lcpf(k_1, k_2)$ to denote the *least common prime factor* of k_1 and k_2, and $lcm(k_1, k_2, \ldots, k_t)$ to denote the *least common multiple* of all such integers. Given a graph $G = (V, E)$, $v \in V$ and $V' \subseteq V$, we use $deg(v, V')$ to denote the number of edges connecting v to any vertex in V'.

The New Impossibility Results. The next theorem provides a sufficient condition for a configuration to be ungatherable, but we first need the following definition of *partitive* configuration.

Definition 4. *Let $G = (V, E)$ be a graph and $C = (G, \lambda)$ be a non-final configuration. C is said partitive on $V' \subseteq V$ if there exists a partition $\mathcal{V} = \{V_1, V_2, \ldots, V_t\}$ of V' into (not necessarily maximal) C-batches where each set V_i fulfills the following conditions:*

1. *$|V_i| \geq 2$;*
2. *if V_i is occupied, then $|Rob(V_i)|$ is a multiple of $|V_j|$ for each $V_j \in \mathcal{V}$.*
3. *if $V_i, V_j \in \mathcal{V}$ are two adjacent batches, then they remain G-batches in the subgraph induced by $V_i \cup V_j$.*

When a configuration C is partitive on the entire set of vertices V we simply say that C is partitive. Notice that when C is both initial and partitive, then the partition in the above definition necessarily requires that each occupied batch V_i has the same size and the same occupancy. In fact, if V_i and V_j are both occupied, then $|Rob(V_i)| = |V_i|$ is a multiple of $|V_j|$ and $|Rob(V_j)| = |V_j|$ is a multiple of $|V_i|$: this implies $|V_i| = |V_j|$.

It is worth to remark that this definition leads to the following observations: if V_i and V_j are two adjacent sets of $\mathcal{V}$, then according to the definition of equivalent vertices in C, we get that $deg(v, V_j)$ is the same for each vertex $v \in V_i$

and, symmetrically, $deg(v, V_i)$ is the same for each vertex $v \in V_j$. This implies that, if we denote by $E_{i,j}$ the set of edges between V_i and V_j, then $|E_{i,j}|$ is a multiple of both $|V_i|$ and $|V_j|$; in particular $|E_{i,j}|$ ranges from $lcm(|V_i|, |V_j|)$ to $|V_i| \cdot |V_j|$. A relevant property arising from this observation - and also showing the rationale of the above definition - is the following.

Lemma 2. *Let $G = (V, E)$ be a graph and $C = (G, \lambda)$ be a non-final configuration. Assume C partitive with respect to the partition $\mathcal{V} = \{V_1, V_2, \ldots, V_t\}$. If an algorithm moves robots from an occupied batch V_i toward V_j then the following properties hold:*

(a) the adversary can equally distribute all robots in V_i on all the vertices of V_j;

(b) after the move, the adversary can lead to configuration C' with the following properties: (1) it is still partitive, (2) if V_j is occupied in C, then C' has one occupied batch less than C, and (3) if V_j is unoccupied in C, then C' has the same number of occupied batches of C.

The next theorem provides a general impossibility result for the gathering problem on graphs of FSYNC (hence also for SSYNC and ASYNC) robots.

Theorem 1. *Let $G = (V, E)$ be any graph and let $C = (G, \lambda)$ be any non-final configuration composed of FSYNC robots. If C is partitive then C is ungatherable.*

Proof. Let $\mathcal{A}$ be any gathering algorithm for C. Since C is partitive, assume that $\mathcal{V} = \{V_1, V_2, \ldots, V_t\}$ is the partition of V fulfilling the conditions of Definition 4. Assume that $\mathcal{A}$ wants to move robots in a set $R \subseteq Rob(V)$ toward a target set $T \subseteq V$ adjacent to the vertices where robots in R reside. Since $\mathcal{A}$ cannot distinguish between equivalent robots and between equivalent vertices, R and T must consist of all robots located on vertices of some batches in $\mathcal{V}$ and all vertices of some batches in $\mathcal{V}$, respectively. In what follows we describe the behavior of the adversary when R and T refer to two single batches (not necessarily distinct) - in case of multiple batches the same behavior is applied to each batch.

If there is only one occupied batch V_i in $\mathcal{V}$, each move can only specify whether robots must move toward an unoccupied batch different from V_i or toward a vertex already occupied within V_i. If the move is toward a vertex already occupied, then the adversary can always make all robots move concurrently toward different destinations, in such a way that the robots just exchange their positions on the same set of occupied vertices. Then $\mathcal{A}$ will always produce a configuration isomorphic to C. If $\mathcal{A}$ moves robots from V_i toward some unoccupied batch $V_j \in \mathcal{V}$ and this move creates a new configuration C', according to Lemma 2 we have: (1) the adversary equally distributes all robots in V_i on the vertices of V_j, (2) C' still fulfills the same conditions of the statement, and (3) C' contains only one occupied C-batch. It is evident that such properties hold regardless the number of moves performed by $\mathcal{A}$. This implies that $\mathcal{A}$ cannot accomplish the gathering.

If there are several occupied C-batches, regardless whether $\mathcal{A}$ moves robots from any V_i toward an occupied or unoccupied batch $V_j \in \mathcal{V}$, the adversary is able to apply the strategy described before (which is based on Lemma 2): it equally distributes all robots in V_i on the vertices of V_j. Let C' be the produced configuration. If V_j is unoccupied in C, the same number of occupied batches remains in C'. If V_j is occupied, C' has one occupied batch less than C. Notice that in any case C' still fulfills the same conditions of the statement. This implies that during the execution of $\mathcal{A}$ the number of occupied C-batches can be eventually reduced to one, but, as observed before, the algorithm is not able to reduce to one the size of this single batch (according to the definition of partitive configuration, each batch has size greater than one) and hence it cannot accomplish the gathering.   □

Notice that this theorem extends the impossibility result provided in [15]. The following corollary states that some configurations can be gathered only at some predetermined vertices.

Corollary 1. *Let $G = (V, E)$ be any graph, $C = (G, \lambda)$ be any non-final configuration, and $V' \subset V$ unoccupied. If C is partitive on $V \setminus V'$, then each gathering algorithm for C (if any) must move robots toward V'.*

The rationale of this corollary is the following: if the algorithm limits the movements of robots within $V \setminus V'$ then Theorem 1 applies which in turn means the obtained configurations always remain partitive on $V \setminus V'$. Hence, in order to gather, the move toward V' must be performed, eventually.

Theorem 1 is a central means for studying the gathering problem in general graphs. We now extend this theorem to specific graph topologies.

Definition 5. *Let $G = (V, E)$ be a graph and $C = (G, \lambda)$ be a non-final configuration. C is said weak-partitive if there exists an unoccupied subset $V' \subseteq V$ such that the following conditions hold:*

1. *C is partitive on $V \setminus V'$, and let $\mathcal{V} = \{V_1, V_2, \ldots, V_t\}$ be the corresponding partition into C-batches of $V \setminus V'$;*
2. *for each maximal C-batch B contained in V', let B' be the maximal G-batch of C containing B. Then, $|B| \geq d \cdot t$, where: (a) d is the least common prime factor of $|Rob(V_i)|$ for any occupied batch $V_i \in \mathcal{V}$; (b) t is the number of occupied maximal C-batches not contained in B'.*

The rationale of this definition is the following. According to Corollary 1, each gathering algorithm for C (if any) must move robots toward V'. Let B be any maximal batch contained into V', and let B' be a G-batch containing B. As soon as the algorithm moves robots into B, the size of B is reduced and the number of occupied vertices in B' increases. We will discuss later what happens to the decomposition into batches after robots moved into B'. The above definition takes care of assuring enough space (i.e., number of unoccupied vertices) to host all robots not in B'. Notice that, in the above definition, the integer d could be defined as the size of the smallest batch in $\mathcal{V}$ - the current definition simply imposes a weaker constraint on d. To see that this definition captures a wider set of configurations with respect to those captured by Definition 4, observe that

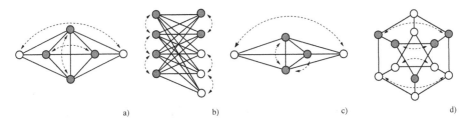

Fig. 1. A gray vertex indicates the presence of one robot; arrows indicate C-equivalent vertices/robots relationship (and its closure generates C-batches). (a) A partitive configuration C_1 with three C-batches, each containing two vertices. (b) A configuration partitive on $V \setminus V'$, where V' is the set containing all the unoccupied vertices. For observing this, the four vertices on the left side must be considered as formed by two G-batches. Notice that the five vertices on the right side form a G-batch B' that fulfills Condition 2 of Definition 5. This implies that this configuration is weak-partitive; (c) and (d) Configurations C_3 and C_4 which are neither partitive nor weak-partitive.

when each maximal batch B contained in V' guarantees that $|B|$ is exactly a multiple of d, then C is partitive on the entire set V.

The following theorem shows that weak-partitive configurations restricted to specific topologies are ungatherable by SSYNC robots.

Theorem 2. *Let $G = (V, E)$ be any graph, $C = (G, \lambda)$ be any non-final configuration composed of SSYNC robots, and C be weak-partitive. If the subgraph induced by each G-batch B' in the Condition 2 of Definition 5 is either a complete graph or a stable set, then C is ungatherable.*

Actually, when all the batches B' stated in the above theorem are stable sets, the result holds even for FSYNC robots.

Corollary 2. *Let $G = (V, E)$ be any complete graph or complete bipartite graph, $C = (G, \lambda)$ be any non-final configuration composed of SSYNC robots. If C is weak-partitive then it is ungatherable*

Figure 1(a) shows a partitive configuration C_1 - by Theorem 1 we deduce that C_1 is ungatherable by means of FSYNC robots. Figure 1(b) shows a weak-partitive configuration C_2 fulfilling the condition of the statement of Theorem 2, hence it is ungatherable by means of SSYNC robots. Figures 1(c) and (d) show two configurations that are neither partitive nor weak-partitive. Actually C_3 is partitive on $V \setminus V'$, where V' is the subset containing the unoccupied vertices - by Corollary 1 any gathering algorithm (if it there exists) must move robots toward V'.

The characterization for the unsolvability of the gathering problem by means of SSYNC robots on complete graphs directly follows from Theorem 2.

Theorem 3. *Let $G = (V, E)$ be a complete graph, and let $C = (G, \lambda)$ be an initial configuration. If C is composed of SSYNC robots, then C is ungatherable.*

3 A Sufficient Condition for Gathering in Arbitrary Graphs

In this section we recall from [6] a sufficient condition for gathering FSYNC robots in arbitrary graphs. This condition is based on the concepts of *recognizable subgraphs* and *d-primality*.

Recognizable Subgraphs. Informally, a subgraph H of a graph G is said *recognizable* if any automorphism of G maps H on itself, that is, H cannot be confused with other subgraphs. Formally:

Definition 6. [6] *A subgraph $H = (V_H, E_H)$ of a graph $G = (V, E)$ is recognizable if $V_H = \{\varphi(v) \mid v \in V_H\}$ for each automorphism $\varphi \in Aut(G)$.*

According to Definition 3, it can be observed that any maximal G-batch induces a recognizable subgraph. Moreover, we have already observed by Lemma 1 that the set containing all the maximal G-batches forms a unique partition of V. This implies that each robot can agree on the elements of the unique partition of V into maximal C-batches.

Let $G = (V, E)$ be a graph and H be a subgraph induced by any maximal batch of G. If H is disconnected, all the connected components are pairwise isomorphic. We denote by $c(H)$ and $s(H)$ the number of connected components of H and the size of each connected component of H, respectively.

*d-**primality.*** Concerning the concept of d-primality, it is useful for characterizing when moving k equivalent robots over d equivalent vertices always produces batches with different orders.

Definition 7. [6] *Let k and d be two positive integers. We say that k is d–prime if $lpf(k) > d$, where $lpf(k)$ denotes the* least prime factor of k.

Note that when k is d–prime the next properties hold: (1) $k > d > 0$; (2) if $d = 1$ then every $k > 1$ is d–prime; (3) for any integer $2 \le d' \le d$, d' does not divide k.

The Sufficient Condition. The notions of recognizable graphs and d-primality are used in the following result to provide a sufficient condition to the solvability of the gathering problem by means of SSYNC robots. This condition is applicable to any graph topology.

Theorem 4. [6] *Let $G = (V, E)$ be a graph with n vertices, and let $C = (G, \lambda)$ be an initial configuration composed of k SSYNC robots, $2 \le k \le n$. If there exists a minimal recognizable subgraph H of G such that k is $\max\{c(H), s(H)\}$–prime, then C is gatherable.*

As observed, configurations in Fig. 1(a) and 1(b) are partitive and weak-partitive, respectively: according to Theorems 1 and 2 they are both ungatherable by SSYNC robots. On the contrary, Fig. 1(c) and 1(d) show configurations where Theorem 4 applies: C_3 is gatherable since $k = 3$ and there exists a maximal batch inducing a subgraph H with $c(H) = 2$ and $s(H) = 1$ (the empty vertices form such a subgraph); C_4 is gatherable since $k = 5$ and there exists a maximal batch inducing a subgraph H with $c(H) = 2$ and $s(H) = 3$. Subgraph H is given by the six internal vertices that form two disjoint triangles.

4 Complete Bipartite Graphs

In this section we study the gathering problem of SSYNC robots on complete bipartite graphs. Throughout the section, we use the following notation. If $G = (V_1 \cup V_2, E)$ is a complete bipartite graph and $C = (G, \lambda)$ is any initial configuration, then:

- n_1 and n_2 denote the number of vertices of V_1 and V_2, resp.; k_1 and k_2 denote the number of robots on V_1 and V_2, resp.
- $(k_1, n_1; k_2, n_2)$ is a compact representation for C; sometimes, instead of a fixed number, in this notation we will use intervals for denoting vertices. For instance, $(13, [13, 15]; 2, 2+)$ denotes all the configurations where $k_1 = 13$, $13 \leq n_1 \leq 15$, $k_2 = 2$, and $n_2 \geq 2$.
- if $k_i = 0$ we say that set V_i, $i = 1, 2$, is *unoccupied, occupied* otherwise.
- $B^1_{max} = B_{max} \cap V_1$ and $B^2_{max} = B_{max} \cap V_2$;
- Λ^1 and Λ^2 denote the number of *occupied vertices* in V_1 and V_2, respectively.

All the notation provided above refers to an initial configuration C, but it can be extended to any configuration $C(t)$ obtained during the execution $\mathbb{E} : C(0), C(1), C(2), \ldots$ of any gathering algorithm by simply fixing the time t: i.e. $\Lambda^1(t)$ denotes the number of occupied vertices in V_1 of configuration $C(t)$.

The Gatherable Cases $\mathcal{G}_1$. The following theorem shows there exists a set $\mathcal{G}_1$ of gatherable configurations. A configuration C belongs to $\mathcal{G}_1$ if at least one of the following conditions holds:

- $k_1 + k_2$ is $\min\{n_1, n_2\}$–prime, but for $(1, 1; 1, 1)$;
- $k_1 + k_2$ is $\min\{k_1, k_2\}$–prime, but for $(1, n/2; 1, n/2)$, for any $n \geq 2$ even.

Theorem 5. *Let $G = (V_1 \cup V_2, E)$ be a complete bipartite graph with $n_1 + n_2$ vertices, and let $C = (G, \lambda)$ be an initial configuration composed of $k_1 + k_2$ SSYNC robots. If $C \in \mathcal{G}_1$ then C is gatherable.*

The proof of Theorem 5 basically concerns the correctness of Algorithm $\mathcal{A}_{bip}$ (Fig. 2).

The Gatherable Cases $\mathcal{G}_2$. In this section we show there are some special configurations, not belonging to $\mathcal{G}_1$, that can be gathered as well. All such configurations form a set denoted as $\mathcal{G}_2$. A configuration C belongs to $\mathcal{G}_2$ if it is one of the following:

- $(2, 2+; k_2, k_2 + 1)$ and $(k_1, k_1 + 1; 2, 2+)$, for each $k_2 \geq 0$;
- $(3, 3+; k_2, k_2 + 2)$ and $(k_1, k_1 + 2; 3, 3+)$, for each $k_2 \geq 0$;
- $(5, 5+; 3, 3)$, $(6, 6+; 2, 3)$, $(7, 7+; 3, 3)$, and $(8, 8+; 2, 3)$.

Before proving that each configuration in $\mathcal{G}_2$ is gatherable, we provide a simple result useful for the current task.

Algorithm: $\mathcal{A}_{bip}$
Input: Configuration $C = (G, \lambda)$, where G is a complete bipartite graph, fulfilling conditions of Theorem 5.

```
1  if n₁ > n₂ and k₁ + k₂ is n₂-prime then
2  |  call Ag // gathering accomplished on a vertex in V₂
3  else if n₂ > n₁ and k₁ + k₂ is n₁-prime then
4  |  call Ag // gathering accomplished on a vertex in V₁
5  else
6  |  if Λ¹ ≤ Λ² then let α = 1 else let α = 2 ;
7  |  let ᾱ = 3 − α // Vᾱ is the side with more robots ;
8  |  if both V₁ and V₂ are occupied then
9  |  |  let Origin = Rob(Vᾱ) ;
10 |  |  if Bmax intersects both V₁ and V₂ then
11 |  |  |  let Dest = Bmax ∩ Vα
12 |  |  else
13 |  |  |  let Dest = Bmax
14 |  else
15 |  |  let Origin = Rob(Vα ∖ Bmax) ;
16 |  |  let Dest = Vᾱ ;
17 |  each robot in Origin moves toward an arbitrary vertex in Dest ;
```

Fig. 2. Algorithm $\mathcal{A}_{bip}$ for gathering SSYNC robots in a complete bipartite graph G. $\mathcal{A}_g$ denotes the algorithm related to Theorem 4.

Lemma 3. *Let $G = (V_1 \cup V_2, E)$ be a complete bipartite graph with $n_1 + n_2$ vertices, let $C = (G, \lambda)$ be a non-initial configuration composed of $k = k_1 + k_2$ SSYNC robots. If $|B^1_{\max}| = 1$ or $|B^2_{\max}| = 1$ or $\Lambda^1 = 1$ or $\Lambda^2 = 1$ then C is gatherable.*

Theorem 6. *Let $G = (V_1 \cup V_2, E)$ be a complete bipartite graph with $n_1 + n_2$ vertices, and let $C = (G, \lambda)$ be an initial configuration composed of $k_1 + k_2$ SSYNC robots. If C belongs to $\mathcal{G}_2$ then C is gatherable.*

Proof. Consider C in the form $(2, 2+; k_2, k_2 + 1)$, with $k_2 \geq 0$ (equivalently we may consider $(k_1, k_1 + 1; 2, 2+)$). Notice that in C there is only one unoccupied vertex $u \in V_2$. By moving robots from V_1 to u we get the following cases. If the adversary moves only one robot, then a configuration $C(1)$ is obtained where condition $\Lambda^1 = 1$ holds. Then, $C(1)$ is gatherable by Lemma 3. Conversely, if from $C(0)$ the adversary moves two robots from V_1 to u, then $|B^2_{\max}| = 1$ and Lemma 3 applies.

Consider C in the form $(3, 3+; k_2, k_2 + 2)$, with $k_2 \geq 0$. Notice that in C there are two unoccupied vertices in V_2. By moving robots from V_1 to such unoccupied vertices we get the following cases. If the adversary moves only one robot, then a configuration $C(1)$ of the form $(2, 2+; k_2, k_2 + 1)$ is obtained. We have already proved it is gatherable. If the adversary moves two robots, $C(1)$ is gatherable by Lemma 3 as $\Lambda^1 = 1$ holds. Finally, if from $C(0)$ the adversary moves all the robots from V_1, then condition $|B^2_{\max}| = 1$ of Lemma 3 applies.

When $C = (6, 6+; 2, 3)$ (or $C = (8, 8+; 2, 3)$) a gathering algorithm $\mathcal{A}$ can allow robots in V_1 to move toward the unoccupied vertex in V_2.

If more than one robot move, Lemma 3 becomes true as $|B_{max}^2| = 1$ and the gathering can be easily finalized. Whereas, if only one robot moves, $C = (5, 5+; 3, 3)$ (or $C = (7, 7+; 3, 3)$, resp.) is obtained.

From $C = (5, 5+; 3, 3)$, algorithm $\mathcal{A}$ allows robots in V_2 to move onto occupied vertices in V_1. If only one robot moves, again Lemma 3 applies $|B_{max}^1| = 1$. If two robots move Lemma 3 applies $\Lambda_2 = 1$. If all the three robots move, then the adversary may create three multiplicities of size 2 in V_1 as otherwise still Lemma 3 applies. The only possible move now is given by the two singletons moving toward V_2. This leaves the three multiplicities in V_1 and two singletons in V_2. Now the algorithm can move the robots in V_1 toward the unoccupied vertex in V_2. If only one robot moves, there is a unique singleton in V_1 which is moved by the algorithm toward V_2, hence creating $|B_{max}^2| = 1$; in more than one robot move, again it is obtained $|B_{max}^2| = 1$.

A similar case analysis can be discussed when $C = (7, 7+; 3, 3)$. □

About Gatherability of Remaining Configurations. In this section we consider configurations of SSYNC robots defined on a complete bipartite graph. In particular, denote by $\mathcal{I}$, $\mathcal{G}_1$, $\mathcal{G}_2$, $\mathcal{P}$, and $\mathcal{W}$ the set containing the initial configurations, the gatherable configurations considered in Theorem 5, the partitive configurations, and the weak-partitive configurations, respectively.

Here we study the gathering problem for configurations in $\mathcal{R} = \mathcal{I} \setminus (\mathcal{G}_1 \cup \mathcal{G}_2 \cup \mathcal{P} \cup \mathcal{W})$. By Theorem 5 and Definition 7, any configuration in $\mathcal{R}$ fulfills: (a) $\min\{k_1, k_2\} \geq lpf(k_1 + k_2)$, and (b) $\min\{n_1, n_2\} \geq lpf(k_1 + k_2)$.

In the remainder of this section we address each configuration $C \in R_e$, where R_e is the subset of $\mathcal{R}$ formed by configuration with an even number of robots. From the first of the above conditions, we derive that in any configuration of R_e, both $k_1 \geq 1$ and $k_2 \geq 1$ holds. Symmetrically, from the second of the above conditions, we derive that in any configuration of R_e, both $n_1 \geq 1$ and $n_2 \geq 1$ holds, otherwise the configuration is clearly gatherable.

Lemma 4. *Let $G = (V_1 \cup V_2, E)$ be a complete bipartite graph with $n_1 + n_2$ vertices, let $C = (G, \lambda)$ be any configuration in R_e composed of $k = k_1 + k_2$ SSYNC robots, $k_1 \geq k_2$. If there are at least three unoccupied vertices in both V_1 and V_2, then C is ungatherable.*

Proof. Let $C(0) = C \in R_e$. By definition of R_e we get $k_2 \geq 2$. We now show that $k_2 > 2$. In fact, since $C \in R_e$ then k is even and hence when $k_2 = 2$ then k_1 is even as well. Moreover, since $n_1 - k_1 \geq 3$ and $n_2 - k_2 \geq 3$, then C is weak-partitive and hence not belonging to R_e. In the remainder of the proof we analyze the cases $k_2 = 3$ and $k_2 > 3$. Let $\mathcal{A}$ be any possible algorithm for C.

If $k_2 = 3$ consider the four possible targets in $C(0)$ given by the occupied and unoccupied set of vertices in V_1 and V_2. We first show that if $\mathcal{A}$ moves the robots towards unoccupied vertices, we can apply Theorem 2 at the resulting configuration $C(1)$. In fact: (1) if robots are moved from V_2 to the unoccupied vertices in V_1, then, since $n_1 - k_1 \geq 3$, each robots can occupy a different vertex;

(2) if robots are moved from V_1 to the unoccupied vertices in V_2, then, the adversary can form a configuration $C(1)$ with four vertices occupied by a single robot and two multiplicities with $k/2 - 2$ robots each. In both cases the resulting configuration $C(1)$ is weak-partitive and hence ungatherable.

We now show that if $\mathcal{A}$ moves the robots from V_1 to the occupied vertices in V_2, then necessarily a weak-partitive configuration $C(2)$ is obtained. If in $C(0)$ the targets are the three occupied vertices of V_2, the adversary creates in $C(1)$ a multiplicity with two robots and two multiplicities with $k/2 - 1$ robots each. From $C(1)$, if $\mathcal{A}$ moves the robots from the multiplicity with two robots a configuration $C(2)$ with two vertices occupied in V_1 and two multiplicities with $k/2 - 1$ in V_2 is created. From $C(1)$, if $\mathcal{A}$ moves the robots from the multiplicities with $k/2 - 1$ then a configuration $C(2)$ with two multiplicities with $k/2 - 2$ robots in V_2 and two multiplicities with two robots in V_2 is created. In both cases, $C(2)$ is weak-partitive and hence ungatherable.

We now show that if $\mathcal{A}$ moves the robots from V_2 to the occupied vertices in V_1, then necessarily a weak-partitive configuration is obtained. If in $C(0)$ the targets are the occupied vertices of V_1, the adversary creates in $C(1)$ three multiplicities with two robots each. From $C(1)$ the only possibility for $\mathcal{A}$ that avoids $C(2) = C(0)$ is to move the robots non belonging to multiplicities (which are $k - 6$ in total). Notice that if $k - 6 = 0$ then $C(1)$ is weak-partitive. When $k - 6 > 0$ the adversary create a configuration C_2 by moving all such robots on two vertices of V_2 with $(k - 6)/2$ robots each. From C_2, if robots on V_1 are moved toward the unoccupied vertices, two further multiplicities with three robots each are created. If robots V_1 are moved on the occupied vertices of V_2, two multiplicities of $k/2$ robots are created. In both cases, a weak-partitive configuration $C(3)$ is obtained. On the other hand, if in $C(2)$ algorithm $\mathcal{A}$ moves robots from V_2 toward the unoccupied vertices then a configuration isomorphic to $C(1)$ is re-created. When $\mathcal{A}$ moves robots from V_2 towards the occupied vertices in V_1, the adversary moves all the robots by creating a configuration $C(3)$ with two multiplicities with $k/2 - 1$ robots each and a multiplicity with two robots. To avoid $C(4)$ isomorphic to $C(2)$ the only possible move is to move the robots in the multiplicity with two robots toward V_2. In this case the adversary moves the two robots on two different vertices of V_2 thus creating again a weak-partitive configuration. This concludes the proof for the case $k_2 = 3$.

If $k_2 > 3$, we can assume k_1 and k_2 both odd in $C(0)$, otherwise the configuration is weak-partitive. In $C(0)$, as in the previous case, we have four targets, but if the initial move is toward a target with unoccupied vertices, the adversary moves a single robot generating a configuration $C(1)$ weak-partitive since k_1 and k_2 become both even.

We now show that if in $C(0)$ algorithm $\mathcal{A}$ moves the robots from V_1 to the occupied vertices in V_2, then necessarily a weak-partitive configuration is obtained. The adversary forms $C(1)$ by moving all the k_1 robots on V_2 forming two multiplicities with $(k_1 - k_2 + 2)/2 + 1$ robots and $k_2 - 2$ multiplicities with two robots. If $(k_1 - k_2 + 2)/2 + 1 = 2$ the configuration is ungatherable since it consists of a single batch. Then we can assume there are two batches in $C(1)$, both entirely contained in V_2. If $\mathcal{A}$ moves the robots from $B_{\min}$, the adversary moves the robots only on two vertices creating two multiplicities on V_1 with the same number of robots. The resulting configuration is weak-partitive. If $\mathcal{A}$ moves the robots in $B_{\max}$, again two multiplicities on V_1 with the same number of robots are created, but two robots of $B_{\max}$ are left on a single vertex of V_2. Again, the resulting configuration is weak-partitive.

We now show that if in $C(0)$ algorithm $\mathcal{A}$ moves the robots from V_2 to the occupied vertices in V_1, then necessarily a weak-partitive configuration is obtained. If in $C(0)$ the targets are the occupied vertices of V_1, all the k_2 robots are moved to form $C(1)$ such that they create k_2 multiplicities with two robots on V_1. Then in $C(1)$, $k_1 - k_2$ vertices of V_1 are occupied by a single robot. Note that $k_1 - k_2 \geq 2$ and that $k_1 - k_2$ is even. To avoid $C(2) = C(0)$, the algorithm $\mathcal{A}$ can move only the robots not in multiplicities. Then the adversary moves all these robots creating a configuration $C(2)$ with two multiplicities on V_2 with $(k_1 - k_2)/2$ robots each and k_2 multiplicities with two robots on V_1. From $C(2)$, if the robots on V_1 are moved towards the unoccupied vertices of V_2, the adversary forms two new multiplicities with k_2 robots each, and the resulting configuration is weak-partitive. If the robots on V_1 are moved towards the occupied vertices of V_2, the adversary forms two multiplicities on V_2 with $k/2$ robots each and the configuration is ungatherable since it is formed by a single batch. From $C(2)$, the only remaining move that avoids $C(3) = C(1)$ is to move the robots form V_2 to the occupied vertices of V_1. In this case, the configuration $C(3)$ created by the adversary consists of two multiplicities with $(k_1 - k2)/2 + 2$ robots and $k_2 - 2$ multiplicities with two robots on V_1, whereas V_2 is empty. From $C(3)$, the only move that avoids $C(4) = C(2)$ is to move the robots in the multiplicities with two robots. The adversary forms a configuration $C(4)$ with two multiplicities with $k_2 - 2$ robots each in V_2 and since in V_1 there are two multiplicities left, the configuration is weak-partitive. □

5   Conclusion

We have considered the gathering problem in graphs. First we have extended the set of ungatherable configurations with respect to what was previously known in the literature for FSync robots. Then we have further extended ungatherability for SSync robots. Still in the SSync context we have focussed on symmetric and dense network topologies such as complete and complete bipartite graphs. The obtained results point out how the problem becomes difficult to be handled by means of a general approach, hence requiring specific strategies for different cases. Our study confirms how the graph environment reveals to be rather hostile with respect to the gathering task.

References

1. Bose, K., Kundu, M.K., Adhikary, R., Sau, B.: Optimal gathering by asynchronous oblivious robots in hypercubes. In: Gilbert, S., Hughes, D., Krishnamachari, B. (eds.) ALGOSENSORS 2018. LNCS, vol. 11410, pp. 102–117. Springer, Cham (2019). https://doi.org/10.1007/978-3-030-14094-6_7

2. Cicerone, S., Di Stefano, G., Navarra, A.: MinMax-distance gathering on given meeting points. In: Paschos, V.T., Widmayer, P. (eds.) CIAC 2015. LNCS, vol. 9079, pp. 127–139. Springer, Cham (2015). https://doi.org/10.1007/978-3-319-18173-8_9

3. Cicerone, S., Di Stefano, G., Navarra, A.: Gathering of robots on meeting-points: feasibility and optimal resolution algorithms. Distrib. Comput. **31**(1), 1–50 (2018)

4. Cicerone, S., Di Stefano, G., Navarra, A.: Asynchronous arbitrary pattern formation: the effects of a rigorous approach. Distrib. Comput. **32**(2), 91–132 (2019)

5. Cicerone, S., Di Stefano, G., Navarra, A.: Asynchronous robots on graphs: gathering. In: Flocchini, P., Prencipe, G., Santoro, N. (eds.) Distributed Computing by Mobile Entities. Lecture Notes in Computer Science, vol. 11340, pp. 184–217. Springer, Cham (2019). https://doi.org/10.1007/978-3-030-11072-7_8

6. Cicerone, S., Di Stefano, G., Navarra, A.: Gathering synchronous robots in graphs: from general properties to dense and symmetric topologies. In: Censor-Hillel, K., Flammini, M. (eds.) SIROCCO 2019. LNCS, vol. 11639, pp. 170–184. Springer, Cham (2019). https://doi.org/10.1007/978-3-030-24922-9_12

7. Cieliebak, M., Flocchini, P., Prencipe, G., Santoro, N.: Distributed computing by mobile robots: gathering. SIAM J. Comput. **41**(4), 829–879 (2012)

8. D'Angelo, G., Di Stefano, G., Klasing, R., Navarra, A.: Gathering of robots on anonymous grids and trees without multiplicity detection. Theor. Comput. Sci. **610**, 158–168 (2016)

9. D'Angelo, G., Di Stefano, G., Navarra, A.: Gathering on rings under the look-compute-move model. Distrib. Comput. **27**(4), 255–285 (2014)

10. D'Angelo, G., Di Stefano, G., Navarra, A.: Gathering six oblivious robots on anonymous symmetric rings. J. Discret. Algorithms **26**, 16–27 (2014)

11. D'Angelo, G., Di Stefano, G., Navarra, A., Nisse, N., Suchan, K.: Computing on rings by oblivious robots: a unified approach for different tasks. Algorithmica **72**(4), 1055–1096 (2015)

12. D'Angelo, G., Navarra, A., Nisse, N.: A unified approach for gathering and exclusive searching on rings under weak assumptions. Distrib. Comput. **30**(1), 17–48 (2017)

13. D'Emidio, M., Di Stefano, G., Frigioni, D., Navarra, A.: Characterizing the computational power of mobile robots on graphs and implications for the Euclidean plane. Inf. Comput. **263**, 57–74 (2018)

14. Di Stefano, G., Navarra, A.: Gathering of oblivious robots on infinite grids with minimum traveled distance. Inf. Comput. **254**, 377–391 (2017)

15. Di Stefano, G., Navarra, A.: Optimal gathering of oblivious robots in anonymous graphs and its application on trees and rings. Distrib. Comput. **30**(2), 75–86 (2017)

16. Guilbault, S., Pelc, A.: Gathering asynchronous oblivious agents with local vision in regular bipartite graphs. Theor. Comput. Sci. **509**, 86–96 (2013)

17. Izumi, T., Izumi, T., Kamei, S., Ooshita, F.: Time-optimal gathering algorithm of mobile robots with local weak multiplicity detection in rings. IEICE Trans. **96-A**(6), 1072–1080 (2013)

18. Klasing, R., Kosowski, A., Navarra, A.: Taking advantage of symmetries: gathering of many asynchronous oblivious robots on a ring. Theor. Comput. Sci. **411**, 3235–3246 (2010)

Brief Announcement: Analysis of a Memory-Efficient Self-stabilizing BFS Spanning Tree Construction

Ajoy K. Datta[1], Stéphane Devismes[2(⊠)], Colette Johnen[3], and Lawrence L. Larmore[1]

[1] Department of Computer Science, University of Nevada, Reno, USA
[2] Université Grenoble Alpes, VERIMAG, UMR 5104, Grenoble, France
stephane.devismes@univ-grenoble-alpes.fr
[3] Université de Bordeaux, LaBRI, UMR 5800, Bordeaux, France

Abstract. We present preliminary results on the last topic we collaborate with our late friend, Professor Ajoy Kumar Datta (1958–2019), who prematurely left us a few months ago. In this work, we shed new light on a self-stabilizing wave algorithm proposed by Colette Johnen in 1997 [12]. This algorithm constructs a BFS spanning tree in any connected rooted network. Nowadays, it is still the best existing self-stabilizing BFS spanning tree construction in terms of memory requirement, *i.e.*, it only requires $\Theta(1)$ bits per edge. However, it has been proven assuming a weakly fair daemon. Moreover, its stabilization time was unknown. Here, we study the slightly modified version of this algorithm, still keeping the same memory requirement. We prove the self-stabilization of this variant under the distributed unfair daemon and show a stabilization time in $O(\mathcal{D} \cdot n^2)$ rounds, where $\mathcal{D}$ is the network diameter and n the number of processes.

Keywords: Self-stabilization · BFS spanning tree · Distributed unfair daemon · Stabilization time · Round complexity

1 Introduction

We consider the problem of constructing a spanning tree in a self-stabilizing manner. Numerous self-stabilizing spanning tree constructions have been studied until now, *e.g.*, the spanning tree may be arbitrary (see [4]), *depth-first* (see [5]), *breadth-first* (see [6]). Deterministic solutions to these problems have been investigated in either fully identified networks [2], or rooted networks [5]. Here, we consider rooted connected networks. By "rooted" we mean that one process, called the *root* and noted r, is distinguished from the others. All other processes are fully anonymous. We focus on the construction of a Breadth-First Search

This study was partially supported by the French ANR projects ANR-16-CE40-0023 (DESCARTES) and ANR-16 CE25-0009-03 (ESTATE).

© Springer Nature Switzerland AG 2019
M. Ghaffari et al. (Eds.): SSS 2019, LNCS 11914, pp. 99–104, 2019.
https://doi.org/10.1007/978-3-030-34992-9_8

(BFS) spanning tree in such a rooted connected network, *i.e.*, a spanning tree in which the (hop-)distance from any node to the root is minimum. Spanning tree construction is a fundamental task in communication networks. Indeed, spanning trees are often involved in the design of *routing* [10] and *broadcasting* tasks [3], for example. Moreover, improving the efficiency of the underlying spanning tree algorithm usually also implies an improvement of the overall solution.

We consider here the *atomic state model*, also called the *locally shared memory model with composite atomicity*. In this model, the daemon assumption accepted by the algorithm is crucial since it captures the asynchrony of the system. More generally, self-stabilizing solutions are also discriminated according to their stabilization time (usually in rounds) and their memory requirement.

Related Work. There are many self-stabilizing BFS spanning tree constructions in the literature. Maybe the first one is that of Chen *et al.* [4]. It is proven in the atomic state model under the central unfair daemon and no time complexity analysis is given. The algorithm of Huang and Chen [11] is proposed in the atomic state model, yet under a distributed unfair daemon. In [8], the stabilization time of this algorithm is shown to be $\Theta(n)$ rounds in the worst case, where n is the number of processes. Another algorithm, implemented in the link-register model, is given in [9]. It uses unbounded process local memories. However, it is shown in [8] that a straightforward bounded-memory variant of this algorithm, working in the atomic state model, achieves an optimal stabilization time in rounds, *i.e.*, $O(\mathcal{D})$ rounds where $\mathcal{D}$ is the network diameter. In [1], Afek and Bremler design a solution for unidirectional networks in the message-passing model, assuming bounded capacity links. The stabilization time of this latter algorithm is $O(n)$ rounds. The algorithm given in [7] has a stabilization time $O(\mathcal{D}^2)$ rounds, assuming the atomic state model and a distributed unfair daemon. All aforementioned solutions [1, 4, 7, 9, 11] also achieve silence. Two other non-silent, *a.k.a. talkative*, self-stabilizing BFS spanning tree constructions have been proposed in the atomic state model. The algorithm in [6] is proven under the distributed unfair daemon and has a stabilization time in $O(n)$ rounds. In [12], the proposed solution assumes a distributed weakly fair daemon and its stabilization time is not investigated. Except for [12], in all these aforementioned algorithms, each process has a *distance* variable which keeps track of the current level of the process in the BFS tree. Thus, these BFS spanning tree constructions have a space complexity in $\Omega(\log(\mathcal{D}))$ bits per process. In contrast, the solution given in [12] does not compute any distance value (actually, the construction is done using synchronization phases). Consequently, the obtained memory requirement only depends on local parameters, *i.e.*, $\Theta(\log(\delta_p))$ bits per process p, where δ_p the local degree of p. In other words, the space complexity of this algorithm is intrinsically $\Theta(1)$ bits per edge. Moreover, the algorithm does not need *a priori* knowledge of any global parameter on the network such as $\mathcal{D}$ or n. It is worth noticing that today it is still the best self-stabilizing BFS spanning tree construction in terms of memory requirement.

Contribution. We fill the blanks in the analysis of the memory-efficient self-stabilizing BFS spanning tree construction given in [12]. Precisely, we study a slightly modified (maybe simpler) version of the algorithm. This variant still achieves a memory requirement in $\Theta(1)$ bits of memory per edge. We prove its self-stabilization under the distributed unfair daemon, the weakest scheduling assumption. Moreover, we establish a stabilization time in $O(\mathcal{D} \cdot n^2)$ rounds, where $\mathcal{D}$ is the network diameter and n the number of processes. All the technical material has been omitted from this brief announcement and is available online at the following address:

https://arxiv.org/abs/1907.07944

2   The Algorithm

Our algorithm executes infinitely many tree constructions in sequence: at the end of each construction, the root of the network (r), called here the *legal root*, initiates a new one. The algorithm alternatively builds 0-colored and 1-colored BFS spanning trees. The color is used to eventually distinguish processes belonging to the legal tree (rooted at r) from those which do not. We denote by r_color the current color of the legal tree.

One of the difficulty to build a BFS tree without using a *distance* variable is to ensure that the path from any process to r in the tree is minimal. To solve this issue, each construction is split into phases synchronized at r: when r detects the end of the current phase, it initiates the next one, if the construction is not yet complete; otherwise, r starts a new tree construction. Once the system is stabilized, each construction is made of at most $\mathcal{D}$ phases: during the kth phase, all processes at distance k from r join the current tree (by choosing as a parent a neighbor at distance $k - 1$ from r). However, during the convergence, a tree construction can contain up to n phases.

Shared Variables. In the following, for every process u, we denote by $N(u)$ the set of u's neighbors and by $dist(u)$ the distance from u to r.

Each non-root process u maintains the following variables:

$TS.u \in N(u) \cup \{\bot\}$: the *output parent pointer* of u. Once the system is stabilized, TS variables are constant and distributedly define a BFS spanning tree rooted at r. More precisely, when the system is stabilized, $dist(u) = dist(TS.u) + 1$ if $u \neq r$, $dist(u) = 0$ otherwise.

$P.u \in N(u) \cup \{\bot\}$: another *parent pointer*. Once the system is stabilized, if $P.u \neq \bot$, then $P.u = TS.u$. Actually, $P.u$ is used to inform the pointed neighbors about whether or not the construction of the subtree rooted at u is terminated. Precisely, if at end of a phase u is *childless, i.e.,* $\forall v \in N(u)$, $P.v \neq u$, then the subtree construction rooted at u is done.

$C.u \in \{0, 1\}$: the *color* of u. Once the system is stabilized, the processes in the current tree have color $C.r = r\_color$, while other processes have color $\overline{r\_color}$.

$S.u \in \{Idle, Working, Power, WeakE, StrongE\}$: the *status* of u. Status $WeakE$ and $StrongE$ are only used to correct errors. Process u is said to have an *Erroneous* status if $S.u \in \{WeakE, StrongE\}$. Only processes of status *Power* can gain new children. Once the system is stabilized, during the kth phase, only processes at distance $k-1$ from r, *i.e.*, leaves of the tree under construction, will acquire the status *Power*. If u is an internal process in the tree under construction and is participating to the current phase, then u has the status *Working*. If u is not involved in the current tree construction (*i.e.*, $P.u = \perp$), or the current phase is either not started or finished in the subtree of u, then u has the status *Idle*.

$ph.u \in \{a, b\}$: the current *phase*. This variable is used to distinguish any two consecutive phases. In particular, it allows to solve the following ambiguity. A process in the tree is *Idle* either when it has completed or has not yet started the current phase. In order to distinguish these two cases, we use the phase variable. If the phase value of an *Idle* process u is the same as that of its parent $P.u$, then u has finished the current phase. Otherwise, it has not yet initiated the current phase.

The root r maintains the same variables, except P and TS: r does not have a parent. Moreover, $S.r \in \{Power, Working, StrongE\}$: Status $WeakE$ does not exist for r.

Tree Construction. This part is identical to [12]. The current tree construction is over when r becomes childless, *i.e.*, $\forall v \in N(r), P.v \neq r$. Then, r starts a new tree construction: it changes its color and initiates the first phase by taking the status *Power*.

A phase is divided into three stages: *forwarding, tree expansion,* and *backward*. The processes in the tree under construction first propagate the phase initiated by r to the leaves of the tree under construction (forwarding part). Then, the neighbors of the leaves that are not in the tree join it (tree expansion part). Finally, non-root tree nodes backtrack to inform r of the end of the tree expansion (backward part).

In the *forwarding* stage of the kth phase, along the tree under construction, internal nodes take the status *Working* and copy the phase value of r in a top-down fashion; then, the leaves (at distance $k-1$ from r) take the status *Power* and also copy the phase value of r. This stage requires h rounds, where $h \leq n$ is the height of the tree rooted at r under construction.

In the *tree expansion* stage of the kth phase, all processes at distance k from r join the tree under construction: they choose a neighbor of status *Power* as a parent (by updating their variables P and TS), copy both the phase value and color of their new parent, and take the status *Idle*. Each *tree expansion* stage requires $O(1)$ round (*n.b.*, this stage may be slow down by a constant number of rounds due to local corrections).

In the *backward* stage of the kth phase, the processes of status *Power* finalize the kth phase by switching their status to *Idle* when the current phase is over in their neighborhood: all their neighbors are in the tree (they have the color

r_color). The $Working$ processes then finalize the current phase by switching their status to $Idle$ bottom-up when their children have finished the current phase (they have the status $Idle$ and the same phase value as them). Moreover, when the subtree rooted at a process u is completely built (*i.e.*, $\forall v \in N(u)$ we have $P.v \neq u$), u additionally sets its P variable to $\perp$. The backward part of tree phase construction requires $h - 1$ rounds.

When the kth phase is over (*i.e.*, all processes of the legal tree have the status $Idle$, the same phase as r, and their color is r_color), r initiates the $k + 1$ phase if the construction is not done (otherwise, it starts a new tree construction): r changes its phase value and takes the status $Working$.

Thus, a phase lasts $O(n)$ rounds and a tree construction costs $O(n^2)$ rounds.

Error Correction. There are two major correction tasks: one is to remove the illegal trees, the other is to break the (parent) cycles. We solve these two tasks using mechanisms that are simpler than those proposed in [12].

The error correction is based on *conflicts, faulty process, and illegal root detection*. There are two kinds of conflicts, the *(weak) conflicts* and the *strong* ones.

When a process detects a conflict, is faulty, or is an illegal root, it switches to an $Erroneous$ status, either $StrongE$ (if it is a strong conflict), or $WeakE$ (in all other cases), and leaves its tree by setting their P variable to $\perp$. Then, its children take status $WeakE$ and leave the tree, and so on: these corrections are propagated toward the leaves. The $Erroneous$ detached processes (*i.e.*, processes without parent and children) recover by changing their status to $Idle$. We have two cases. If the process has the status $WeakE$, it directly recovers. Otherwise, it waits until none of its neighbors has the status $Power$ (these latter, if any, are enabled to take the status $WeakE$ and leave their tree, as explained later).

Let u be a non-root process. First, if u has children but no parent, it is an *illegal root*. Then, u is *faulty* if it has a parent, but its state is not consistent with that of its parent. In either case, u takes the status $WeakE$ and leaves its tree, as previously explained.

The major difficulty for the error correction is to remove parent cycles while no distance value is available. A process having a parent assumes that it is in the legal tree and its color is equal to r_color (even if it is inside a cycle). Based on this assumption, it detects a *conflict* when one of its neighbors does not have its color but has the status $Power$ (both processes cannot be inside the legal tree). Once a *conflict* detected by a process $u \neq r$, it takes $WeakE$ and leaves its tree. Hence, the potential cycle is transformed into an illegal tree.

If a process $u \neq r$ has in its closed neighborhood two $Power$ processes, but not of the same color, then u detects a *strong conflict*: both processes cannot be in the legal tree, u takes the status $StrongE$, and leaves its tree. The root r detects a strong conflict if one of its neighbors v has the status $Power$ but either v has not its color, or r is childless. When the root detects a strong conflict it simply takes the status $StrongE$ (recall that r has no P variable and so cannot leave its tree). All $Power$ neighbors of a $StrongE$ process take the status $WeakE$

and leave their tree. So, a *StrongE* process has to wait until the status *Power* vanishes from its neighborhood before recovering.

Overview of the Round Complexity. After $O(n)$ rounds, no process in the legal tree may detect a conflict, and so create a new illegal branch that may contain *Power* processes. After one more $O(n)$ rounds, only process in the legal tree may have the status *Power*. Let γ be a configuration where only processes in the legal tree may have the *Power* status. After at most d complete tree constructions from γ, processes at distance less than d from r never more belong to any cycle or illegal tree, *i.e.*, they are stabilized. So after at most $\mathcal{D} + 1$ complete tree construction from γ, there is no cycle or illegal tree at all: all processes are stabilized. Any complete tree construction requires $O(n^2)$ rounds. So, the round complexity of our algorithm is $O(\mathcal{D} \cdot n^2)$ rounds.

References

1. Afek, Y., Bremler-Barr, A.: Self-stabilizing unidirectional network algorithms by power supply. Chicago J. Theor. Comput. Sci. (1998)
2. Afek, Y., Kutten, S.,Yung, M. : Memory-efficient self-stabilizing protocols for general networks. In: WDAG 1990, pp. 15–28 (1990)
3. Bui, A., Datta, A.K., Petit, F., Villain, V.: Optimal PIF in tree networks. In: WDAS 1999, pp. 1–16 (1999)
4. Chen, N., Yu, H., Huang, S.: A self-stabilizing algorithm for constructing spanning trees. IPL **39**, 147–151 (1991)
5. Collin, Z., Dolev, S.: Self-stabilizing depth-first search. IPL **49**(6), 297–301 (1994)
6. Cournier, A., Devismes, S., Villain, V.: Light enabling snap-stabilization of fundamental protocols. ACM TAAS. **4**(1), 6 (2009)
7. Cournier, A., Rovedakis, S., Villain, V.: The first fully polynomial stabilizing algorithm for BFS tree construction. In: OPODIS 2011, pp. 159–174 (2011)
8. Devismes, S., Johnen, C.: Silent self-stabilizing BFS tree algorithms revisited. JPDC **97**, 11–23 (2016)
9. Dolev, S., Israeli, A., Moran, S.: Self-stabilization of dynamic systems assuring only read/write atomicy. Distrib. Comput. **7**, 3–16 (1993)
10. Glacet, C., Hanusse, N., Ilcinkas, D., Johnen, C.: Disconnected components detection and rooted shortest-path tree maintenance in networks. In: SSS 2014, pp. 120–134 (2014)
11. Huang, S.-T., Chen, N.-S.: A self-stabilizing algorithm for constructing breadth-first trees. IPL **41**, 109–117 (1992)
12. Johnen, C.: Memory-efficient self-stabilizing algorithm to construct BFS spanning trees. In: WSS 1997, pp. 125–140 (1997)

Brief Announcement: Distributed Computing in the Asynchronous LOCAL Model

Carole Delporte-Gallet[1], Hugues Fauconnier[1], Pierre Fraigniaud[1(✉)], and Mikaël Rabie[2]

[1] IRIF, CNRS and Université de Paris, Paris, France
pierre.fraigniaud@irif.fr
[2] LIP6, Sorbonne Université, Paris, France

Abstract. We show that, for any task T associated to a locally checkable labeling (LCL), if T is solvable in t rounds by a deterministic algorithm in the LOCAL model, then T remains solvable by a deterministic algorithm in $O(t)$ rounds in an asynchronous variant of the LOCAL model whenever $t = O(\text{polylog } n)$.

1 The LOCAL model

Distributed network computing [10] deals with the power and limitation of a collection of computing entities (a.k.a. processes) occupying the nodes of a network, and exchanging messages along the links of this network. In this framework, a primary interest has been placed on *locality*, that is, determining what tasks can be solved whenever every process has to output after having exchanged information with processes in its vicinity only, i.e., at bounded distance in the network. The LOCAL model [8] has been extensively used for studying locality in network computing over the last 25 years [13]. In this model, the network is modeled as a connected simple graph $G = (V, E)$, with processing nodes occupying the vertices of G, and communicating through the edges of G. Initially, every process is aware solely of its identity, which is supposed to be unique in the network. The LOCAL model is *synchronous*: computation proceeds as a sequence of *rounds*, with all nodes starting at the same round. At each round, every node sends messages to its neighbors in G, receives messages from its neighbors, and performs some individual computation. The round complexity of an algorithm is the number of rounds until all nodes output. For instance, a celebrated result in this context is Linial's lower bound [8] stating that 3-coloring the n-node ring requires at least $\frac{1}{2} \log^* n - O(1)$ rounds. This bound is tight up to additive constants, thanks to Cole and Vishkin's algorithm [4], which 3-colors the n-node ring in at most $\frac{1}{2} \log^* n + O(1)$ rounds (see [14], and [12] for the exact constant additive factors).

This research is supported by the ANR project DESCARTES (ref. DS0702-2016). Additional support from the INRIA project GANG.

M. Ghaffari et al. (Eds.): SSS 2019, LNCS 11914, pp. 105–110, 2019.
https://doi.org/10.1007/978-3-030-34992-9_9

Moreover, the LOCAL model can be further simplified. Indeed, as pointed out in [8], a t-round algorithm in the LOCAL model can be simulated by another t-round algorithm which proceeds in two phases: first, each node collects all the data present at the nodes at distance at most t around it, and, second, each node individually simulates the behavior of the original algorithm, without communication. In other words, a t-round algorithm in the LOCAL model can simply be viewed as a function from the ball $B_G(v, t)$ of radius t around every node v in G to the output set. This vision of the LOCAL model considerably simplifies the design of algorithms, and the analysis of the complexity of the problems.

2   Criticisms of the LOCAL model

Despite all its positive aspects, the LOCAL model is subject to pertinent criticisms. One such criticism is that the model assumes no bound on the computing power of the nodes, and on the throughput of the links. While this criticism is valid, it must be underlined that this apparent weakness of the model insures that lower bounds such as the one in [8] are non-conditional, i.e., they hold even if processes have infinite computing power, and even for full information protocols—where every node is forwarding at each round all the knowledge that it accumulated during the previous rounds. Moreover, most of (but indeed not all) the upper bounds do not abuse this power, that is, most algorithms involve polynomial-time computation at the nodes, like the Cole and Vishkin's algorithm [4]. Furthermore, the CONGEST model [10] has been designed especially for measuring the impact of limiting the bandwidth of the links for tasks involving high throughput, whose study in the context of the LOCAL model would be inappropriate. For instance, C_4 detection, i.e., determining whether the given network contains a cycle of length 4, is a trivial task in the LOCAL model, but it requires $\Theta(\sqrt{n})$ rounds to be solved in the CONGEST model [6]. The conclusion is that research in the context of the LOCAL model does not (too frequently) abuse of infinite computation power at the nodes, or of infinite link bandwidth.

Another criticism is the fact that all nodes start at the same time, and proceed in locksteps. Indeed, in practice, the processes may proceed at different speed depending on various factors including heterogeneity of the CPUs, clock drifts, cache misses, poor load balancing, etc. Moreover, the speeds of the different processes may vary with time, and processes may even be subject to all kinds of failures. In this context, the celebrated FLP theorem [7] states that binary consensus cannot be solved in the asynchronous message passing systems, even if at most one process can crash. The argument opposed to the criticism about synchrony in the LOCAL model is the ability to use *synchronizers* [1,2,11] for implementing synchronous algorithms in an asynchronous environment. However, while synchronizers are well suited to handle delays in the communications, they use *waiting* mechanisms allowing each node to figure out when a round finishes. Such mechanisms are not suited for an environment in which processes can vary in speed and eventually crash. Indeed, waiting can cause deadlocks occurring when a process forever waits for a process that has crashed. Instead, in asynchronous computing with an unbounded number of crashes, algorithms are required to be *wait-free*, that is, the algorithms must guarantee that every process can terminate and output correctly, independently from the behavior of the other processes.

To sum up, one must admit that there is still a gap between the study of asynchronous crash-prone computing and the study of locality in network computing. The main issue addressed in the paper is therefore: *Is making the LOCAL model slightly more realistic, e.g., by introducing some form of asynchrony and failures, resulting in a weaker model, for which stronger lower bounds could be derived?* Surprisingly, the answer to this question is shown to be essentially negative: we show that introducing asynchrony in the computation (while keeping synchronous communication between all nodes) does not reduce much the power of the LOCAL model, at least as far as the large class of tasks are concerned.

3   Decoupling Computations from Communications

Castañeda et al. [3] has initiated a line of work aiming at bridging the asynchrony-locality gap, by demonstrating that one can study locality in network computing even in the framework of asynchronous crash-prone processes. For this purpose, they introduced an asynchronous variant of the LOCAL model, called DECOUPLED, applied to symmetric networks (rings, toruses, etc.). This latter model decouples the computing entities (processes) from the communicating entities (routers). The communications remain synchronous, that is, there is still a notion of rounds. However, the processes are fully asynchronous and subject to crash failures. In particular, the processes may wake up at different times. It is shown in [3] that 3-coloring the n-node ring can still be done in $\frac{1}{2}\log^* n + O(1)$ rounds in the DECOUPLED model. I.e., it is sufficient that every node v waits for at most $t = \frac{1}{2}\log^* n + O(1)$ rounds after it wakes up for 3-coloring the n-node ring, even if processes are fully asynchronous and subject to crash failures.

In this paper, we simplify the *operational* DECOUPLED model into an *abstract* model, called ABST-DECOUPLED, that will be shown not stronger than the DECOUPLED model (in symmetric networks), but easier to handle, and defined for all kinds of networks. In a nutshell, one can view an algorithm in the ABST-DECOUPLED model as performing in two phases, like in the standard LOCAL model. The first phase consists, for every awake process v, of taking a snapshot of the ball $B_G(v, t)$ of radius t around v in G. This snapshot returns the structure of the ball $B_G(v, t)$, and the identifiers of some processes in $B_G(v, t)$, depending on the wake up times of these processes. The second phase consists of an individual computation at v eventually resulting in the output of node v. The main difference between the ABST-DECOUPLED model and the (synchronous) LOCAL model is the following. In the ABST-DECOUPLED model, if $w \in B_G(v, t)$ is not awake when v is making its snapshot, then the identity of w and its input remain unknown to v. Instead, in the LOCAL model, a snapshot of $B_G(v, t)$ systematically returns the identifiers and inputs of all nodes in $B_G(v, t)$.

Before presenting our results, we want to stress the fact that, in practice, the communication links as well as the memory registers can be viewed as synchronous, but the *access* to the links and to the registers is asynchronous, due to, e.g., scheduling and contention issues. The DECOUPLED and ABST-DECOUPLED models are aiming at taking this phenomenon into account, as least some aspects of it. We also want to stress the fact that, as we shall discuss further in the text, although the ABST-DECOUPLED model includes a strong snapshot functionality, it is not stronger than the realistic operational DECOUPLED model, at least as far as symmetric networks are concerned.

4   Our Results

We extend the results in [3] to the entire class of *locally checkable labeling* (LCL) tasks [9], which is of primary interest for the research in the framework for local computing in networks. Many classical graph problems, e.g., vertex or edge-coloring, maximal matching, maximal independent set, minimal dominating set, etc., are LCL tasks. In a nutshell, an (input-free) LCL task is specified by a set L of labels, and a family $\mathcal{F}$ of balls with constant radius $r \geq 0$, in which every node is labeled by a label in L. For instance, c-coloring corresponds to the LCL task with $L = \{1, \ldots, c\}$ and $\mathcal{F}$ is the family of balls with radius 1, such that the label of the center is different from the labels of all its neighbors.

In the ABST-DECOUPLED model, solving an LCL task $(L, \mathcal{F})$ of radius r in a graph $G = (V, E)$ asks every correct process $v \in V$ to output a label in L such that every ball $B_G(v, r)$ in which nodes corresponding to correct processes are labeled by their outputs, and nodes corresponding to processes that crashed are unlabeled, can be extended to a ball in $\mathcal{F}$, by assigning labels in L to the unlabeled nodes.

We prove the following general result, which shows that asynchronous crash-prone processes are essentially as efficient as reliable synchronous processes. A particular case of the statement below is when processes are initially aware of the size n of the network, and that the identifiers are in $[0, n-1]$.

Lemma 1. *Let $(L, \mathcal{F})$ be an LCL task. Assume that, in the LOCAL model, $(L, \mathcal{F})$ can be solved by a deterministic algorithm in $t(N)$ rounds in n-node graphs whenever the processes are initially aware of an upper bound N for n, and that the identifiers are in the range $[0, N)$. Then, in the ABST-DECOUPLED model, $(L, \mathcal{F})$ can be solved by a deterministic algorithm in at most $3\,t(N^2)$ rounds in n-node graphs, whenever the processes are initially aware of an upper bound N for n, and that the identifiers are in the range $[0, N)$.*

In particular, every LCL task that can be solved in a polylogarithmic number of rounds in the LOCAL model, whenever the processes are initially aware of an upper bound $N = O(poly(n))$ on the number of nodes and the range of IDs, can also be solved in a polylogarithmic number of rounds in the ABST-DECOUPLED model.

Since the ABST-DECOUPLED model will be shown not to be stronger than the DECOUPLED model in symmetric graphs, we get the following, as a direct consequence of Lemma 1.

Theorem 1. *Let* $(L, \mathcal{F})$ *be an LCL task. Assume that, in the LOCAL model,* $(L, \mathcal{F})$ *can be solved by a deterministic algorithm in* $t(N)$ *rounds in symmetric n-node graphs whenever the processes are initially aware of an upper bound* N *for* n, *and that the identifiers are in the range* $[0, N)$. *Then, in the DECOUPLED model,* $(L, \mathcal{F})$ *can be solved by a deterministic algorithm in at most* $3\,t(N^2)$ *rounds in symmetric n-node graphs whenever the processes are initially aware of an upper bound* N *for* n, *and that the identifiers are in the range* $[0, N)$.

In particular, 3-coloring the n-node ring whose processes are given the initial knowledge of n and that identifiers are in the range $[1, n]$, and where processes are sharing a consistent notion of clockwise and counterclockwise directions, can be solved in $\frac{3}{2} \log^* n + O(1)$ rounds in the DECOUPLED model. The algorithm in [3] performs in $\frac{1}{2} \log^* n + O(1)$ rounds, but it is specific to 3-coloring the ring, while our approach is generic, and applies to all LCL tasks.

The full version of this paper, including the proofs of the above claims, can be found in [5].

References

1. Awerbuch, B.: Complexity of network synchronization. J. ACM **32**(4), 804–823 (1985)
2. Awerbuch, B., Peleg, D.: Network synchronization with polylogarithmic overhead. In: 31st Symposium on Foundations of Computer Science (FOCS), pp. 514–522 (1990)
3. Castañeda, A., Delporte, C., Fauconnier, H., Rajsbaum, S., Raynal, M.: Making local algorithms wait-free: the case of ring coloring. Theory Comput. Syst. **63**(2), 344–365 (2019)
4. Cole, R., Vishkin, U.: Deterministic coin tossing with applications to optimal parallel list ranking. Inf. Control **70**(1), 32–53 (1986)
5. Delporte-Gallet, C., Fauconnier, H., Fraigniaud, P., Rabie, M.: Distributed computing in the asynchronous LOCAL model. CoRR abs/1904.07664 (2019)
6. Drucker, A., Kuhn, F., Oshman, R.: On the power of the congested clique model. In 33rd ACM Symposium on Principles of Distributed Computing (PODC), pp. 367–376 (2014)
7. Fischer, M.J., Lynch, N.A., Paterson, M.: Impossibility of distributed consensus with one faulty process. J. ACM **32**(2), 374–382 (1985)
8. Linial, N.: Locality in distributed graph algorithms. SIAM J. Comput. **21**(1), 193–201 (1992)
9. Naor, M., Stockmeyer, L.J.: What can be computed locally? SIAM J. Comput. **24**(6), 1259–1277 (1995)
10. Peleg, D.: Distributed Computing: A Locality-Sensitive Approach. SIAM, Philadelphia (2000)
11. Peleg, D., Ullman, J.D.: An optimal synchronizer for the hypercube. SIAM J. Comput. **18**(4), 740–747 (1989)

12. Rybicki, J., Suomela, J.: Exact bounds for distributed graph colouring. In 22nd International Colloquium on Structural Information and Communication Complexity (SIROCCO), pp. 46–60 (2015)
13. Suomela, J.: Survey of local algorithms. ACM Comput. Surv. **45**(2), 24:1–24:40 (2013)
14. Szegedy, M., Vishwanathan, S.: Locality based graph coloring. In: 25th ACM Symposium on Theory of Computing (STOC), pp. 201–207 (1993)

An Environment for Specifying and Model Checking Mobile Ring Robot Algorithms

Ha Thi Thu Doan[1](✉)[iD], Adrián Riesco[2][iD], and Kazuhiro Ogata[1]

[1] Japan Advanced Institute of Science and Technology, Ishikawa, Japan
{doanha,ogata}@jaist.ac.jp
[2] Universidad Complutense de Madrid, Madrid, Spain
ariesco@fdi.ucm.es

Abstract. An environment for specifying and model checking mobile robot algorithms on rings (or mobile ring robot algorithms) is proposed. We have developed the Maude Ring Specification Enviaude RSE), a specification environment that explicitly supports ring-shaped networks. Maude RSE is implemented on top of Maude, a rewriting logic-based specification language. The underlying key behind the tool is pattern matching between ring patterns and ring instances, called "ring pattern matching." Because rings are not commonly available data structures in any existing specification language, we encode ring patterns as sets of sequence patterns and simulate ring pattern matching by pattern matching between sets of sequence patterns and sequence instances, which is proven correct and transparent to Maude RSE users. The advantages of Maude RSE are demonstrated by case studies analyzing exploration and gathering algorithms.

Keywords: Distributed mobile robot system · Ring discrete model · Specification environment · Formal verification · Model checking

1 Introduction

The past two decades, theoretical computer science has seen the rapid growth and development of distributed computing by mobile entities. Recent developments focus on models and algorithms for autonomous mobile robots that self-organize and cooperate in order to achieve global goals. Autonomous mobile robots have been proposed for several important applications, such as rescue activities in disaster areas and outer space activities. The seminal model proposes a distributed system of k robots that have low capacities: they are identical

This research was partially supported by JSPS KAKENHI Grant Number JP19H04082, Comunidad de Madrid project BLOQUES-CM (S2018/TCS-4339) co-funded by EIE Funds of the European Union, and MINECO project *TRACES* (TIN2015-67522-C3-3-R).

M. Ghaffari et al. (Eds.): SSS 2019, LNCS 11914, pp. 111–126, 2019.
https://doi.org/10.1007/978-3-030-34992-9_10

(they are indistinguishable and all execute the same algorithm), oblivious (they have no memory of their past actions), and disoriented (they share no common orientation). Moreover, the robots do not communicate by sending or receiving messages, but have the ability to sense their environment and see the relative positions of the other robots.

Various models and algorithms [20, 25] have been proposed to solve particular problems for autonomous mobile robots. This paper focuses on ring discrete models [4, 5, 10], in which robots perform their activities in a ring-shaped network. What and how problems can be solved by a group of autonomous mobile robots on ring-shaped networks is an important topic in the area, as shown by the large number of algorithms that have been proposed: e.g. the papers [4, 12–14, 18, 27] propose algorithms for ring exploration, robot gathering on rings is solved in [5, 9, 11, 24, 26, 30], and some other problems are solved in [10, 19]. It is possible to make virtual rings over arbitrary-shaped network topologies and then mobile ring robot algorithms can be essentially applied to such topologies. Therefore, mobile ring robot algorithms are generic and worth investigating.

In the literature, the correctness of such algorithms relies on handmade mathematical proofs, which are error-prone. The untrustfulness of handmade mathematical proofs has been pointed out in [1, 3, 15, 16]. Formal, automatic techniques could help us increase the confidence of the existing algorithms/proofs, as shown in [1, 3, 8, 15, 16]. For discrete models, model-checking has been proven useful to find errors in the proposed algorithms [3, 15, 16]. However, ring discrete models are not well supported by any existing specification language, such as DVE [2], SPIN [22], and Maude [7]. This is because of the particular symmetries owned by *rings*. Consequently, the specifiers, such as Berard *et al.* in [3] and Doan *et al.* in [15, 16], need to specify rings by adapting other defined structures, such as *sequences*. It, therefore, makes the specification task tedious as well as time-consuming, while the specifications obtained are complicated and lengthy.

Context. Because rings cannot be directly supported by any existing specification language, we defined rings as associative sequences that satisfy two properties: rotative and reversible. We used Maude [7] as specification language because it allows us to use associative sequences. Now, the Maude Ring Specification Environment (Maude RSE), which explicitly supports ring-shaped networks, has been implemented on top of Maude. One key behind the tool is pattern matching between ring patterns and ring instances, called "ring pattern matching." Because of the above-mentioned reason, however, we encode ring patterns as sets of sequence patterns and simulate ring pattern matching by standard pattern matching between sets of sequence patterns and sequence instances, which is proven correct and transparent to Maude RSE users.

Contributions. Maude RSE itself and its theoretical foundations are the main achievements. Our research illustrates the power of rewriting logic in that Maude RSE can be implemented by extending Maude, more precisely Full Maude. That is, we do not need to implement such formal tools from scratch but we can do so by extending Maude and/or new formal tools on top of Maude. The case studies conducted in Maude RSE demonstrate that, because Maude RSE supports ring structures, mobile ring robot algorithm specifications in Maude RSE are more concise and compact than those in Maude, while the time overhead incurred by handling rings is almost irrelevant. From a theoretical point of view, we prove that ring pattern matching can be simulated by pattern matching between sets of sequence patterns and sequence instances. Therefore, Maude RSE will benefit researchers in both the formal methods community and the distributed computing community.

Outline. Section 2 overviews mobile robots on ring architectures and the problems of specifying mobile ring robot algorithms. Section 3 introduces Maude RSE and outlines the theory of ring-pattern matching. It, then, presents how to specify mobile ring robot algorithms in Maude RSE. Section 4 evaluates Maude RSE. Finally, Sect. 5 concludes the paper. The source code of the tool and three case studies, the detailed descriptions of the ring pattern match theory, and more information in Sect. 3 are available at [29].

2   Problems

In this paper, we restrict our attention to discrete models, and more specifically to the ring topology. About timing assumption, we consider the more general asynchronous model ASYNC. In addition, we take into account multiplicities, which make much harder to formalize mobile robot algorithms. Multiplicities appear in robot algorithms when more than one robot is allowed in one node; in the following, we will call *multiplicities* to these nodes.

Robots follow a three-phase behavior: *Look*, *Compute*, and *Move*. During its Look phase a robot takes a snapshot of other robots' positions. The collected information is used in the Compute phase during which the robot decides whether to move or stay idle. There may be lag between the Compute phase and the subsequent Move phase and then some other movements by other robots may be done in-between. A move that has been decided by a robot in a Compute phase but has not yet been conducted by the robot in the subsequent Compute phase is called a pending move. In the Move phase, the robot may move to one of the two adjacent nodes, as computed in the previous phase. Rings are *anonymous*, that is, there are neither node nor edge labeled.

Anonymous rings have *rotative* and *reversible* characteristics, which cannot be directly handled by any existing specification language. Let us illustrate these problems with a simple example. Assume that we specify the ring (the system state on a ring) shown in Fig. 1(a), in which robots are *disoriented*. Such a system state can be expressed as a sequence $q_0 \ q_1 \ \dots \ q_{j-1} \ q_j$ of intervals, where an interval q_i is the number of consecutive empty nodes between two non-empty nodes, in a view starting from any robot and traversing the ring in one arbitrary direction. System state representations are called *configurations*.

The system state as shown in Fig. 1(a) could be expressed as 2 1 0 3 1 in the (clockwise) view starting from the one at the bottom. Because it is a ring, the state could be also expressed, starting from other robots, as 1 0 3 1 2, 0 3 1 2 1, 3 1 2 1 0, and 1 2 1 0 3. Since robots are disoriented, the state

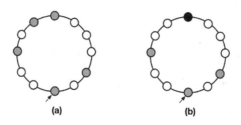

Fig. 1. (a) A system with two adjacent robots and (b) The system after the movement.

could be expressed as 1 3 0 1 2, 2 1 3 0 1, 1 2 1 3 0, 0 1 2 1 3, and 3 0 1 2 1 by reversing (i.e., considering counterclockwise) these sequences. All these configurations should be considered the same. Generally, given a sequence q_0 $q_1 \cdots q_{j-1}$ q_j, the state it expresses is equivalent, in a ring, to all sequences obtained by rotating it — $q_0 q_1 \cdots q_{j-1} q_j$, $q_1 \cdots q_{j-1} q_j q_0$, $\ldots$, $q_j q_0$, $q_1 \cdots$ q_{j-1} — and by reversing them — $q_j q_{j-1} \cdots q_1 q_0$, $q_{j-1} \cdots q_1 q_0 q_j$, $\ldots$, $q_0 q_j$ $q_{j-1} \cdots q_1$. Unfortunately, it is impossible to directly specify this in any existing specification language. Actually, all the expressions above are considered totally different from any existing specification language point of view, so specifiers are required to implement their own strategies to handle them. Consequently, specifiers need to specify rings by adapting other defined structures, such as *sequences*. For instance, in [16], Doan et al. use associative operators in Maude.

To illustrate the idea used in [16], let us show how to specify a mobile ring robot algorithm in Maude. Given a ring on which there are two robots located at two adjacent nodes, respectively (such two robots are called adjacent robots), we want to put them together (i.e., create a multiplicity) by moving one of them to the node at which the other is located, where there is a non-empty node closer to the node at which the former is located than to the other node. For example, in Fig. 1(a) (where we define nodes with respect to the bottom node, with an arrow) we have two adjacent robots on the top, the one on the left (the fifth, clockwise, from the bottom) is separated from the rest of nodes by one empty node, while the one on the right is separated by three empty nodes. Hence, we would move the one on the left to the node at which the other one is located, as shown Fig. 1(b), where the black node indicates a multiplicity. Assuming we use -1 to denote multiplicities, we can use a rewrite rule to specify this transition. The source state would use (i) 0 to indicate that two robots are adjacent, (ii) variables I1 and I2 to denote the intervals next to the adjacent robots, and (iii) a variable S to denote the remaining sequences. Assuming I2 is larger than I1, we will increment the smaller interval (I1) and replace 0 (robots are adjacent) by -1 (two robots are in the same node):

```
crl 0 I2 S I1 => -1 I2 S (I1 + 1) if I2 > I1.
```

In the particular case depicted in Fig. 1(a) the state could be expressed, clockwise and starting from the fifth node clockwise from the bottom, as 0 3 1 2 1. This configuration matches (the left-hand side of) the rule[1] by substituting I2 with 3, I1 with 1, and S with 1 2. The state is rewritten to −1 3 1 2 2, which expresses the configuration in Fig. 1(b).

However, the state shown in Fig. 1(a) could be also expressed, clockwise from the top, as 3 1 2 1 0. In this case, there is no substitution such that the sequence can match the rule. For this reason we need another rule to handle it:

```
crl I2 S I1 0 => -1 I2 S (I1 + 1) if I2 > I1.
```

The configuration 3 1 2 1 0 matches this rule by substituting I2 with 3, I1 with 1, and S with 1 2.

Splitting Problem. The state in Fig. 1(a) could be also expressed, clockwise from the bottom, as 2 1 0 3 1, but it is impossible to apply any of the rules above to this configuration. The rest of the sequence is split into two sub-sequences at both sides of the whole sequence. Thus, it is necessary to split the variable S into two variables S1 and S2 that denote the remaining sequences at the left side and the right side, respectively.

```
crl S2 I1 0 I2 S1 => -1 I2 S1 S2 (I1 + 1) if I2 > I1.
```

In our theoretical framework we need to formally define and work on splitting and joining (which puts together two sub-sequences that substitute two sequence variables obtained from the splitting before) functions that deal with these cases.

Reversing Problem. Let us take a look at the state in Fig. 1(a), which could be expressed, counter-clockwise from the fifth node (clockwise) from the bottom, as 1 2 1 3 0. We need the following rule for this case:

```
crl I1 S I2 0 => -1 I2 rev(S) (I1 + 1) if I2 > I1.
```

When the configuration matches the rule, what substitutes S is 2 1. We need to reverse 2 1, the sequence that substitutes S because otherwise what is obtained by applying the rule to the configuration is −1 3 2 1 2, which is different from −1 3 1 2 2. The function rev reverses a sequence, e.g rev(2 1) is 1 2. The configuration 1 2 1 3 0 matches this rule by substituting I2 with 3, I1 with 1, and S with 2 1. The state is rewritten to −1 3 1 2 2, the configuration in Fig. 1(b).

Hence, we need to have all the rules by rotating and reversing the left-hand side of the first rule to handle all possible sequences. We need 10 rules to specify the above-mentioned transition. Note that we name the rules RL1 to RL10.

```
crl[RL1]  0 I2 S I1      => -1 I2 S (I1 + 1)        if I2 > I1.
crl[RL2]  I2 S I1 0      => -1 I2 S (I1 + 1)        if I2 > I1.
crl[RL3]  S I1 0 I2      => -1 I2 S (I1 + 1)        if I2 > I1.
crl[RL4]  I1 0 I2 S      => -1 I2 S (I1 + 1)        if I2 > I1.
crl[RL5]  S2 I1 0 I2 S1  => -1 I2 S1 S2 (I1 + 1)    if I2 > I1.
```

---

[1] In the actual specification, we need an operator enclosing the sequence, such as {_} to avoid rewriting sub-sequences. However, to make the explanation as close as possible to mathematical description, we omit it here.

```
crl[RL6]   I1 S I2 0      => -1 I2 rev(S) (I1 + 1)           if I2 > I1.
crl[RL7]   0 I1 S I2      => -1 I2 rev(S) (I1 + 1)           if I2 > I1.
crl[RL8]   I2 0 I1 S      => -1 I2 rev(S) (I1 + 1)           if I2 > I1.
crl[RL9]   S I2 0 I1      => -1 I2 rev(S) (I1 + 1)           if I2 > I1.
crl[RL10]  S1 I2 0 I1 S2 => -1 I2 rev(S1) rev(S2) (I1 + 1) if I2 > I1.
```

This makes the specification complicated and lengthy and specifiers exhausted. If a ring is not faithfully specified, the formal verification of a mobile ring robot algorithm may overlook cases.

3   Maude Ring Specification Environment (Maude RSE)

One possible way to solve these problems is to develop a specification environment in which rings are explicitly supported. It is reasonable, and saves time and effort, if the environment is built on top of an existing specification system. For this reason, Maude Ring Specification Environment (Maude RSE) is implemented on top of Maude, a rewriting logic-based programming and specification language, taking advantage of its meta-programming features. This section outlines a theory of pattern matching on rings ("ring-pattern matching") that guarantees that our way of dealing with ring-pattern matching makes sense and briefly describes how Maude RSE is built, its architecture, and how to define a ring topology in it.

3.1   Ring Pattern Match Theory

Sequences. Let sequence patterns be in the form $ES_1\ ES_2\ \ldots\ ES_n$, where each ES_i is an element, an element variable, or a sequence variable. We suppose that the juxtaposition operator used as the constructor in sequence patterns is associative and the empty sequence, denoted ϵ, is its identity. Sequence instances are sequence patterns that do not contain variables. Let **SP** and **Seq** be the sets of sequence patterns and sequence instances, respectively. Let **Elt** be the set of (concrete) elements, **EV** be the set of element variables, and **SV** be the set of sequence variables.

Definition 1 (Sequence pattern match). *Pattern match between $sp \in$ **SP** & $seq \in$ **Seq** is to find all substitutions σ such that $\sigma(sp) = seq$. Let $sp =?= seq$ be the set of all such substitutions.*

Definition 2 (Split sequence patterns). *For $sp \in$ **SP**, $\mathrm{split}(sp)$ is a sequence pattern such that each sequence variable S in sp is replaced with $\mathrm{sv}(S,0)\ \mathrm{sv}(S,1)$. Then, the inductive definition of split is $\mathrm{split}(\varepsilon) = \varepsilon$, $\mathrm{split}(e) = e$ for $e \in$ **Elt**, $\mathrm{split}(E) = E$ for $E \in$ **EV**, $\mathrm{split}(S) = \mathrm{sv}(S,0)\ \mathrm{sv}(S,1)$ for $S \in$ **SV** and $\mathrm{split}(SP_1\ SP_2) = \mathrm{split}(SP_1)\ \mathrm{split}(SP_2)$ for $SP_1, SP_2 \in$ **SP**.*

Definition 3 (Joining split sequence variables). *For $sp \in$ **SP** and $seq \in$ **Seq**, let σ be in $(\mathrm{split}(sp) =?= seq)$. $\mathrm{join}(\sigma)$ is the substitution σ' such that for each sequence variable S in sp $\sigma'(S) = \sigma(\mathrm{sv}(S,0))\ \sigma(\mathrm{sv}(S,1))$ and for any* *other variables X $\sigma'(X) = \sigma(X)$. The domain of* join *can be naturally extended to the set of substitutions such that* join(split(sp) =?= seq) *is* {join(σ) | $\sigma \in$ (split(sp) =?= seq)}.

Rings

Definition 4 (Rings). *For $sp \in$ SP, $[sp]$ is called a ring pattern and satisfies (1) the rotative law ($[sp] = [\text{rtt}(sp)]$) and (2) the reversible law ($[sp] = [\text{rev}(sp)]$). When sp is a sequence $seq \in$ Seq, $[seq]$ is called a ring.* rtt(sp) *rotates sp rightward;* rev(sp) *reverses sp.*

Definition 5 (Ring pattern match). *For $sp \in$ SP and $seq \in$ Seq, pattern match between $[sp]$ and $[seq]$ is to find all substitutions σ such that $[\sigma(sp)] = [seq]$. Let $[sp]$ =?= $[seq]$ be the set of all such substitutions.*

Definition 6 (Sequences rotated and/or reversed). *For $sp \in$ SP, $[[sp]]$ is the set of sequences inductively defined as follows: (1) $sp \in [[sp]]$ and (2) if $sp' \in [[sp]]$, then* rtt(sp') $\in [[sp]]$ *and* rev(sp') $\in [[sp]]$.

The intuitive idea is that $[sp]$ is an implicit notation indicating that sp behaves as a ring while $[[sp]]$ is a explicit notation that lists all possible combinations after applying rotation and reverse to sp. In this way, given a particular sequence seq it is possible to use standard pattern matching between the elements in $[[sp]]$ and seq, so ($[[sp]]$ =?= seq) = {$\sigma \mid \sigma = (sp'$ =?= seq), for $sp' \in [[sp]]$}. The following theorem shows that it is possible to use $[[sp]]$ as an effective implementation of $[sp]$ (see Ring pattern match theory in [29] for details):

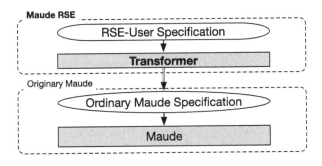

Fig. 2. Architecture of Maude RSE.

Theorem 1. *For any sequence pattern $sp \in$ SP and any sequence $seq \in$ Seq,* join($[[\text{split}(sp)]]$ =?= seq) = ($[sp]$ =?= $[seq]$).

3.2   Extending Maude with Ring Attributes

It has been demonstrated in [16,17,21,28] that Maude allows programmers to specify distributed algorithms/systems more succinctly than others programming languages. In particular, we extend Full-Maude [7], which is an extension of Maude written in Maude itself that provides extra features to extend Maude. The specification environment is built as depicted in Fig. 2. A specification in Maude RSE is considered as a user specification, which may contain specifications of a ring topology that would not be supported by the standard Maude engine. The main player in the system is *Transformer* that takes a user specification and transforms it into an ordinary Maude specification. Technically, a user specification is represented as a term at the meta-level. Transformer, then, analyzes and modifies it by adding extra equations/rules that handle rings. Pattern matching is a key functionality in Maude. Because pattern matching between ring patterns and ring instances is not supported by Maude and any other existing specification languages, we need to simulate it. There are two possible ways to simulate ring pattern matching. For a sequence pattern SP and a sequence instance SI, (1) we generate all sequence instances that denote the ring instance denoted by SI and model check each sequence instance with SP, and (2) we generate all sequence patterns that denote the ring pattern denoted by SP and model check SI with each ring pattern. We have adopted (2) because Maude automatically matches one sequence instance with many sequence patterns, while (1) would force us to manually handle a collection of sequence instances. It is nontrivial, however, to decide whether a matching between a ring pattern and a ring instance can be simulated by pattern matching between a collection of sequence patterns and a sequence instance. We have formally proved that the former can be simulated by the latter, see Sect. 3.1. The main idea is that given a user ring specification as a ring pattern, Maude RSE generates all corresponding sequence patterns to deal with the "ring" characteristic. Intuitively, given a ring pattern $[ES_1 \ldots ES_i \ldots ES_n]$, Transformer generates as the left-hand side of a rule: n rotative patterns $[ES_1 \ldots ES_i \ldots ES_n]$, $\ldots$, $[ES_i \ldots ES_n \ ES_1]$, $\ldots$, $[ES_n \ ES_1 \ldots ES_i]$ and n reversible patterns $[ES_1 \ ES_n \ldots ES_i]$, $\ldots$, $[ES_i \ldots ES_1 \ldots ES_n]$, $\ldots$, $[ES_n \ldots ES_i \ldots ES_1]$. When ES_i (i = 1, 2, $\ldots$, n) is a variable, it is split and jointed afterwards. We can basically use the right-hand side of the given rule as the right-hand side for the other $2n-1$ patterns generated as the left-hand side. We, however, need to reverse sequences that substitute sequence variables occurring in the right-hand side for the n reversible patterns. For example, in the problem in Sect. 2, users only need to specify the first rule RL1 while all other rules are automatically generated by Transformer by **1. Splitting:** All sequence variables are splitted. e.g. variable S is splitted into $S1$ and $S2$.

```
crl 0 I2 S1 S2 I1 => -1 I2 S1 S2 (I1 + 1)   if I2 > I1.
```

2. Rotating and **3. Joining:** All elements in the sequence of the left-hand side are rotated. After that, some pairs of sequence variables that are splitted from one sequence variable and appeared in the splitted order are joined. We get the rules RL2 to RL5. **4. Reversing:** All sequences on the left-hand sides are reversed (rules RL6 to RL10). As the result, we get all 10 rules. Users do not need to deal with the "ring" characteristic, which is handled transparently by Maude RSE.

In fact, Transformer needs to handle more complicated user specification rules/equations, all of them guarateed correct by Theorem 1. Because the result of the transformation is a standard Maude specification, we can guarantee that all Maude facilities, such as the LTL model checker, can be directly used.

3.3   Syntax Declaration

We consider two kinds of rings: *oriented* rings in which the orientation of the ring (clockwise and anti-clockwise order) is taken into account, and *disoriented* rings in which there is no orientation. In Maude, types are called *sorts*. A sort denotes the set of elements of the same type. For example, the sort Nat denotes the set of natural numbers. A sort is a *subsort* of another sort if and only if the set denoted by the former is a subset of the one denoted by the latter, and the latter is called a supersort of the former. Keywords sort and subsort are used to declare sort and subsort relation, respectively. Elements of a given sort are built by constructors, with keyword op, together with the keyword ctor, given the arity and the coarity. Moreover, operators can have equational axioms, such as associativity (assoc) and identity (id:).

We first consider *disoriented* rings, implemented by the ring attribute. In particular, rings are constructed as a sequence of elements with this attribute. Let us assume Elem and Seq the sorts for elements and sequences, respectively. The configurations of a system as rings could be defined as:

```
subsort Elem < Seq.
op emp : -> Seq [ctor].
op __ : Seq Seq -> Seq [ctor assoc id: emp].
op [_] : Seq -> Config [ring ctor].
```

An operator without any argument is called a *constant*, such as emp, which stands for the empty sequence. Underscores are placeholders where arguments are placed. Similarly, id: emp indicates that operator emp is the identity element of the juxtaposition (empty syntax) operator __. Seq is a supersort of Elem, which means that each Elem is treated as the singleton sequence only consisting of this element. The operator __ is used to construct sequences of elements: for s_1 and s_2 of sort *Seq*, s_1 s_2 has sort *Seq*. The structure __ is presented just as an example; it could be replaced by any other structure that depends on the user's preferences, such as _,_ and _|_. Likewise, the structure [_] is an optional preference. A configuration is defined as a ring structure that is specified based on a sequence of elements. Because a ring is disoriented, the mirror image of a ring represents the same state as the original state. When we use intervals as ring elements, we could use Int for Elem, where Int is the sort for integers. The system shown in Fig. 1(a) is expressed using this syntax as [0 3 1 2 1], [3 1 2 1 0], [1 2 1 0 3], [1 2 1 3 0], and so on.

For *oriented* rings, Maude RSE provides the `r-ring` attribute that could be considered as a sub-class attribute of the `ring` attribute. The *oriented* ring and its mirror image do not necessarily represent the same state.

3.4   Applications

We have specified and model checked three algorithms for exploration with stop, exploration, and gathering. The two last algorithms have been also specified in standard Maude. We compare our new specifications with existing ones (see Sect. 4). Due to page limitation, this section only presents how to formalize and specify the algorithm for exploration with stop in Maude RSE.

Robots Exploration with Stop on Ring Under ASYNC. The ASYNC (or asynchronous) model is considered. Each robot can distinguish whether a node is empty, occupied by one robot, or occupied by more than one robot. The problem of exploring with stop requires that, regardless of the initial placement of the robots, each node must be visited by at least one robot and the robots must be in a configuration in which they all remain idle.

Exploration Algorithm [18]. The algorithm works following a sequence of three distinct phases: Set-Up, Tower-Creation, (towers are the equivalent notion to multiplicities in our notation) and Exploration. The Set-Up phase transforms any initial configuration into one that is in a predetermined set of configurations (called *no-towers-final*) with special properties. After that, the Tower-Creation phase and then the Exploration phase are executed. Finally, all nodes are visited and no robot will make any further moves.

Formal Specification of Exploration Algorithm. Let us suppose that there are n nodes denoted u_0 u_1 ... u_{n-1} and each node may be occupied by more than one robot. The multiplicity of robots located at node u_i is denoted d_i: $d_i = 0$, $d_i = 1$, and $d_i = 2$ indicate that there are no robots, there is exactly one robot, and there are more than one robot, respectively.

State Expressions. So far the state of a system on a ring has been represented as a sequence of elements, e.g, for the system in Fig. 1(a), elements are intervals. For this system, each element is a node of the ring. Remember from Sect. 2 that robots move in two phases: first they decide where to move and then the movement is performed; when the movement has been decided but not applied yet it is called *pending move*. A pending move is represented as a snapshot of the ring from the node the robot will move; to avoid ambiguities due to the symmetries of the ring, the same snapshot is stored in the target node.

Once the pending move is completed a new movement can be computed, so we have at most one pending move at a time (this is true because of the particular structure of the algorithm given above, that first creates multiplicities and does not move them afterwards); on the other hand, many target-pending moves can be stored in each node, because robots from different nodes may want to move

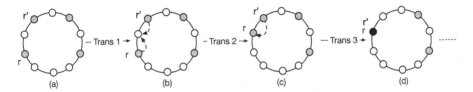

Fig. 3. One possible transitions from the initial configuration: a dashed arrow represents a pending move and a black node represents a tower.

to the same node. because robots from different nodes may want to move to the same node. In this way, executing a movement is as simple as finding two nodes, one with the pending move and another one with the same pending move as a target. Note that, in this stage, robots just know that they have to move, but they cannot access other robots' pending moves.

Hence, we denote a node as a tuple $\langle d_i, p_i, ps_i \rangle$, where d_i denotes the multiplicity of the node, p_i denotes a pending move, and ps_i denotes the set of pending moves that will be done by other robots. p_i is either one pending move or nil meaning that there are no pending moves. Given this definition of node, a snapshot (i.e., the pending move) for a robot at node u_i is the sequence d_i $d_{i+1} \ldots d_{i-1}$ of the multiplicities taken from that node. ps_i is either a set of pending moves or emp, the empty set. The sort `Pending` denotes pending moves, `PendingSet` pending move sets, and `Node` nodes. A configuration, of sort `Config`, is expressed as a ring of nodes with the `ring` attribute:

```
subsort Node < Seq.
op <_,_,_> : Nat Pending PendingSet -> Node [ctor].
op emp : -> Seq [ctor].
op __ : Seq Seq -> Seq [ctor assoc id: emp].
op [_] : Seq -> Config [ctor ring].
```

The structure `<_,_,_>` is used to construct nodes. For $d \in Nat$, $p \in Pending$, $ps \in PendingSet$, we have $\langle d, p, ps \rangle \in Node$. A configuration is defined as a ring `[_]`, which takes as argument a sequence of nodes. For example, let v and v' be the pending moves 1 0 1 0 1 0 1 0 0 0 and 1 0 1 0 1 0 0 0 1 0 for r and r', respectively. Note that each snapshot is taken from the node making the movement in clockwise order, although the anti-clockwise order would be valid as well. The configuration of the system with two pending moves as shown in Fig. 3(b) could be expressed from robot r' clockwise as $[\langle 1, v', emp \rangle \langle 0, nil, emp \rangle \langle 1, nil, emp \rangle \langle 0, nil, emp \rangle \langle 1, nil, emp \rangle \langle 0, nil, emp \rangle \langle 0, nil, emp \rangle \langle 0, nil, emp \rangle \langle 1, v, emp \rangle \langle 0, nil, (v; v') \rangle]$. We see that two nodes (standing for r and r', respectively) have pending moves in the second component of the tuple while the target node has both pending moves in the third component.

State Transitions. When either (1) a robot takes the snapshot of the system and then computes a move, or (2) a robot executes its pending move, the current configuration of the system changes. Such changes, called (state) transitions, are specified by rewrite rules. For example, one possible transition path is as shown in Fig. 3. The following rewrite rule describes the action when a robot performs its pending move (note that variable P appears twice, once as pending move and then as part of the set in the target node; it disappears once the rule is applied):

```
rl [S1 < 1, P, PS > < D, P', (P; PS') > S2] =>
   [S1 < 0, nil, PS > < D + 1, P', PS' > S2] .
```

where S1 and S2 are variables of sort Seq, P and P' are variables of sort Pending, PS and PS' are variables of sort PendingSet, and D is a variable of sort Nat.

The configuration [S1 ⟨ 1, P, PS ⟩⟨ D, P', (P; PS') ⟩ S2] represents any state such that the robot ⟨ 1, P, PL ⟩ has a pending move P and the next node is ⟨ D, P', (P; PL') ⟩ in which P is in the set of pending moves. In addition to these two nodes, such a state may have some more nodes that are expressed as $S1$ and $S2$. The term $[\langle 1, v', emp \rangle\langle 0, nil, emp \rangle\langle 1, nil, emp\rangle\langle 0, nil, emp \rangle\langle 1, nil, emp\rangle\langle 0, nil, emp\rangle \langle 0, nil, emp\rangle\langle 0, nil, emp\rangle\langle 1, v, emp\rangle\langle 0, nil, (v; v')\rangle]$ expresses the state of Fig. 3(b). The left-hand side of the above rewrite rule matches this term by using the substitution $S1 \mapsto \langle 1, v', emp\rangle\langle 0, nil, emp\rangle\langle 1, nil, emp\rangle\langle 0, nil, emp\rangle\langle 1, nil, emp\rangle \langle 0, nil, emp\rangle\langle 0, nil, emp\rangle \langle 0, nil, emp\rangle$, $P \mapsto v$, $PS \mapsto emp$, $D \mapsto 0$, $P' \mapsto nil$, $PS' \mapsto v'$ and $S2 \mapsto emp$, and the rewrite rule can be applied to the term, creating the state $[\langle 1, v', emp\rangle\langle 0, nil, emp\rangle\langle 1, nil, emp\rangle \langle 0, nil, emp\rangle\langle 1, nil, emp\rangle \langle 0, nil, emp\rangle \langle 0, nil, emp\rangle\langle 0, nil, emp\rangle\langle 0, nil, emp\rangle\langle 1, nil, v' \rangle]$, which corresponds to the state in Fig. 3(c). Note that the size of the ring is not fixed.

Model Checking the Algorithm. We use the Maude LTL model checker to verify that the algorithm enjoys desired properties. The authors in [18] give some important theorems, such as *Theorems 3.1* and *3.2*, that must hold to guarantee the correctness of the algorithm. For example, *Theorem 3.1* states a property that must be satisfied at the end of the Set-Up phase: any initial configuration is transformed into a *no-towers-final* configuration. We have formally expressed these theorems as LTL formulas [23]. For example, *Theorem 3.1* then is expressed as the LTL formula:

```
theorem3-1 = [] (endOf -> SetUp) /\ <> endOf .
```

where [] is the always operator and <> is the eventually operator. The proposition endOf holds if and only if the Set-Up phase has finished and the proposition SetUp holds if and only if the state does not have any towers and all robots are located adjacent to each other in one or two groups.

As the result of the model checking, no counterexamples are found for the LTL formula. This makes us more confident on the correctness of the algorithm.

4   Evaluation

We compare the sizes and performance of specifications in plain Maude [15,16] and in Maude RSE and report on the overheads (which is almost nothing) introduced by Maude RSE for model checking. We consider two algorithms solving two main problems on ring: the perpetual exploration algorithm, which was defined in [4] and specified in [15], and the gathering algorithm, designed in [9] and specified in [16]. Note that Maude RSE successfully reproduces the model checking experiments reported in [15,16], finding the counterexamples demonstrating that the algorithms do not enjoy some properties.

4.1   A Perpetual Exploration Algorithm [4]

In [15], the ring is represented as the set of all non-empty nodes. The ring features, namely rotation and reversibility, are dealt with by using associative and commutative sets. However, the commutative attribute makes it impossible to keep the order in the ring. Specifiers, thus, are forced to use complex constraints to specify the algorithms because the order of elements might change. For this reason, several functions are defined to handle these constrains.

By using Maude RSE, we do not need to handle ring characteristics and we do not need to use commutativity as the attribute of the constructor used to construct configurations. Our specification gets rid of all these extra functions. In total, we reduce over 50% of the code.

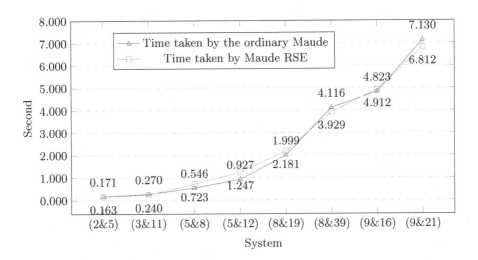

Fig. 4. Maude RSE preserves the performance of the ordinary Maude environment.

4.2   A Gathering Algorithm [9]

In [16], the authors use 44 rules to specify the system and 53 equations to handle some constraints about configurations. Many of these rules and equations are defined to handle rings. In Maude RSE, the specification requires 17 rules and 18 equations, that is, we obtain a code reduction of more than 60%.

Performance Analysis. We conducted model checking experiments for the gathering algorithm to compare the performances. 8 different systems in terms of the number of robots and the size of the ring, e.g. System 1 with 2 robots and 5 nodes denoted as (2&5), are taken. Experiments were conducted on a 4 GHz Intel Core i7 processor with 32 GB of RAM. The results are shown in Fig. 4. Based on these experiments, we can conclude that Maude RSE preserves the performance of the ordinary Maude environment; no extra time consuming.

5 Conclusion

Because mobile robot systems are a new form of distributed system, the existing specification methods (and tools) do not support these systems appropriately. In this case, a new language or an extension of an existing language is needed. This paper introduces an extension of Maude to mobile ring robot algorithms: Maude RSE. Maude RSE makes it possible to specify ring structures, which need to be used to specify mobile ring robot algorithms. As extensions of ring topology, recent research on rings consider dynamic rings where edges may appear and disappear unpredictably [6]. Furthermore, more kinds of robots, such as myopic luminous robots [31] are proposed to work on rings. As future work, we try to tackle other kinds of robots on rings, such as myopic luminous robots. We then consider extending Maude RSE in the following directions: (1) to support other features on ring, such that rings are dynamic, and (2) to support other topologies by making virtual rings over them.

References

1. Balabonski, T., Delga, A., Rieg, L., Tixeuil, S., Urbain, X.: Synchronous gathering without multiplicity detection: a certified algorithm. In: Bonakdarpour, B., Petit, F. (eds.) SSS 2016. LNCS, vol. 10083, pp. 7–19. Springer, Cham (2016). https://doi.org/10.1007/978-3-319-49259-9_2
2. Barnat, J., Brim, L., Češka, M., Ročkai, P.: Divine: parallel distributed model checker. In: Parallel and Distributed Methods in Verification and High Performance Computational Systems Biology. IEEE (2010)
3. Bérard, B., Lafourcade, P., Millet, L., Potop-Butucaru, M., Thierry-Mieg, Y., Tixeuil, S.: Formal verification of mobile robot protocols. Distrib. Comput. **29**(6), 459–487 (2016)
4. Blin, L., Milani, A., Potop-Butucaru, M., Tixeuil, S.: Exclusive perpetual ring exploration without chirality. In: Lynch, N.A., Shvartsman, A.A. (eds.) DISC 2010. LNCS, vol. 6343, pp. 312–327. Springer, Heidelberg (2010). https://doi.org/10.1007/978-3-642-15763-9_29
5. Bonnet, F., Potop-Butucaru, M., Tixeuil, S.: Asynchronous gathering in rings with 4 robots. In: Mitton, N., Loscri, V., Mouradian, A. (eds.) ADHOC-NOW 2016. LNCS, vol. 9724, pp. 311–324. Springer, Cham (2016). https://doi.org/10.1007/978-3-319-40509-4_22
6. Bournat, M., Dubois, S., Petit, F.: Computability of perpetual exploration in highly dynamic rings. In: IEEE 37th International Conference on Distributed Computing Systems, pp. 794–804 (2017)

7. Clavel, M., et al.: All About Maude - A High-Performance Logical Framework. LNCS, vol. 4350. Springer, Heidelberg (2007). https://doi.org/10.1007/978-3-540-71999-1

8. Courtieu, P., Rieg, L., Tixeuil, S., Urbain, X.: Certified universal gathering in $\mathbb{R}^2$ for oblivious mobile robots. In: Gavoille, C., Ilcinkas, D. (eds.) DISC 2016. LNCS, vol. 9888, pp. 187–200. Springer, Heidelberg (2016). https://doi.org/10.1007/978-3-662-53426-7_14

9. D'Angelo, G., Di Stefano, G., Navarra, A.: Gathering on rings under the look-compute-move model. Distrib. Comput. **27**, 255–285 (2014)

10. D'Angelo, G., Di Stefano, G., Navarra, A., Nisse, N., Suchan, K.: Computing on rings by oblivious robots: a unified approach for different tasks. Algorithmica **72**(4), 1055–1096 (2015)

11. D'Angelo, G., Navarra, A., Nisse, N.: A unified approach for gathering and exclusive searching on rings under weak assumptions. Distrib. Comput. **30**(1), 17–48 (2017)

12. Datta, A.K., Lamani, A., Larmore, L.L., Petit, F.: Enabling ring exploration with myopic oblivious robots. In: IEEE International Parallel and Distributed Processing Symposium Workshop, Hyderabad, pp. 490–499 (2015)

13. Devismes, S.: Optimal exploration of small rings. In: Proceedings of the Third International Workshop on Reliability, Availability, and Security, pp. 91–96 (2010)

14. Devismes, S., Petit, F., Tixeuil, S.: Optimal probabilistic ring exploration by semi-synchronous oblivious robots. Theor. Comput. Sci. **498**, 10–27 (2013). https://doi.org/10.1016/j.tcs.2013.05.031

15. Doan, H.T.T., Bonnet, F., Ogata, K.: Model checking of a mobile robots perpetual exploration algorithm. In: Liu, S., Duan, Z., Tian, C., Nagoya, F. (eds.) SOFL+MSVL 2016. LNCS, vol. 10189, pp. 201–219. Springer, Cham (2017). https://doi.org/10.1007/978-3-319-57708-1_12

16. Doan, H.T.T., Bonnet, F., Ogata, K.: Model checking of robot gathering. In: Proceedings of The 21th Conference on Principles of Distributed Systems, pp. 12:1–12:16 (2017)

17. Doan, H.T.T., Bonnet, F., Ogata, K.: Specifying a distributed snapshot algorithm as a meta-program and model checking it at meta-level. In: Proceedings of The 37th IEEE International Conference on Distributed Computing Systems, pp. 1586–1596 (2017)

18. Flocchini, P., Ilcinkas, D., Pelc, A., Santoro, N.: Computing without communicating: ring exploration by asynchronous oblivious robots. Algorithmica **65**(3), 562–583 (2013)

19. Flocchini, P., Kranakis, E., Krizanc, D., Santoro, N., Sawchuk, C.: Multiple mobile agent rendezvous in a ring. In: Farach-Colton, M. (ed.) LATIN 2004. LNCS, vol. 2976, pp. 599–608. Springer, Heidelberg (2004). https://doi.org/10.1007/978-3-540-24698-5_62

20. Flocchini, P., Prencipe, G., Santoro, N.: Distributed Computing by Oblivious Mobile Robots. Morgan & Claypool Publishers (2012)

21. Grov, J., Ölveczky, P.C.: Formal modeling and analysis of Google's megastore in real-time maude. In: Iida, S., Meseguer, J., Ogata, K. (eds.) Specification, Algebra, and Software. LNCS, vol. 8373, pp. 494–519. Springer, Heidelberg (2014). https://doi.org/10.1007/978-3-642-54624-2_25

22. Holzmann, G.J.: The SPIN Model Checker: Primer and Reference Manual. Addison-Wesley, Boston (2004)

23. Huth, M., Ryan, M.: Logic in Computer Science: Modelling and Reasoning about Systems. Cambridge University Press, Cambridge (2004)

24. Izumi, T., Izumi, T., Kamei, S., Ooshita, F.: Mobile robots gathering algorithm with local weak multiplicity in rings. In: Patt-Shamir, B., Ekim, T. (eds.) SIROCCO 2010. LNCS, vol. 6058, pp. 101–113. Springer, Heidelberg (2010). https://doi.org/10.1007/978-3-642-13284-1_9

25. Kawamura, A., Kobayashi, Y.: Fence patrolling by mobile agents with distinct speeds. Distrib. Comput. **28**(2), 147–154 (2015)

26. Klasing, R., Markou, E., Pelc, A.: Gathering asynchronous oblivious mobile robots in a ring. Theor. Comput. Sci. **390**(1), 27–39 (2008)

27. Lamani, A., Potop-Butucaru, M.G., Tixeuil, S.: Optimal deterministic ring exploration with oblivious asynchronous robots. In: Patt-Shamir, B., Ekim, T. (eds.) SIROCCO 2010. LNCS, vol. 6058, pp. 183–196. Springer, Heidelberg (2010). https://doi.org/10.1007/978-3-642-13284-1_15

28. Liu, S., Ölveczky, P.C., Wang, Q., Meseguer, J.: Formal modeling and analysis of the walter transactional data store. In: Rusu, V. (ed.) WRLA 2018. LNCS, vol. 11152, pp. 136–152. Springer, Cham (2018). https://doi.org/10.1007/978-3-319-99840-4_8. https://sites.google.com/ site/siliunobi/walter

29. Our Maude source files. https://goo.gl/6AnwHE

30. Millet, L., Potop-Butucaru, M., Sznajder, N., Tixeuil, S.: On the synthesis of mobile robots algorithms: the case of ring gathering. In: Felber, P., Garg, V. (eds.) SSS 2014. LNCS, vol. 8756, pp. 237–251. Springer, Cham (2014). https://doi.org/10.1007/978-3-319-11764-5_17

31. Ooshita, F., Tixeuil, S.: Ring exploration with myopic luminous robots. In: Izumi, T., Kuznetsov, P. (eds.) SSS 2018. LNCS, vol. 11201, pp. 301–316. Springer, Cham (2018). https://doi.org/10.1007/978-3-030-03232-6_20

Brief Announcement: Self-stabilizing LCM Schedulers for Autonomous Mobile Robots Using Neighborhood Mutual Remainder

Shlomi Dolev[1], Sayaka Kamei[2(✉)], Yoshiaki Katayama[3], Fukuhito Ooshita[4], and Koichi Wada[5]

[1] Department of Computer Science, Ben-Gurion University of the Negev,
Beersheba, Israel
dolev@cs.bgu.ac.il

[2] Graduate School of Engineering, Hiroshima University, Higashihiroshima, Japan
s-kamei@se.hiroshima-u.ac.jp

[3] Graduate School of Engineering, Nagoya Institute of Technology, Nagoya, Japan
katayama@nitech.ac.jp

[4] Nara Institute of Science and Technology, Ikoma, Japan
f-oosita@is.naist.jp

[5] Faculty of Science and Engineering, Hosei University, Tokyo, Japan
wada@hosei.ac.jp

Abstract. A vast number of algorithms for robots assume a given schedule that ensures the simultaneous LOOK (observing the location of all other non-moving robots), only then simultaneous COMPUTE (computing the target for moving) and only then simultaneous MOVE (LCM). The assumption on the existence of synchronization signal for movements, when other robots do not LOOK is questionable. We present several self-stabilizing techniques to implement such a scheduler using global/local clock pulses and thus converting the vast numbers of robot algorithms that assume the LCM schedule, to eliminate this assumption at once. Namely, we present implementation of a move-atomic property scheduler, where robots possess an independent clock that is advanced at the same speed. We realize it by applying the neighborhood mutual remainder. This research presents the first self-stabilizing implementations of the LCM synchronization.

Keywords: Self-stabilization · LCM robot system · Move-atomic · Neighborhood mutual remainder

This work was supported in part by JSPS KAKENHI No. 17K00019, 18K11167, 19K11823 and 19K11828, Rita Altura Trust Chair in Computer Science, the Ministry of Science and Technology, Israel &Japan Science and Technology Agency (JST) SICORP (Grant#JPMJSC1806) and the German Research Funding Organization (DFG, Grant#8767581199).

M. Ghaffari et al. (Eds.): SSS 2019, LNCS 11914, pp. 127–132, 2019.
https://doi.org/10.1007/978-3-030-34992-9_11

1   Introduction

There is a vast literature on algorithms for swarms of robots that are scheduled to perform steps simultaneously, abstracting away the mechanism used to synchronize the simultaneous operations. An abstraction in which the operations between two successive robot configurations consists of a global period for LOOK (observing the location of the other robots) after which a global period for COMPUTE (computing the next move step) and at the end a global period for MOVE (moving to the computed location). The LCM abstraction simplifies the algorithm design and proof processes.

In distributed computing, synchronizers are used to "compile" such synchronous distributed algorithm to work in a more realistic asynchronous distributed systems. The synchronizer supplies a local indication on a clock pulse, that triggers the next operation. The synchronizer is designed to preserve the invariants and proofs of the synchronous algorithm. Thus, automatic convergence of a synchronous algorithm works in asynchronous settings. In the case of LCM, one needs to supply a local indication on three pulses, one for LOOK, one for COMPUTE and one for MOVE. In fact, LOOK and COMPUTE can be regarded as one period in which no robot moves.

Requiring all robots not to move at a certain period may be too restrictive. Our LCM synchronizer compiler obtains the effect of an atomic move (preserving the invariants and proofs of atomic LCM algorithm) by local means. Namely, when a robot looks no other robot in its neighborhood moves (thus, avoiding race conditions).

To implement such synchronizers, we introduce a new synchronization concept, called the *neighborhood mutual remainder* (NMR), and give a simple self-stabilizing NMR algorithm in the full paper [1]. A distributed algorithm that satisfies the NMR requirement should satisfy *global fairness*, *l-exclusion*, and *local rendezvous* requirements. Global fairness is satisfied when each participant executes the critical section (CS) infinitely often, *l*-exclusion is satisfied when at most *l* neighboring processes enter the CS at the same time, and local rendezvous is satisfied when for each participant infinitely often no participant in the closed neighborhood is in the CS or a trying section.

In this BA, we give a self-stabilizing synchronization algorithm for an LCM robot system by using the NMR. We realize the move-atomic property in a self-stabilizing manner on the assumption that robots repeatedly receive clock pulses at the same time, where the move-atomic property guarantees that, while some robot executes LOOK and COMPUTE phases, no robot in its sight executes a MOVE phase. This research presents the first self-stabilizing implementation of the LCM synchronization, allowing the implementation in practice of any self-stabilizing or stateless robot algorithm.

In the full paper [1], we extend the above algorithm to the assumption that robots receive clock pulses at different times but the duration between two pulses is identical for all robots. Then, on the same assumption, we implement the fully synchronous (FSYNC) model.

2 Preliminaries for Robot Systems

By applying NMR, we implement traditional LCM robot models on an ordinary distributed system composed of mobile terminals (Fig. 1). In the rest of this section, we describe an underlying mobile terminal model where NMR is executed, and a simulated robot model where algorithms for LCM robot model can be executed.

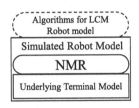

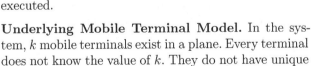

Fig. 1.   Structure of robot system.

Underlying Mobile Terminal Model. In the system, k mobile terminals exist in a plane. Every terminal does not know the value of k. They do not have unique IDs and execute the same deterministic algorithm. Every terminal has no direct communication means except lights which can emit a color to other terminals. They also have an observation device to obtain other terminals' positions and colors of lights within a fixed distance ϕ from its current position.

The terminal can move up to y in one time unit (one move). If the terminal moves toward the target position, it may stop moving before arriving at the target position, however it precedes the distance of at least σ. If the distance from the current position to the target position is at most σ, the terminal always reaches the target position.

A communication graph is defined as $G = (V, E)$ where V is a set of terminals and E is a set of terminal pairs that can observe each other. Note that the communication graph may change when terminals move. We say terminal r_i is a neighbor of r_j if $(r_i, r_j) \in E$ holds. A neighborhood of terminal r_i is denoted by $N(i) = \{r_j | (r_i, r_j) \in E\}$, and a closed neighborhood of r_i is denoted by $N[i] = N(i) \cup \{r_i\}$.

Every terminal operates based on pulses. When a terminal receives a pulse, it instantaneously takes a snapshot by using its observation device and obtains positions and colors of neighboring terminals. And then, it executes an algorithm based on the snapshot before the next pulse. We consider *global pulses* as external pulses. All terminals receive these pulses at the same time, and the duration between two successive pulses is identical for all robots. We regard the duration between two successive pulses as one time unit.

Simulated Robot Model. Next, we describe simulated robot models that we will implement on the underlying mobile terminal model. Our objective robot models are in the traditional LCM (Look-Compute-Move) robot model.

In the traditional LCM robot model, each robot repeats three-phase cycles: LOOK, COMPUTE, and MOVE. During the LOOK phase, the robot obtains positions and colors of its neighboring robots using its observation device. During the COMPUTE phase, the robot computes its next state, color, and movement according to the observation in the LOOK phase. The robot may change its state and color at the end of the COMPUTE phase. If the robot decides to move, it moves toward the target position during the MOVE phase. The robot may stop moving before arriving at the target position, however it precedes the distance of at least σ. If the distance from the current position to the target position is at most σ, the robot always reaches the target position. In the following, we simply describe that a robot executes LOOK, COMPUTE, and MOVE instead of executing the LOOK, COMPUTE, and MOVE phase, respectively.

The simulated robot also operates based on pulses as the underlying terminal model. When the terminal receives a pulse, the simulated robot also operates one or more of the phases among LOOK, COMPUTE or MOVE immediately after the underlying terminal finished the operation. The underlying terminal tells the simulated robot which phase(s) can be executed. When the simulated robot executes LOOK, it obtains positions and colors of other robots (terminals) within the distance $\phi - y$. The reason why this view restriction is necessary will be described later.

In literature, some synchronization models are considered in the LCM model. In this paper, we focus on the move-atomic model. The move-atomic model guarantees that, while a robot executes MOVE, none of its neighbors executes LOOK or COMPUTE.

Dynamic Graph Reduction. Let ϕ be the distance of underlying terminal views, namely, the NMR algorithm of a terminal is executed with all terminals within ϕ distance from the terminal. At this point, we must assume that each simulated robot should move up to $y < \phi$ in a single time unit and uses neighbors within distance up to $\phi - y$ when it executes LOOK, the only neighborhood dynamism is from another simulated robot r_j that is not viewed by r_i in the NMR algorithm, therefore is not included in the local synchronization, but penetrates to be in the COMPUTE zone of r_i while r_i executes LOOK. Hence having a y-tier eliminates such a scenario by any r_j.

Self-stabilizing LCM Implementations. We aim to implement the move-atomic model on the underlying system model in a self-stabilizing manner. To do this, we assign some time units to execute LOOK, COMPUTE, and MOVE, and trigger the phases upon pulses. We assume that each phase does not last beyond the next pulse. This implies that the duration from a pulse to the next pulse is sufficiently long so that robots precede the distance of at least σ.

Definition 1. *The system implements a self-stabilizing move-atomic model if there exists some time t such that, after time t, (1) every robot repeats three-phase cycles infinitely and (2) while a robot executes MOVE, none of its neighbors executes LOOK or COMPUTE.*

3   Self-stabilizing Move-Atomic Algorithm

The main idea of the implementation is to apply the NMR algorithm [1] to terminals that simulate robots. Let $robot(r_i)$ be the robot simulated by terminal r_i. Terminal r_i can make $robot(r_i)$ execute LOOK and COMPUTE only when r_i

---

Algorithm 1. Self-Stabilizing Move-Atomic Algorithm with Global Pulses for r_i.

Upon a global pulse
 $MaxN_i := \max\{Nlight_j \mid r_j \in N[i]\}$
 if $Light_i = 0 \wedge LC_i = true$ **then**{
  // Enter the CS
  make $robot(r_i)$ execute LOOK and COMPUTE
  $LC_i := false$
 } **else if** $\forall r_j \in N[i][Light_j \neq 0] \wedge LC_i = false$ **then**{
  // Rendezvous (No closed neighbors enter the CS)
  make $robot(r_i)$ execute MOVE
  $Nlight_i := |N[i]|$
  $LC_i := true$
 }
 $Clock_i := (Clock_i + 1) \mod (MaxN_i + 1)$
 $Light_i := Clock_i$

---

enters the CS, and it can make $robot(r_i)$ execute MOVE only when the neighborhood of r_i is in rendezvous, namely, $rendezvous_i$. When $rendezvous_i$ takes place, no neighbor of r_i is in the CS (i.e., no neighbor r_j of r_i makes $robot(r_j)$ execute LOOK or COMPUTE), and thus r_i can make $robot(r_i)$ execute MOVE. By this behavior, we can achieve the move-atomic property: While a robot executes MOVE, none of its neighbors executes LOOK or COMPUTE.

Each terminal r_i has following two lights: $Nlight_i \in \{1, \ldots, k\}$ is the color representing $|N[i]|$, and $Light_i \in \{0, \ldots, k\}$ is the color representing the local clock phase based on global pulses. Additionally, r_i maintains the following variables: $MaxN_i \in \{1, \ldots, k\}$ is the maximum value of $Nlight$ among the closed neighborhood of r_i, LC_i is a Boolean which represents whether the next operation is LOOK or not, and $Clock_i \in \{0, \ldots, k\}$ is a local counter of global pulses, not necessarily identical among robots.

When r_i detects a global pulse, r_i obtains visible neighbors' $Nlight$ values and updates $MaxN_i$[1]. The local counter of global pulses $Clock_i$ is bounded by $MaxN_i$, and maintained by each terminal r_i, that is, they are not necessarily the same. By the value of its counter, each terminal decides its color of $Light_i$. When $Light_i$ is 0, r_i can make $robot(r_i)$ execute LOOK and COMPUTE (*i.e.*, r_i enters the CS). Only immediately after all values of $Light$ of r_i's closed neighbors become not 0, meaning none is planning to make its robot execute LOOK and COMPUTE in the next (long) global pulse, r_i can make $robot(r_i)$ execute MOVE (*i.e.*, no closed neighbors enter the CS and hence rendezvous is satisfied). Then, if the

---

[1] Because the maximal number of neighboring terminals is typically much less than the total number of terminals, the number of colors is typically much smaller than the number of terminals. Moreover, when a global upper bound on the number of neighbors is known and used, only two colors of light suffices (indicating whether $Clock$ value is 0 or not) for the entire operation of the algorithm.

visible graph changes, $|N[i]|$ also changes. Thus, after a MOVE execution, r_i updates the color of $Nlight_i$.

Reference

1. Dolev, S., Kamei, S., Katayama, Y., Ooshita, F., Wada, K.: Neighborhood mutual remainder: Self-stabilizing implementation of look compute move. arXiv:1903.02843 (2019)

Reducing the Number of Messages
in Self-stabilizing Protocols

Anaïs Durand[1]([⊠]) and Shay Kutten[2]

[1] Sorbonne Université, CNRS, LIP6, 75005 Paris, France
anais.durand@lip6.fr
[2] Technion - Israel Institute of Technology, Haifa, Israel
kutten@ie.technion.ac.il

Abstract. Self-stabilizing algorithms recover from sever faults, such as inconsistent initialization. Traditionally, when designing a self-stabilizing message-passing algorithm, the main goal was to reduce the time until stabilization. The message cost was neglected. In this work, we strive to reduce the number of messages sent on the average per time period. As a tool, we present a stabilizing module that can message-efficiently determine when a task (from a wide family of tasks) is terminated. False positive detection is possible, but only when faults occurred. This module can then be used in the transformation of non self-stabilizing algorithms into self-stabilizing ones.

Keywords: Fault-tolerance · Self-stabilization · Message complexity · Quiescence detection · Termination detection

1 Introduction

In 1974, Dijkstra [11] introduced the *self-stabilization* as a property of distributed algorithms that withstand sever faults. If a self-stabilizing system is led by faults into any incorrect global state, it eventually recovers a correct behavior. For example, the token circulation algorithms proposed in [11] can recover from an arbitrary initial configuration where several processes hold a token instead of only one. After recovery, exactly one token remains. Self-stabilizing protocols for various problems have been devised: leader election, synchronization, *etc.* However, when designing a self-stabilizing algorithm, the message complexity is traditionally neglected and the designers only aim at reducing the stabilization time, *i.e.*, the time before recovering a correct behavior. This happened probably because a self-stabilizing message-passing algorithm cannot stop. It needs to continuously send messages in order to check whether faults occurred and recovery is needed.

In particular, multiple transformers that convert a non self-stabilizing algorithm $\mathcal{A}$ into a self-stabilizing one have been designed [1,3–5]. Most of those

---

This study has been partially supported by the ANR project ESTATE (ANR-16-CE25-0009).

M. Ghaffari et al. (Eds.): SSS 2019, LNCS 11914, pp. 133–148, 2019.
https://doi.org/10.1007/978-3-030-34992-9_12

transformers work roughly as follows. First, $\mathcal{A}$ is executed. Then, once the execution of $\mathcal{A}$ terminates, a local checking algorithm is executed (called a "local detection" algorithm [1] or the local *verifier* of a *Proof Labeling Scheme* [19]). This checking detects when the state is illegal (because a fault occurred). For example, if $\mathcal{A}$ is an algorithm to construct a routing tree of shortest paths (SPT), the verifier checks that the state of every node (but the root) includes a parent pointer, and that the collection of parent pointers forms a tree of shortest paths. If a fault is actually detected, a self-stabilizing *reset* algorithm, *e.g.*, [2], is executed in order to bring all the nodes to a legal initial state of $\mathcal{A}$. Then, the process starts again, *i.e.*, $\mathcal{A}$ is executed, termination detected, *etc.*[1]

In most transformers, being able to detect termination is necessary in order to know when the verifier should be activated. Otherwise, if the verification is done before the output is computed, a fault would be signaled. For example, before algorithm $\mathcal{A}$ of the above example terminates, the SPT is not yet computed so the verifier may interpret that as a fault. The above transformers assume a *synchronous* network to detect that enough time has passed so $\mathcal{A}$ must have terminated. However, we do not want such an assumption. Alternatively [18], such transformers can use a self-stabilizing synchronizer [2,6]. Unfortunately, a self-stabilizing synchronizer is very costly in number of exchanged messages. It uses $\Omega(m)$ messages per round (where m is the number of edges). For example, if $\mathcal{A}$'s time complexity is $\Omega(n)$, its self-stabilizing version (using such a transformer), would need $\Omega(nm)$ messages till stabilization (and would continue using $\Omega(m)$ messages forever). An earlier transformer uses even more messages [17]. (It assumed a self stabilizing leader election, which was then provided by [1]).

In this work, we present a method for reducing the number of messages sent on the average per time period (compared to using synchronizers), at least for the termination detection of a wide class of tasks called *diffusing computations* [12] (*e.g.*, DFS, broadcast and echo, two-phase commit, token circulation). In a diffusing computation, only a single process, the *initiator*, can spontaneously send a message to one or more of its neighbors and only once. Whenever receiving a message, any process can send messages to its neighbors. Indeed, we propose a snap-stabilizing[2] quiescence detection algorithm tailored to detect when the execution of $\mathcal{A}$ is terminated by proposing a termination detection method more message-efficient than a self-stabilizing synchronizer. Note that both detection methods may have a one sided error. That is, *if* faults occurred, the detector may detect termination even though $\mathcal{A}$ has not terminated; such a false detection is still useful for a transformer, since triggering *reset* to rerun $\mathcal{A}$ would be the right thing to do in the case faults occurred.

Another component is needed for the transformer. If the execution of $\mathcal{A}$ starts in an arbitrary configuration because of faults, it may never terminate since $\mathcal{A}$ is not self-stabilizing. Thus, the transformer needs a mechanism to enforce

---

[1] A proof labeling scheme has to be designed especially for $\mathcal{A}$, and some changes to $\mathcal{A}$ may be needed in order to generate the specific "label" for the proof labeling scheme.

[2] A *snap-stabilizing* [7] algorithm is a self-stabilizing algorithm that recovers immediately after faults occurred.

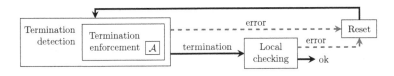

Fig. 1. Schematic overview of the proposed transformer.

termination. We can use a very simple enforcer as follows. Assume that an upper bound x on the number of messages that a node sends in $\mathcal{A}$ when there is no fault is known. For example, in broadcast and echo, the number of messages each node sends is bounded by twice the number of its neighbors. To implement the enforcer, each node just refuses to send more than x messages. Figure 1 proposes a schematic overview of the transformer.

Quiescence Detection. *Quiescence* [8] is a global property of distributed systems. A distributed system is quiet when the communication channels are empty and a local indicator of stability holds at every process. Termination is an example of quiescence property. Detecting quiescence is a known fundamental problem in distributed computing. For example, in addition to its usefulness in the self-stabilizing transformer, detecting the termination of a task allows the system to know the computed result is ready for output. Moreover, termination detection simplifies the design of a complex task. The task is broken into modules, such that some module m_2 must wait until some other module m_1 terminates. It is easier to design a terminating m_1, and then couple it with a termination detection protocol [15].

Since the seminal works of Dijkstra and Scholten [12] and Francez [14] on termination detection in distributed systems, the quiescence detection and its sub-problems have been extensively studied. For a survey, see [21]. Two main kinds of quiescence detection algorithms can be distinguished. *Ongoing detection* algorithms monitor the execution since its beginning and eventually detects quiescence when it is reached, *e.g.*, [12]. A different approach is the *immediate detection* algorithms that answer whether the system has reached quiescence by now or not, *e.g.*, [14]. Ongoing quiescence detection is needed for the transformer, and for most other applications. Ongoing detection can be designed using an immediate detection algorithm by repeatedly executing the detection algorithm until it actually detects quiescence, however this may be highly inefficient.

A self-stabilizing *Propagation of Information and Feedback (PIF)* algorithm [10,20,22] can be used to design an immediate termination detection algorithm, see [9]. Varghese [22] proposes a self-stabilizing PIF algorithm in the message-passing model. Snap-stabilizing PIF algorithms are proposed in [10,20]. Such an application would have high communication and memory complexity even without the need to convert this farther to an ongoing detection.

Contributions. First, we propose a new measure for message efficiency for *asynchronous* networks, where we count the number of messages in executions that are "reasonable" synchronous, *i.e.*, *k-synchronous executions*. Then, we propose a self-stabilizing and snap-stabilizing ongoing quiescence detection algorithm $\mathcal{Q}$ for *diffusing computations*. Using $\mathcal{Q}$, one can implement a message-efficient self-stabilizing transformer. When $\mathcal{Q}$ monitors an algorithm $\mathcal{A}$, it detects quiescence or signals an error in $O(t_{\mathcal{A}} + n)$ rounds, where $t_{\mathcal{A}}$ is the round complexity of $\mathcal{A}$ and n is the number of processes. Its memory complexity is $O(\Delta \log n)$ bits per process, where Δ is the maximum degree. If the execution is k-synchronous, the message complexity of the quiescence detection algorithm is $O\big(k(m + n(t_{\mathcal{A}} + n) + M_{\mathcal{A}})\big)$, where m is the number communication links and $M_{\mathcal{A}}$ is the message-complexity of $\mathcal{A}$.

Roadmap. In the next section, we detail the considered computational model and the specification of the quiescence detection problem. Section 3 presents the new snap-stabilizing quiescence detection algorithm $\mathcal{Q}$ and an analysis of its correctness and complexities is given in Sect. 4.

2   Preliminaries

Consider connected *distributed systems* of n processes operating in the asynchronous message-passing model. The topology of the system is represented by a graph $\mathcal{G} = (V, E)$ where V is the set of processes and E is the set of communication links. Each process can send messages to and receive messages from a subset of other process called *neighbors*. $\mathcal{N}_p$ denotes the set of neighbors of process p, *i.e.*, $(p, q) \in E \Leftrightarrow q \in \mathcal{N}_p$. Communications are bidirectional, *i.e.*, $p \in \mathcal{N}_q \Leftrightarrow q \in \mathcal{N}_p$. We assume reliable (no message is lost) and FIFO (messages are received in the order they are sent) channels of bounded capacity c. Messages are received in finite time. The size of a message is restricted to $\Theta(\log n)$ bits.

Variables and Executions. Every process has a finite number of variables. Let us denote $p.x$ the variable x of process p. Assume a unique process is distinguished as the initiator of the diffusing computation, *i.e.*, every process p has a constant $p.\text{init}$ that evaluates to **true** at a unique process. The *state* of a process is the vector of the values of its variables. The *state* of a channel is the list of messages it contains. A *configuration* is a vector of states, one for every process or channel in the network. Denote by $\gamma(p).x$ the value of variable $p.x$ in configuration γ.

Let $\mapsto$ be a binary relation over configurations such that $\gamma \mapsto \gamma'$ is a *step*, *i.e.*, γ' can be obtained from γ by the *activation* of one or more processes (some messages are received and/or sent, some internal computation is done). It is required that during a step, a process receives and sends at most one message. An *execution* is a sequence of configurations $\Gamma = \gamma_0, \gamma_1, \ldots, \gamma_i, \ldots$, such that $\forall i \geq 0$, $\gamma_i \mapsto \gamma_{i+1}$. Configuration γ_0 is the *initial configuration* of Γ. Infinitely often during an execution, a process triggers a timeout and processes it to do some internal computation and/or to send some messages. The diffusing computation algorithms monitored by the quiescence detection algorithm do not use timeouts but are only message-driven. In a stabilizing context, we consider executions starting from an arbitrary configuration that may be caused by faults, yet we assume that no faults occur during the execution.

A *round* is a unit of complexity measure and is defined as follows. The first round of an execution $\Gamma = \gamma_0, \gamma_1, \ldots, \gamma_i, \ldots$ is the minimal prefix Γ' of Γ such that every message in transit (*i.e.*, inside the channels) in γ_0 is received (and processed) by its recipient and every process triggers (and processes) a timeout. Let γ_j be the last configuration of Γ'. The second round of Γ is the first round of Γ'' and so on, where Γ'' is the suffix of Γ starting from γ_j.

Quiescence Detection. A *(global) quiescent* property is characterized by a *local quiescence-indicator* $Quiet(p)$ at each process p such that:

- *Quiescence:* If $Quiet(p)$ holds, p does not send messages and, as long as p does not receive a message, $Quiet(p)$ continues to hold.
- *Local Indicativity:* The channels are empty and $Quiet(p)$ holds at every process p if and only if quiescence is reached.

For example, for the termination property, $Quiet(p)$ holds when p is disabled.

A distributed algorithm is *snap-stabilizing* [7] *w.r.t.* some specification S if any execution starting from an arbitrary configuration satisfies S. In this context, we define the set of *regular* initial configurations of the quiescence detection algorithm, *i.e.*, initial configurations where the detection algorithm is well initialized. Notice that, since the initial configuration is arbitrary, it can be non-regular. (The definition of regular initial configurations for the quiescence detection algorithm is given in Definition 7.)

The goal of a quiescence detection algorithm $\mathcal{Q}$ is to detect when the execution of another algorithm $\mathcal{A}$ that $\mathcal{Q}$ monitors reaches quiescence.

Definition 1. $\mathcal{Q}$ is a snap-stabilizing quiescence detection algorithm for diffusing computations *if for every execution Γ where $\mathcal{Q}$ monitors an algorithm $\mathcal{A}$ since the beginning of its execution the following holds:*

- Eventual Detection: *If the execution of $\mathcal{A}$ reaches quiescence, some process eventually calls SigQuiet() or SigError().*
- Soundness: *If SigQuiet() is called, either the execution of $\mathcal{A}$ reached quiescence or was not a diffusing computation, or the initial configuration of $\mathcal{Q}$ was not regular.*
- Relevance: *If the execution of $\mathcal{A}$ is a diffusing computation and the initial configuration of $\mathcal{Q}$ is regular, no process ever calls SigError().*

$SigQuiet()$ and $SigError()$ are two output signals. When such a signal is emitted, it can be used to trigger an external response from the system, *e.g.*, a reset [1–3]. Notice that there is no hypothesis on $\mathcal{A}$, *i.e.*, we do not require $\mathcal{A}$ to be self-stabilizing or even to compute a correct result.

3   Quiescence Detection Algorithm $\mathcal{Q}$

In this section, we propose a self-stabilizing ongoing quiescence detection algorithm $\mathcal{Q}$ for diffusing computations written in the message-passing model. The code of $\mathcal{Q}$ is presented in Algorithm 1.

Algorithm 1. Algorithm $\mathcal{Q}$ for Process p

```
 1: upon PIF_rcv(q, ⟨pckt, dist⟩) do
 2:     if ¬Error(p) then
 3:         p.status := ACT ;   Deliver(q, pckt) ;
 4:         if ¬p.init ∧ p.par = ⊥ then  p.par := q ;   p.dist := dist + 1 ;
 5:     if p.par = q then  PIF_send_fbck(q, ⟨true⟩);
 6:     else PIF_send_fbck(q, ⟨false⟩);

 7: upon PIF_fbck(q, ⟨isChild⟩) do
 8:     if ¬Error(p) then
 9:         p.pckt[q] := ⊥ ;
10:         if isChild then                                    // q is a child of p
11:             if p.status = ACT ∧ (p.init ∨ p.par ≠ ⊥) then  p.child[q] := true ;
12:             else p.status := ERR ;   SigError() ;
13:         else p.child[q] := false ;

14: upon rcv(q, ⟨PAR⟩) do                      // q thinks it is the parent of p
15:     if ¬Error(p) ∧ p.par ≠ q then  send(q, ⟨NOCHILD⟩) ;

16: upon rcv(q, ⟨CHILD, dist⟩) do                       // q is a child of p
17:     if ¬Error(p) then
18:         if p.status = ACT ∧ (p.init ∨ p.par ≠ ⊥) ∧ dist = p.dist + 1 then
19:             p.child[q] := true;
20:         else p.status := ERR ;   SigError() ;

21: upon rcv(q, ⟨NOCHILD⟩) do                    // q is not a child of p
22:     if ¬Error(p) then p.child[q] := false;

23: upon timeout do
24:     if Error(p) then p.status := ERR ;   SigError() ;
25:     else if Passive(p) then
26:         p.status := PASS ;
27:         if p.par ≠ ⊥ then  send(p.par, ⟨NOCHILD⟩) ;   p.par := ⊥ ;
28:         if p.init then  SigQuiet() ;
29:     else
30:         foreach q ∈ 𝒩ₚ : p.pckt[q] ≠ ⊥ do
31:             p.status := ACT ;
32:             PIF_send(q, ⟨p.pckt[q], p.dist⟩) ;
33:         foreach q ∈ 𝒩ₚ : p.child[q] = true do  send(q, ⟨PAR⟩) ;
34:         if p.par ≠ ⊥ then  send(p.par, ⟨CHILD, p.dist⟩) ;
```

Overview. The idea of $\mathcal{Q}$ adapts the algorithm of Dijkstra and Scholten [12] to the stabilizing context using local checking [1,3]. To monitor an algorithm $\mathcal{A}$ and detect its quiescence, $\mathcal{Q}$ handles the sending and reception of messages of $\mathcal{A}$, that we will call *packets* in order to avoid confusion. $\mathcal{Q}$ builds the *tree of the execution* defined as follows. The initiator of the diffusing computation is the root of the tree. When a process p receives a packet *pckt*, it joins the tree by choosing the sender of *pckt* as its parent by updating variable p.par. Each process p has also a Boolean variable p.child$[q]$ for each of neighbor q, stating if q is a child of p. When a process p has no children and predicate *Passive*(p) holds, p leaves the tree by notifying its parent and removing its parent pointer, if p is not the initiator. Otherwise, it signals quiescence.

Handling the messages of $\mathcal{A}$. In order to allow $\mathcal{Q}$ to manage the packets of the monitored algorithm $\mathcal{A}$, we assume that the functions of $\mathcal{A}$ to send and receive packets are slightly altered as shown in Algorithm 2. Every process p has a variable p.pckt$[q]$ for each neighbor q used to communicate between $\mathcal{A}$ and $\mathcal{Q}$. Indeed, p.pckt$[q]$ contains the packet of $\mathcal{A}$ that p wants to send to q, or $\bot$ if no packet

---

Algorithm 2. Macro of Modification of Algorithm $\mathcal{A}$

```
    /* Replace every:        */
 1: send(q, pckt);
    /* by:                   */
 2: wait until p.pckt[q] = ⊥;
 3: p.pckt[q] := pckt;
```

---

has to be sent. When p needs to send some packet *pckt* to q in $\mathcal{A}$, p must wait until the previous packet is processed, *i.e.*, until $\mathcal{Q}$ sets p.pckt$[q]$ to $\bot$. On the other hand, when a process p receives a packet *pckt* from q in $\mathcal{Q}$, this packet is delivered to $\mathcal{A}$ at p using *Deliver*$(q, pckt)$, *i.e.*, it triggers a $rcv(q, pckt)$ in $\mathcal{A}$.

In order to ensure that the packets of $\mathcal{A}$ are delivered and quiescence is not signaled when some messages are in transit, $\mathcal{Q}$ uses a snap-stabilizing *Propagation of Information with Feedback (PIF)* algorithm [10], denoted here $\mathcal{PIF}$. A PIF allows a process to send a messages to other processes (*propagation*) and to receive in return an acknowledgment from those other processes (*feedback*). Let us use $\mathcal{PIF}$ as follows. To send packets to a neighbor q, a process p uses a dedicated instance of $\mathcal{PIF}$ involving only p and q. Notice that one instance of $\mathcal{PIF}$ between only two processes requires a constant number of bits per process and so $O(\Delta)$ bits per process to send and receive packets from all neighbors (where Δ is the maximum degree). Primitives of the PIF algorithm are prefixed with $PIF\_$. When a message *msg* is sent from p to q through the $\mathcal{PIF}$ protocol (*i.e.*, using $PIF\_send(q, msg)$), it triggers a $PIF\_rcv(p, msg)$ at process q. Then q sends a feedback to p using $PIF\_send\_fbck(p, ack)$ that triggers a $PIF\_fbck(q, ack)$. Notice that the messages of $\mathcal{Q}$ used to do the detection are not transmitted through $\mathcal{PIF}$ since no feedback on those message is required.

Those three protocols – $\mathcal{A}$, $\mathcal{Q}$, and $\mathcal{PIF}$ – are composed using a fair composition [13]. Figure 2 illustrates how the packets of $\mathcal{A}$ are handled and the interactions between the three protocols.

Quiescence and Error Detection. To check whether the execution tree is correctly built and to update the knowledge of a process about its children, every process p regularly send control messages along the tree: $\langle\text{PAR}\rangle$ to its children (Line 33) and $\langle\text{CHILD}, p.\text{dist}\rangle$ to its parent (Line 34), where p.dist is the distance from p to the root. In particular, if a process q receives a message $\langle\text{CHILD}, dist\rangle$ from

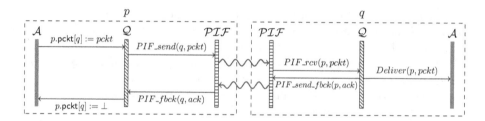

Fig. 2. Schematic view of how the packets of $\mathcal{A}$ from process p to process q are handled. The wavy arrows illustrate the communications between p and q through protocol $\mathcal{PIF}$.

one of its children p and $dist \neq q.\text{dist} + 1$, the distances in the tree are not well computed (*e.g.*, the tree contains a cycle) so q signals an error. If a process q receives a message $\langle \text{PAR} \rangle$ from a neighbor p that is not its parent, it sends back a message $\langle \text{NOCHILD} \rangle$ and p can update its variable $p.\text{child}[q]$.

In addition, process p locally checks the correctness of the tree, *i.e.*, if predicate $LCorrect(p)$ holds. For example, p can verify that it has no children if it is not attached to the tree. (Notice that the formal definition of $LCorrect(p)$ is given in Definition 3.) If $Error(p) \equiv \left(\neg LCorrect(p) \lor p.\text{status} = \text{ERR} \right)$ holds, p signals an error (Lines 24). A process p which calls $SigError()$ also sets $p.\text{status}$ to ERR.

A process p leaves the tree when it becomes *passive*, *i.e.*, the following predicate $Passive(p)$ holds.

Definition 2. *Predicate $Passive(p)$ holds if $Quiet(p)$ holds,[3] p has no packets to send and no received packets to process* (i.e., there is no incoming event PIF_rcv *and* $\forall q \in \mathcal{N}_p$, $p.\text{pckt}[q] = \bot$), *and p has no children (that it knows of), i.e.,* $\forall q \in \mathcal{N}_p, \neg p.\text{child}[q]$.

4   Analysis

In this section, we show that $\mathcal{Q}$ is a snap-stabilizing ongoing quiescence detection algorithm and we analyze its complexities. Due to the lack of space, some straightforward proofs are omitted. Notice that since the three algorithms $\mathcal{A}$, $\mathcal{Q}$, and $\mathcal{PIF}$ are fairly composed, a process executes a round of one of these algorithms every 3 rounds.

4.1   Properties of the PIF Protocol

First, let us state some useful property of the snap-stabilizing PIF protocol $\mathcal{PIF}$ from [10].

---

[3] Predicate $Quiet(p)$ is defined in Sect. 2.

Proposition 1. *If a process p initiates a PIF to send a message m to its neighbor q through the protocol $\mathcal{PIF}$ (i.e., using $PIF\_send(q,m)$), m is received by q ($PIF\_rcv(p,m)$) in at most 8 rounds of $\mathcal{PIF}$. Then, p receives the feedback from q in at most 1 additional round of $\mathcal{PIF}$ ($PIF\_fbck(q,fbck)$). Moreover, under such condition, p cannot receive a feedback from q before q received m.*

Notice that it does not prevent situations where the arbitrary initial configuration generates faulty communications of the $\mathcal{PIF}$ protocol leading to some process p receiving feedback without initiating any PIF. However, once p actually initiated some PIF with a neighbor q, *i.e.*, the call to function $PIF\_send(q,m)$ is terminated and $\mathcal{PIF}$ will manage the communications between p and q to ensure the transmission of m, p cannot receive any feedback that is sent by q before q receives m. Let *tainted messages* be messages of $\mathcal{PIF}$ present in the arbitrary initial configuration or generated afterwards and that are not related to a PIF *actually* initiated by some process.

Proposition 2. *If there are no tainted messages in the channels (i.e., in transit or in the incoming mailboxes of processes but not yet processed by their recipient), a process p cannot receive any feedback from a neighbor q if p does not initiate any PIF to send a packet to q.*

Since one PIF lasts at most 9 rounds of $\mathcal{PIF}$ (by Proposition 1), we can deduce the following corollary.

Corollary 1. *After $9 \times 3 = 27$ rounds, there are no more tainted messages.*

4.2   Execution Trees

Now, let us show that par-variables actually define trees. If the structure of the tree is incorrect, process p locally detects the errors using predicate $LCorrect(p)$ defined as follows.

Definition 3. *Predicate $LCorrect(p)$ holds at some process p if all of the four following conditions are satisfied:*

$C_1(p)$. $(p.\mathsf{status} \neq \text{ERR} \wedge p.\mathsf{init}) \Rightarrow (p.\mathsf{par} = \perp \wedge p.\mathsf{dist} = 0)$
$C_2(p)$. $(p.\mathsf{status} \neq \text{ERR} \wedge \neg p.\mathsf{init} \wedge p.\mathsf{par} = \perp) \Rightarrow (\forall q \in \mathcal{N}_p : \neg p.\mathsf{child}[q])$
$C_3(p)$. $(p.\mathsf{status} \neq \text{ERR} \wedge \neg p.\mathsf{init} \wedge p.\mathsf{par} = \perp) \Rightarrow (\forall q \in \mathcal{N}_p : p.\mathsf{pckt}[q] = \perp)$
$C_4(p)$. $p.\mathsf{status} = \text{PASS} \Rightarrow p.\mathsf{par} = \perp$

Lemma 1. *Let $p \in V$. If the execution of $\mathcal{A}$ is a diffusing computation, the following two claims hold.*

1. Let $\gamma \mapsto \gamma'$. If $LCorrect(p)$ holds in γ then $LCorrect(p)$ holds in γ'.
2. In at most 3 rounds, $LCorrect(p)$ holds.

Now, let us prove that par-variables actually define trees. We define three conditions on the relationship between p and its parent $q \in \mathcal{N}_p$.

Definition 4.   *Let $p \in V$ and $q \in \mathcal{N}_p$.*

$C_5(p)$. $(p.\text{status} \neq \text{ERR} \land q.\text{status} \neq \text{ERR} \land p.\text{par} = q) \Rightarrow (q.\text{child}[p] \lor q.\text{pckt}[p] \neq \bot)$
$C_6(p)$. $(p.\text{status} \neq \text{ERR} \land q.\text{status} \neq \text{ERR} \land p.\text{par} = q) \Rightarrow (q.\text{init} \lor q.\text{par} \neq \bot)$
$C_7(p)$. $(p.\text{status} \neq \text{ERR} \land q.\text{status} \neq \text{ERR} \land p.\text{par} = q) \Rightarrow p.\text{dist} = q.\text{dist} + 1$

Lemma 2.   *Let $p \in V$ and $q \in \mathcal{N}_p$.*

1. *Let $\gamma \mapsto \gamma'$ s.t. no tainted messages are in the channels. If $C_5(p,q)$, $C_6(p,q)$, and $C_7(p,q)$ hold in γ then $C_5(p,q)$, $C_6(p,q)$, and $C_7(p,q)$ hold in γ'.*
2. *In at most 33 rounds, $C_5(p,q)$, $C_6(p,q)$, and $C_7(p,q)$ hold.*

Definition 5.   *The* execution graph $\mathcal{G}_{ex} = (V_{ex}, E_{ex})$ *is the subgraph of non-error processes induced by* par*-variables, i.e., $V_{ex} = \{p \in V : p.\text{status} \neq \text{ERR}\}$ and $E_{ex} = \{(p,q) \in E : p.\text{par} = q \land p, q \in V_{ex}\}$.*

From Lemmas 1 and 2, one can deduce the following corollary.

Corollary 2.   *In at most 33 rounds, $\mathcal{G}_{ex}$ is a forest.*

4.3   Detection of Quiescence

In this subsection, we show that $\mathcal{Q}$ fulfills the three properties of quiescence detection: eventual detection (Theorem 1), soundness (Theorem 2), and relevance (Theorem 3). Let γ_0 be the initial configuration and let $\gamma_0(\mathcal{A})$ be the projection of γ_0 on the variables and messages of $\mathcal{A}$.

Theorem 1 (Eventual Detection).   *If an execution of $\mathcal{A}$ starting from $\gamma_0(\mathcal{A})$ reaches quiescence, then a process calls $SigError()$ or $SigQuiet()$ in $O(t_{\mathcal{A}} + n)$ rounds, where n is the number of processes and $t_{\mathcal{A}}$ is the maximum number of rounds for $\mathcal{A}$ to reach quiescence from $\gamma_0(\mathcal{A})$.*

Proof. First, notice that if the execution is not a diffusing computation, *i.e.*, if some process p that is not the initiator of $\mathcal{A}$ spontaneously requires the sending of some packet of $\mathcal{A}$ in $\mathcal{Q}$ (*i.e.*, $p.\text{pckt}[q]$ becomes different than $\bot$ with $q \in \mathcal{N}_p$), p signals an ERR (see Line 24). If $\mathcal{Q}$ signals an error, we are done. Now, assume that $SigError()$ is never called.

Since every packet of $\mathcal{A}$ is eventually delivered through $\mathcal{PIF}$ and thanks to the fair composition, $\mathcal{Q}$ does not block the execution of $\mathcal{A}$. Let γ be the first configuration after which $\mathcal{G}_{ex}$ is a forest (see Corollary 2) and $\mathcal{A}$ reached quiescence, *i.e.*, at every process p, $Quiet(p)$ holds and p does not require the sending of a packet of $\mathcal{A}$ ($\forall q \in \mathcal{N}_p, p.\text{pckt}[q] = \bot$).

When some process p requires the sending of a packet in $\mathcal{A}$ to some neighbor q, *i.e.*, $p.\text{pckt}[q] \neq \bot$, p initiates a PIF (Line 32) in at most 3 rounds. The PIF lasts 27 rounds (Proposition 1) before p sets $p.\text{pckt}[q]$ to $\bot$ and allows the sending of the next packet of $\mathcal{A}$. Thus, in at most $30t_{\mathcal{A}}$ rounds, every packet of $\mathcal{A}$ has been delivered. So γ is reached in at most $33t_{\mathcal{A}} + 27$ rounds. Notice that no process has status ERR or it would have signaled an error in the meantime (Lines 24).

Now, let us show that the height of the trees decreases in at most 12 rounds and that eventually every process has status PASS. First, since $\forall p \in V$, $\forall q \in \mathcal{N}_p$, $p.\mathsf{pckt}[q] = \bot$, no PIF is initiated after γ. Hence, no process can get a PIF_rcv anymore. Let p be a leaf in $\mathcal{G}_{ex}$. By definition, $\forall q \in \mathcal{N}_p$, $\gamma(q).\mathsf{par} \neq p$. Since p will never send a packet to q, $q.\mathsf{par}$ cannot become equal to p anymore. Now, let us show that $p.\mathsf{child}[q]$ becomes false in finite time.

If $\gamma(p).\mathsf{child}[q] = \mathsf{true}$, p sends a message $\langle \mathrm{PAR} \rangle$ to q in at most 3 rounds (Line 33). At most 3 rounds later, q receives this message (Line 14) and sends back a message $\langle \mathrm{NOCHILD} \rangle$ since $q.\mathsf{par} \neq p$ (Line 15). When p receives this message at most 3 rounds later (Line 16), it sets $p.\mathsf{child}[q]$ to false (Line 21). Since $q.\mathsf{par}$ remains different than p afterwards, q can never send a feedback containing $\langle \mathsf{true} \rangle$ or a message $\langle \mathrm{CHILD}, * \rangle$ to p. Thus, p never sets $p.\mathsf{child}[q]$ to true again. Hence, in at most 9 rounds after γ, $\forall q \in \mathcal{N}_p$, $p.\mathsf{child}[q] = \mathsf{false}$ so $Passive(p)$ continuously holds. In at most 3 additional rounds, p gets status PASS and leaves the tree by setting $p.\mathsf{par}$ to $\bot$ (Lines 25–28). So the height of the tree that contained p in γ decreases in at most 12 rounds.

By repeating this argument, eventually every tree is composed of only one process of status PASS. In particular, the initiator signals quiescence when it gets status PASS (Line 28). Since the height of the trees is at most $n - 1$, the initiator signals quiescence at most $12n$ rounds after γ. □

Let us define a last property on $p \in V$ and its neighbors $q \in \mathcal{N}_p$.

Definition 6. *Let $p \in V$ and $q \in \mathcal{N}_p$.*

$C_8(p)$. $p.\mathsf{status} = \mathrm{PASS} \wedge q.\mathsf{status} \neq \mathrm{ERR} \Rightarrow q.\mathsf{par} \neq p$

Lemma 3. *Let $p \in V$ and $q \in \mathcal{N}_p$.*

1. *Let $\gamma \mapsto \gamma'$ s.t. no tainted messages are in the channels. If $C_5(p,q)$ and $C_8(p,q)$ hold in γ then $C_8(p,q)$ holds in γ'.*
2. *In at most 33 rounds, $C_8(p,q)$ holds.*

Proof.

1. Let $\gamma \mapsto \gamma'$ s.t. there is no tainted messages in the channels, and $C_5(p,q)$ and $C_8(p,q)$ hold in γ. A process with status ERR cannot change it. Assume $\gamma'(p).\mathsf{status} = \mathrm{PASS}$ and $\gamma'(q).\mathsf{status} \neq \mathrm{ERR}$.
   (a) If $\gamma(p).\mathsf{status} = \mathrm{PASS}$ then, by C_8, $\gamma(q).\mathsf{par} \neq p$. Assume by contradiction that q sets $q.\mathsf{par}$ to p during $\gamma \mapsto \gamma'$, this means that it receives a packet $pckt$ from p (Line 4). However, when p initiates the PIF to send $pckt$ (Line 32), $p.\mathsf{pckt}[q] \neq \bot$ and $p.\mathsf{status} = \mathrm{ACT}$. Moreover, p cannot get status PASS without receiving a feedback from q, which must happen after $\gamma \mapsto \gamma'$ in which $pckt$ is assumed to be received by q (Proposition 2), a contradiction. Hence $\gamma'(q).\mathsf{par} \neq p$.

(b) If p gets status PASS during $\gamma \mapsto \gamma'$, $\gamma(p).\mathsf{pckt}[q] = \bot$ and $\gamma(p).\mathsf{child}[q] =$ false. Thus, by the contrapositive of $C_5(p,q)$, $\gamma(q).\mathsf{par} \neq p$. Similarly to case 1a, q cannot set $q.\mathsf{par}$ to p.

2. Let γ_T be the first configuration after 27 rounds. By Corollary 1, no tainted messages are in the channels. If $\gamma_T(p).\mathsf{status} = $ PASS but $\gamma_T(q).\mathsf{par} = p$ then q sends a message $\langle\text{CHILD}, *\rangle$ to p in at most 3 rounds (Line 34). When p receives this message at most 3 rounds later, it gets status ERR (Line 20). □

Definition 7. *An initial configuration γ_0 is regular if:*

R_1 $\forall p \in V$, $\gamma_0(p).\mathsf{init} \Rightarrow (\gamma_0(p).\mathsf{dist} = 0 \wedge \gamma_0(p).\mathsf{par} = \bot)$
R_2 $\forall p \in V$, $\neg\gamma_0(p).\mathsf{init} \Rightarrow (\gamma_0(p).\mathsf{par} = \bot \wedge \forall q \in \mathcal{N}_p, \neg\gamma_0(p).\mathsf{child}[q])$
R_3 $\forall p \in V$, $\gamma_0(p).\mathsf{status} = $ PASS $\wedge \forall q \in \mathcal{N}_p, \gamma_0(p).\mathsf{pckt}[q] = \bot$
R_4 *There is no messages in the channels in γ_0.*

By Definition 7 and Lemmas 1, 2, and 3, we have the following corollary.

Corollary 3. *In any configuration γ of an execution starting from a regular initial configuration γ_0 such that every execution of $\mathcal{A}$ starting from $\gamma_0(\mathcal{A})$ is a diffusing computation, $\forall p \in V$, $LCorrect(p)$ holds and $\mathcal{G}_{ex}$ is a forest in γ. Moreover, $\forall p \in V$, $\forall q \in \mathcal{N}_p$, $C_5(p,q)$, $C_6(p,q)$, $C_7(p,q)$, and $C_8(p,q)$ hold in γ.*

Theorem 2 (Soundness). *If $SigQuiet()$ is called, either the execution of $\mathcal{A}$ actually reached quiescence or was not a diffusing computation, or the initial configuration of $\mathcal{Q}$ was not regular.*

Proof. We prove this theorem by the contrapositive. Assume the execution of $\mathcal{A}$ never reaches quiescence (*i.e.*, there is always some enabled process in $\mathcal{A}$ or a process that needs to send or receive a packet of $\mathcal{Q}$) and is a diffusing computation, and the initial configuration of $\mathcal{Q}$ is regular. Assume by contradiction that $\mathcal{Q}$ signals quiescence.

By hypothesis, quiescence is signaled before any error signal. Moreover, recall that a process gets status ERR before signaling an error and the initial status of every process is PASS. Hence no process ever gets status ERR. Let r be the only initiator and thus the process that signaled quiescence. Let $\gamma_i \mapsto \gamma_{i+1}$ be the step where r calls $SigQuiet()$ (Line 28). Thus, $Passive(r)$ holds in γ_i. Since the execution of $\mathcal{A}$ has not reached quiescence, there are three cases:

1. $\exists p \in V$ s.t. $\neg Quiet(p)$ holds in γ_i. Thus, $\neg Passive(p)$ holds in γ_i.
2. A packet of $\mathcal{A}$ is required to be sent, *i.e.*, $\exists p \in V$, $\exists q \in \mathcal{N}_p$, s.t. $\gamma_i(p).\mathsf{pckt}[q] \neq \bot$. Thus, $\neg Passive(p)$ holds in γ_i.
3. A PIF has been initiated by some process p to send a packet $pckt$ to some neighbor q, yet q did not received $pckt$ yet. In this latter case, when p initiated the PIF (Line 32), $p.\mathsf{pckt}[q] = pckt \neq \bot$. Moreover, p cannot set $p.\mathsf{pckt}$ to $\bot$ until p receives a feedback from q. However, q cannot send such a feedback before receiving $pckt$. Thus, $\gamma_i(p).\mathsf{pckt}[q] \neq \bot$ and $\neg Passive(p)$ holds in γ_i.

In those three cases, $\exists p \in V$ s.t. $\neg Passive(p)$ holds in γ_i. Notice that $p \neq r$.

The initial configuration is regular, $\neg Passive(p)$ holds in γ_i, and p is not the initiator. Thus, p received a first packet from some neighbor x (Line 1) at some point in the execution. Again, since the initial configuration is regular, p set $p.\mathsf{par}$ to x and got status ACT. Consider the last time p changes its status from PASS to ACT (say in $\gamma_j \mapsto \gamma_{j+1}$) before γ_i, i.e., $j < i$, $\gamma_j(p).\mathsf{status} = $ PASS, and $\forall j' \in \{j+1,\ldots,i\}$, $\gamma_{j'}(p).\mathsf{status} = $ ACT. Again, in order to get status ACT, p had received a packet from some neighbor y (Line 1) in $\gamma_j \mapsto \gamma_{j+1}$. Hence, $\gamma_{j+1}(p).\mathsf{par} = y$. Process p cannot change the value of $p.\mathsf{par}$ after γ_{j+1} without getting status PASS (Line 27). Hence, $\gamma_i(p).\mathsf{par} = y \neq \perp$.

By Corollary 3, Property C_6, and since $\mathcal{G}_{ex}$ is a forest, there is a sequence of distinct processes $seq = p_0, p_1, \ldots, p_k$, $0 < k \geq n$, such that $p_0 = p$, $\forall i \in \{0, \ldots, k-1\}$, $\gamma(p_i).\mathsf{par} = p_{i+1}$, and $p_k = r$. Now, by recursively applying $C_5(p_i, p_{i+1})$ for $i \in \{0, \ldots, k-1\}$, we have $\gamma(p_{i+1}).\mathsf{pckt}[p_i] \neq \perp \vee \gamma(p_{i+1}).\mathsf{child}[p_i]$. Thus, $\neg Passive(p_{i+1})$ holds in γ_i. In particular, $\neg Passive(r)$ holds in γ_i, a contradiction.   □

Theorem 3 (Relevance). *If the initial configuration of $\mathcal{Q}$ is regular and the execution of $\mathcal{A}$ is a diffusing computation, no process ever calls SigError().*

Proof. Let $p \in V$. By Corollary 3, $LCorrect(p)$ holds throughout the execution so p cannot call $SigError()$ executing Lines 24 without already being in status ERR. Initially, $p.\mathsf{status} \neq$ ERR since the initial configuration is regular. Except by executing Lines 24, p can get status ERR in two cases.

1. If p gets status ERR and calls $SigError()$ executing Line 12, it has just received a feedback $fbck$ containing $\langle \mathtt{true} \rangle$ from some $q \in \mathcal{N}_p$ while $p.\mathsf{status} = $ PASS or $\neg p.\mathsf{init} \wedge p.\mathsf{par} = \perp$. When q sent $fbck$ (say in $\gamma \mapsto \gamma'$), $q.\mathsf{par} = p$. By Corollary 3, $C_5(q,p)$, $C_6(q,p)$, and the contrapositive of $C_8(q,p)$, $\gamma(p).\mathsf{status} = $ ACT, $\gamma(p).\mathsf{pckt}[q] \neq \perp \vee \gamma(p).\mathsf{child}[q]$, and $\gamma(p).\mathsf{init} \vee \gamma(p).\mathsf{par} \neq \perp$. So p cannot get status PASS or change the value of its par-variable before receiving $fbck$ (Proposition 2), a contradiction.

2. If gets status ERR and p calls $SigError()$ executing Line 20, it received some message $msg = \langle$CHILD$, dist\rangle$ from a neighbor $q \in \mathcal{N}_p$ while (a) $p.\mathsf{status} = $ PASS, or (b) $\neg p.\mathsf{init} \wedge p.\mathsf{par} = \perp$, or (c) $dist \neq p.\mathsf{dist} + 1$. When q sent msg (say in $\gamma \mapsto \gamma'$), $q.\mathsf{par} = p$ and $q.\mathsf{dist} = dist$. Thus, situations (a) and (b) are similar to case 1. Now, consider situation (c). By Corollary 3, $C_5(q,p)$, $C_7(q,p)$, and the contrapositive of $C_8(q,p)$, we have $\gamma(p).\mathsf{status} = $ ACT, $\gamma(p).\mathsf{pckt}[q] \neq \perp \vee \gamma(p).\mathsf{child}[q]$, and $\gamma(p).\mathsf{dist} = \gamma(q).\mathsf{dist} + 1 = dist + 1$. Process p cannot change its distance without getting status PASS and so, without $q.\mathsf{par}$ becoming different than p. If $q.\mathsf{par}$ remains equal to p in between γ and the reception of m by p, p never changes its distance, a contradiction. Otherwise, $q.\mathsf{par}$ becomes different than p at some point between γ and the reception of m by p. To allow p to get status PASS, q must send a message $msg' = \langle$NOCHILD$\rangle$ or a feedback $fbck$ containing $\langle\mathtt{false}\rangle$ to p after changing its par-variable, i.e., after sending msg. Moreover, p must receive msg' or $fbck$ before msg even if the channels are FIFO, a contradiction.   □

4.4   Message Complexity

Finally, we study the message complexity of Q in k-*synchronous* execution. We adapt the definition of k-synchronous executions from [16] to message-passing systems.

Definition 8. *An execution Γ is* k-synchronous *if the following conditions hold.*

(a). *The ratio of speed between the slowest and fastest messages is at most k. More precisely, let $\gamma_{s_m} \mapsto \gamma_{s_m+1}$ and $\gamma_{r_m} \mapsto \gamma_{r_m+1}$ be the steps during which the message m is sent and received, respectively. For every pair of messages m, m' sent during Γ, $(r_m - s_m) \le k(r_{m'} - s_{m'})$.*

(b). *The ratio of speed between the slowest and fastest processes is at most k. More precisely, for any $\Gamma_0, \Gamma_1, \Gamma'$ such that $\Gamma = \Gamma_0\Gamma_1\Gamma'$, and for any two processes p and q, if q triggers at least $k + 1$ timeout during Γ_1 then p triggers at least one timeout during Γ_1.*

Let Γ be an execution of the composition of Q and $\mathcal{A}$.

Theorem 4 (Message Complexity). *If Γ is k-synchronous of and any execution of $\mathcal{A}$ starting from $\gamma_0(\mathcal{A})$ reaches quiescence, then $O\big(k(m + n(t_{\mathcal{A}} + n) + M_{\mathcal{A}})\big)$ messages are sent before some process calls $SigQuiet()$ or $SigError()$, where n is the number of processes, m is the number of edges, $t_{\mathcal{A}}$ (respectively, $M_{\mathcal{A}}$) is the maximum number of rounds (respectively, of exchanged messages in $\mathcal{A}$) for $\mathcal{A}$ to reach quiescence from $\gamma_0(\mathcal{A})$.*

Proof. Let p be a process. Γ is k-synchronous and a process sends at most one message per neighbor at each activation, so p sends at most k messages per neighbor during one round. By Corollary 2, $\mathcal{G}_{ex}$ becomes a forest in at most 33 rounds. During these 33 rounds, up to $O(k\,m)$ messages are exchanged. The remaining of the computation before a process signals quiescence or an error lasts $O(t_{\mathcal{A}}+n)$ rounds. During this part of the computation, messages $\langle\text{CHILD}, *\rangle$ and $\langle\text{PAR}\rangle$ are only exchanged along the trees, i.e., a total of $O(k(n-1)(t_{\mathcal{A}} + n))$ messages. Moreover, $\mathcal{PIF}$ requires 27 rounds to transmit a packet and the corresponding feedback (Proposition 1), i.e., a total of $O(kM_{\mathcal{A}})$ messages. Hence $O\big(k(m + (n-1)(t_{\mathcal{A}} + n) + M_{\mathcal{A}})\big)$ messages are exchanged before a signal.   □

Remark 1. Notice that, if the initial configuration of Q is regular and the execution of $\mathcal{A}$ is a diffusing computation, $\mathcal{G}_{ex}$ is always a forest. Thus, in this case, $O\big(k(n(t_{\mathcal{A}} + n) + M_{\mathcal{A}})\big)$ messages are exchanged before a signal.

5   Discussion and Future Work

We proposed the first self-stabilizing and snap-stabilizing ongoing quiescence detection algorithm Q. This algorithm works for diffusing computations. One can use Q to detect termination before safely re-starting a task or starting a new one, for example to transform a non self-stabilizing algorithm into a self-stabilizing one. This transformer is more message-efficient than the Awerbuch and Varghese transformation. For example, let us consider a non self-stabilizing algorithm $\mathcal{A}$ whose time and message complexity are x and y, respectively. If no faults hit the system (in particular, the initial configuration of $\mathcal{Q}$ is regular), the transformer using $\mathcal{Q}$ requires $O(k(n\,x + n^2 + y))$ messages before detecting the termination of $\mathcal{A}$, if the execution is k-synchronous (where k is a constant). In a similar context, the transformer of Awerbuch and Varghese uses m messages per time unit for the synchronizer. Hence, $\Omega(k\,m\,x + y)$ messages are exchanged.

In addition to the improved performance, the resulting transformer has other advantages over the method with a synchronizer. Indeed, it does not require to know the time complexity of $\mathcal{A}$ contrary to the methods that use this bound to know when the execution of $\mathcal{A}$ is terminated. Moreover, the time complexity of $\mathcal{A}$ is an upper bound on the time before $\mathcal{A}$ terminates, but an execution of $\mathcal{A}$ can actually terminate (far) earlier. In this case, our method stabilizes faster since it detects termination when it happens. These advantages hold only if the execution of $\mathcal{A}$ terminates. Otherwise, it requires a termination enforcement method like the one proposed above.

One natural open problem is the generalization of this quiescence detection algorithm to non-diffusing computations. Another open problem would be to design more general and more efficient methods for termination enforcement.

References

1. Afek, Y., Kutten, S., Yung, M.: The local detection paradigm and its application to self-stabilization. Theor. Comput. Sci. **186**(1–2), 199–229 (1997)
2. Awerbuch, B., Kutten, S., Mansour, Y., Patt-Shamir, B., Varghese, G.: Time optimal self-stabilizing synchronization. In: STOC 1993, pp. 652–661 (1993)
3. Awerbuch, B., Patt-Shamir, B., Varghese, G.: Self-stabilization by local checking and correction (extended abstract). In: FOCS 1991, pp. 268–277 (1991)
4. Awerbuch, B., Patt-Shamir, B., Varghese, G., Dolev, S.: Self-stabilization by local checking and global reset. In: WDAG 1994, pp. 326–339 (1994)
5. Awerbuch, B., Varghese, G.: Distributed program checking: a paradigm for building self-stabilizing distributed protocols. In: FOCS 1991, pp. 258–267 (1991)
6. Boulinier, C., Petit, F., Villain, V.: When graph theory helps self-stabilization. In: PODC 2004, pp. 150–159 (2004)
7. Bui, A., Datta, A.K., Petit, F., Villain, V.: State-optimal snap-stabilizing PIF in tree networks. In: WSS 1999, pp. 78–85 (1999)
8. Chandy, K.M., Misra, J.: An example of stepwise refinement of distributed programs: quiescence detection. ACM TOPLAS **8**(3), 326–343 (1986)
9. Cournier, A., Datta, A.K., Devismes, S., Petit, F., Villain, V.: The expressive power of snap-stabilization. Theor. Comput. Sci. **626**, 40–66 (2016)
10. Delaët, S., Devismes, S., Nesterenko, M., Tixeuil, S.: Snap-stabilization in message-passing systems. J. Parallel Distrib. Comput. **70**(12), 1220–1230 (2010)
11. Dijkstra, E.W.: Self-stabilizing systems in spite of distributed control. Commun. ACM **17**(11), 643–644 (1974)
12. Dijkstra, E.W., Scholten, C.S.: Termination detection for diffusing computations. Inf. Process. Lett. **11**(1), 1–4 (1980)
13. Dolev, S.: Self-stabilization. MIT press, Cambridge (2000)

14. Francez, N.: Distributed termination. ACM TOPLAS **2**(1), 42–55 (1980)
15. Francez, N., Rodeh, M., Sintzoff, M.: Distributed termination with interval assertions. In: Díaz, J., Ramos, I. (eds.) ICFPC 1981. LNCS, vol. 107, pp. 280–291. Springer, Heidelberg (1981). https://doi.org/10.1007/3-540-10699-5_105
16. Hendler, D., Kutten, S.: Bounded-wait combining: constructing robust and high-throughput shared objects. Distrib. Comput. **21**(6), 405–431 (2009)
17. Katz, S., Perry, K.J.: Self-stabilizing extensions for message-passing systems. Distrib. Comput. **7**(1), 17–26 (1993)
18. Korman, A., Kutten, S., Masuzawa, T.: Fast and compact self-stabilizing verification, computation, and fault detection of an MST. In: PODC 2011, pp. 311–320 (2011)
19. Korman, A., Kutten, S., Peleg, D.: Proof labeling schemes. Distrib. Comput. **22**(4), 215–233 (2010)
20. Levé, F., Mohamed, K., Villain, V.: Snap-stabilizing PIF on arbitrary connected networks in message passing model. In: SSS 2016, pp. 281–297 (2016)
21. Matocha, J., Camp, T.: A taxonomy of distributed termination detection algorithms. J. Syst. Softw. **43**(3), 207–221 (1998)
22. Varghese, G.: Self-stabilization by counter flushing. SIAM J. Comput. **30**(2), 486–510 (2000)

A Loosely Self-stabilizing Protocol for Randomized Congestion Control with Logarithmic Memory

Michael Feldmann$^{(\boxtimes)}$, Thorsten Götte, and Christian Scheideler

Department of Computer Science, Paderborn University, Paderborn, Germany
{michael.feldmann,thorsten.goette,scheideler}@upb.de

Abstract. We consider congestion control in peer-to-peer distributed systems. The problem can be reduced to the following scenario: Consider a set V of n peers (called *clients* in this paper) that want to send messages to a fixed common peer (called *server* in this paper). We assume that each client $v \in V$ sends a message with probability $p(v) \in [0, 1)$ and the server has a capacity of $\sigma \in \mathbb{N}$, i.e., it can receive at most σ messages per round and excess messages are dropped. The server can modify these probabilities when clients send messages. Ideally, we wish to converge to a state with $\sum p(v) = \sigma$ and $p(v) = p(w)$ for all $v, w \in V$.

We propose a *loosely* self-stabilizing protocol with a slightly relaxed legitimate state. Our protocol lets the system converge from *any* initial state to a state where $\sum p(v) \in [\sigma \pm \epsilon]$ and $|p(v) - p(w)| \in O(\frac{1}{n})$. This property is then maintained for $\Omega(n^c)$ rounds in expectation. In particular, the initial client probabilities and server variables are not necessarily well-defined, i.e., they may have arbitrary values.

Our protocol uses only $O(W + \log n)$ bits of memory where W is length of node identifiers, making it very lightweight. Finally we state a lower bound on the convergence time an see that our protocol performs asymptotically optimal (up to some polylogarithmic factor) in certain cases.

1 Introduction

Consider a set of n nodes (called *clients* in this paper) that want to continuously send messages to a fixed node (called *server*) with a certain probability in each round. The server is not aware of its connections and has limited capabilities with regard to the number of messages it is able to receive in each round and its internal memory. The task for the server is to use a *congestion control* protocol to modify the client probabilities such that the server receives only a constant amount of messages in each round (on expectation). As client probabilities may be arbitrary at the beginning, we further require the protocol to be

This work was partially supported by the German Research Foundation (DFG) within the Collaborative Research Center On-The-Fly Computing (GZ: SFB 901/3) under the project number 160364472.

© Springer Nature Switzerland AG 2019
M. Ghaffari et al. (Eds.): SSS 2019, LNCS 11914, pp. 149–164, 2019.
https://doi.org/10.1007/978-3-030-34992-9_13

self-stabilizing, i.e., it should be able to reach its goal starting from any arbitrary initial state. Self-stabilization comes with the advantage that the protocol is able to recover from transient faults like message loss or blackout of processes automatically. As the system grows larger, these kinds of faults occur more often, which makes self-stabilization as a concept very desirable.

At first glance, one may think that this setting only applies to client/server-architectures. However, we believe that solving this problem is quite important for distributed systems where nodes constantly have to communicate with their neighbors. Also there are distributed systems where nodes are not aware of their incoming connections, e.g. in rooted trees, random graphs [19] or linearized de Bruijn networks [20]. On these networks one is able to effectively perform many important techniques relevant to distributed computing such as aggregation, sampling, or broadcast which are important for applications like distributed data structures (e.g. hash tables [17], queues [11] or heaps [10]). Also nodes with limited capabilities can be found in internet of things applications like wireless networks [23].

In this paper we present a loosely self-stabilizing protocol for congestion control. In contrast to classical self-stabilization, loose self-stabilization relaxes the closure property. Our protocol guarantees that the server only receives a constant amount of messages on expectation in each round while only using a logarithmic amount of bits for its internal protocol variables for a period of $O(n^c)$ rounds (and not forever as classical self-stabilization would require). Furthermore we can guarantee *fairness*, i.e., the probabilities of all clients are the same (up to some small constant deviation). By slightly weakening the definition for a legitimate state, we are able to analyze the runtime of our protocol and show that it is able to quickly reach a state that is already practical for both, the clients and the server.

2   Model and Definitions

2.1   System Model

Network Model. Since we only consider communication of nodes with their direct neighborhood in the overlay network, we consider the following directed graph $G = (V \cup \{s\}, E)$. $V = \{v_1, \ldots, v_n\}$ represents the set of n clients and s represents the server. We assume n to be fixed. The set of edges is defined by $E = \{(v, s) \mid v \in V\}$, i.e., all clients know the server, but the server does not know which client is connected to it. More particularly, the server does not know the value n. All clients and the server can be identified via their unique *reference*, represented by values $v_i.id \in \mathbb{N}$ for all $i \in \{1, \ldots, n\}$ and $s.id \in \mathbb{N}$ respectively. We assume that identifiers can be stored by at most W bits, where $W \geq \log n$ is known to the server. If a node v knows the reference of another node w, then v is allowed to send messages to w.

Each client $v \in V$ maintains a probability $p(v) \in (0, \hat{p}]$, where $\hat{p} \leq 1$ is a protocol-specific constant. Denote by $p_{min} \in (0, \hat{p}]$ the minimum client probability, i.e., $p_{min} = \min_{v \in V}\{p(v)\}$ and denote the sum of all client probabilities by P, i.e., $P = \sum_{v \in V}^{n} p(v)$. We assume that the probability $p(v)$ for a client v cannot become smaller than $1/2^{b \cdot W}$ for some fixed constant $b > 0$, i.e., it can be encoded by $O(W)$ bits. This means that all probabilities are multiples of $1/2^{b \cdot W}$.

Computational Model and Definition of a Round. We divide time into synchronous *rounds*, where a single round consists of the following steps:

(i) Each client v tosses a biased coin that shows 'heads' with probability $p(v)$. If v's coin shows 'heads', v sends a message $m = (v.id, p(v))$ to the server s. Otherwise v stays idle for the rest of the round. We assume that the server is only able to receive up to σ messages from clients per round for a fixed constant $\sigma \in \Theta(1)$ that is known to the server. If more than σ clients decide to send a message to the server in this step, then exactly σ of those messages are determined uniformly at random to arrive at the server, while the other ones are dropped.

(ii) The server makes some internal computation based on the messages it received in the previous step.

(iii) For each message $m = (v.id, p(v))$ that the server received, it may send a message $m' = (p(v)')$ back to v.

(iv) Each client $v \in V$ that received a message $m = (p(v)')$ in the previous step sets $p(v)$ to $p(v)'$.

A message sent by a client to the server in step (i) is denoted as a *ping* or *ping message* and we may also just say that the client *pings* the server in this case. We say that a client *successfully pings* the server (in round t) if it sends a ping message to the server (in round t) that is actually being processed by the server, i.e., that is not dropped. We may use $p_t(v)$ to refer to the probability of client v in round t. Note that the server s is able to answer v in step (iii) because v sent its reference $v.id$ to s in step (i). Once the round is over, the server forgets about $v.id$. Also observe that the server is not required to send an answer to each message it received in (iii).

Last, the *state* S_t of the system before round t is defined by the assignment of variables $p(v)$ at each client $v \in V$ and internal variables at the server. The system transitions from S_t to S_{t+1} by performing the steps (i) to (iv) mentioned above.

2.2 Problem Statement

We wish to state a protocol that reaches a state with the following two conditions, namely *Busyness* and *Fairness*. They are defined as follows:

Definition 1 (Busyness). *Let $L, R \in O(\sigma)$ be protocol-specific constants. We say that the server is* busy *in some state S of the system if $P \in [L, R]$ holds in S. We say that a state is* busy *for short.*

Definition 2 (Fairness). *The system satisfies* fairness *in some state S, if $\sum_{v \in V} \left(p(v) - \frac{P}{n} \right)^2 \leq \frac{1}{n^c}$ holds in S for some constant $c > 0$. We say that a state is* fair *for short.*

We believe these to be natural and reasonable safety properties given our problem and model setup. With the first property we ensure that the server operates close to its limits and is not under- or overutilized. Note that $L, R \geq 1$ can be chosen freely by the server, so it can adjust these values depending on its computational power in practice. Note that this is not fully precise in a sense that P does converge to some desired fixed value, but we can guarantee that P will eventually converge to some value within the interval $[L, R]$. Moreover, the notion of busyness prevents the trivial solution of letting all clients send with probability 1. Fairness assures that all clients (roughly) send the same amount of data to the server and every client will eventually send. This prevents the trivial solution of letting σ clients send with probability 1 and all others with 0.

Note that in a distributed setting errors are the norm rather than the exception, which means that the probabilities of the clients and the variables can be corrupted through malicious messages, crashes, and memory faults. Thus, we are specifically interested in a *self-stabilizing* protocol that reaches a safe state even if all probabilities and server variables are corrupted.

In the classical sense, a protocol is *self-stabilizing* w.r.t. a set of legitimate states if it satisfies *Convergence* and *Closure*: Convergence means that the protocol is guaranteed to arrive at a legitimate state in a finite amount of time when starting from an arbitrary initial state. Closure means that if the protocol is in a legitimate state, it remains in legitimate states thereafter as the set of clients does not change and no faults occur. However, our protocol will *not* meet these strong requirements of classical self-stabilization due to the clients' probabilistic nature. To account for this, we will instead show that our protocol is *loosely self-stabilizing*.

The notion of *probalistic loose self-stabilization* was introduced by Sudo et al. in [21] to deal with probabilistic protocols that violate the Closure with *very small* probability. Instead of a set of legitimate states that are never left, a loosely self-stabilizing protocol maintains a safety condition for a sufficiently long time. More precisely, a protocol is (α, β)-loose self-stabilizing, if it fulfills the following two properties: First, it reaches a legitimate state after α rounds (in expectation) starting from *any* possible initial state. Second, given that the execution starts in a legitimate state, the protocol fulfills a safety condition for at least β rounds (in expectation). That means for β consecutive rounds, all states fulfill a certain condition if their execution started in a legitimate state. We call this the holding time. To put it more formally, let $\mathfrak{S}$ be the set of all possible system states and $\mathfrak{L} \subset \mathfrak{S}$ be the set of all legitimate states. Then the random variable $C(s, \mathfrak{L})$ denotes the convergence time if the algorithm started in $s \in \mathfrak{S}$. Likewise, let $\mathfrak{L}^*$ the be set of all states that fulfill the safety condition, then $H(\ell, \mathfrak{L}^*)$ denotes the holding time given that we start in $\ell \in \mathfrak{L}$. Thus, for a (α, β)-loose self-stabilizing protocol, it holds

$$\max_{\mathfrak{s} \in \mathfrak{S}} \mathbb{E}\left[C(\mathfrak{s}, \mathfrak{S})\right] \leq \alpha \quad \text{and} \quad \min_{\ell \in \mathfrak{L}} \mathbb{E}\left[H(\ell, \mathfrak{L}^*)\right] \geq \beta$$

Note that for an efficient protocol it should hold $\alpha << \beta$, i.e, we quickly reach a legitimate state and then stay safe for a long time.

2.3   Technical Contributions

Our goal is to construct a self-stabilizing protocol for the server that converges the system into a state where busyness (Definition 1) and fairness (Definition 2) hold. In the following we discuss the most major obstacles that we have to overcome when constructing a solution.

Dealing with Arbitrary Initial States. In initial states the variables at both the clients and the server may contain arbitrary values. Particularly, each client probability may initially be an arbitrary value out of $(0, \hat{p}]$. Due to the restrictions on the message size this may lead to P being as low as $O(1/poly(n))$ initially which means that it may take a long time until the server receives the first ping message. This means that our protocol needs to be designed in a way such that for initially low values of P we make significant progress in reaching a legitimate state once the probability of a client is modified.

Knowledge of $\Theta(\log n)$**.** Our algorithm requires the server to estimate $\Theta(\log n)$. The problem of approximating $\Theta(\log n)$ can be non-trivial when additionally requiring a self-stabilizing solution for this, i.e., the server may think of any value to be $\log n$ initially. Our loosely self-stabilizing solution for approximating $\Theta(\log n)$ at the server may be of independent interest.

2.4   Our Contribution

We propose a congestion control protocol that is loosely self-stabilizing. It converges to a legitimate state that is busy and fair within $\tilde{O}(\mathfrak{c}\left(p_{min}^{-1} + n^3\right))^1$ rounds starting from any initial state where clients may have arbitrary probabilities. Then all following states are also busy and fair for at least another $O(n^\mathfrak{c})$ rounds in expectation. Here, $\mathfrak{c}$ is a parameter and can be chosen depending on the context. Note that even for small $\mathfrak{c}$ the system stays stable long enough for practical purposes. Furthermore, the server uses only $O(W + \log n)$ bits in legitimate states. This makes the protocol very lightweight and ideal for servers with strong memory constraints, e.g., in sensor networks.

The rest of the paper is structured as follows: First, we review some related work in Sect. 3. Then, we present our protocol in Sect. 4. Last, in Sect. 5 we rigorously analyze our protocol and show that it is loosely self-stabilizing. Due to space constraints, the full proofs are deferred to the full version of this paper [9].

3   Related Work

Congestion Control. There exists a wealth of literature on *congestion control* in the internet. Classical approaches that have been considered are MIMD (Multiplicative Increase, Multiplicative Decrease [15]) and AIMD (Additive Increase, Multiplicative Decrease [5]). Many other researchers studied congestion control for the AIMD model, which resulted in various extensions of the original work,

---

[1] We use $\tilde{O}$ to hide polylogarithmic factors.

see for example [6,16,18]. Although these protocols work for arbitrary initial probabilities, their auxiliary variables are always assumed to be well-initialized. In contrast, our protocol also tolerates completely arbitrary initial states including auxiliary variables, making it truly self-stabilizing. Also, to the best of our knowledge, prior congestion control protocols do not provide a rigorous theoretical analysis on their convergence time.

Flow Control. Close to congestion control problems are flow control problems (see [12] for a survey). These protocols differ from our setting in the sense that they operate on a continuous data stream, whereas we consider discrete rounds where only small self-contained control-messages are exchanged between the server and multiple clients, so flow control strategies are not applicable here.

Contention Resolution. Close but different to congestion control protocols is the area of contention resolution in multiple access channels (see for example [1, 2,4] or [13] for a survey). A multiple access channel (MAC) is a medium shared among all nodes through which they can send messages. In each round a node may either send a message or sense the channel. Messages that have been sent in the same round by two or more nodes *collide* and are not transmitted. By sensing the channel a node gets informed whether the channel is *idle* (no message has been sent), *busy* (a collision occurred) or it receives a message (in case there has been exactly one message sent). Contention resolution differs from congestion control in a sense that once two or more messages are sent in the same round there already is a collision, whereas in congestion control multiple messages are allowed to be processed by the receiver. Also the MAC allows clients to only receive binary feedback, making it less powerful compared to our server.

Distributed Consensus and Load Balancing. Further related areas on a technical level are *distributed average consensus* (see [14] for a survey) and (discrete) load balancing (see [3,22] and the references therein). In both problems, multiple agents try to find the arithmetic mean of a given set of initial values. Our protocol tries the same in order to achieve fairness. However, we need to deal with dynamically changing probabilities as the adaption of the nodes' values directly influences their sampling probabilities. In other settings the probabilities may be arbitrary but are fixed in advance.

Self-stabilization. Self-stabilization was first proposed in [8]. Since inventing self-stabilizing protocols can be quite difficult, people came up with relaxed versions for the convergence property like probabilistic self-stabilization or weak-stabilization [7]. The notion of loose-stabilization [21] that is used in this paper relaxes the closure property instead of the convergence property.

4   Protocol Description

Intuitively our protocol works as follows: We constantly let the server count the number of pings it received in each round for an interval of Δ rounds. Probabilities of clients that ping are averaged in these rounds. Once an interval of Δ

rounds ends, the server is able to precisely approximate P in case $\Delta \in \Theta(\log n)$ and decide whether to either raise the probability of a client that has pinged in that round (if P is too small), decrease the probability of a client (if P is too large) or adjust the probabilities of clients by computing the average (if P lies within a desired interval).

We describe the protocol in greater detail now starting with the introduction of variables and constants. Afterwards we describe how the approximation for $\Theta(\log n)$ and P at the server works, followed by the description of the core protocol.

4.1 Variables and Constants

Table 1 shows the variables and constants that are maintained by the server.

Table 1. Variables and constants used by our algorithm

$\varepsilon > 0$	A constant used for the approximation of P
$L, R \in \Theta(1)$	Constants for the left and right border of the desired interval $[L, R]$ to which P should converge. In order to guarantee that eventually $P \in [L, R]$, we require that $\|R - L\| > \hat{p} + 2\varepsilon$. Note that L, R are chosen such that $1 \le L < R \le \sigma$, i.e., on expectation, the server receives at least L, but no more than R messages in legitimate states
$\Delta \in \Theta(\log n)$	A variable indicating the interval of rounds in which the server counts the number of incoming pings
$\delta \in [0, \Delta]$	A counter that is incremented each round and reset to 0 once it is equal to Δ
$X \in \mathbb{N}_0$	A counter that sums up the number of incoming pings within a period of Δ rounds

Note that the constants L, R and ε are protocol-based constants, which means they are chosen preemptively by the server and thus are fixed while the stabilization process of the system is going on. On the other side the variables δ, Δ and X may contain arbitrary values out of their domains in initial states.

4.2 Approximating $\Theta(\log n)$ at the Server

In order to work properly, our protocol needs an approximation of $\Theta(\log n)$. In the following we sketch a protocol to obtain such an approximation given that we have one server and n clients.

We let the server maintain a table of $\log \log N$ columns where each column i represents a value $c_i = \sqrt[2^i]{N}$ and a timestamp $t_i \ge 0$ (see Table 2). The first column c_0 represents the value N, which may be arbitrary large in initial states. Therefore the table along with its timestamps may initially be completely

Table 2. Table maintained at the server.

$c_0 = N$	$c_1 = \sqrt{N}$	$c_2 = \sqrt[4]{N}$	...	$c_{\log \log N - 1} = 2$
t_0	t_1	t_2	...	$t_{\log \log N - 1}$

arbitrary. The table is maintained as follows by the server: We map the identifiers of the server and the clients to the interval $[0, 1)$ via a uniform hash function $h : \mathbb{N} \to [0, 1)$. Whenever a client v with $|h(s.id) - h(v.id)| \leq \frac{1}{c_i}$ successfully pings the server, the server resets all timestamps $t_i, \ldots, t_{\log \log N - 1}$ to 0. Aside from this, each timestamp t_i gets incremented by one in each round. Once the entry t_i for column c_i gets larger than $O(c_i \cdot polylog(c_i))$, all columns $c_0, \ldots, c_i$ are deleted from the table and the value N is set to the column c_{i+1}. On the other side, once a client v pings for which $|h(s.id) - h(v.id)| \leq \frac{1}{c_0^2}$ holds we update the table by adding that many columns to the left until $\frac{1}{c_0^2} < |h(s.id) - h(v.id)| \leq \frac{1}{c_0}$ holds. The server always sets $\Delta = \Theta(\log c_0)$ to approximate $\Theta(\log n)$.

This protocol will run in parallel to anything described in the remainder of this section.

4.3 Approximating P at the Server

At the end of an interval of rounds of size Δ, the server checks whether P is (approximately) less than L, larger than R or within $[L, R]$. We use the operator $\prec$ to indicate the result of the approximation, for example if P is approximately less than L we say $P \prec L$ and otherwise $P \succ L$. In order to check whether $P \prec L$ or $P \succ L$, the server checks whether $X/\Delta < L$ holds. If that is the case then the server decides on $P \prec L$, otherwise it decides $P \succ L$. By comparing X/Δ to R the server can do the same to decide whether $P \prec R$ or $P \succ R$ holds.

4.4 Core Protocol

The server executes Algorithm 2 in each round after each client has decided whether to ping the server or not (Algorithm 1, Line 3). Here $v_1, \ldots, v_k$ are the clients that successfully pinged the server in round t.

Algorithm 1. Pseudocode executed at each client v in each round

---
1: Toss a coin that shows 'heads' with probability $p(v)$
2: **if** Coin shows 'heads' **then**
3:     Send $m = (v.id, p(v))$ to s
4: **if** v received $p'(v)$ from s **then**
5:     $p(v) \leftarrow p'(v)$

Algorithm 2. Pseudocode executed at the server in each round

1: Let $v_1, \ldots, v_k$ be the clients that successfully pinged the server in ascending order of their probabilities, i.e., $p(v_1) \leq \ldots \leq p(v_k)$
2: $X \leftarrow X + k$
3: $\delta \leftarrow (\delta + 1) \mod \Delta$
4: **if** $\delta = 0$ **then**
5:      **if** $P \prec L$ **then**
6:          Send $\hat{p}$ to v_1 ▷ Increase minimum probability
7:      **else if** $P \succ R$ and $k \geq 2$ **then**
8:          Send $p(v_k)/(1 + 1/\sigma)$ to v_k ▷ Decrease maximum probability
9:      $X \leftarrow 0$
10: **else**
11:      **for all** $i \in \{1, \ldots, k\}$ **do**
12:          Send $\lfloor \sum_{i=1}^{k} p(v_i)/k \rfloor + r_i$ to v_i ▷ Average probabilities

The protocol given by Algorithm 2 works as follows: At the beginning of each round we let clients ping the server with their corresponding probabilities. Assume that k clients $v_1, \ldots, v_k$ pinged the server ordered by their probabilities, i.e., $p(v_1) \leq \ldots \leq p(v_k)$. The server first increments X by k (Line 2) and then sets δ to $(\delta + 1) \mod \Delta$ (Line 3). In case $\delta \neq 0$, the server sets each probability $p \in \{p(v_1), \ldots, p(v_k)\}$ to the average of these probabilities (Line 12). In a round where $\delta = 0$ holds the server instead approximates P based on X and Δ. Using the approximation for P, the server checks whether $P \prec L$, i.e., whether P is currently too low. If that is the case, then the server raises the minimum probability $p(v_1)$ to $\hat{p}$ (Line 6). On the other hand, if P is too large ($P \succ R$) and at least $k \geq 2$ clients pinged, the server sets the maximum probability $p(v_k)$ to $p(v_k)/(1 + 1/\sigma)$ (Line 8). Once this has been done, the server resets X to 0 (Line 9).

Notice that parts of our algorithm (specifically the way we choose client probabilities to be decreased) are related to the well-known *two-choice process* where we (greedily) choose the process with minimum probability to have its probability reduced (Line 8). As it turns out in the analysis, we can make use of this by modelling our setting as a balls-and-bins process for which we can apply a result from [22].

Due to messages being restricted to only $O(W)$ bits it may happen that we lose accuracy on the overall sum of probabilities P if we were to simply compute the averages of client probabilities and round it up or down. To overcome this problem, we use the following rounding approach when computing average client probabilities (Line 12): In a round where k clients ping the server and the average of these clients has to be computed, we initially set the probabilities to the average rounded down on W bits, i.e., the least significant bit is set to 0. As the real average value leaves some residue value of the form $r \cdot \frac{1}{2^{b \cdot W}}$ for an integer $r < k$, we set the least significant bit of r clients (chosen randomly among the v_i's) to 1. This is indicated by the values $r_i \in \{0, \frac{1}{2^{b \cdot W}}\}$. By doing so we ensure that P does not get modified when only computing averages and all the client probabilities remain multiples of $\frac{1}{2^b \cdot w}$. For the analysis we assume for simplicity that we compute the average value without rounding and only consider the rounding approach when it actually influences a proof.

5   Analysis

We analyze our algorithm in this section and show that it is loosely self-stabilizing. Therefore, we need to give a formal definition for a legitimate state and a safety condition. Obviously, we want our system to be in a *busy* and *fair* state, but moreover, in order to guarantee a long holding time, we need a correct estimate of $\Theta(\log n)$. Therefore, we introduce the notion of stability.

Definition 3 (Stability). *A state $s \in S$ fulfills the stability property, if c_0, the biggest entry in the table, is in $\Omega(n^{\frac{1}{2}})$ and all t_i are 0. We call such a state s stable for short.*

As we will see, this ensures that the protocol correctly estimates $\Theta(\log n)$ for at least $\Omega(n^c)$ rounds in expectation.

Furthermore, we need to weaken the fairness property a bit to get more practical results. This comes from the fact that the algorithm may erroneously increase or decrease the probabilities, even if $\Delta \in \Theta(\log n)$. We wish to acknowledge that our protocol does reach an arbitrarily fair state after $O(poly(n))$ rounds and then stays that way for another $O(poly(n))$ rounds (both in expectation), i.e., it would hold $\alpha \approx \beta$. We sketch this in the full version of the paper [9] and focus here on the so-called weakly fair state as we deem it more practical. It is defined as follows:

Definition 4 (Weakly Fairness). *A state S of the system is a weakly fair state if $\forall v \in V : p(v) \in \Omega\left(\frac{P}{n}\right)$.*

Given this definition, we can now simply define the legitimate state. Over the course of this chapter, we will show that the following holds:

Theorem 1. *Let c be a big enough constant. Further, let $L, R \in O(\sigma)$ and $\varepsilon > 0$ be protocol-specific constants. Then it holds:*

- *A state $\ell \in \mathfrak{L}(L, R, \varepsilon)$ of the system is a legitimate state if it is busy, weakly fair, and stable.*
- *A state $\ell \in \mathfrak{L}^*(L, R, \varepsilon)$ of the system fulfills the safety condition if it is busy and weakly fair.*

Then, our protocol is $\left(\tilde{O}(p_{min}^1 + n^3), \Omega(n^c)\right)$-loosely self-stabilizing with regard to the legal states $\mathfrak{L}(L, R, \varepsilon)$ and safe states $\mathfrak{L}^*(L, R, \varepsilon)$.

5.1 Convergence Time

Now we show that the system converges to a legitimate state after $\tilde{O}(p_{min}^{-1} + n^3)$ rounds w.h.p. We split the analysis into three phases: First we analyze the time it takes until $\Delta \in \Theta(\log n)$ is fixed. In the second phase we analyze the time it takes for P to reach a value within $[L, R]$. Finally we show a bound on the time it takes until weak fairness is reached, i.e., until all probabilities are in $\Omega(P/n)$. Note that these phases exist purely for analytical purposes and the algorithm itself is oblivious of them.

Phase I: Approximating $\Theta(\log n)$. We start by showing that there exists a appropriate self-stabilizing approximation algorithm for $\Theta(\log n)$ given that the communication graph is a star graph of $\Theta(n)$ nodes. In particular, the following holds:

Theorem 2. *Our protocol provides a fixed estimation of $\Theta(\log n)$ for the server within $O(p_{min}^{-1} + n^2 \cdot polylog(n))$ rounds w.h.p. starting from any configuration, and reaches a stable state every $O(n^2)$ rounds with probability $1 - o(n^{-c})$.*

Proof (Sketch). For the analysis of this approach we first show that after $O(n^2 \cdot polylog(n))$ rounds all superfluous columns that may exist in initial states have been deleted and thus $\Delta \leq \Theta(\log n)$ holds. Afterwards we show that after $\Theta(n/\log n)$ clients have successfully pinged the server at least once (which needs $O(p_{min}^{-1} + n \cdot \log^2 n)$ rounds), at least $\Theta(n/\log n)$ clients are *visible*, i.e., they have a probability of at least $\Omega\left(\frac{P}{n \cdot polylog(n)}\right)$. This suffices to show convergence for our strategy. For the second property we show that no columns gets added or deleted w.h.p. and that a visible client remains visible throughout the algorithm via a slight adaptation of the server's behavior. This leads to the timestamp t_0 of the first column c_0 being reset to 0 after at most $O(n^2)$ rounds w.h.p.   □

Phase II: Convergence for P. In the following we bound the time until we arrive at a configuration with $P \in [L, R]$ once $\Delta \in \Theta(\log n)$ has stabilized. Here, we need to take into account that in the first phase all probabilities could be arbitrarily adapted by the algorithm. In particular, through negative feedback the smallest probability p_{min} could be further reduced. This could potentially delay the stabilization of our algorithm ad infinitum. However, recall that the minimal probability is only decreased when two nodes of (almost) minimal probability successfully ping the server. Thus, the smaller p_{min} gets, the more unlikely it is for it to be reduced further. Formally, we can show the following:

Lemma 1. *During the execution of the first phase, no node will be assigned a probability smaller than $O\left(\frac{\min\{p_{min}, n^{-2}\}}{\log n}\right)$ w.h.p.*

Proof (Sketch). The proof works similar to the analysis of a ball-into-bins process with d choices. Whenever the probabilities are reduced through the algorithm, this can be seen as throwing a ball to the biggest of the d randomly chosen nodes that pinged in that round. Through a careful adaption of the corresponding proof, we see that the minimal node's probability is reduced at most $\log \log n$ times if the protocol runs for $O(p_{min}^{-1})$ rounds. This corresponds to reducing the probability by a factor $\left(1 + \frac{1}{\sigma}\right)^{-\log \log n}$. Since σ is constant, this is within $O\left(\log n^{-1}\right)$. □

Given this insight, we can now show the following:

Theorem 3 (Convergence Time for P). *Let the system be in any state where $\Delta \in \Theta(\log n)$ is already fixed. After $O((p_{min}^{-1} + n) \log^2 n)$ rounds, the system reaches a state where $P \in [L, R]$ w.h.p.*

Proof (Sketch). We need to consider the cases $P < L$ and $P > R$. In case $P < L$ we can show that it takes $O(p_{min}^{-1} \log n)$ rounds until $P \in [L, R]$ w.h.p. This follows from the time needed to set the probabilities of at least α different clients to $\hat{p}$ for a constant $\alpha \in \mathbb{N}$ with $\alpha \cdot \hat{p} > L$. For $P > R$ we can conclude that, with at least constant probability, at least one client out of the set $V' = \{v \in V \mid p(v) \geq \frac{P}{2n}\}$ pings the server successfully in a round where $\delta = 0$ and thus gets its probability reduced. It follows by calculation that after $O(n \log n)$ of these reductions $P < R$ holds. These reductions can be achieved within $O(n \log^2 n)$ rounds w.h.p. □

Phase III: Convergence to Weak Fairness. Finally, we show that we reach (weakly) fair state after at most $\tilde{O}(p_{min}^{-1} + n^3)$ rounds w.h.p. given that the initial state is already busy and stable. Our definition of a weakly fair state requires that all probabilities are close to $\frac{P}{n}$ (and moreover will *stay* close for $O(n^c)$ rounds). To achieve this, the protocol must not in- or decrease the client probabilities too often. On the first glance, one might think that $\Delta \in \Theta(\log n)$ and $P \in [L, R]$ are sufficient to ensure that. A closer look reveals that in cases where P is close to the borders of the interval $[L, R]$ this might not be the case. However, if we assume that P only changes very infrequent, then we can adapt the results of Berenbrink et al. [3] and obtain the following result.

Theorem 4. *Let the system be in a legitimate state where $P \in [L, R]$. Then it holds:*

1. *After at most $O(n^3 \log n)$ rounds, P changes only with prob $o(n^{-2})$.*
2. *After $O(p_{min}^{-1} \cdot \log n)$ rounds the system reaches a weakly fair state w.h.p.*

Proof (Sketch). For the first claim note that the probability to decrease P depends on P itself. The main idea is that after n reductions (which take n^3 rounds in expectation), P is so small that further reductions are very unlikely.

For the second part, we model our system as a balls-and-bins process. The clients represent the bins and the client probabilities represent balls, where the number of balls depends on P, i.e., if the probability of a client v is $p(v) = \frac{c}{2^{b \cdot W}}$ (recall that client probabilities are multiples of $1/2^{b \cdot W}$), then we say that v has c balls. At the beginning the $P \cdot 2^{b \cdot W}$ balls are arbitrarily distributed among all clients. Then, we use an adaption of the potential function analysis from [3]. As potential, we/they use the sum of squared differences, i.e., $\Phi_t := \sum_{v \in V} \left(p_t(v) - Pn^{-1}\right)^2$. In particular, we need to adapt the following:

1. The probabilities are not uniform *and* change dynamically during the process. We solve this by observing that with constant probability, the sampled values are close to the arithmetic mean. Therefore, clients with small probability are increased quickly once they send.
2. The probabilities can be reduced. However, since the probability for a change is small, i.e., $o(n^{-2})$, we can amortize it through the balancing.

Given these adaptations, we can show that after $O(p_{min}^{-1} \cdot \log n)$ rounds the sum of the squared differences between all clients and the average is at most n. This corresponds to the sum of the squared differences between all client probabilities and the average probability P/n being at most $1/n$, which suffices to show fairness. Together with the time it takes for probabilities to be small enough, the theorem follows. $\qquad\square$

5.2   Holding Time

It remains to bound the holding time. However, this simply follows from the observations we made so far.

Theorem 5. *Let c be an arbitrary constant. Suppose the system is in a legitimate state $\ell \in \mathfrak{L}(L, R, \epsilon)$, then it will remain in a safe state for $\Omega(n^c)$ rounds in expectation.*

Proof (Sketch). We show that both busyness and fairness are maintained with probability $1 - o(n^c)$ if we start in a stable state. First, note that starting in a stable state, the system maintains $\Delta \in \Theta(\log n)$ until the first entry in the table is deleted. For a deletion, a node (which pings with probability $\frac{P}{n}$) must not ping for consecutive $O(cn \log n)$ rounds. The probability for this is $O(n^{-c})$ and hence this holds for $O(n^c)$ round in expectation. Given that $\Delta \in \Theta(\log n)$ remains fixed, we can show the following.

1. The system remains in a busy state. We violate busyness if and only if P leaves the interval $[L, R]$. Therefore, the probabilities need to be de- or increased at least $\omega(n)$ times. This only happens if the server (wrongly) predicts $P \succ L$ or $P \prec R$, which happens with prob. $1 - o(n^c)$ given that $\Delta \in \Theta(\log n)$.
2. The system remains in a weakly fair state. By a similar argument, we see that fairness is violated if few nodes are decreased too often. This also only happens if the server (wrongly) predicts $P \succ L$ or $P \prec R$ and is therefore evenly unlikely. In particular, the times between two decreases are so long that the nodes can re-balance themselves and thus stay weakly fair. $\qquad\square$

5.3   Tightness

Last, we observe the tightness of our convergence time. One can easily see that *any* self-stabilizing protocol needs $\Omega(p_{min}^{-1} \log n + n)$ rounds to reach a legitimate state. This follows from the fact that each client need to ping the server at least once to get a probability in $O(\frac{P}{n})$. As we see, our protocol is indeed optimal if

$p_{min} \in O(\frac{1}{n^3})$, but is slower otherwise. However, note that the slowdown only happens because of two important properties that our protocol fulfills. First, it takes an additional $O(n^3)$ rounds in phase I, i.e., during approximation of $\Theta(\log n)$. Given that the protocol has a stable estimation of $\Theta(\log n)$ (which is reasonable in many contexts) the convergence time is asymptotically optimal in this phase. Second, it takes $O(n^3)$ rounds until the probability for a decrease is so low that the protocol converges to a weakly fair state. For an even notion of fairness (e.g. at most $o(n)$ nodes may have very low probability) this could be improved.

6   Conclusion

We proposed a self-stabilizing protocol for congestion control in overlay networks that performs reasonably well in our model. Finally we want to make a remark on the system's performance in arbitrary topologies.

Remark 1. Consider an overlay network $G = (V, E)$ with indegree at most ζ. Further, let each node know a (probably rough) estimation N of n. Assume we apply our protocol for loose self-stabilizing congestion control such that each node acts as a server for its incoming connections and as a (separate) client for each of its outgoing connections. This way we obtain $(O((p_{min}^{-1}+\zeta^3)\cdot polylog(N)),$ $N^c)$ loosely self-stabilizing protocols for all servers.

This follows from Theorem 1, if we assume that all nodes $v \in V$ run our algorithm with neighbors as clients. However note that in cases where ζ is constant our results would hold only with probability in $\Theta(e^{-\zeta})$ and not w.h.p. To circumvent this we just use the estimation N of n for nodes and let the value for Δ at each node be in $\Theta(\log N)$ instead of $\Theta(\log \zeta)$. Given that all nodes know $\Theta(\log N)$, the approximation algorithm is obsolete and all states are stable. Note that all other bounds only depend on the number of client. Thus, we plug in the maximum degree ζ of a node instead of n. This gives us the $polylog(N)$-factor in the runtime above.

References

1. Bender, M.A., Farach-Colton, M., He, S., Kuszmaul, B.C., Leiserson, C.E.: Adversarial contention resolution for simple channels. In: SPAA, pp. 325–332. ACM (2005). https://doi.org/10.1145/1073970.1074023
2. Bender, M.A., Fineman, J.T., Gilbert, S., Young, M.: Scaling exponential backoff: constant throughput, polylogarithmic channel-access attempts, and robustness. J. ACM **66**(1), 6:1–6:33 (2019). https://doi.org/10.1145/3276769
3. Berenbrink, P., Friedetzky, T., Kaaser, D., Kling, P.: Tight & simple load balancing. In: IPDPS, pp. 718–726. IEEE (2019). https://doi.org/10.1109/IPDPS.2019.00080
4. Chang, Y., Jin, W., Pettie, S.: Simple contention resolution via multiplicative weight updates. In: SOSA@SODA, OASICS, vol. 69, pp. 16:1–16:16. Schloss Dagstuhl - Leibniz-Zentrum fuer Informatik (2019). https://doi.org/10.4230/OASIcs.SOSA.2019.16

5. Chiu, D., Jain, R.: Analysis of the increase and decrease algorithms for congestion avoidance in computer networks. Comput. Netw. **17**, 1–14 (1989). https://doi.org/10.1016/0169-7552(89)90019-6

6. Corless, M.J., Shorten, R.: Deterministic and stochastic convergence properties of AIMD algorithms with nonlinear back-off functions. Automatica **48**(7), 1291–1299 (2012). https://doi.org/10.1016/j.automatica.2012.03.014

7. Devismes, S., Tixeuil, S., Yamashita, M.: Weak vs. self vs. probabilistic stabilization. Int. J. Found. Comput. Sci. **26**(3), 293–320 (2015)

8. Dijkstra, E.W.: Self-stabilizing systems in spite of distributed control. Commun. ACM **17**(11), 643–644 (1974). https://doi.org/10.1145/361179.361202

9. Feldmann, M., Götte, T., Scheideler, C.: A loosely self-stabilizing protocol for randomized congestion control with logarithmic memory. CoRR abs/1909.04544 (2019). https://arxiv.org/abs/1909.04544

10. Feldmann, M., Scheideler, C.: Skeap & seap: scalable distributed priority queues for constant and arbitrary priorities. In: SPAA, pp. 287–296. ACM (2019). https://doi.org/10.1145/3323165.3323193

11. Feldmann, M., Scheideler, C., Setzer, A.: Skueue: a scalable and sequentially consistent distributed queue. In: IPDPS, pp. 1040–1049. IEEE Computer Society (2018). https://doi.org/10.1109/IPDPS.2018.00113

12. Gerla, M., Kleinrock, L.: Flow control: a comparative survey. IEEE Trans. Commun. **28**(4), 553–574 (1980). https://doi.org/10.1109/TCOM.1980.1094691

13. Goldberg, L.A.: Notes on contention resolution (2002). https://www.cs.ox.ac.uk/people/leslieann.goldberg/contention.html

14. Guerraoui, R., Hurfinn, M., Mostefaoui, A., Oliveira, R., Raynal, M., Schiper, A.: Consensus in asynchronous distributed systems: a concise guided tour. In: Krakowiak, S., Shrivastava, S. (eds.) Advances in Distributed Systems. LNCS, vol. 1752, pp. 33–47. Springer, Heidelberg (2000). https://doi.org/10.1007/3-540-46475-1_2

15. Kelly, T.: Scalable TCP: improving performance in highspeed wide area networks. Comput. Commun. Rev. **33**(2), 83–91 (2003). https://doi.org/10.1145/956981.956989

16. Kesselman, A., Mansour, Y.: Adaptive AIMD congestion control. Algorithmica **43**(1–2), 97–111 (2005). https://doi.org/10.1007/s00453-005-1160-3

17. Kniesburges, S., Koutsopoulos, A., Scheideler, C.: CONE-DHT: a distributed self-stabilizing algorithm for a heterogeneous storage system. In: Afek, Y. (ed.) DISC 2013. LNCS, vol. 8205, pp. 537–549. Springer, Heidelberg (2013). https://doi.org/10.1007/978-3-642-41527-2_37

18. Lahanas, A., Tsaoussidis, V.: Performance evaluation of τ-AIMD over wireless asynchronous networks. In: Braun, T., Carle, G., Koucheryavy, Y., Tsaoussidis, V. (eds.) WWIC 2005. LNCS, vol. 3510, pp. 86–96. Springer, Heidelberg (2005). https://doi.org/10.1007/11424505_9

19. Mahlmann, P., Schindelhauer, C.: Distributed random digraph transformations for peer-to-peer networks. In: SPAA, pp. 308–317. ACM (2006). https://doi.org/10.1145/1148109.1148162

20. Richa, A., Scheideler, C., Stevens, P.: Self-stabilizing De Bruijn networks. In: Défago, X., Petit, F., Villain, V. (eds.) SSS 2011. LNCS, vol. 6976, pp. 416–430. Springer, Heidelberg (2011). https://doi.org/10.1007/978-3-642-24550-3_31

21. Sudo, Y., Nakamura, J., Yamauchi, Y., Ooshita, F., Kakugawa, H., Masuzawa, T.: Loosely-stabilizing leader election in a population protocol model. Theor. Comput. Sci. **444**, 100–112 (2012). https://doi.org/10.1016/j.tcs.2012.01.007

22. Talwar, K., Wieder, U.: Balanced allocations: a simple proof for the heavily loaded case. In: Esparza, J., Fraigniaud, P., Husfeldt, T., Koutsoupias, E. (eds.) ICALP 2014. LNCS, vol. 8572, pp. 979–990. Springer, Heidelberg (2014). https://doi.org/10.1007/978-3-662-43948-7_81
23. Tang, B., Gupta, H., Das, S.R.: Benefit-based data caching in ad hoc networks. IEEE Trans. Mob. Comput. **7**(3), 289–304 (2008). https://doi.org/10.1109/TMC.2007.70770

Exploration of Dynamic Ring Networks by a Single Agent with the H-hops and S-time Steps View

Tsuyoshi Gotoh[1][✉], Yuichi Sudo[1], Fukuhito Ooshita[2],
and Toshimitsu Masuzawa[1]

[1] Graduate School of Information Science and Technology, Osaka University,
1-5 Yamadaoka, Suita, Osaka 565-0871, Japan
{t-gotoh,y-sudou,masuzawa}@ist.osaka-u.ac.jp
[2] Nara Institute of Science and Technology,
8916-5 Takayamacho, Ikoma, Nara 630-0101, Japan
f-oosita@is.naist.jp

Abstract. The researches about a mobile entity (called agent) on dynamic networks have attracted a lot of attention in recent years. Exploration which requires an agent to visit all the nodes in the network is one of the most fundamental problems. While the exploration with complete information or with no information about network changes is proposed, despite its practical scenario and applicability, an agent with partial information about the network changes has not been considered yet. In this paper, we consider the exploration of 1-interval connected rings by a single agent with the H-hops and S-time steps view such that the agent can see not all but a part of network changes, i.e., the network changes of links within H-hops for the next S-time steps. In the setting, we show that $H + S \geq n$ and $S \geq \lceil n/2 \rceil$ (n is the size of networks) is necessary and sufficient condition to explore 1-interval connected rings by a single agent. Moreover, we investigate the upper-bounds and the lower-bounds of the exploration time. It is proven that the exploration time is $O(n^2)$ for $S < n - 1$, $O(n^2/H + n \log H)$ for $S \geq n - 1$, and $\Omega(n^2/H)$ for any S.

1 Introduction

More applications of dynamic networks have arisen in recent years, for example, wireless mobile ad hoc, transportation, inter vehicle, or social networks and so on, more important the researches about the dynamic networks have got. In such networks, the topology changes with time due to faults or movements of nodes and the existing method for static networks (which do not change with time)

This work was supported by JSPS KAKENHI Grant Numbers 17K19977, 18K11167, 18K18000 and 19H04085 and JST SICORP Grant Numbers JPMJSC1606 and JPMJSC1806.

M. Ghaffari et al. (Eds.): SSS 2019, LNCS 11914, pp. 165–177, 2019.
https://doi.org/10.1007/978-3-030-34992-9_14

might no longer work. For this reason, the researchers have started to consider several problems on dynamic networks [4].

The exploration which requires a mobile entity called an agent (e.g., a software agent, a robot, or a vehicle) to visit all the nodes of the network is one of the most fundamental problems. The exploration is useful for solving fundamental tasks on the networks such as broadcast or network maintenance. It has been well-studied for static networks [7] and recently been studied for dynamic networks. In the previous works about the exploration of dynamic networks, two extreme cases are considered: an agent has the complete knowledge about changes of all the links for all the future time steps a priori [7,9,10]; or an agent can only see whether the links adjacent to its current node are present or not at the moment [2,3,6,8]. The former one models the situation where the network changes are completely predictable as the public transportation networks in which the network changes are introduced by totally scheduled movements of the nodes. The latter one models the situation where the network changes are caused by unscheduled events, for example, faults or unscheduled movements of the nodes.

Although the above two models are plausible and also theoretically important, the intermediate model, i.e., an agent with partial information or, in other words, capability to know link changes within some distance in the near future should be considered due to the following reason: even in the totally scheduled situation (if exists), computing all the future changes often costs computation time and it is desirable to compute only the necessary information to solve a problem to save computing time or memories; the visibility of an agent to monitor whether there are faults or environmental changes roughly depends on the quality (or costs) of its sensor and it can save some costs to compute only the necessary information needed for a problem. Moreover, such a model is so interesting from a theoretical viewpoint: how the amount of information available for an agent influences the solvability or the time complexity of problems.

To formalize such a concept and analyze its influences, in this paper, we first introduce the *H-hops and S-times view* such that the agent with the view can see the link scheduling (when and which links disappear or appear) of the links within H hops from its location for S time steps from the current time. Then, we consider how the value of H or S influences the solvability or the time complexity of the exploration of 1-interval connected rings by a single agent.

1.1 Related Works

To see various settings and exploration algorithms on static networks, there is a good survey [5].

The following works consider the exploration of dynamic networks. As a randomized approach, Avin et al. [1] or Lamprou et al. [11] use a random walk to explore dynamic networks using only the local degree to decide the destination in each time step. The deterministic exploration of dynamic networks by an agent with the full knowledge of a *link scheduling* (the information about when and which links disappear or appear) is considered in [7] for *carrier networks*,

[10] for *T-interval connected rings* or [9] for *T-interval connected cactuses*. The deterministic exploration of dynamic networks by multiple agents (or robots) without the knowledge of a link scheduling (or only with the ability to detect whether the adjacent links are present or not at the moment) is considered in [6] for 1-interval connected rings. The *perpetual exploration* (i.e., the exploration without termination) on *connected-over-time rings* in the setting is considered in [2,3]. The difference between with or without the ability to detect whether the adjacent links are present or not at the moment (called the *link presence detection*) is considered in [8] for $n \times m$ *dynamic tori* which consist of n horizontal rings and m vertical rings each of which is a 1-interval connected ring. It is shown that the optimal number of agents with the link presence detection to explore the networks is a half of the optimal number of agents without the one to explore.

1.2 Our Contributions

In this paper, we consider the exploration of 1-interval connected rings by a single agent with the H-hops and S-time steps view where the agent can see not all but a part of network changes, i.e., a link scheduling of the links within H-hops from its location for S-time steps from the current time. With our best knowledge, this is the first work to generalize the agent capacity to see a link scheduling.

The results are summarized in Table 1. For the proposed model, we show that $H + S \geq n$ and $S \geq \lceil n/2 \rceil$ (n is the size of networks) is the necessary and sufficient condition to explore 1-interval connected rings by a single agent. The result is proved by showing the impossibility of the exploration for $H + S < n$ or $S < \lceil n/2 \rceil$ and proposing an exploration algorithm for $H + S \geq n$ and $S \geq \lceil n/2 \rceil$. The algorithm also gives an upper-bound of the exploration time, $O(n^2)$. Moreover, we show a lower-bounds of the exploration time is $\Omega(n^2/H)$ for any S. This implies that the upper and lower bounds for $H + S \geq n$ and $\lceil n/2 \rceil \leq S < n - 1$ are tight ($\Theta(n^2)$) when H is constant. Finally, we gives an algorithm for $S \geq n - 1$ by which a single agent can explore a 1-interval connected ring within $O(n^2/H + n \log n)$ time. This also leads to $O(n \log n)$ time when $H = \Theta(n)$ and indicates the upper and lower bounds for $S \geq n - 1$ are tight ($\Theta(n \log n)$) when $H = O(n/\log n)$.

Table 1. Upper and lower bound of the exploration time in 1-interval connected rings.

H and S	Upper bound	Lower bound
$H + S < n$ or $S < \lceil n/2 \rceil$	The exploration is **impossible**	
$H + S \geq n$ and $\lceil n/2 \rceil \leq S < n - 1$	$O(n^2)$	$\Omega(n^2/H)$
$S \geq n - 1$	$O(n^2/H + n \log H)$	

2   Models and Terminologies

We consider *a time variant ring* $\mathcal{R} = (V, E, \rho)$ where $G = (V, E)$ is a ring network, i.e., $V = \{v_0, v_1, \ldots, v_{n-1}\}$ is a set of n nodes, $E = \{e_0, e_1, \ldots, e_{n-1}\}$ is a set of n links such that $e_i = (v_i, v_{i+1 \bmod n})$. The nodes of the network is anonymous. For simplicity, we omit mod n in the following. A function $\rho : E \times \mathbb{N} \leftarrow \{0, 1\}$ is called a *link presence function* such that $\rho(e, t)$ is 1 (resp. 0) if link e is present (resp. missing) at time step (or *step*) $t \in \mathbb{N}$. A network at each step t is denoted as $R_t = (V, E_t)$ where $E_t = \{e_i \in E \mid \rho(e_i, t) = 1\}$. We assume that $\mathcal{R}$ is 1-interval connected, i.e., at each step t, a network R_t is connected. In other words, at each step t, there is at most one missing link $e_i \in E$ such that $e_i \notin E_t$.

We say the ascending (resp. descending) order of node indices is the right (resp. left) direction. Each port of e_i has a globally consistent label at v_i and v_{i+1} which gives an entity on the ring a global direction (the right direction at v_i and the left direction at v_{i+1}) of the ring. Given a connected component $V' \subsetneq V$, the right (resp. left) extremity of V' is the node $v_i \in V'$ such that $v_{i+1} \notin V'$ (resp. $v_{i-1} \notin V'$). If $|V'| = 1$, the unique node in V' is both the right extremity and the left extremity of V'.

In the network, a single agent A is operational. Agent A knows the network size n, has computation capacity and its own memory, and can traverse at most one link in each step. In addition to them, A can get the *view* which contains information of presence of nearby links in near future as defined later. In a step t, A at a node, say v_i, first computes and decides which direction it moves depending on its memory and the view from v_i. If the corresponding link is present at t, A succeeds to move and reaches a neighbor of v_i. Otherwise, A fails to move and stays at v_i.

Informally speaking, the *H-hops and S-time steps view* that agent A can get shows which link is missing within H hops from the current node and within S steps from the current step. Formally speaking, for $\lceil n/2 \rceil \geq H > 0$ and $S > 0$, A gets the H-hops and S-time steps view $\beta_{H,S}(i, s) = \{(e_j, t, \rho(e_j, t)) \mid i - H \leq j \leq i + H - 1, s \leq t \leq s + S - 1\}$ when A exists on v_i at step s. For example, when $H = 2$, $S = 2$, and A exists on v_0 at step 5, A can see $\beta_{2,2}(0, 5) = \{(e_1, 5, 0), (e_0, 5, 1), (e_{n-1}, 5, 1), (e_{n-2}, 5, 1), (e_1, 6, 1), (e_0, 6, 0), (e_{n-1}, 6, 1), (e_{n-2}, 6, 1)\}$. When no confusion arises, we simply write *the view* instead of writing the "H-hops and S-time steps view."

It is assumed that a link scheduling (or $\rho(e_i, t)$ for every e_i and every step $t > 0$) is decided by the adversary. The adversary knows the algorithm of A, has infinite computation capacity, and tries to prevent A from exploring the ring.

In this paper, we consider the exploration problem by a single agent A: A is required to visit all the nodes in the ring. A node is said to be *explored* when it is visited by A for the first time. The set of explored (resp. unexplored) nodes at the start of step t is denoted by V^t (resp. $\overline{V^t}$). In the following, without loss of generality, we assume A starts the exploration from v_0.

3   Impossibility Result

We show an impossibility result in this section. Specifically, we show that the exploration is impossible when $H + S < n$ or $S < \lceil n/2 \rceil$ holds.

Lemma 1. *If $H + S < n$ or $S < \lceil n/2 \rceil$, a deterministic single agent with the H-hops and S-time steps view cannot explore 1-interval connected rings.*

Proof. We first consider the condition $S < \lceil n/2 \rceil$. Note that $H \leq \lceil n/2 \rceil$. It suffices to show that the exploration is impossible when $S = \lceil n/2 \rceil - 1$. We assume for contradiction, that there is an algorithm by which A can explore any ring under any link scheduling when $S = \lceil n/2 \rceil - 1$. Since A can explore the ring, A starting from v_0 eventually reaches v_{n-1} (no matter whether the exploration is completed or not).

The adversary can decide a link scheduling so that e_{n-1} (resp. e_{n-2}) is missing when A exists on v_0 (resp. v_{n-2}). The adversary first keeps showing a link scheduling where e_{n-1} is kept deleted for S steps from the current step until A moves to $v_{n-\lceil n/2 \rceil}$. If A does not move to $v_{n-\lceil n/2 \rceil}$ and stays v_i for $0 \leq i < n - \lceil n/2 \rceil$, e_{n-1} is kept deleted and A cannot reach v_{n-1} (A must pass through e_{n-1} or $e_{n-\lceil n/2 \rceil - 1}$ to reach v_{n-1} from v_0), which is a contradiction. Thus, A eventually moves to $v_{n-\lceil n/2 \rceil}$ at some step, say t. Then, the adversary deletes e_{n-2} from the $(t + S)$-th step (the $(t + \lceil n/2 \rceil - 1)$-th step) until A moves to $v_{n-\lceil n/2 \rceil - 1}$. By the scheduling, since A reaches v_{n-2} at earliest at the end of the $(t + n - 2 - (n - \lceil n/2 \rceil))$-th step (or at the start of the $(t + \lceil n/2 \rceil - 1)$-th step) from $v_{n-\lceil n/2 \rceil}$, e_{n-2} starts to disappear when (or before) A reaches v_{n-2} and keeps disappearing unless A moves to $v_{n-\lceil n/2 \rceil - 1}$. Thus, if A does not move to $v_{n-\lceil n/2 \rceil - 1}$, A cannot reaches v_{n-1}. This is a contradiction. This means that A moves to $v_{n-\lceil n/2 \rceil - 1}$ after the $(t + 1)$-th round. However, by the similar way, the adversary can prevent A from reaching v_{n-1}. This is a contradiction. Hence, when $S < \lceil n/2 \rceil$, a single agent cannot explore 1-interval connected rings.

Secondly, we consider the condition $H + S < n$ and $S \geq \lceil n/2 \rceil$. It is sufficient to show that A cannot explore the ring when $S = n - H - 1$ for $1 \leq H \leq \lfloor n/2 \rfloor - 1$ since $H < \lfloor n/2 \rfloor$ from the conditions. Again, we assume for contradiction, that there is an algorithm by which A can explore any ring under any link scheduling. Since A can explore the ring, A starting from v_0 eventually reaches v_{n-1} (no matter whether the exploration is completed or not).

The adversary first keeps showing a link scheduling where e_{n-1} is kept deleted for S steps from the current step until A moves to v_H. If A does not move to v_H and stays at v_i for $0 \leq i \leq H - 1$, e_{n-1} is kept deleted and A cannot reach v_{n-1}, which is a contradiction. Thus, A eventually moves to v_H at some step, say t. Then, depending on whether A visits v_{H-1} before v_{n-H-1} after t or not, the missing link is decided (see Fig. 1). Note that A can see neither e_{n-1} nor e_{n-2} in its view when existing at v_i for $H \leq i \leq n - H - 2$ and such nodes v_i ($H \leq i \leq n - H - 2$) always exist since $H \leq \lfloor n/2 \rfloor - 1$.

If A visits v_{H-1} before v_{n-H-1}, the adversary keeps deleting e_{n-1}. By the link scheduling, unless A decides to reach v_{n-H-1} from v_H, e_{n-1} is kept deleted

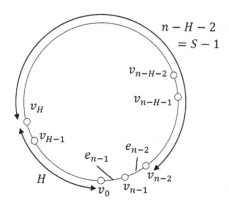

Fig. 1. Illustrating the proof of Theorem 1.

and A cannot reach v_{n-1}, which is a contradiction. This means that A eventually reaches v_{n-H-1}. Let t' be the last step before A reaches v_{n-H-1} such that A exists at v_{H-1} at the start of t'.

When A leaves v_{H-1} at the t'-th step, the adversary makes a scheduling so that e_{n-2} starts and keeps disappearing from the $(t'+n-H-1)$-th step until A comes back to v_{n-H-2}. This does not conflict with the link scheduling in the past view of A since at the t'-th step, e_{n-1} is scheduled to be deleted for the next $S = n - H - 1$ steps and for the next $n - H - 1 - x$ steps at the $(t'+x)$-th step. Since it takes at least $n - H - 2$ steps to reach v_{n-2} from v_H, A reaches v_{n-2} at earliest at the end of the $(t'+n-H-2)$-th step. However, at the $(t'+n-H-1)$-th step, e_{n-2} is missing and the adversary keeps deleting e_{n-2} until A moves to v_{n-H-2}. Then, A cannot reach v_{n-1} unless moving to v_{n-H-2}. However, by the similar way, the adversary can prevent A from reaching v_{n-1}. This is a contradiction. Hence, when $H + S < n$ or $S < \lceil n/2 \rceil$, a single agent cannot explore 1-interval connected rings.   □

4   Solvability Result and Upper Bound of Exploration Time

In this section, we prove the exploration is possible when $H + S \geq n$ and $S \geq \lceil n/2 \rceil$ by proposing an exploration algorithm by a single agent. The algorithm also gives an upper bound of the exploration time, $O(n^2)$. Note that $S \geq H$ since $S \geq \lceil n/2 \rceil$ and $H \leq \lceil n/2 \rceil$. Moreover, to simplify the time complexity analysis, when $H + S > n$ or $S > n - 1$, A uses the H'-hops and S'-time steps view where $S' = \min(S, n - 1)$ and $H' + S' = n$ (clearly, $\beta_{H',S'}(i,s) \subseteq \beta_{H,S}(i,s)$ for every v_i and every step s). Thus, we assume $H + S = n$ in this section.

The algorithm is described in Algorithm 1. The algorithm consists of $n - 1$ phases. In each phase i ($0 \leq i \leq n - 2$), A starts from v_i and ends at v_{i+1}. Let t_i be the first step of phase i. Agent A starting from v_i moves to v_{i+1} through e_i if

A sees e_i appear in its view (see Fig. 2). Otherwise, A moves one hop in the left direction and sees whether e_i appears in its view or not. This movement in the left direction continues as long as e_i does not appear in its view until A reaches v_{i-H} where e_i is no longer included in the view of A. If A reaches v_{i-H} at the $(t_i + H - 1)$-th step, A keeps moving the left direction from v_{i-H} until reaching v_{i+1} and the exploration finishes at the time (see Fig. 3). If A sees e_i appear in its view before reaching v_{i-H}, A moves in the right direction and reaches v_{i+1} through e_i.

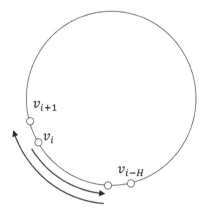

Fig. 2. The move in the right direction of Algorithm 1.

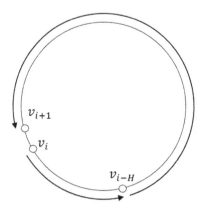

Fig. 3. The move in the left direction of Algorithm 1.

Theorem 1. *For $H + S \geq n$ and $S \geq \lceil n/2 \rceil$, the exploration time of 1-interval connected rings by a single agent with the H-hops and S-time steps view is upper-bounded by $O(n^2)$.*

Algorithm 1. Exploration algorithm for $S + P \geq n$

1: $n_{exp} \rightarrow 1$ //v_0 has been explored before the first step.
2: **while** $(n_{exp} < n)$ **do**
3:    $d \rightarrow 0$
4:    let v_i be the current node
5:    $tmp = i$
6:    //trying to reach v_{i+1} from v_i in the following loop.
7:    **while** $(d < H)$ **do**
8:       **if** $((v_{tmp}, v_{tmp+1})$ is always missing in the next S steps) **then**
9:          move one hop in the left direction.
10:          $d \rightarrow d + 1$
11:       **else**
12:          move d hops in the right direction //returning v_i
13:          wait for e_{tmp} to appear and pass through e_{tmp} as soon as it appears
14:          exit from the inner while loop (at lines 7-14)
         //A does not reach v_{i+1} in the above loop
15:    **if** $(d \geq H)$ **then**
16:       move $n - 1 - H$ hops in the left direction //reaching v_{i+1} in the left direction
17:       exit from the outer while loop (at lines 2-18) //the exploration ends
18:    $n_{exp} \rightarrow n_{exp} + 1$

Proof. It suffices to show that A with the H-hops and S-time steps view completes exploration within $O(n^2)$ steps by Algorithm 1. We show that, in each phase i, A starting from v_i can reach v_{i+1} within $H + S - 1$ $(= n - 1)$ steps by Algorithm 1.

We first show the claim for the case e_i appears in the view of A at v_j for $i \geq j > i - H$. For the first step of phase i, A can clearly reach v_{i+1} within S steps from the current step if e_i appears in its view. When A sees e_i appear in its view at step t at v_j for $i > j > i - H$, e_i must appear at the $(t+S-1)$-th step and be missing at step t' for $t \leq t' \leq t + S - 2$ by the construction. This means that all the other links than e_i are present at step t' $(t \leq t' \leq t+S-2)$, and thus A can move for $S - 1$ steps from v_j in the right direction without interference by missing links. Since $i - j < H$ and $H \leq S$, A always reaches v_i by the start of the $(t + S - 1)$-th step. Then, A reaches v_{i+1} as soon as e_i appears. Since A moves at most $H - 1$ hops in the left direction and e_i appears S steps after A starts to move in the right direction, it takes $H + S - 1$ steps to reach v_{i+1} from v_i through e_i.

When A reaches v_{i-H} at step t', e_i must be deleted for at least $S - 1$ steps from t' and all the other links than e_i are present in the $S - 1$ steps and thus, A can move for $S - 1$ steps from v_{i-H} in the left direction without interference by missing links. Since $H + S - 1 = n - 1$, A reaches $v_{i-(n-1)} = v_{i+1}$ after $S - 1$ steps and the exploration is completed.

The exploration time of Algorithm 1 is $O(n^2)$ since each phase consists of at most $H + S - 1 = n - 1$ steps and the number of phases is at most $n - 1$. $\qquad \square$

From Lemma 1 and Theorem 1, the following theorem holds.

Theorem 2. *If and only if $H + S \geq n$ and $S \geq \lceil n/2 \rceil$, a single agent with the H-hops and S-time steps view can explore of 1-interval connected rings within finite time steps.*

5 Lower Bound of Exploration Time

The lower bound of the exploration time for any S is considered in this section. The following theorem holds.

Theorem 3. *The exploration time of 1-interval connected rings by a single agent with the H-hops and S-time steps view is lower-bounded by $\Omega(n^2/H)$.*

Proof. The adversary separates the whole steps into phases, which are defined as follows: the first phase starts from step 1 and ends at the t-th step such that $|V^t| = H$ and $|V^{t+1}| = H + 1$. Letting t_p be the first step of phase p, phase p for $2 \leq p \leq \lfloor (n-1)/H \rfloor - 1$ directly follows the previous phase and ends at the $(t_{p+1} - 1)$-th step such that $|V^{t_{p+1}} - V^{t_p}| = H$. In other words, A newly explores H nodes in each phase. Clearly, $|V^{t_p}|$ is $(p-1)H + 1$. Note that $(\lfloor (n-1)/H \rfloor - 1)H + 1 < n$ and some nodes remain unexplored at the $(\lfloor (n-1)/H \rfloor)$-th phase. We will handle the nodes later in this proof. Let v_{i_p} be the node where A exists at the start of phase p.

The adversary decides a link scheduling so that each phase includes at least pH steps. For this goal, the adversary deletes $e_{i_p + H - 1}$ (resp. $e_{i_p - H}$) for $|V^{t_p}| + H - 2 = pH - 1$ steps from t_p including t_p if v_{i_p} is the right (resp. left) extremity of V^{t_p}. Without loss of generality, we assume v_{i_p} is the right extremity of V^{t_p}.

Under the link scheduling, A explores H nodes from v_{i_p}. Let n_r be the number of the explored nodes on the right of V^{t_p} and n_l be the number of the explored nodes on the left of V^{t_p} in phase p. We show that, in phase p, it takes at least pH steps to explore the H nodes for A by contradiction. We assume that A explores the H nodes in phase p within less than pH steps. If A traverses $e_{i_p + H - 1}$, it takes at least pH steps ($pH - 1$ steps by appearance of $e_{i_p + H - 1}$ and one step to pass the link). This is a contradiction. Then, $n_r < H$ and thus $n_l > 0$. Since A does not traverse $e_{i_p + H - 1}$ and $n_l > 0$, it takes at least $|V^{t_p}| - 1 + H = pH$ steps (not $pH + 1$ steps when A always moves in the left direction) to explore H nodes. This is a also contradiction. Hence, each phase p includes at least pH steps. Thus, the exploration takes at least $\Sigma_{p=1}^{\lfloor (n-1)/H \rfloor - 1} pH = \Omega(n^2/H)$.   □

In total, the exploration takes at least $\Sigma_{p=1}^{\lfloor (n-1)/H \rfloor - 1} pH + n - 1 + \lfloor ((n-1) \bmod H)/2 \rfloor = \Omega((n-H)^2/H + n) = \Omega(n^2/H)$.   □

6 Upper Bound of Exploration Time for $S \geq N - 1$

In this section, we show the upper bound of exploration for $S \geq n - 1$ by proposing an exploration algorithm described in Algorithm 2.

The exploration algorithm is separated into phases. There are $\lfloor (n-1)/H \rfloor - 1 + \lceil \log(H + 1 + ((n-1) \bmod H)) \rceil$ phases. We call the first $\lceil (n-1)/H \rceil - 1$ phases *poly phases* and the following $\lceil \log(H + 1 + ((n-1) \bmod H)) \rceil$ phases *log phases*. Let t_p be the first step of phase p and v_{i_p} be the node where A exists at t_p. In each poly phase, A explores H nodes, i.e., $|V^{t_p}| = (p-1)H + 1$. In each log phase p, a half of the nodes in $\overline{V^{t_p}}$ are explored. Without loss of generality, we assume v_{i_p} be the right extremity of V^{t_p} in the following.

In each poly phase p, the agent first sees its view from v_{i_p}. If A can reach v_{i_p+H} by the $(t_{p+1} - 1)$-th step by moving in the right direction where $t_{p+1} - 1 = t_p + 2H + |V^{t_p}| - 3 = t_p + (p+1)H - 2$ is the last step of phase p, A moves in the right direction during the p-th phase. Otherwise, A moves in the left direction during the p-th phase and reaches $v_{i_p-|V^{t_p}|+1-H}$ by $(t_{p+1} - 1)$-th step (see Fig. 4).

A log phase includes $n-1$ steps. Also in each log phase p, the agent first sees its view from v_{i_p}. If A can explore at least $\lceil |\overline{V^{t_p}}|/2 \rceil$ within $n-1$ steps by moving in the right direction, A moves in the right direction during the p-th phase. Otherwise, A moves in the left direction during the p-th phase and explores at least $\lceil |\overline{V^{t_p}}|/2 \rceil$ (see Fig. 5).

The following theorem holds.

Theorem 4. *For $S \geq n-1$, the exploration time of 1-interval connected rings by a single agent with the H-hops and S-time steps is upper-bounded by $O(n^2/H + n \log H)$.*

Proof. It suffices to show that A completes exploration within $O(n^2/H + n \log H)$ steps by Algorithm 2 when $S \geq n-1$.

We first consider poly phases and show that A can explore H nodes in each poly phase, which leads to that the total exploration time of poly phases is $O(n^2/H)$. By the construction, each poly phase p lasts for $(p+1)H - 1$ steps. Since $p \leq \lfloor (n-1)/H \rfloor - 1$, $(p+1)H - 1$ is always less than $n-1$ and S during the poly phases. If A can reach v_{i_p+H} by the $(t_p + (p+1)H - 2)$-th step, A moves in the right direction during the p-th phase and explores H nodes within $t_p + (p+1)H - 2$ steps. Otherwise, A can move at least $pH \; (= |V^{t_p}| - 1 + H)$ steps in the left direction and reaches $v_{i-|V^{t_p}|+1-H}$ since moving in the right direction succeeds at most $H - 1$ steps and thus fails at least $(p+1)H - 1 - (H-1) = pH$ steps. This means that the total exploration time of the poly phases is $\Sigma_{p=1}^{\lfloor (n-1)/H \rfloor - 1}((p+1)H - 1) = O((n-H)^2/H) = O(n^2/H)$.

Then, we consider log phases and show that A can explore at least $\lceil |\overline{V^{t_p}}|/2 \rceil$ nodes, which leads to that the total exploration time of poly phases is $O(n \log H)$. By the construction, each log phase lasts for $n-1$ steps. If A can explore $\lceil |\overline{V^{t_p}}|/2 \rceil$ by moving in the right direction, A moves in the right direction and the claim holds. Otherwise, letting v_{mid_p} the middle node of $\overline{V^{t_p}}$ (when there are two middle nodes, v_{mid_p} is the one closer to v_{i_p} in the right direction), A can move less than $mid_p - i_p$ steps in the right direction. This means A can move at least $n - 1 - (mid_p - i_p) + 1$ in the left direction during the p-th phase. Combining this with $((mid_p - i_p) \bmod n) + ((i_p - mid_p) \bmod n) = n$, we have $n - 1 - (mid_p -$

$i_p) + 1 = i_p - mid_p$. In the expression, $i_p - mid_p$ corresponds to the distance from v_{i_p} to v_{mid_p} in the left direction. Thus, A reaches v_{mid_p} in the p-th phase. Since each log phase includes $n - 1$ steps, the number of the remaining nodes at the first log phase is at most $2H - 1$ and at least a half of the remaining nodes is explored in each log phase, the total exploration time of the log phases is $(n - 1)\lceil \log(2H - 1 + 1) \rceil = O(n \log H)$.

By combining the above steps, we obtain the total exploration steps as $O(n^2 + n \log H)$. $\qquad \square$

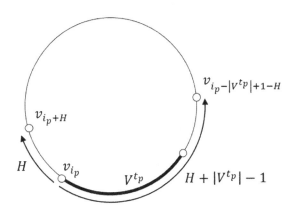

Fig. 4. The move in a poly phase of Algorithm 2.

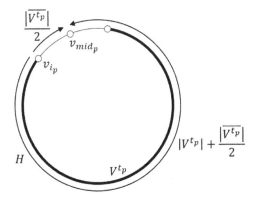

Fig. 5. The move in a log phase of Algorithm 2.

Algorithm 2. Exploration algorithm for $S + P \geq n$

```
1:  p → 1 //starting poly phases
2:  while (p < ⌊(n − 1)/H⌋ − 1) do
3:      let t be the current step and vᵢ be the current node
4:      if vᵢ is the right extremity of Vᵗ then
5:          if A can reach v_{i+H} in the right direction by the (t+(p+1)H−1)-th step then
6:              move H hops in the right direction
7:          else
8:              move pH hops in the left direction
9:      else
10:         if A can reach v_{i−H} in the left direction by the (t+(p+1)H−1)-th step then
11:             move H hops in the left direction
12:         else
13:             move pH hops in the right direction
14:     p → p + 1
15: p → 1 //starting log phases
16: while (p ≤ ⌈log(H + ((n − 1) mod H) + 1)⌉) do
17:     let t be the current step and vᵢ be the current node
18:     if vᵢ is the right extremity of Vᵗ then
19:         if A can reach v_{i+⌈|V̄ᵗ|/2⌉} in the right direction by the (t+n−2)-th step then
20:             move in the right direction until the (t+n−2)-th step
21:         else
22:             move in the left direction until the (t+n−2)-th step
23:     else
24:         if A can reach v_{i−⌈|V̄ᵗ|/2⌉} in the left direction by the (t+n−2)-th step then
25:             move in the right direction until the (t+n−2)-th step
26:         else
27:             move in the left direction until the (t+n−2)-th step
28:     p → p + 1
```

7 Conclusions and Future Works

In this paper, we introduced the H-hops and S-time steps view which can be used to model some situations where an agent (or robot) can partly see their nearby environment or can predict the near-future changes of the environment. This is the first paper considering such a model to the best of our knowledge. Then, for a single agent with the H-hops and S-time steps view, we studied the exploration of 1-interval connected rings. We give some fundamental results, i.e., impossibility of the exploration for $H + S < n$ or $S < \lceil n/2 \rceil$, solvability of the exploration for $H + S \geq n$ and $S \geq \lceil n/2 \rceil$, and the upper bound and the lower bound of the exploration time for some cases. Although these results themselves are important, there remain a few works to do in the future, for example, the efficient exploration in the case $H + S \geq n$ and $S < n - 1$, the tight bound for the exploration time for all the cases. We will work for these problems.

It is also interesting to consider the exploration of other topologies with the H-hops and S-time steps view. Although in dense networks the view of a few diameter gives an agent almost global information (thus, it may be improper for modeling an ability of the agent to get a partial information of a dynamic network), our model can be applied to sparse networks.

References

1. Avin, C., Koucký, M., Lotker, Z.: How to explore a fast-changing world (cover time of a simple random walk on evolving graphs). In: Aceto, L., Damgård, I., Goldberg, L.A., Halldórsson, M.M., Ingólfsdóttir, A., Walukiewicz, I. (eds.) ICALP 2008. LNCS, vol. 5125, pp. 121–132. Springer, Heidelberg (2008). https://doi.org/10.1007/978-3-540-70575-8_11
2. Bournat, M., Datta, A.K., Dubois, S.: Self-stabilizing robots in highly dynamic environments. In: Bonakdarpour, B., Petit, F. (eds.) SSS 2016. LNCS, vol. 10083, pp. 54–69. Springer, Cham (2016). https://doi.org/10.1007/978-3-319-49259-9_5
3. Bournat, M., Dubois, S., Petit, F.: Computability of perpetual exploration in highly dynamic rings. In: IEEE 37th International Conference on Distributed Computing Systems, pp. 794–804. IEEE (2017)
4. Casteigts, A., Flocchini, P., Quattrociocchi, W., Santoro, N.: Time-varying graphs and dynamic networks. Int. J. Parallel Emergent Distrib. Syst. 27(5), 387–408 (2012)
5. Das, S.: Mobile agents in distributed computing: network exploration. Bull. EATCS 1(109), 54–69 (2013)
6. Di Luna, G.A., Dobrev, S., Flocchini, P., Santoro, N.: Distributed exploration of dynamic rings. Distrib. Comput. 1–27 (2018)
7. Flocchini, P., Mans, B., Santoro, N.: On the exploration of time-varying networks. Theoret. Comput. Sci. 469, 53–68 (2013)
8. Gotoh, T., Sudo, Y., Ooshita, F., Kakugawa, H., Masuzawa, T.: Group exploration of dynamic Tori. In: IEEE 38th International Conference on Distributed Computing Systems, pp. 775–785. IEEE (2018)
9. Ilcinkas, D., Klasing, R., Wade, A.M.: Exploration of constantly connected dynamic graphs based on cactuses. In: Halldórsson, M.M. (ed.) SIROCCO 2014. LNCS, vol. 8576, pp. 250–262. Springer, Cham (2014). https://doi.org/10.1007/978-3-319-09620-9_20
10. Ilcinkas, D., Wade, A.M.: Exploration of the t-interval-connected dynamic graphs: the case of the ring. Theory Comput. Syst. 62(5), 1144–1160 (2018)
11. Lamprou, I., Martin, R., Spirakis, P.: Cover time in edge-uniform stochastically-evolving graphs. Algorithms 11(10), 149 (2018)

IPERFTZ: Understanding Network Bottlenecks for TrustZone-Based Trusted Applications

Christian Göttel, Pascal Felber, and Valerio Schiavoni(✉)

University of Neuchâtel, Rue Emile-Argand 11, 2000 Neuchâtel, Switzerland
{christian.gottel,pascal.felber,valerio.schiavoni}@unine.ch

Abstract. The growing availability of hardware-based trusted execution environments (TEEs) in commodity processors has recently advanced support (*i.e.,* design, implementation and deployment frameworks) for network-based secure services. Examples of such TEEs include ARM TRUSTZONE or Intel SGX, largely available in embedded, mobile and server-grade processors. TEEs shield services from compromised hosts, malicious users or powerful attackers. TEE-enabled devices are largely being deployed on the edge of the network, paving the way for large-scale deployments of trusted applications. These applications allow processing and disseminating sensitive data without having to trust cloud providers. However, uncovering network performance limitations of such trusted applications is difficult and currently lacking, despite the interest and reliance by developers and system deployers.

IPERFTZ is an open-source tool to uncover network performance bottlenecks rooted at the design and implementation of trusted applications for ARM TRUSTZONE and underlying runtime systems. Our evaluation based on micro-benchmarks shows current trade-offs for trusted applications, both from a network as well as an energy perspective; an often overlooked yet relevant aspect for edge-based deployments.

Keywords: Network · Performance · Bottleneck · Measurement · ARM TrustZone · OP-TEE

1 Introduction

Services are being moved from the cloud to the edge of the network. This migration is due to several reasons: lack of trust in the cloud provider [7], energy savings [19,24] or reclaiming control over data and code. Edge devices are used to accumulate, process and stream data [20,30]. The nature of such data can be very sensitive: edge devices can be used to process health-based data emitted by body sensors (*e.g.,* cardiac data [26]), data originated by smart home sensors indicating the presence of humans inside a household, or even financial transactions [16,28]. In this context, applications using this information must be protected against powerful attackers, potentially even with physical access to

© Springer Nature Switzerland AG 2019
M. Ghaffari et al. (Eds.): SSS 2019, LNCS 11914, pp. 178–193, 2019.
https://doi.org/10.1007/978-3-030-34992-9_15

the devices. Additionally, communication channels for inter-edge device applications must also be secured to prevent attacks such as man-in-the-middle attacks. Edge devices are low-energy units with limited processing and storage capacity. As such, it is unpractical to rely on sophisticated software-based protection mechanisms (*e.g.,* homomorphic encryption [22]), currently due to their high processing requirements and low performance [12]. Alternatively, new hardware-based protection mechanisms can be easily leveraged by programmers to provide prior protection guarantees. Specifically, *trusted execution environments* (TEEs) are increasingly made available by hardware vendors in edge-devices [29]. Several ARM-based devices, such as the popular Raspberry Pi[1], embed native support for TEEs called TRUSTZONE [4,23]. TRUSTZONE can be leveraged to deploy *trusted applications* (TAs) with additional security guarantees.

There exist several programming frameworks and runtime systems to develop TAs for TRUSTZONE with varying capabilities and different degrees of stability and support (*e.g.,* SierraTEE[2], OP-TEE[3], and [21]). While a few studies look at the interaction between TEEs and the corresponding untrusted execution environments [2,14], little is known on the network performance bottlenecks experienced by TAs on ARM processors. We fill this gap by contributing IPERFTZ, a tool to measure accurately the network performance (*e.g.,* latency, throughput) of TAs for TRUSTZONE. IPERFTZ consists of three components, namely *(1)* a client application, *(2)* a TA, and *(3)* a server. Our tool can be used to guide the calibration of TAs for demanding workloads, for instance understanding the exchanges with untrusted applications or for secure inter-TEE applications [28]. In addition, IPERFTZ can be used to study the impact of network and memory performance on the energy consumption of running TAs. By adjusting IPERFTZ's parameters, users evaluate the network throughput of their TAs and can quickly uncover potential bottlenecks early in the development cycle. For instance, internal buffer sizes affect the achievable network throughput rates by a factor of 1.8×, almost halving throughput rates.

The rest of the paper is organized as follows. Section 2 motivates the need for tools analyzing TAs. We provide an in-depth background on TRUSTZONE in Sect. 3, as well as covering details on the TRUSTZONE runtime system OP-TEE. In Sect. 4 we present the architecture of IPERFTZ and some implementation details in Sect. 5. We report our evaluation results in Sect. 6. We cover related work in Sect. 7 before concluding in Sect. 8.

2   Motivating Scenario

We consider scenarios with simple yet practical services deployed as TAs. For instance, in [13] authors deploy key-value stores inside a TRUSTZONE runtime system. Benchmarks show a 12×–17× slowdown when compared to plain (yet

---

[1]   https://www.raspberrypi.org, accessed on 30.07.2019.

[2]   https://www.sierraware.com/open-source-ARM-TrustZone.html,   accessed   on
      30.07.2019.

[3]   https://www.op-tee.org, accessed on 30.07.2019.

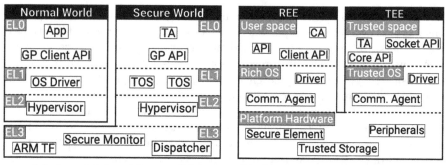

Fig. 1. Block diagrams highlighting relevant software components

unsecure) deployments, due to shared memory mechanisms between the trusted and untrusted environments. As further detailed in Sect. 4, networking in Op-Tee is supported by similar shared memory mechanisms. Yet, we observe the lack of tools to clearly highlight the root causes of such bottlenecks. Further, in the TRUSTZONE ecosystem, there is a lack of proper tools to evaluate network bottlenecks contrary to untrusted environments (*e.g.*, iperf3[4], netperf[5], nuttcp[6]). The overhead originating from the shared memory mechanism can be identified by comparing the measured network throughput inside and outside the TEE. Measuring such overheads is of particular relevance in embedded, mobile and IoT environments. In those scenarios, devices are often battery powered, limited both in time and capacity. Hence, network performance tools should further highlight energy costs, pointing users to specific bottlenecks.

3 Background

This section provides a background on ARM TRUSTZONE (Sect. 3.1), the GLOB-ALPLATFORM specifications (Sect. 3.2) and OP-TEE, the TRUSTZONE runtime system used for IPERFTZ (Sect. 3.3). This background helps understanding technical challenges in our context and how IPERFTZ addresses them.

3.1 ARM TrustZone in a Nutshell

TRUSTZONE is a security architecture designed for ARM processors and was introduced in 2003 [3]. It partitions hardware and software resources into two worlds, *i.e., secure* and *normal* world, as shown in Fig. 1a. A dedicated NS bit [4] drives this world separation and allows to execute secure (NS bit set low) or

---

[4] https://software.es.net/iperf/, accessed on 30.07.2019.
[5] https://hewlettpackard.github.io/netperf/, accessed on 30.07.2019.
[6] https://www.nuttcp.net/Welcome%20Page.html, accessed on 30.07.2019.

non-secure (NS bit set high) transactions on the system bus. In general, non-secure transactions cannot access system resource secured by a low NS bit. The TRUSTZONE architecture spans beyond the system bus, including peripherals (*e.g.,* GPUs [31] and I/O). Every TRUSTZONE-enabled processor is logically split into a secure and a non-secure (virtual) core, executing in a time-shared manner. Hence, accessible system resources are determined by the executing core: secure cores can access all system resources, while non-secure cores can only access non-secure ones. ARM processors embed one *memory management unit* (MMU) per virtual core in charge of mapping virtual addresses to physical addresses. The *translation lookaside buffer* (TLB) in the MMU is used to maintain the mapping translations from virtual to physical memory addresses. Tagging TLB entries with the identity of the world allows secure and non-secure address translation entries to co-exist. With tags the TLB no longer has to be flushed making fast world switches possible.

The implementation of TRUSTZONE is organized into four *exception levels* (EL) with increasing privileges [5] (Fig. 1a). EL0, the lowest one, executes unprivileged software. EL1 executes operating systems, while EL2 provides support for virtualization. Finally, ARM Trusted Firmware is running at EL3 dispatching boot stages at boot time and monitoring secure states. Switches between the two worlds are supervised by a secure monitor [6]. It is invoked in two ways: *(1)* by executing a *secure monitor call* (SMC), or *(2)* by a subset of *hardware exception mechanisms* [4]. When invoked, the secure monitor saves the state of the currently executing world, before restoring the state of the world being switched to. After dealing with the worlds' state, the secure monitor returns from exception to the restored world.

3.2   The GlobalPlatform Standard

GLOBALPLATFORM[7] publishes specifications for several TEEs (*e.g.,* OP-TEE and [21]). We provide more details on OP-TEE in Sect. 3.3 (an implementation of such specifications), while briefly explaining the terminology in the remainder to understand Fig. 1b. An *execution environment* (EE) provides all components to execute applications, including hardware and software components. A *rich execution environment* (REE) runs a rich OS, generally designed for performance. However, it lacks access to any secure component. In contrast, TEEs are designed for security, but programmers have to rely on a reduced set of features. A trusted OS manages the TEE under constrained memory and storage bounds. TEE and REE run alongside each other. In recent ARM releases (since v8.4), multiple TEEs can execute in parallel [3], each with their own trusted OS. TAs rely on system calls usually implemented by the trusted OS as specific APIs [10]. *Client applications* (CA) running in the rich OS can communicate with TAs using the *TEE Client API*. Similarly, TAs can access resources such as *secure elements* (*i.e.,* tamper-resistant devices), *trusted storage*, and *peripherals*, or send messages outside the TEE. *Communication agents* in the TEE and REE

---

[7] https://globalplatform.org, accessed on 30.07.2019.

mediate exchanges between TAs and CAs. Finally, the *TEE Socket API* can be used by TAs to setup network connections with remote CAs and TAs.

3.3   OP-TEE: Open Portable Trusted Execution Environment

OP-TEE is an open-source implementation of several GLOBALPLATFORM specifications [8–11] with native support for TRUSTZONE. The OP-TEE OS manages the TEE resources, while any Linux-based distribution can be used as rich OS alongside it. OP-TEE supports two types of TAs: *(1)* regular TAs [11] running at EL0, and *(2)* *pseudo TAs* (PTAs), statically linked against the OP-TEE OS kernel. PTAs run at EL1 as secure privileged-level services inside OP-TEE OS's kernel. Finally, OP-TEE provides a set of client libraries to interact with TAs and to access secure system resources from within the TEE.

4   Networking for Trusted Applications

For networked TAs, *i.e.,* generating or receiving network traffic respectively from and to TAs, runtime systems must provide support for sockets and corresponding APIs. To do so, either *(1)* the TEE borrows the network stack from the REE, or *(2)* the TEE relies on *trusted device drivers*. The former solution implies leveraging *remote procedure calls* (RPC) to a `tee-supplicant` (an agent which responds to requests from the TEE), and achieves a much smaller *trusted computing base*. The latter allows for direct access to the network device drivers for much lower network latencies. Furthermore, it simplifies confidential data handling as the data does not have to leave the TEE. The former requires developers to provide data confidentiality before network packets leave the TEE, for instance by relying on encryption.

IPERFTZ leverages `libutee`[8] and its socket API, supporting streams or datagrams. The socket interface exposes common functions: `open`, `send`, `recv`, `close`, `ioctl` and `error`. The GLOBALPLATFORM specification allows TEE implementations to extend protocol-specific functionalities via command codes and `ioctl` functions. For example, it is possible to adjust the receiving and sending socket buffer sizes with TCP socket or changing the address and port with UDP sockets.

The `libutee` library manages the lifecycle of sockets via a TA session to the socket's PTA. The socket PTA handles the RPC to the `tee-supplicant`, in particular allocating the RPC parameters and assigning their values. Afterwards, a SMC instruction is executed to switch back to the normal world. The `tee-supplicant` constantly checks for new service requests from the TEE. Once a new request arrives, its arguments are read by the `tee-supplicant` and the specified command is executed. Finally, when the data is received by the `tee-supplicant`, it is relayed over POSIX sockets to the rich OS. In essence, when data is sent or received over a socket, it traverses all exception levels, both secure (from EL0 up to EL3) and non-secure (from EL2 to EL0 and back up).

---

[8] https://optee.readthedocs.io/architecture/libraries.html#libutee,   accessed   on 30.07.2019.

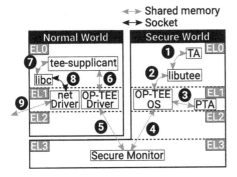

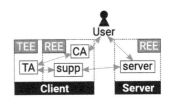

Fig. 2. Execution flow inside OP-TEE.

Fig. 3. Interaction of IPERFTZ's components in the client-server model.

Figure 2 summarizes the previous paragraphs and shows the interaction between the secure and normal worlds in OP-TEE. The secure world hosts the TA, which interacts directly with `libutee` (Fig. 2-❶). When using GLOB-ALPLATFORM's Socket API, `libutee` does a system call (Fig. 2-❷) to OP-TEE. OP-TEE then delegates the request to the socket PTA (Fig. 2-❸). The secure monitor is invoked through a SMC (Fig. 2-❹), which maps the data from the TEE to the REE's address space. From there execution switches into the normal world and the OP-TEE driver (Fig. 2-❺) resumes operation. Requests are then handled by the `tee-supplicant` (Fig. 2-❻) over `ioctl` system calls. The agent executes system calls using `libc` (Fig. 2-❼) to directly relate the underlying network driver (Fig. 2-❽) over the POSIX interface. Once data reaches the network driver, it can be sent over the wire (Fig. 2-❾).

4.1 Threat Model

For our threat model we consider a malicious user that has physical access or is able to obtain remote access on the devices used to deploy IPERFTZ as depicted in Fig. 3. By gaining access to the network or devices connected to it, the malicious user can break security by either compromising these devices or exploiting IPERFTZ for *denial-of-service* (DoS) attacks. We assume that the REE, which includes the rich OS and the user space, cannot be trusted. However, we consider that the devices and the TEE, which includes dispatcher, OP-TEE, and secure monitor, can be trusted. As also stated in [4], side-channel attacks are out of scope of our threat model. We also point out that some ARM *systems on a chip* (SoCs) are affected by the Meltdown [18] and Spectre [15] attacks[9].

For use of IPERFTZ in production, we recommend hardcoding network test parameters in the TA and disabling any argument passing to reduce the potential of DoS attacks. Furthermore, the signing key used for TAs should be kept

---

[9] https://developer.arm.com/support/arm-security-updates/speculative-processor-vulnerability, accessed on 30.07.2019.

confidential as it allows the malicious user to modify TA binaries and create authentic TA binaries. Assuming the TRUSTZONE-enabled device is equipped with an *embedded MultiMediaCard* (eMMC), then TAs can be securely stored on the eMMC and the malicious user cannot tamper with a TA's binary. In development use, manipulation of the CA's parameters by the malicious user to exploit a buffer overflow can be excluded. During a network bandwidth measurement, the malicious user can run a (distributed) DoS attack to reduce the network bandwidth, such that a lower network throughput is measured and reported by IPERFTZ. At the time of writing, OP-TEE does not provide support for the TLS protocol which renders secure connections unusable. Although irrelevant to IPERFTZ but applicable in general to networked TAs, the malicious user could run a man-in-the-middle attack, either directly within the REE or on the network, and intercept the traffic exchanged between the two devices.

5   Implementation

We describe the implementation challenges of the three components included in IPERFTZ,[10] namely *(1)* a CA acting as proxy for IPERFTZ's *(2)* TA, and *(3)* the server component which the TA is interfacing. All components are implemented in the C language, and consists of 927 lines of code: 243 for the client, 314 for IPERFTZ's TA, and 430 for the server.[11]

5.1   IPERFTZ: Client Application

When the CA starts, the TEE context is initialized (`TEEC_InitializeContext`) using the file descriptor fetched from the OP-TEE driver. Two distinct dynamic shared-memory areas are allocated (`TEEC_AllocateSharedMemory`) at this time, to *(1)* exchange arguments passed over the *command line interface* with the TA (see Sect. 5.2) and *(2)* to retrieve metrics gathered by the TA during the network measurement. Several arguments (*e.g.*,, IP of the target server node, dummy data size, socket buffer size) are written in the shared memory area. The dummy data size is used by the TA to read/write data to the interface socket. Both shared memory areas get registered with the operation data structure (`TEEC_OpenSession`) before calling the `TEEC_InvokeCommand` function. The executing thread in the CA is blocked until the TA completes. The execution inside the TEE is resumed at the TA's main entry point upon world switch. Once the TA completes, an SMC instruction drives the CPU core to switch back into the normal world, where execution is resumed. The metrics gathered from the TA are available to the user as persistent files.

5.2   IPERFTZ: Trusted Application

The IPERFTZ TA is the primary executing unit. It takes the role of the client in the client-server model. The TA allocates a buffer for the dummy data on the

---

[10] https://github.com/ChrisG55/iperfTZ.

[11] Numbers for individual components include local header lines of code.

Table 1. Comparison of evaluation platforms.

Device	QEMU	Raspberry
CPU Model	Intel Xeon E3-1270 v6	Broadcom BCM2837
CPU Frequency	3.8 GHz	1.2 GHz
Memory Size	63 GiB DDR4	944 MiB LPDDR2
Memory data rate	2400 MT/s	800 MT/s
Disk Model	Samsung MZ7KM480HMHQ0D3	Transcend micro SDHC UHI-I Premium
Disk Size	480 GB	16 GB
Disk Read Speed	528.33 MB/s	90 MB/s
Network Bandwidth	1 Gbit/s	100 Mbit/s

heap, filled with random data generated by OP-TEE's Cryptographic Operations API [10]. With the information from the arguments, the TA finally sets up a TCP interface socket and opens a client connection before assigning the socket buffer sizes. Our implementation relies on the Time API [10] to measure the elapsed time during the network throughput measurement inside the TEE. OP-TEE computes the time value from the physical count register and the frequency register. The count register is a single instance register shared between normal and secure world EL1. The network throughput measurement is then started while either maintaining a constant bit rate, transmitting a specific number of bytes or running for 10 seconds. During the measurement, the TA gathers metrics on the number of transmit calls, *i.e.,* recv and send, bytes sent, time spent in the transmit calls and the total runtime. Upon completion, results are written to the shared memory area and the execution switches back to the normal world.

5.3   IPERFTZ: Server

The server component is deployed and executed inside the normal world. This is used to wait for incoming TCP connections (or inbound UDP datagrams) from IPERFTZ's TA. While executing, it gathers similar network metrics as the other components. Additionally, this component collects TCP specific metrics, such as the smoothed *round trip time* or the *maximum segment size*. This TCP specific data is not accessible for TAs and can only be retrieved on the server side using a getsockopt system call.

6   Evaluation

In this section we will demonstrate how IPERFTZ can measure the network throughput. We further draw conclusions regarding hardware and software implementation designs. We report that it is particularly challenging to assess network throughput, given the remarkable diversity one can find on embedded and mobile ARM systems.

Evaluation Settings. We deploy IPERFTZ on the Raspberry Pi platform. Due to the limited network bandwidth of Raspberry Pi devices supported by OP-TEE, we also include results under emulation using QEMU.[12] With QEMU we can run the same evaluation as on the Raspberry Pi and we also profit from a higher network bandwidth. Table 1 compares in detail the two setups. For both setups we use the same machine as server, on which we collect power consumptions and run the IPERFTZ server component.

Server. The server is connected to a Gigabit switched network, with access to power meter measurements. The nodes being measured are at a single-hop from the server. During the micro-benchmarks server components will be deployed on the server with fixed dummy buffer and socket buffer sizes of 128 KiB. This allows creating an accurate time series of the recorded throughput, latency and power metrics by concentrating the data acquisition on a single node.

QEMU. We deploy OP-TEE with QEMU v3.1.0-rc3 running on a Dell PowerEdge R330 server. The OP-TEE project has built-in support for QEMU and uses it in system emulation mode. In system emulation mode QEMU emulates an entire machine, dynamically translating different hardware instruction sets when running a virtual machine with a different architecture. In order to provide full network capability, we replace the default SLiRP network[13] deployed with OP-TEE by a bridged network with a tap device.

Raspberry Pi. OP-TEE only supports the Raspberry Pi 3B. We deploy OP-TEE on a Raspberry Pi 3B v1.2 equipped with a Broadcom BCM2837 SoC. The SoC implements an ARM Cortex-A53 with ARMv8-A architecture. The BCM2837 chip lacks support for cryptographic acceleration instructions and is not equipped with TRUSTZONE *Protection Controller* (TZPC), TRUST-ZONE *Address Space Controller* (TZASC), *Generic Interrupt Controller* (GIC) or any other proprietary security control interfaces on the bus [27]. The Raspberry Pi 3B lacks an on-chip memory or eMMC to provide a securable memory. We take these limitations into account in our evaluation, and leave further considerations once a more mature support for the Raspberry Pi platform is released.

Power Measurement. To measure the power consumption of the two platforms, we connect the Dell PowerEdge server to a LINDY iPower Control 2 × 6M *power distribution unit* (PDU) [17] and the Raspberry Pi 3B to an Alciom PowerSpy2 [1]. The LINDY PDU provides a HTTP interface queried up to every second with a resolution of 1 W and a precision of 1.5%. Alciom PowerSpy2 devices rely on Bluetooth channels to transfer the collected metrics. Both measuring devices collect voltage, current and power consumption in real time.

---

[12] https://www.qemu.org, accessed on 30.07.2019.

[13] https://wiki.qemu.org/Documentation/Networking#User_Networking_.28SLIRP.
29, accessed on 30.07.2019.

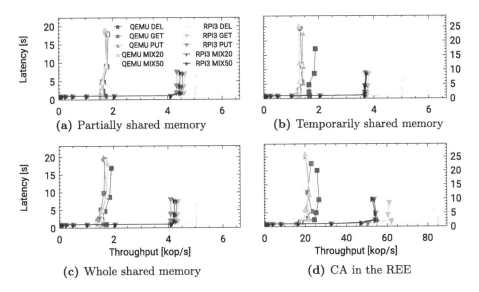

Fig. 4. Throughput-latency plots for different kinds of shared memory.

Memory Bandwidth. We use an existing key-value store TA [13] to evaluate the overhead of the different types of shared memory. The hash-table at the core of the key-value store uses separate chaining for collision resolution and implements modular hashing. The GLOBALPLATFORM specification defines three different types of shared memory: *whole* (an entire memory area), *partial* (a subset of an entire memory area with a specified offset), and *temporarily* (a memory area within the REE with an optional offset). The temporarily shared memory area is only shared with the TA for the duration of the TEE method invocation; the two others get registered and unregistered with the TEE session. The key-value store supports common operations such as DEL, GET and PUT on key-value pairs. We benchmark each operation in isolation as well as combining GET and PUT operations (MIXed benchmark). The benchmarks operate as follows: for whole and partially shared memory, the CA will request a shared memory region of 512 KiB from the TEE and fills it with random data from /dev/urandom. With temporarily shared memory, the CA will allocate a 512 KiB buffer and initialize it similarly with random data. Before invoking a key-value operation a chunk size of 1 KiB is selected as data object at a random offset in the shared memory respectively buffer. The random offset is then used as key and every operation is timed using CLOCK_MONOTONIC.[14] During the benchmark 256 operations are issued at a fixed rate between 1 and 32768 operations per second. Figure 4 shows the throughput-latency plots for each type of shared memory as well as for running the key-value store as a CA in the REE.

Compared to the Raspberry Pi, the results on QEMU are predominantly superposed and only achieve about half the throughput. We believe this is due

---

[14] Manual page: man time.h.

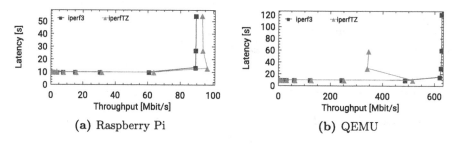

Fig. 5. TCP network throughput measurements for 128 KiB buffer sizes.

to an I/O bound from the ARM instruction and TRUSTZONE emulation using QEMU. We further observe with QEMU that the DEL benchmark for temporarily shared memory (Fig. 4b) and as CA (Fig. 4d) is clearly distinguishable from the other benchmarks. On the Raspberry Pi platform the graphs are well separated and ranked according to our expectations (lowest to highest throughput): PUT, MIX50, MIX20, GET, and DEL. The PUT operation has the lowest throughput because of memory allocation, memory copy and object insertion in the TA. The GET operation looks up the data object and copies it to the shared memory resulting in a higher throughput than the PUT operation. The mixed benchmarks show a similar behavior: the higher the PUT ratio, the lower the throughput. Hence, the MIX50 (50% PUT operations) has a lower average throughput than MIX20. The DEL operation avoids any time intensive memory operation and only has to free a data object after looking it up in the store. An interesting observation is made when comparing the memory throughput of the benchmarks executed in the REE against the benchmarks executed in the TEE. Key-value store operations executed inside TAs experience a 12×-14× overhead with QEMU and a 12×-17× overhead on the Raspberry Pi. This overhead is due to the world and context switches associated to TA method invocations.

Network Bandwidth. This micro-benchmark compares the network throughput measured with IPERFTZ in OP-TEE to the network throughput measured with iperf3 in Linux. We deploy both programs with the same set of parameters, *i.e.,* 128 KiB socket and dummy buffer sizes. Upon each iteration the data send is doubled starting at 1 MiB up to 10 GiB. We allocate not more than 512 KiB for the dummy data on the TA's heap, since TAs are by default limited in OP-TEE to 1 MiB in size. Linux has two kernel parameters which limit the maximum size of read and write socket buffers: /proc/sys/net/core/rmem_max and /proc/sys/net/core/wmem_max. These kernel parameters can be changed at runtime using sysctl, in order to allocate larger socket buffers.

As shown in Fig. 5, IPERFTZ generally exceeds the network throughput of iperf3 in both setups. On the Raspberry Pi 3B we cannot observe any degradation of the network throughput due to an overhead from frequent world switches. This result does not come as a surprise. The memory bandwidth benchmark operates at a throughput of several hundred MB/s, while the network bandwidth benchmark operates at about 10 MB/s. There is a gap of one order of magnitude

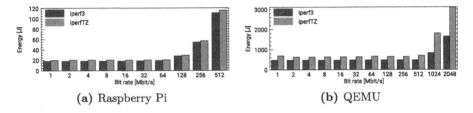

(a) Raspberry Pi   **(b)** QEMU

Fig. 6. Energy consumption during TCP network throughput measurements. Bit rates on the x-axis are given in logarithm to base 2.

in throughput between the two benchmarks, which we assume to be sufficient for the overhead not to arise. However, on QEMU we observe a serious degradation of the network throughput, when trying to achieve Gbit/s bit rate with OP-TEE. Remarkably, high throughput rates are strongly affected by the world switching overhead, even degrading beyond unaffected throughput rates. Our measurements indicate that network throughput beyond 500 Mbit/s is affected by a 1.8× world switching overhead, almost halving the network throughput.

Energy. During the network bandwidth benchmark, we recorded the power consumed by both setups. The LINDY iPower Control and the Alciom PowerSpy2 both record the timestamp as Unix time in seconds and the instantaneous power in watts. We use those units to execute a numerical integration over time using the trapezoidal method to obtain the total energy consumed by both setups during a benchmark run. Figure 6 shows these results. The total energy on the y-axis (in joule) is consumed by the device while executing a benchmark run for a specific bit rate on the x-axis (as binary logarithmic scale in Mbit/s). On the Raspberry Pi (Fig. 6a) we observe that before reaching saturation, IPERFTZ is consuming about 2 J (11%) more than `iperf3`. In the highly saturated range, the energy doubles with the throughput. However, with QEMU (Fig. 6b), the energy difference between the execution in the REE and the TEE is significant. Given that QEMU is running on an energy-demanding and powerful server, IPERFTZ consumes about 173 J (36%) more before the overhead arises than `iperf3` in the REE. We can clearly attribute this additional energy consumption observed on both setups to the execution of IPERFTZ in the TEE. Certainly, the world switching overhead also contributes to an increase of the energy consumption with QEMU. By assuming a similar behavior for the energy consumption on QEMU as in the saturated range on the Raspberry Pi, we obtain a 1.6× energy overhead due to world switching.

7   Related Work

There exists a plethora of network benchmarking and tuning tools. We note that the implementation of IPERFTZ is heavily inspired by the well-known `iperf` tool.

In this sense, IPERFTZ supports a subset of its command-line parameters, for instance to facilitate the execution of existing benchmarking suites.[15]

The ttcp (Test TCP) tool was one of first programs implemented to measure the network performance over TCP and UDP protocols. Lately, it has been superseded by nuttcp.[16] A tool with similar features is netperf.[17] Unlike the aforementioned tools, tcpdump[18] is a packet analyzer that captures TCP packets being sent or received over a network. IPERFTZ does not provide packet analysis tools. Instead, it does offer client and server-side measurements both for TCP and UDP data flows. More recently, iperf integrated most of the functionalities of ttcp, extending it with multi-threading capabilities (since iperf v2.0) and allowing bandwidth measurements of parallel streams. While it would be possible to provide similar support in IPERFTZ, the execution of code inside the TAs is currently single-threaded, hence limiting the achievable outbound throughput. The most recent version of iperf (v3.0) ships a simplified (yet single-threaded) implementation specifically targeting non-parallel streams. Flowgrind[19] is a distributed TCP traffic generator. In contrast, IPERFTZ follows a client-server model, with traffic generated between a server and a TA. StreamBox-TZ [25] is a stream analytics engine, which processes large IoT streams on the edge of the cloud. The engine is shielded from untrusted software using TRUSTZONE. Similar to IPERFTZ, StreamBox-TZ runs on top of OP-TEE in a TA. Yet, IPERFTZ does not process data streams but can generate and measure network performance of those streams.

To summarize and to the best of our knowledge, IPERFTZ is the first tool specifically designed to run as a TA for TRUSTZONE that can measure the achievable network throughput for such applications.

8   Conclusion and Future Work

The deployment of TAs is becoming increasingly pervasive for the management and processing of data over the network. However, due to constraints imposed by the underlying hardware and runtime system, network performance of TAs can be affected negatively. IPERFTZ is a tool to measure and evaluate network performance of TAs for ARM TRUSTZONE, a widely available TEE on embedded, IoT and mobile platforms. We implemented the IPERFTZ prototype on top of OP-TEE and we evaluated it on the Raspberry Pi platform. Our experimental results highlight performance and energy trade-offs deployers and programmers are confronted with both on hardware and emulated environments. We believe the insights given by our work can be exploited to improve design and configuration of TEEs for edge devices handling real-world workloads for TAs.

---

[15] Full compatibility with iperf would require substantial engineering efforts that we leave out of the scope of this work.

[16] See footnote 6.

[17] See footnote 5.

[18] https://www.tcpdump.org, accessed on 30.07.2019.

[19] www.flowgrind.net, accessed on 30.07.2019.

We intend to extend our work to support different types of sockets (*e.g.*, datagram sockets) and to leverage on-chip cryptographic accelerators. This would allow us to provide TLS-like channels for TAs, a feature that has not yet been implemented in OP-TEE. Finally, we aim for supporting various kinds of TEEs, especially in the context of embedded platforms and SoC, such as Keystone[20] for RISC-V processors.

Acknowledgments. The authors would like to thank the anonymous reviewers for their helpful comments and suggestions. The research leading to these results has received funding from the European Union's Horizon 2020 research and innovation programme under the LEGaTO Project (legato-project.eu), grant agreement No. 780681.

References

1. Alciom: PowerSpy2, 1.01 edn, 4 March 2013
2. Amacher, J., Schiavoni, V.: On the performance of ARM TrustZone. In: Pereira, J., Ricci, L. (eds.) DAIS 2019. LNCS, vol. 11534, pp. 133–151. Springer, Cham (2019). https://doi.org/10.1007/978-3-030-22496-7_9
3. Arm Limited: Isolation using virtualization in the Secure world. https://developer.arm.com/-/media/Files/pdf/Isolation_using_virtualization_in_the_Secure_World_Whitepaper.pdf?revision=c6050170-04b7-4727-8eb3-ee65dc52ded2
4. Arm Limited: ARM Security Technology: Building a Secure System using TrustZone Technology, April 2009. http://infocenter.arm.com/help/topic/com.arm.doc.prd29-genc-009492c/PRD29-GENC-009492C_trustzone_security_whitepaper.pdf. Accessed 30 July 2019
5. Arm Limited: Arm Cortex-A53 MPCore Processor: Technical Reference Manual, 8 February 2016. https://developer.arm.com/docs/ddi0500/g. Revision: r0p4
6. Arm Limited: Fundamentals of ARMv8-A, March 2017. https://static.docs.arm.com/100878/0100/fundamentals_of_armv8_a_100878_0100_en.pdf. Accessed 30 July 2019
7. Baumann, A., Peinado, M., Hunt, G.: Shielding applications from an untrusted cloud with Haven. ACM Trans. Comput. Syst. **33**(3), 8:1–8:26 (2015). https://doi.org/10.1145/2799647
8. GlobalPlatform, Inc.: TEE Client API Specification Version 1, July 2010
9. GlobalPlatform, Inc.: TEE Sockets API Specification Version 1.0.1, January 2017
10. GlobalPlatform, Inc.: TEE Internal Core API Specification 1.1.2.50, June 2018
11. GlobalPlatform, Inc.: TEE System Architecture Version 1.2, November 2018
12. Göttel, C., et al.: Security, performance and energy implications of hardware-assisted memory protection mechanisms on event-based streaming systems. In: 2018 IEEE 37th Symposium on Reliable Distributed Systems (SRDS), pp. 264–266 (2018). https://doi.org/10.1109/SRDS.2018.00042
13. Göttel, C., Felber, P., Schiavoni, V.: Developing secure services for IoT with OP-TEE: a first look at performance and usability. In: Pereira, J., Ricci, L. (eds.) DAIS 2019. LNCS, vol. 11534, pp. 170–178. Springer, Cham (2019). https://doi.org/10.1007/978-3-030-22496-7_11
14. Jang, J.S., Kong, S., Kim, M., Kim, D., Kang, B.B.: SeCReT: secure channel between rich execution environment and trusted execution environment. In: NDSS, pp. 1–15 (2015)

---

[20] https://keystone-enclave.org, accessed on 30.07.2019.

15. Kocher, P., et al.: Spectre attacks: exploiting speculative execution. In: 40th IEEE Symposium on Security and Privacy (S&P 2019) (2019)

16. Lind, J., Eyal, I., Pietzuch, P., Sirer, E.G.: Teechan: payment channels using trusted execution environments. ArXiv preprint arXiv:1612.07766 (2016)

17. Lindy Electronics Ltd.: iPower Control 2x6M/2x6XM, 1 edn, June 2015

18. Lipp, M., et al.: Meltdown: reading kernel memory from user space. In: 27th USENIX Security Symposium (USENIX Security 2018) (2018)

19. Lyu, X., et al.: Selective offloading in mobile edge computing for the green internet of things. IEEE Netw. **32**(1), 54–60 (2018). https://doi.org/10.1109/MNET.2018.1700101

20. Mäkinen, O.: Streaming at the edge: local service concepts utilizing mobile edge computing. In: 2015 9th International Conference on Next Generation Mobile Applications, Services and Technologies, pp. 1–6 (2015). https://doi.org/10.1109/NGMAST.2015.35

21. McGillion, B., Dettenborn, T., Nyman, T., Asokan, N.: Open-TEE - an open virtual trusted execution environment. In: 2015 IEEE Trustcom/BigDataSE/ISPA, TRUSTCOM 2015, vol. 1, pp. 400–407. IEEE Computer Society, Washington, DC (2015). https://doi.org/10.1109/Trustcom.2015.400

22. Naehrig, M., Lauter, K., Vaikuntanathan, V.: Can homomorphic encryption be practical? In: Proceedings of the 3rd ACM Workshop on Cloud Computing Security Workshop, CCSW 2011, pp. 113–124. ACM, New York (2011). https://doi.org/10.1145/2046660.2046682

23. Ngabonziza, B., Martin, D., Bailey, A., Cho, H., Martin, S.: TrustZone explained: architectural features and use cases. In: 2016 IEEE 2nd International Conference on Collaboration and Internet Computing (CIC), pp. 445–451 (2016). https://doi.org/10.1109/CIC.2016.065

24. Ning, Z., Kong, X., Xia, F., Hou, W., Wang, X.: Green and sustainable cloud of things: enabling collaborative edge computing. IEEE Commun. Mag. **57**(1), 72–78 (2019). https://doi.org/10.1109/MCOM.2018.1700895

25. Park, H., Zhai, S., Lu, L., Lin, F.X.: StreamBox-TZ: secure stream analytics at the edge with TrustZone. In: 2019 USENIX Annual Technical Conference (USENIX ATC 2019), pp. 537–554. USENIX Association, Renton, July 2019. https://www.usenix.org/conference/atc19/presentation/park-heejin

26. Segarra, C., Delgado-Gonzalo, R., Lemay, M., Aublin, P.-L., Pietzuch, P., Schiavoni, V.: Using trusted execution environments for secure stream processing of medical data. In: Pereira, J., Ricci, L. (eds.) DAIS 2019. LNCS, vol. 11534, pp. 91–107. Springer, Cham (2019). https://doi.org/10.1007/978-3-030-22496-7_6

27. Sequitur Labs Inc.: Easing Access to ARM TrustZone - OP-TEE and Raspberry Pi 3, 26 September 2016

28. Shepherd, C., Akram, R.N., Markantonakis, K.: Establishing mutually trusted channels for remote sensing devices with trusted execution environments. In: Proceedings of the 12th International Conference on Availability, Reliability and Security, ARES 2017, pp. 7:1–7:10. ACM, New York (2017). https://doi.org/10.1145/3098954.3098971

29. Shepherd, C., et al.: Secure and trusted execution: past, present, and future - a critical review in the context of the internet of things and cyber-physical systems. In: 2016 IEEE Trustcom/BigDataSE/ISPA, pp. 168–177 (2016). https://doi.org/10.1109/TrustCom.2016.0060

30. Varghese, B., Wang, N., Barbhuiya, S., Kilpatrick, P., Nikolopoulos, D.S.: Challenges and opportunities in edge computing. In: 2016 IEEE International Confer-

ence on Smart Cloud (SmartCloud), pp. 20–26 (2016). https://doi.org/10.1109/SmartCloud.2016.18

31. Volos, S., Vaswani, K., Bruno, R.: Graviton: trusted execution environments on GPUs. In: Proceedings of the 12th USENIX Conference on Operating Systems Design and Implementation, pp. 681–696, OSDI 2018. USENIX Association, Berkeley (2018). http://dl.acm.org/citation.cfm?id=3291168.3291219

Atomic Cross-Chain Swaps
with Improved Space and Local Time
Complexity

Soichiro Imoto[1]($\boxtimes$), Yuichi Sudo[1], Hirotsugu Kakugawa[2],
and Toshimitsu Masuzawa[1]

[1] Osaka University, Suita, Japan
{s-imoto,y-sudou,masuzawa}@ist.osaka-u.ac.jp
[2] Ryukoku University, Kyoto, Japan
kakugawa@rins.ryukoku.ac.jp

Abstract. An effective atomic cross-chain swap protocol is introduced
by Herlihy [Herlihy, 2018] as a distributed coordination protocol in order
to exchange assets across multiple blockchains among multiple parties.
An atomic cross-chain swap protocol guarantees; (1) if all parties conform
to the protocol, then all assets are exchanged among parties, (2)even if
some parties or coalitions of parties deviate from the protocol, no party
conforming to the protocol suffers a loss, and (3) no coalition has an
incentive to deviate from the protocol. Herlihy [Herlihy, 2018] invented
this protocol by using hashed timelock contracts.

A cross-chain swap is modeled as a directed graph $D = (V, A)$. Vertex set V denotes a set of parties and arc set A denotes a set of proposed asset transfers. Herlihy's protocol uses the graph topology and
signature information to set appropriate hashed timelock contracts. The
space complexity of the protocol (i.e., the total number of bits written
in the blockchains in a swap) is $O(|A|^2)$. The local time complexity of
the protocol (i.e., the maximum execution time of a contract in a swap
to transfer the corresponding asset) is $O(|V| \cdot |L|)$, where L is a feedback
vertex set computed by the protocol.

We propose a new atomic cross-chain swap protocol which uses only
signature information and improves the space complexity to $O(|A| \cdot |V|)$
and the local time complexity to $O(|V|)$.

1 Introduction

1.1 Motivation

The seminal work [2] by Nakamoto in 2008 for developing bitcoins has attracted
many researchers to the research of blockchains. However, the blockchain has
problems in privacy level, increased transaction time and scalability. In order

This work was supported by JSPS KAKENHI Grant Numbers 17K19977, 18K18000,
19H04085, and 19K11826 and JST SICORP Grant Number JPMJSC1606.

M. Ghaffari et al. (Eds.): SSS 2019, LNCS 11914, pp. 194–208, 2019.
https://doi.org/10.1007/978-3-030-34992-9_16

to overcome them, new cryptocurrencies with a wide variety of advantages are developed. There are also blockchains that handle physical rights as well as virtual currency (e.g., ownership of cars, copyrights of songs, proof of circulation and so on) [3]. It is a great advantage of blockchains that it allows us to exchange them in the absence of any trusted third parties.

As trading on blockchains becomes popular, demands for trading across multiple blockchains increase [3]. As a specific example of exchanging assets across multiple blockchains among multiple parties, consider the case that Alice wants to sell the copyrights of her songs for bitcoins. Bob is willing to buy her copyrights with alt-coins. Carol wants to exchange alt-coins for bitcoins. An *atomic cross-chain swap* protocol is a mechanism by which multiple parties exchange their assets managed by multiple blockchains. It is common that some parties do not know and do not trust each other, thus the protocol must guarantee that no party conforming to the protocol suffers from a loss in their trading. Specifically, this protocol guarantees the following three conditions. (1) if all parties conform to the protocol, then all assets are exchanged among parties, (2) even if some parties or coalitions of parties deviate from the protocol, no party conforming to the protocol suffers a loss, and (3) no coalition has an incentive to deviate from the protocol. The more blockchain users have request to trade as blockchain technology develops in the future, the more important atomic cross-chain swaps will be [10–17].

In many blockchains, assets are transferred from one party to another party by using *smart contracts*. A smart contract is a program that runs on a blockchain and has its correct execution enforced by the consensus protocol [4,5]. In this paper, *hashed timelock contracts* (HTLCs) [6,7] are used. In HTLC, recipients of a transaction have to acknowledge payment by generating cryptographic proof within a certain timeframe. Otherwise, the transaction does not take place. For example, consider the case Alice wants to send an asset to Bob by using HTLCs in gratitude for taking money from Bob. Alice first generates random secret data s, called a *secret*, and produces *hashlock* $h = H(s)$, where H is a cryptographic hash function. Next, Alice publishes the contract with hashlock h. After that, if Alice takes money form Bob, Alice reveals the secret s to Bob. When Bob sends the secret s to the contract, the contract irrevocably transfers Alice's asset to Bob. Alice also sets *timelock* t so that her escrowed asset can be returned if Bob does not give money to Alice within the time.

In this paper, we consider the case that more than two parties exchange their assets, as shown in Fig. 1.

In the following, we quote a simple protocol presented by Herlihy [1] for exchanging assets among Alice, Bob, and Carol. In that exchanging, Alice sends the copyrights of her songs to Bob, Bob sends alt-coins to Carol and Carol sends bitcoins to Alice. Let Δ be time enough for one party to publish a smart contract on any of the blockchains, or to change the state of a contract, and for the other party to detect the change:

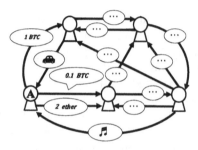

Fig. 1. Several parties exchange their assets by using (possibly) different blockchains. For example, the party A gets ownership of a car and the copyright of a song for 1.1 BTC and 2 ether.

1. Alice creates a secret s, hashlock $h = H(s)$, and publishes a contract on the music copyright blockchain with hashlock h and timelock 6Δ in the future, to transfer her music copyrights to Bob.
2. When Bob confirms that Alice's contract has been published on the copyright blockchain, he publishes a contract on the alt-coin blockchain with the same hashlock h but with timelock 5Δ in the future, to transfer his alt-coins to Carol.
3. When Carol confirms that Bob's contract has been published on the alt-coin blockchain, she publishes a contract on the Bitcoin blockchain with the same hashlock h, but with timelock 4Δ in the future, to transfer her bitcoins to Alice.
4. When Alice confirms that Carol's contract has been published on the Bitcoin blockchain, she sends the secret s to Carol's contract, acquiring the bitcoins and revealing s to Carol.
5. Carol then sends s to Bob's contract, acquiring the alt-coins and revealing s to Bob.
6. Bob sends s to Alice's contract, acquiring the copyrights and completing the swap.

Everyone can stop the swap if published contracts are different from predetermined ones. There is no possibility that Alice transfers to Bob the copyrights without acquiring the bitcoins because only Alice initially knows secret s. If Carol's bitcoins have been transferred, this guarantees that she can get s and acquire the alt-coins as well because she publishes her contract after confirming publication of Bob's contract and the timeout specified by the time lock of Bob's contract is one Δ greater than that of her contract. Bob also acquires the copyrights if his alt-coins have transferred. Even if Alice and Bob conspire to deceive Carol, they can not get the bitcoins without payment for Carol, thus Carol never suffers from a loss. As seen from these facts, every party should publish his contracts only after all the contracts for assets transferred to him are published unless he generates a secret for a swap. We call a party who generates a secret for a swap a *leader*, and denote the set of leaders by L. From the above discussion, L must be a feedback vertex set. This simple protocol works if and only if we have exactly one leader l such that $\{l\}$ is a feedback set.

Generally, a cross-chain swap is modeled as a directed graph $D = (V, A)$. Vertex set V denotes a set of parties and arc set A denotes a set of proposed asset transfers. When there are multiple cycles, this simple protocol does not work. The leader can publish his two contracts without anxiety about a result of the swap. However, each of other two parties cannot publish his contracts until he confirms the other publishes a contract to him. Therefore, at least one leader is necessary on each cycle of a swap graph. Moreover, if there are multiple cycles, it is not possible to assign consistent timeouts across cycles in a way that guarantees a gap of at least Δ between contracts for giving assets and those for taking assets. That is, it is not easy to realize a mechanism such that each leader reveals its secret with confidence when there are multiple leaders.

To circumvent the problem, Herlihy [1] sets complicated conditions for each smart contract to design an atomic cross-chain swap protocol. In the Herlihy's protocol, each leader creates a secret and one hashlock for each of the $|L|$ secrets is set in each contract in order to publish contracts safely. Each leader sends his own secret with his signature to the contracts corresponding the asset which he wants to get. The party who published the contract looks at it and gets the secret with the leader's signature. Each party unlocks the hashlock created by the leader by sending all secrets with the signature and his own signature. Each hashlock has its timelock, which depends on the length of the path along which the secret is transferred from the leader generating the secret. The sequence of signatures of parties indicates the path. It is necessary to check whether the path actually exists in a given swap. Therefore, topology information of a swap graph is stored in every contract, which requires $O(|A|)$ bits. A contract is triggered when all secrets of each leader are sent into its timelock to the contract and the network nodes of the blockchain of the corresponding contract verify that these paths exist.

1.2   Our Contributions

Let $D(V, A)$ be a given directed graph that represents a cross-chain swap to general graphs. We propose an atomic cross-chain swap protocol, which improves the space complexity (bits stored on all blockchains) of Herlihy's protocol [1] from $O(|A|^2)$ to $O(|A| \cdot |V|)$, and the local time complexity (the maximum execution time of a contract in a swap to transfer the corresponding asset) from $O(|V| \cdot |L|)$ to $O(|V|)$. The protocol does not store the swap topology in a contract. Instead, it simply assigns each contract the timelock depending on the number of signatures of parties. Specifically, we set the timelock of a contract as follows: the contract is triggered if and only if the contract is given signatures of distinct x parties in a swap before time $t + x\Delta$ for some $x \in [1, n]$ where t is the time instance we will explain later (the starting time of the execution plus alpha). Suppose that an asset of party v is transferred to another party by executing a contract. This means that the signature of x parties are published in the corresponding blockchain before time $t + x\Delta$. Then, party v can trigger all the contracts that transfer an asset to v because v can obtain the signature of x parties from that blockchain and can give those contracts $x + 1$ signatures by adding its own signature before time $t + (x + 1)\Delta$. As a result, it is not necessary to store the topology information of a swap in any contract, thus the space complexity is improved.

Let *local time complexity* be how long a contract takes to be triggered from executing a first action for triggering, which is sending a secret to the contract from a party at first. The local time complexity is much smaller than the execution time of the protocol. However, it is worthwhile to mention that each contract is executed by *all* network nodes that host the copies of the ledger of the corresponding blockchain. Therefore, the local time complexity is an index independent of the execution time and reducing local time complexity of a smart contract is of practical importance. In Herlihy's protocol, a contract must verify at most $|V|$ signatures for the hashlock of each leader. Thus, we need $\Theta(|V| \cdot |L|)$ time to trigger one contract. On the other hand, in the proposed protocol, we only need to verify $|L|(\leq |V|)$ secrets and at most $|V|$ signatures. That is why the proposed protocol improves local time complexity from $|V| \cdot |L|$ to $|V|$. We summarize our contribution to Table 1.

Table 1. Comparison between the performances of Herlihy's protocol and our protocol.

	Local time complexity	Execution time	Space complexity
Herlihy's protocol	$O(|V| \cdot |L|)$	$O(|V|\Delta)$	$O(|A|^2)$
Our protocol	$O(|V|)$	$O(|V|\Delta)$	$O(|A| \cdot |V|)$

2   Model

2.1   Graph Model

$D = (V, A)$ is a *directed graph*, where V is a finite set of *vertexes*, and A is a finite set of ordered pairs of distinct vertexes called *arcs*. The number of vertexes $|V|$ is denoted by n. An arc (u, v) has tail u and head v. An arc *leaves* its tail and *enters* its head. Therefore, an arc (u, v) is *leaving arc* from u and *entering arc* to v. A digraph $D' = (V', A')$ is a *subgraph* of $D = (V, A)$ if $V' \subseteq V$, $A' \subseteq A \cap (V' \times V')$.

A *path* p in D is a sequence of vertexes $(v_0, v_1, ..., v_k)$ with $(v_i, v_{i+1}) \in A$ for every $i = 0, 1, \ldots, k-1$, where we define k as the length of the path. The *distance* $dist(u, v)$ from vertex u to vertex v is the length of the shortest path from u to v in D. D's *diameter* $diam(D)$ is the largest distance between two vertexes in D. A vertex v is *reachable* from vertex u if there is a path from u to v. Directed graph D is *strongly connected* if every vertex is reachable from any other vertex in D. A path $(u_0, ..., u_k)$ with $k \geq 1$ is a *cycle* if $u_0 = u_k$. A *feedback vertex set* of D is a set of vertexes whose removal makes a graph acyclic. By definition, any feedback vertex set contains at least one vertex in any cycle in D.

2.2   Blockchain and Smart Contract Model

A *blockchain* is a distributed ledger that can record transactions between parties in a verifiable and permanent way. We do not assume any specific blockchain algorithm. We only assume that every blockchain used in a cross-chain swap supports *smart contracts*. An owner of an asset can transfer ownership of the asset to a *counterparty* by creating a smart contract. The owner specifies conditions in the contract to transfer the asset. A counterparty can get ownership of the asset when the conditions are satisfied. We say that a contract is *published* when the owner of the corresponding asset releases the contract on a blockchain. A contract is *triggered* when all conditions of the contract are satisfied and the ownership of the asset on the contract is transferred to the counterparty. The owner of an asset can also specify conditions to get back the asset on a smart contract. If the conditions are satisfied before the contract is triggered, the asset is returned to the owner. Thereafter, the asset is never transferred to the counterparty by this contract. Typically, the owner can specify the condition to get back the asset in the form of *timeout condition*. The owner can regain the asset anytime after the specified time in a contract. Smart contracts are immutable, which means smart contracts can never be changed and no one can tamper with or break the contract. We assume that every operation on any blockchain can be completed within a known amount of time Δ. In particular, every party can publish a contract within Δ unit time and every party can trigger a contract within Δ unit time (if he has all information to satisfy all conditions of the contract). Blockchains require each party v to create the public key $pkey_v$ and the private key $skey_v$ for transactions. A *signature* $sign(x, v)$ is the signature on data x signed with $skey_v$. Every party can verify the signature $sign(x, v)$ with the public key $pkey_v$.

2.3   Swap Model

Parties $v_1, v_2, \ldots, v_n$ are given a digraph $D = (V, A)$, where each vertex in V represents a party, and each arc $(u, v) \in A$ represents an asset on a blockchain to be transferred from party u to party v. When ownership of an asset on an arc (u, v) transfers from u to v, we say the asset transfer happens, or the arc is triggered.

We denote the corresponding asset for arc $(u, v) \in A$ as $a_{u,v}$. The value of each asset may vary from one party to another; for example, v may find a higher value on the asset than u. Otherwise, that is, if all the parties see the same value about an asset, exchanging their assets following $D = (V, A)$ is meaningless (no one can gain unless someone loses.). The value of $a_{u,v}$ for u (resp. v) is denoted by $value_{u,v}^-$ (resp. $value_{u,v}^+$). We define $value_{D,v}^+$ as the sum of the values for v of all entering arcs to v in D, i.e., $value_{D,v}^+ = \sum_{(u,v) \in A} value_{u,v}^+$. We also define $value_{v,D}^- = \sum_{(v,u) \in A} value_{v,u}^-$ as the sum of the values for v of all leaving arcs from v in D.

We assume that D meets the following condition because otherwise party v should not participate in a swap D.

$$\forall v \in V : value^+_{D,v} > value^-_{v,D}$$

Moreover, we assume that there exists no connected subgraph $D' = (V', A)$ of D such that

$$\exists C \subset V' : (\sum_{v' \in C} (value^+_{D',v'} - value^-_{v',D'}) > \sum_{v' \in C} (value^+_{D,v'} - value^-_{v',D})) \quad (1)$$

$$\wedge (\forall v'' \notin C : value^+_{D',v''} - value^-_{v'',D'} \geq value^+_{D,v''} - value^-_{v'',D}). \quad (2)$$

Inequality (1) means that one coalition C in D' gets more benefits in D' than D, and (2) means that any party in $D' - C$ gets equal to or more benefits in D' than D. If swap D has such a subgraph D', we can say that D forces C to perform disadvantage exchanges because C gets larger benefits in swap D'. Therefore, D is not an appropriate swap. This is the reason why we exclude such a swap. As we will explain in Sect. 5, this assumption is weaker than the assumption that Herlihy [1] makes to guarantee that no party and coalition can get more benefits by deviating from the protocol than following the protocol. Even without this assumption, the proposed protocol is an atomic cross-chain swap protocol under another specific assumption about value function which Herlihy's protocol requires. We discuss that in the Sect. 5.

A protocol is a strategy for a party, that is, a set of rules that determines which action the party takes. Ideally, all parties in a swap $D = (V, A)$ follows the common protocol P. However, we must consider the case that some party does not follow (i.e, deviates from) the common protocol to get larger benefits. To make the matter worse, those parties may make a coalition and take actions cooperatively to get larger benefits in total (some party in the coalition willingly loses aiming at larger total benefits of the coalition). We must design a protocol such that any party following the protocol does not suffer from a loss even if such selfish parties or coalitions exist. We assume that every party in a swap $D = (V, A)$ can send any message to any party in the swap. *Space complexity* of a protocol is measured by the total number of bits required store information on all blockchains in a swap. *Local time complexity* of a protocol is measured by the maximum execution time of a contract in a swap to transfer the corresponding asset.

Definition 1. *A swap protocol P is* uniform *if it guarantees the followings:*

- *If all parties follow P, all arcs in $D = (V, A)$ are triggered.*
- *Even if there are parties arbitrarily deviating from P, every party v following P gets all assets of entering arcs to v or regains all assets of leaving arcs from v.*

Definition 2. *A swap protocol P is* Nash equilibrium *if no party can get more benefits by deviating from P than following P unless he joins a coalition.*

Definition 3. *A swap protocol P is* strong Nash equilibrium *if it guarantees that no party and coalition can get more benefits by deviating from P than following P.*

Definition 4. *A swap protocol P is* atomic *if it guarantees to be uniform and strong Nash equilibrium.*

3 Proposed Protocol

3.1 Outline of Proposed Protocol

The proposed protocol P consists of four phases. In Phase 1, every party finds a common feedback vertex set L of $D = (V, A)$ locally by using the same algorithm. Although finding a minimum feedback vertex set is NP-hard [8], we do not need the minimum set, thus we can use any approximate solution. For example, there exists an algorithm to find a 2-approximate solution [9]. The parties belonging to the vertex set $L = \{l_1, l_2, \ldots, l_k\}$ are called leaders. We call the other parties $f_1, f_2, \ldots, f_{n-k}$ followers. In this phase, every leader generates a secret which is a random bit string and calculates a hashlock based on the secret. In Phase 2, a smart contract corresponding to each arc of D is published. Each leader spontaneously publishes contracts for all his leaving arcs. Each follower publishes contracts for all his leaving arcs after confirming that the contracts for all his entering arcs are already published. In Phase 3, each of the leaders $l_2, l_3, \ldots, l_k$ sends its secret to l_1 after confirming that the contracts for all its entering arcs have been published. The leader l_1 starts Phase 4 after l_1 confirms that the contracts for all the entering arcs to l_1 have been published and all secret have been sent from each other leaders. In Phase 4, each arc of D is sequentially triggered, which starts from leader l_1.

As described in Sect. 3.2, we design the contracts in Phase 2 so as to guarantee the following three properties. (i) For any arc, no party can trigger the corresponding contract without knowing the secrets of all leaders. (ii) For any party v, if a contract on leaving arc from v is triggered, v can trigger the contracts published on all entering arcs to v. (iii) For any arc (u, v), party v can regain $a_{u,v}$ if the contract on (u, v) are not triggered during a certain period after it was published. If a leader l_i follows the proposed protocol, l_i sends its secret to the leader l_1 after confirming that the contracts on all entering arcs have been published. Therefore, from Property (i) and (ii), it is guaranteed that no leaving arc from l_i is triggered unless all entering arcs to l_i are triggered. Moreover, from property (iii), after a certain period of time, l_i gets assets of all entering arcs to l_i, or regains assets of all leaving arcs from l_i. If a follower f_i follows the proposed protocol, f_i publishes the contracts for all leaving arcs after confirming that the contracts for all entering arcs have been published. Therefore, according to the same argument, after a certain period of time, f_i gets assets of all entering arcs to f_i, or regains assets of all leaving arcs from f_i.

In the following, we describe the trigger condition and regain condition of the smart contract in Sect. 3.2, and the detailed operation in Phases 1, 2, 3 and 4 in Sects. 3.3, 3.4, 3.5 and 3.6, respectively.

3.2   The Conditions of Smart Contracts

In this subsection, we specify what each party writes in a contract. In a contract, its publisher specifies the conditions for transferring and regaining the asset. Each leader l_i makes hashlock $H(s_i) = h_i$ with random secret data s_i, called a secret, where $H()$ is a cryptographic hash function common to all parties. In any blockchain, as mentioned in Sect. 2.2, time Δ is enough to publish and trigger a contract. The condition of the contract on arc (u, v) to transfer asset $a_{u,v}$ is as follows: Asset $a_{u,v}$ is transferred to counterparty v if v sends the secrets $s_1, s_2, \ldots, s_k$ of all leaders and x signatures[1] on k-tuple $(s_1, s_2, \ldots, s_k)$ by arbitrary x parties to the contract before time $t_s + ((diam(D)+1)\Delta + 2\varepsilon) + x\Delta$, where t_s is the starting time of the protocol. That is, the more a party collects the signatures of the parties, the later he can trigger the contracts on his entering arcs. Let $\varepsilon(<< \Delta)$ be the time required to complete each of Phases 1 and 3. We set the deadline of the smart contract to time $t_s + ((diam(D) + 1)\Delta + \varepsilon) + n\Delta$, so that the publisher can regain his asset on the contract if the corresponding contract is not triggered before that time.

3.3   Phase 1: Preparation

Every party takes $D = (V, A)$ as input. At first, he chooses leaders $l_1, l_2, \ldots, l_k$ such that the leaders form a feedback vertex set of D. Since all parties are given the same swap $D = (V, A)$ as input, they can independently find the same leaders $l_1, l_2, \ldots, l_k$ without communication. We call l_1 the *top leader* and the other leaders $l_2, l_3, \ldots, l_k$ *sub-leaders*.

Next, each leader l_i generates a sequence of random bits s_i called a secret and computes hashlock $h_i = H(s_i)$, after which it sends only the hashlock to all parties. Finally, all parties send their public keys $pkey_{v_1}, pkey_{v_2}, \ldots, pkey_{v_n}$ for verifying their signatures. We assume that all of these can be done in time $\varepsilon \ll \Delta$.

3.4   Phase 2: Publication

Every leader spontaneously publishes contracts for all their leaving arcs with the conditions described in Sect. 3.2.

When a follower finds that the contracts on all his entering arcs are already published, he checks whether the contents of the contracts are consistent with the conditions of Sect. 3.2. Especially, he checks whether the public keys of all the parties and the hashlocks of all leaders that he receives in Phase 1 match the public keys of all parties and the hashlocks of all leaders specified in the contracts on entering arcs. He quits the swap without publishing any contract if those public keys or hashlocks do not match. Thus, even if the party deviating from P may send a fake hashlock or a public key to the other parties in Phase 1, no other party following P suffers from a loss. If all the published contracts are

---

[1] $sign((s_1, s_2, \ldots, s_k), v_1), sign((s_1, s_2, \ldots, s_k), v_2), \ldots, sign((s_1, s_2, \ldots, s_k), v_x)$.

consistent, he publishes contracts for all his leaving arcs. As will be described later, each party v reveals the signature on $s_1, s_2, \ldots, s_k$ with v's secret key $skey_v$ only when v triggers all entering arcs. Therefore, all entering arcs of v can be triggered if any leaving arc of v is triggered. Hence, no follower suffers from a loss, that is, every follower gets assets of all his entering arcs or regains assets of all his leaving arcs.

As described in Sect. 4, Phase 2 can be completed within at most $(diam(D) + 1)\Delta$ time.

3.5   Phase 3: Share Secrets

Every sub-leader reveals his secret to the top-leader if the contracts are published on all his entering arcs. This is to ensure that the top-leader can trigger contracts at first in Phase 4.

The top-leader confirms whether or not the secrets acquired from the sub-leaders are correct using hashlocks shared at Phase 1. If the top-leader finds an incorrect secret, it quits the swap without going to Phase 4.

We assume that Phase 3 can be done in time $\varepsilon \ll \Delta$.

3.6   Phase 4: Trigger

The top-leader starts Phase 4 if he acquires the secrets of all the sub-leaders within time $(diam(D) + 1)\Delta + 2\varepsilon$ which is enough to complete Phases 1, 2 and 3. This is because the top-leader only need to send secrets of all leaders $s_1, s_2, \ldots, s_k$ and its signature within the time $t_s + ((diam(D) + 1)\Delta + 2\varepsilon) + \Delta$ to all his entering arcs in order to get the assets of all his entering arcs. The top-leader sends these secrets and his signature to all his entering arcs, by which he triggers the contracts and acquires all assets of the entering arcs.

If the top-leader deviates from P, he may not trigger some (or all) contracts that transfer assets to the top leader. However, those actions are irrational. We explain the reason in Lemma 8, (Sect. 4).

Next, we describe the behavior of the sub-leaders and followers in Phase 4. Each party v of them waits until any of his leaving arcs is triggered. Consider that one of his leaving arcs is triggered with the information of the secrets $s_1, s_2, \ldots, s_k$ and the signature of x distinct parties $(1 \le x < n)$ on k-tuple $(s_1, s_2, \ldots, s_k)$. By definition of contracts, the leaving arc must be triggered before $t_s + (diam(D) + 1)\Delta + 2\varepsilon + x\Delta$. Then, party v acquires from the contract the secrets $s_1, s_2, \ldots, s_k$ and those x signatures within time $t_s + (diam(D) + 1)\Delta + 2\varepsilon + x\Delta$. He immediately sends $s_1, s_2, \ldots, s_k$ and the x signatures in addition to his signature (thus $x + 1$ signatures in total) to all entering arcs. As a result, v sends all necessary information to the contracts on all his entering arcs before time $t_s + ((diam(D) + 1)\Delta + 2\varepsilon + (x+1)\Delta$, which guarantees that all his entering arcs are triggered. We describe these in Fig. 2.

Every party regains the asset of each of leaving arcs if it is not triggered by the deadline (specified by the timeout) $t_s + (diam(D) + n + 1)\Delta + 2\varepsilon$.

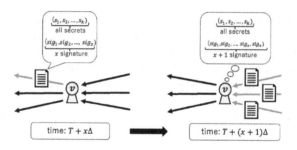

Fig. 2. The party v can trigger all his entering arcs when a leaving arc is triggered.

4   Correctness and Complexity of Protocol

We prove that proposed protocol P is atomic (i.e., uniform and strong Nash equilibrium). First, we prove that P is uniform.

Lemma 1. *Assume that top-leader l_1 follows P. If any leaving arc of l_1 is triggered, all of entering arcs to l_1 are triggered.*

Proof. Assume by contradiction that there is an execution of P such that a leaving arc (l_1, v) of the top-leader l_1 is triggered and an entering arc (u, l_1) is not triggered. After (l_1, v) is triggered, l_1 immediately obtains secrets of all leaders, which is before the time $t_s + diam(D) + 1)\Delta + 2\varepsilon$. Thus, l_1 sends all the secrets and its own signature to the contract on (u, l_1) before the time $t_s + diam(D) + 1)\Delta + 2\varepsilon$, which means that l_1 triggers (u, l_1) in the execution. This contradicts the assumption.

Lemma 2. *Assume that sub-leader $l_i (i \neq 1)$ follows P. If any leaving arc of l_i is triggered, all of entering arcs to l_i are triggered.*

Proof. Assume for contradiction that there is an execution of P such that a leaving arc (l_i, v) of sub-leader l_i is triggered but an entering arc (u, l_i) is not triggered. All secrets $s_1, s_2, ..., s_k$ and the signature of $x(< n)$ parties are revealed in the contract of (l_i, v) by time $t_s + (diam(D) + x + 1)\Delta + 2\varepsilon$. By the assumption, those x signatures do not include the signature of l_i because l_i reveals its signature only when it triggers the contracts of all entering arcs. The contract on (l_i, v) requires the secret of all leaders and l_i does not reveal its secret s_i until the contracts of all its entering arcs are published. Therefore, the contracts of all entering arcs of l_i are published before time $t_s + (diam(D) + x + 1)\Delta + 2\varepsilon$. From the above, l_i sends the secrets $s_1, s_2, \ldots, s_k$ and the x signatures in addition to its signature on $s_1, s_2, \ldots, s_k$ to the contract on (u, l_i) by time $t_s + (diam(D) + x + 1)\Delta + 2\varepsilon$. This implies that edge (u, l_i) is triggered in the execution. This is a contradiction.

Lemma 3. *Assume that follower f follows P. If any leaving arc of f is triggered, all of entering arcs to f are triggered.*

Proof. Follower f publishes a contract for a leaving arc from f only after the contracts of all entering arcs to f are published. Therefore, all entering arcs to f are already published when any leaving arc from f is triggered. We can prove the lemma in the same way as the proof of Lemma 2.

Lemma 4. *Assume that party v follows P. If any leaving arc of v is triggered, all of entering arcs to v are triggered.*

Proof. The lemma immediately follows from Lemmas 1, 2, and 3

Lemma 5. *Let SC be the smart contract published on any leaving arc from party v that follows P. It happens before time $t_s + (diam(D) + n + 2)\Delta + 2\varepsilon$ that contract SC is triggered or the asset of SC is regained by v.*

Proof. If a leaving arc from v is not triggered before time $t_s + (diam(D) + n + 1)\Delta + 2\varepsilon$, party v calls the timeout function of the contract on the arc within at most Δ unit time after the deadline $t_s + (diam(D) + n + 1)\Delta + 2\varepsilon$.

Lemma 6. *If every party follows P, all arcs are triggered within the time $t_s + 2(diam(D) + 1)\Delta + 2\varepsilon$.*

Proof. If all parties follow P, each leader l immediately publishes contracts for all leaving arcs from l after the start of Phase 2, and each follower f publishes contracts for all leaving arcs from f immediately after the contracts for all entering arcs to f are published. Since the time required to publish each contract is at most Δ, we can see that all parties complete Phase 2 in $(diam(D) + 1)\Delta$ time. Therefore, since the upper bound ε of the time is required for the local computation and the message transmission of Phases 1 and 3, Phase 3 is completed by time $t_s + (diam(D) + 1)\Delta + 2\varepsilon$. Thus, top-leader l_1 will immediately begin Phase 4.

After Phase 4 has started, the leader l_1 immediately triggers the contracts of all entering arcs to l_1. Each party v other than l_1 immediately triggers the contracts of all entering arcs to v, once a contract on any leaving arc from v has been triggered. The signature on $(s_1, s_2, \ldots, s_k)$ by v is never used to trigger the contract on the arc which is triggered for the first time among all leaving arcs from v. Therefore, v triggers all entering arcs to v by using the signatures used to trigger the leaving arc and his own signatures. Since the time required to trigger each contract is at most Δ, we can show, by the induction on distance from top-leader l_1, that the time required to trigger the contracts on all arcs in a given swap in Phase 4 is at most $(diam(D) + 1)\Delta$. Summing up these time for Phases 1 to 4, we can show that all arcs are triggered by time $t_s + 2(diam(D) + 1)\Delta + 2\varepsilon$.

Lemma 7. *Protocol P is uniform.*

Proof. The lemma immediately follows from Lemmas 4, 5, and 6

Lemma 8. *Protocol P is strong Nash equilibrium.*

Proof. Assume by contradiction that, in an execution of P, a coalition C formed from some parties gets more profits by deviating from protocol P when the other parties follow P. That is, for C, the profits gained by D' is bigger than that C gets with D, where $D' = (V', A')$ is the subgraph of D induced by all arcs triggered in the execution. In other words, the following inequality holds:

$$\sum_{v \in C}(value^+_{D',v} - value^-_{v,D'}) > \sum_{v \in C}(value^+_{D,v} - value^-_{v,D})$$

Every party in $V' \setminus C$ triggers all his entering arcs by Lemma 4. Therefore, all entering arcs to v' in $V' \setminus C$ are included in A'. Therefore, the following inequality holds for every party $v' \in V' \setminus C$.

$$\forall v' \in V' \setminus C : value^+_{D',v'} - value^-_{v',D'} = value^+_{D,v'} - value^-_{v',D'} \geq$$
$$value^+_{D,v'} - value^-_{v',D}$$

However, these condition contradicts the assumption on a swap $D = (V, A)$ introduced in Sect. 2.3.

Theorem 1. *Protocol P is an atomic cross-chain swaps protocol. If every party follows P, the swap can be completed in time $2(diam(D)+1)\Delta + 2\varepsilon$. Even if any set of parties deviates from P, the swap finishes in at most time $(diam(D)+n+1)\Delta + 2\varepsilon$. Protocol P requires space complexity (the total number of bits on all the blockchains) of $O(|A| \cdots |V|)$. Protocol P requires local time complexity (the maximum execution time of a contract in a swap to transfer the corresponding asset) of $O(|V|)$.*

Proof. Protocol P is atomic by Lemmas 7 and 8. The execution time of P follows from Lemmas 5 and 6. Space complexity of P is $O(|A| \cdot |V|)$ because each arc has one contract and each contract requires public keys of $|V|$ parties and hashlocks of $|L| \leq |V|$ leaders. Local time complexity of P is $O(|V|)$ because we only need to verify $|L|(\leq |V|)$ secrets and at most $|V|$ signatures.

5   Discussion

We made two assumptions, in Sect. 2.3, on the value function specifying the values of assets to parties. Remind that, we assume that for any subgraph D' of D, no coalition C in D' gets more benefits in D' than D when every other party in D' gets benefits in D' no less than in D.

The first alternative is to replace coalition C in the above assumption with just a party v. That is, we assume that for any subgraph D' of D, no party v in D' gets more benefits in D' than D when every other party in D' gets benefits in D' no less than in D. Under this assumption, the proposed protocol P is not strong Nash equilibrium but is Nash equilibrium, which can be proved in the same way as Lemma 8.

The second alternative is the assumption that Herlihy made for his protocol [1]. Specifically, He classifies the parties $v \in V$ into four groups as follows, based on the status of the arcs entering and leaving v in the end of an execution of a swap algorithm.

DEAL: All arcs entering and leaving v are triggered.
NO_DEAL: No arc entering or leaving v is triggered.
FREE_RIDE: Some arc entering v is triggered, but no arc leaving v is triggered.
DISCOUNT: All arcs entering v are triggered, but some arc leaving v is not triggered.
UNDER_WATER: Some arcs entering v is triggered and some arc leaving v is not triggered.

He assumes that no party accepts to be UNDER_WATER, more generally, every coalition tries to avoid the case that any member of the coalition becomes UNDER_WATER, with the highest priority. He also assumes that every party prefer DEAL to NO_DEAL and some parties or coalitions may deviate from the swap protocol because they prefer FREE_RIDE and DISCOUNT to DEAL. Our model has a sufficient power to represent this assumption. It suffices to define the benefit of v in D' as $value^+_{D',v} - value^-_{v,D'} + exp(D',v)$ where the *exception value* $exp(D',v)$ is defined as follows: $exp(D',v) = -\infty$ if there exist arc $(u,v) \in A \setminus A'$ and arc $(v,w) \in A'$, otherwise $exp(D',v) = 0$. If a party (or some party of a coalition) ends up with UNDER_WATER, then the benefit of the party (or the coalition) is $-\infty$, thus they try to avoid the situation with the highest priority. The situation ending up with FREE_RIDE or DISCOUNT brings a party or a coalition a larger benefit than DEAL.

In what follows, we show that our protocol P is strong Nash equilibrium under the second assumption, equivalent to the Herlihy's assumption [1]. Assume by contradiction that, in an execution of P, some coalition C gets more benefits by deviating from protocol P than by conforming to P. Let $D' = (V', A')$ be the subgraph of D induced by the set of the arcs triggered in the execution. Since we introduce the exception value, every party in C must trigger all his entering arcs in D if one of his leaving arc in D is triggered. Since D is strongly connected, if there exists an arc $(u,v) \in A'$, there exists an arc $(v,w) \in A'$, thus all entering arcs to v are included in A'. Therefore, every entering arc of every party in V' is included in A', which implies $D = D'$. This contradicts the assumption that C gets more benefits in D' than in D.

6 Conclusions

In this paper, we proposed an atomic cross-chain protocol to improve space complexity and local time complexity. Herlihy's protocol [1] requires to store the swap topology in each contract and set timelocks to each secret by using the topology. The proposed protocol need not store the swap topology in any contract. Instead, we set the time condition to trigger a contract depending on the number of signatures sent to the contract. Therefore, every party v can immediately trigger the contracts of all entering arcs to v, once a contract on some leaving arc from v is triggered. This is because the signature of v is not included in the signatures used to trigger the leaving contract.

References

1. Herlihy, M.: Atomic cross-chain swaps. In: Proceedings of the 2018 ACM Symposium on Principles of Distributed Computing. ACM (2018)
2. Nakamoto, S.: Bitcoin: A peer-to-peer electronic cash system (2008)
3. Underwood, S.: Blockchain beyond bitcoin. Commun. ACM **59**(11), 15–17 (2016)
4. Szabo, N.: The idea of smart contracts (1997). http://szabo.best.vwh.net/smartcontractsidea.html
5. Luu, L., Chu, D.H., Olickel, H., Saxena, P., Hobor, A.: Making smart contracts smarter. In: Proceedings of the 2016 ACM SIGSAC Conference on Computer and Communications Security, 24–28 October 2016. ACM (2016)
6. Bowe, S., Hopwood, D.: Hashed time-locked contract transactions. https://github.com/bitcoin/bips/blob/master/bip-0199.mediawiki. Accessed 9 Jan 2018
7. BitCoin Wiki. Hashed timelock contracts Timelock Contracts, October 2018
8. Karp, R.M.: Reducibility among combinatorial problems. In: Proceedings of a Symposium on the Complexity of Computer Computations, Held 20–22 March 1972, at the IBM omas J. Watson Research Center, Yorktown Heights, New York, pp. 85–103 (1972)
9. Becker, A., Geiger, D.: Optimization of Pearl's method of conditioning and greedy-like approximation algorithms for the vertex feedback set problem. Artif. Intell. **83**(1), 167–188 (1996)
10. Abraham, D.J., Blum, A., Sandholm, T.: Clearing algorithms for barter exchange markets: enabling nationwide kidney exchanges. In: Proceedings of the 8th ACM Conference on Electronic Commerce, EC 2007, pp. 295–304. ACM, New York (2007)
11. Dickerson, J.P., Manlove, D.F., Plaut, B., Sandholm, T., Trimble, J.: Position indexed formulations for kidney exchange. CoRR, abs/1606.01623 (2016)
12. Jia, Z., Tang, P., Wang, R., Zhang, H.: Efficient near-optimal algorithms for barter exchange. In: Proceedings of the 16th Conference on Autonomous Agents and MultiAgent Systems, AAMAS 2017, pp. 362–370. International Foundation for Autonomous Agents and Multiagent Systems, Richland (2017)
13. Shapley, L., Scarf, H.: On cores and indivisibility. J. Math. Econ. **1**(1), 23–37 (1974)
14. Kaplan, R.M.: An improved algorithm for multi-way trading for exchange and barter. Electron. Commer. Res. Appl. **10**(1), 67–74 (2011). Special Section: Service Innovation in E-Commerce
15. Herlihy, M., Liskov, B., Shrira, L.: Cross-chain Deals and Adversarial Commerce. CoRR, abs/1905.09743 (2019)
16. Borkowski, M., Sigwart, M., Frauenthaler, P., Hukkinen, T., Schulte, S.: Deterministic Cross-Blockchain Token Transfers. CoRR, abs/1905.06204 (2019)
17. Anta, A.F., Georgiou, C., Nicolaou, N.: Atomic Appends: Selling Cars and Coordinating Armies with Multiple Distributed Ledgers. CoRR, abs/1812.08446 (2018)

Achieving Starvation-Freedom with Greater Concurrency in Multi-Version Object-based Transactional Memory Systems

Chirag Juyal[1]([⊠]), Sandeep Kulkarni[2]([⊠]), Sweta Kumari[1]([⊠]), Sathya Peri[1]([⊠]), and Archit Somani[1]([⊠])

[1] Department of Computer Science and Engineering, IIT Hyderabad, Kandi, Telangana, India
{cs17mtech11014,cs15resch01004,sathya_p,cs15resch01001}@iith.ac.in
[2] Department of Computer Science, Michigan State University, East Lansing, MI, USA
sandeep@cse.msu.edu

Abstract. To utilize the multi-core processors properly concurrent programming is needed. The main challenge is to design a correct and efficient concurrent program. Software Transactional Memory Systems (STMs) provide ease of multithreading to the programmer without worrying about concurrency issues as deadlock, livelock, priority inversion, etc. Most of the STMs work on read-write operations known as RWSTMs. Some STMs work at higher-level operations and ensure greater concurrency than RWSTMs. Such STMs are known as Single-Version Object-based STMs (SVOSTMs). The transactions of SVOSTMs can return *commit* or *abort*. Aborted SVOSTMs transactions retry. But in the current setting of SVOSTMs, transactions may *starve*. So, we propose a *Starvation-Freedom* in *SVOSTM* as *SF-SVOSTM* that satisfies the correctness criteria *conflict-opacity*.

Databases and STMs say that maintaining multiple versions corresponding to each shared data-item (or key) reduces the number of aborts and improves the throughput. So, to achieve greater concurrency further, we propose *Starvation-Freedom* in *Multi-Version OSTM* as *SF-MVOSTM* algorithm. The number of versions maintains by SF-MVOSTM either be unbounded with garbage collection as SF-MVOSTM-GC or bounded with latest K-versions as SF-KOSTM. SF-MVOSTM satisfies the correctness criteria as *local opacity* and shows the performance benefits as compared with state-of-the-art STMs.

Keywords: Software Transactional Memory Systems · Concurrency control · Starvation-Freedom · Multi-version · Opacity · Local opacity

A. Somani—Author sequence follows the lexical order of last names.
This research is partially supported by IMPRINT India project 6918F & gift from Intel, USA.

© Springer Nature Switzerland AG 2019
M. Ghaffari et al. (Eds.): SSS 2019, LNCS 11914, pp. 209–227, 2019.
https://doi.org/10.1007/978-3-030-34992-9_17

1   Introduction

In the era of multi-core processors, we can exploit the cores by concurrent programming. But developing an efficient concurrent program while ensuring the correctness is difficult. Software Transactional Memory Systems (STMs) are a convenient programming interface to access the shared memory concurrently while removing the concurrency responsibilities from the programmer. STMs ensure that consistency issues such as deadlock, livelock, priority inversion, etc will not occur. It provides a high-level abstraction to the programmer with the popular correctness criteria opacity [1], local opacity [2] which consider all the transactions (a piece of code) including aborted one as well in the equivalent serial history. This property makes it different from correctness criteria of database serializability, strict-serializability [3] and ensures even aborted transactions read correct value in STMs which prevent from divide-by-zero, infinite loop, crashes, etc. Another advantage of STMs is composability which ensures the effect of multiple operations of the transaction will be atomic. This paper considers the optimistic execution of STMs in which transactions are writing into its local log until the successful validation.

A traditional STM system invokes following methods:(1) $STM\_begin()$: begins a transaction T_i with unique timestamp i. (2) $STM\_read_i(k)$ (or $r_i(k)$): T_i reads the value of key k from shared memory. (3) $STM\_write_i(k, v)$ (or $w_i(k, v)$): T_i writes the value of k as v locally. (4) $STM_t ryC_i()$: on successful validation, the effect of T_i will be visible to the shared memory and T_i returns commit otherwise (5) $STM_t ryA_i()$: T_i returns abort. These STMs are known as *read-write STMs (RWSTMs)* because it is working at lower-level operations such as read and write.

Herlihy et al. [4], Hassan et al. [5], and Peri et al. [6] have shown that working at higher-level operations such as insert, delete and lookup on the linked-list and hash table gives better concurrency than RWSTMs. STMs which work on higher-level operations are known as *Single-Version Object-based STMs (SVOSTMs)* [6]. It exports the following methods: (1) $STM\_begin()$: begins a transaction T_i with unique timestamp i same as RWSTMs. (2) $STM_lookup_i(k)$ (or $l_i(k)$): T_i lookups key k from shared memory and returns the value. (3) $STM_insert_i(k, v)$ (or $i_i(k, v)$): T_i inserts a key k with value v into its local memory. (4) $STM_delete_i(k)$ (or $d_i(k)$): T_i deletes key k. (5) $STM_t ryC_i()$: the actual effect of $STM\_insert()$ and $STM\_delete()$ will be visible to the shared memory after successful validation and T_i returns commit otherwise (6) $STM_t ryA_i()$: T_i returns abort.

Motivation to Work on SVOSTMs: Figure 1 represents the advantage of SVOSTMs over RWSTMs while achieving greater concurrency and reducing the number of aborts. Figure 1(a) depicts the underlying data structure as a hash table (or ht) with M buckets and bucket 1 stores three keys k_1, k_4 and k_9 in the form of the list. Thus, to access k_4, a thread has to access k_1 before it. Figure 1(b) shows the tree structure of concurrent execution of two transactions T_1 and T_2 with RWSTMs at layer-0 and SVOSTMs at layer-1 respectively. Consider the execution at layer-0, T_1 and T_2 are in conflict because write operation of T_2 on key

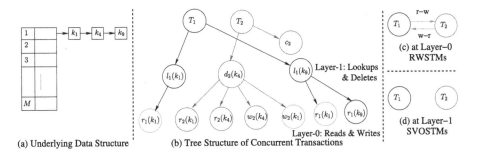

Fig. 1. Advantage of SVOSTMs over RWTSMs

k_1 as $w_2(k_1)$ is occurring between two read operations of T_1 on k_1 as $r_1(k_1)$. Two transactions are in conflict if both are accessing the same key k and at least one transaction performs write operation on k. So, this concurrent execution cannot be atomic as shown in Fig. 1(c). To make it atomic either T_1 or T_2 has to return abort. Whereas execution at layer-1 shows the higher-level operations $l_1(k_1)$, $d_2(k_4)$ and $l_1(k_9)$ on different keys k_1, k_4 and k_9 respectively. All the higher-level operations are isolated to each other so the tree can be pruned [7, Chap 6] from layer-0 to layer-1 and both the transactions return commit with equivalent serial schedule T_1T_2 or T_2T_1 as shown in Fig. 1(d). Hence, some conflicts of RWSTMs does not matter at SVOSTMs which reduce the number of aborts and improve the concurrency using SVOSTMs.

Starvation-Freedom: For long-running transactions along with high conflicts, starvation can occur in SVOSTMs. So, SVOSTMs should ensure the progress guarantee as *starvation-freedom* [8, chap 2]. SVOSTMs is said to be *starvation-free*, if a thread invoking a transaction T_i gets the opportunity to retry T_i on every abort (due to the presence of a fair underlying scheduler [9] with bounded termination) and T_i is not *parasitic*, i.e., if scheduler will give a fair chance to T_i to commit then T_i will eventually return commit. If a transaction gets a chance to commit, still it is not committing because of the infinite loop or some other error such transactions are known as Parasitic transactions [10].

We explored another well known non-blocking progress guarantee *wait-freedom* for STM which ensures every transaction commits regardless of the nature of concurrent transactions and the underlying scheduler [11]. However, Guerraoui and Kapalka [10,12] showed that achieving wait-freedom is impossible in dynamic STMs in which data-items (or keys) of transactions are not known in advance. So in this paper, we explore the weaker progress condition of *starvation-freedom* for SVOSTM while assuming that the keys of the transactions are not known in advance.

Related Work on Starvation-Free STMs: Some researchers Gramoli et al. [13], Waliullah and Stenstrom [14], Spear et al. [15], Chaudhary et al. [9] have explored starvation-freedom in RWSTMs. Most of them assigned priority to the transactions. On conflict, higher priority transaction returns commit

212   C. Juyal et al.

whereas lower priority transaction returns abort. On every abort, a transaction retries a sufficient number of times, will eventually get the highest priority and returns commit. We inspired by this research and propose a novel *Starvation-Free SVOSTM (SF-SVOSTM)* which assigns the priority to the transaction on conflict. In SF-SVOSTM whenever a conflicting transaction T_i aborts, it retries with T_j which has higher priority than T_i. To ensure the starvation-freedom, this procedure will repeat until T_i gets the highest priority and eventually returns commit.

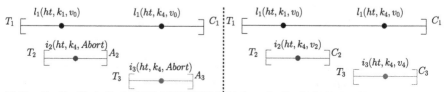

(a). Starvation-Free Single-Version OSTM (SF-SVOSTM) : (b). Starvation-Free Multi-Version OSTM (SF-MVOSTM)

Fig. 2. Benefits of Starvation-Free multi-version OSTM over SF-SVOSTM

Motivation to Propose Starvation-Freedom in Multi-version OSTM: In SF-SVOSTM, if the highest priority transaction becomes slow (for some reason) then it may cause several other transactions to abort and bring down the progress of the system. Figure 2(a) demonstrates this in which the highest priority transaction T_1 became slow so, it is forcing the conflicting transactions T_2 and T_3 to abort again and again until T_1 commits. Database, RWSTMs [16–19] and SVOSTMs [20] say that maintaining multiple versions corresponding to each key reduces the number of aborts and improves throughput.

So, this paper proposes the first *Starvation-Free Multi-Version OSTM (SF-MVOSTM)* which maintains multiple versions corresponding to each key. Figure 2(b) shows the benefits of using SF-MVOSTM in which T_1 lookups from the older version with value v_0 created by transaction T_0 (assuming as initial transaction) for key k_1 and k_4. Concurrently, T_2 and T_3 create the new versions for key k_4. So, all the three transactions commit with equivalent serial schedule $T_1T_2T_3$. So, SF-MVOSTM improves the concurrency than SF-SVOSTM while reducing the number of aborts and ensures the starvation-freedom.

Contributions of the Paper: We propose two Starvation-Free OSTMs as follows:

- Initially, we propose Starvation-Freedom for Single-Version OSTM as SF-SVOSTM which satisfies correctness criteria as *conflict-opacity (or co-opacity)* [6].
- To achieve the greater concurrency further, we propose Starvation-Freedom for Multi-Version OSTM as SF-MVOSTM in Sect. 3 which maintains multiple versions corresponding to each key and satisfies the correctness as *local opacity* [2].

- We propose SF-SVOSTM and SF-MVOSTM for *hash table* and *linked-list* data structure describe in SubSect. 3.2 but its generic for other data structures as well.
- SF-MVOSTM works for unbounded versions with *Garbage Collection (GC)* as SF-MVOSTM-GC which deletes the unwanted versions from version list of keys and for bounded/finite versions as SF-KOSTM which stores finite say latest K number of versions corresponding to each key k. So, whenever any thread creates $(K + 1)^{th}$ version of key, it replaces the oldest version of it. The most challenging task is achieving starvation-freedom in bounded version OSTM because say, the highest priority transaction relies on the oldest version that has been replaced. So, in this case, highest priority transaction has to return abort and hence make it harder to achieve starvation-freedom unlike the approach follow in SF-SVOSTM. Thus, in this paper, we propose a novel approach SF-KOSTM which bridges the gap by developing starvation-free OSTM while maintaining bounded number of versions.
- Section 4 shows that SF-KOSTM is best among all proposed Starvation-Free OSTMs (SF-SVOSTM, SF-MVOSTM, and SF-MVOSTM-GC) for both hash table and linked-list data structure. Proposed hash table based SF-KOSTM (HT-SF-KOSTM) performs 3.9x, 32.18x, 22.67x, 10.8x and 17.1x average speedup on *max-time* for a transaction to commit than state-of-the-art STMs HT-KOSTM [20], HT-SVOSTM [6], ESTM [21], RWSTM [7, Chap. 4], and HT-MVTO [16] respectively. Proposed list based SF-KOSTM (list-SF-KOSTM) performs 2.4x, 10.6x, 7.37x, 36.7x, 9.05x, 14.47x, and 1.43x average speedup on *max-time* for a transaction to commit than state-of-the-art STMs list-KOSTM [20], list-SVOSTM [6], Trans-list [22], Boosting-list [4], NOrec-list [23], list-MVTO [16], and list-KSFTM [9] respectively.

2 System Model and Preliminaries

This section follows the notion and definition described in [12,20], we assume a system of n processes/threads, $th_1, \ldots, th_n$ that run in a completely asynchronous manner and communicate through a set of *keys* $\mathscr{K}$ (or *transaction-objects*). We also assume that none of the threads crash or fail abruptly. In this paper, a thread executes higher-level methods on $\mathscr{K}$ via atomic *transactions* $T_1, \ldots, T_n$ and receives the corresponding response.

Events and Methods: Threads execute the transactions with higher-level methods (or operations) which internally invoke multiple read-write (or lower-level) operations known as *events* (or *evts*). Transaction T_i of the system at read-write level invokes $STM\_begin()$, $STM\_read_i(k)$, $STM\_write_i(k,v)$, $STM\_tryC_i()$ and $STM\_tryA_i()$ as defined in Sect. 1. We denote a method m_{ij} as the jth method of T_i. Method *invocation* (or *inv*) and *response* (or *rsp*) on higher-level methods are also considered as an event.

A thread executes higher-level operations on $\mathscr{K}$ via transaction T_i are known as *methods* (or *mths*). T_i at object level (or higher-level) invokes $STM\_begin()$, $STM\_lookup_i(k)$ (or $l_i(k)$), $STM\_insert_i(k,v)$ (or $i_i(k,v)$), $STM\_delete_i(k)$ (or

$d_i(k))$, $STM\_tryC_i()$, and $STM\_tryA_i()$ methods described in Sect. 1. Here, $STM\_lookup()$, and $STM\_delete()$ return the value from underlying data structure so, we called these methods as *return value methods (orrv_methods)*. Whereas, $STM\_insert()$, and $STM\_delete()$ are updating the underlying data structure after successful $STM\_tryC()$ so, we called these methods as *update methods (or upd_methods)*.

Transactions: We follow multi-level transactions [7] model which consists of two layers. Layer 0 (or lower-level) composed of read-write operations whereas layer 1 (or higher-level) comprises of object-level methods which internally calls multiple read-write events. Formally, we define a transaction T_i at higher-level as the tuple $\langle evts(T_i), <_{T_i} \rangle$, here $<_{T_i}$ represents the total order among all the events of T_i. Transaction T_i cannot invoke any more operation after returning commit ($\mathscr{C}$) or abort ($\mathscr{A}$). Any operation that returns $\mathscr{C}$ or $\mathscr{A}$ are known as $Term(T_i)$ represented as $Term(T_i)$. The transaction which neither committed nor aborted is known as *live transactions* (or *trans.live*).

Histories: A history H consists of multiple transactions, a transaction calls multiple methods and each method internally invokes multiple read-write events. So, a history is a collection of events belonging to the different transactions is represented as $evts(H)$. Formally, we define a history H as the tuple $\langle evts(H), <_H \rangle$, here $<_H$ represents the total order among all the events of H. If all the method invocation of H match with the corresponding response then such history is known as *complete history* denoted as $\overline{H}$. Suppose total transactions in H is *H.trans*, in which number of committed and aborted transactions are *H.committed* and *H.aborted* then the *incomplete history* or *live history* is defined as: *H.incomp = H.live = (H.trans - H.committed - H.aborted)*. This paper considers only *well-form history* which ensures (1) the response of the previous method has received then only the transaction T_i can invoke another method. (2) transaction can not invoke any other method after receiving the response as $\mathscr{C}$ or $\mathscr{A}$.

Due to lack of space, we define other useful notions and definitions used in this paper such as sequential histories [2], real-time order and serial history [3], valid and legal history [20], sub-histories [9], conflict-opacity [6], opacity [1], strict-serializability [3], local opacity [2] formally in accompanying technical report [24].

3   The Proposed SF-KOSTM Algorithm

In this section, we propose *Starvation-Free K-version OSTM (SF-KOSTM)* algorithm which maintains K number of versions corresponding to each key. The value of K is application dependent and may vary from 1 to ∞. When K is equal to 1 then SF-KOSTM boils down to *Starvation-Free Single-Version OSTM (SF-SVOSTM)*. When K is ∞ then SF-KOSTM maintains unbounded versions corresponding to each key known as *Starvation-Free Multi-Version OSTM (SF-MVOSTM)* algorithm. To delete the unused version from the version list of

SF-MVOSTM, we develop a separate Garbage Collection (GC) method [16] and propose SF-MVOSTM-GC. In this paper, we propose SF-SVOSTM and all the variants of SF-KOSTM *(SF-MVOSTM, SF-MVOSTM-GC, SF-KOSTM)* for two data structures *hash table* and *linked-list* but it is generic for other data structures as well.

SubSection 3.1 describes the definition of *starvation-freedom* followed by our assumption about the scheduler that helps us to achieve starvation-freedom in SF-KOSTM. SubSection 3.2 explains the design and data structure of SF-KOSTM. SubSection 3.3 shows the working of SF-KOSTM algorithm.

3.1 Description of Starvation-Freedom

Definition 1. *Starvation-Freedom:* *An STM system is said to be starvation-free if a thread invoking a non-parasitic transaction T_i gets the opportunity to retry T_i on every abort, due to the presence of a fair scheduler, then T_i will eventually commit.*

Herlihy and Shavit [11] defined the fair scheduler which ensures that none of the thread will crash or delayed forever. Hence, any thread Th_i acquires the lock on the shared data-items while executing transaction T_i will eventually release the locks. So, a thread will never block other threads to progress. To satisfy the starvation-freedom for SF-KOSTM, we assumed bounded termination for the fair scheduler.

Assumption 1 *Bounded-Termination:* *For any transaction T_i, invoked by a thread Th_i, the fair system scheduler ensures, in the absence of deadlocks, Th_i is given sufficient time on a CPU (and memory, etc) such that T_i terminates ($\mathscr{C}$ or $\mathscr{A}$) in bounded time.*

In the proposed algorithms, we have considered *TB* as the maximum time-bound of a transaction T_i within this either T_i will return commit or abort in the absence of deadlock. Approach for achieving the *deadlock-freedom* is motivated from the literature in which threads executing transactions acquire the locks in increasing order of the keys and releases the locks in bounded time either by committing or aborting the transaction. We consider an assumption about the transactions of the system as follows.

Assumption 2. *We assume, if other concurrent conflicting transactions do not exist in the system then every transaction will commit. i.e. (a) If a transaction T_i is executing in the system with the absence of other conflicting transactions then T_i will not self-abort. (b) Transactions of the system are non-parasitic as explained in Sect. 1.*

If transactions self-abort or parasitic then ensuring starvation-freedom is impossible.

3.2   Design and Data Structure of SF-KOSTM Algorithm

In this subsection, we show the design and underlying data structure of SF-KOSTM algorithm to maintain the shared data-items (or keys).

To achieve the *Starvation-Freedom* in *K-version Object-based STM (SF-KOSTM)*, we use chaining hash table (or *ht*) as an underlying data structure where the size of the hash table is M buckets as shown in Fig. 3(a) and we propose HT-SF-KOSTM. Hash table with bucket size *one* becomes the linked-list data structure for SF-KOSTM represented as list-SF-KOSTM. The representation of SF-KOSTM is similar to MVOSTM [20]. Each bucket stores multiple nodes in the form of linked-list between the two sentinel nodes *Head*(-∞) and *Tail*(+∞). Figure 3(b) illustrates the structure of each node as ⟨*key, lock, mark, vl, nNext*⟩. Where *key* is the unique value from the range of [1 to $\mathscr{K}$] stored in the increasing order between the two sentinel nodes similar to linked-list based concurrent set implementation [25, 26]. The *lock* field is acquired by the transaction before updating (inserting or deleting) on the node. *mark* is the boolean field which says anode is deleted or not. If *mark* sets to true then node is logically deleted else present in the hash table. Here, the deletion is in a lazy manner similar to concurrent linked-list structure [25]. The field *vl* stands for version list. SF-KOSTM maintains the finite say latest K-versions corresponding to each key to achieving the greater concurrency as explained in Sect. 1. Whenever $(K+1)^{th}$ version created for the key then it overwrites the oldest version corresponding to that key. If K is equal to 1, i.e., version list contains only one version corresponding to each key which boils down to *Starvation-Free Single-Version OSTM (SF-SVOSTM)*. So, the data structure of SF-SVOSTM is same as SF-KOSTM with one version. The field *nNext* points to next available node in the linked-list. From now onwards, we will use the term key and node interchangeably.

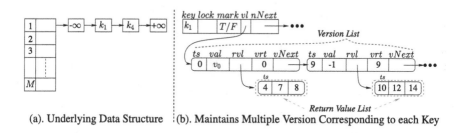

(a). Underlying Data Structure ┊ (b). Maintains Multiple Version Corresponding to each Key

Fig. 3. Design and data structure of SF-KOSTM

The structure of the *vl* is ⟨*ts, val, rvl, vrt, vNext*⟩ as shown in Fig. 3(b). *ts* is the unique timestamp assigned by the *STM_begin()*. If the value (*val*) is *nil* then version is created by the *STM_delete()* otherwise *STM_insert()* creates a version with not *nil* value. To satisfy the correctness criteria as *local opacity*, *STM_delete()* also maintains the version corresponding to each key with *mark* field as *true*. It allows the concurrent transactions to lookup from the older

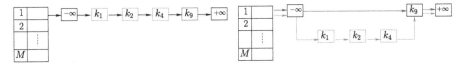

Fig. 4. Searching k_9 over *lazy-list* (Color figure online)

Fig. 5. Searching k_9 over *rblazy-list* (Color figure online)

version of the marked node and returns the value as not *nil*. *rvl* stands for *return value list* which maintains the information about lookup transaction that has lookups from a particular version. It maintains the timestamp (*ts*) of rv_methods (*STM_lookup()* or *STM_delete()*) transaction in it. *vrt* stands for *version real-time* which helps to maintain the *real-time order* among the transactions. *vNext* points to the next available version in the version list.

Maintaining the deleted node along with the live (not deleted) node will increase the traversal time to search a particular node in thelist. Consider Fig. 4, where red color depicts the deleted node $\langle k_1, k_2, k_4 \rangle$ and blue color depicts the live node $\langle k_9 \rangle$. When any method of SF-KOSTM searches the key k_9 then it has to traverse the deleted nodes $\langle k_1, k_2, k_4 \rangle$ as well before reach to k_9 that increases the traversal time.

This motivated us to modify the lazy-list structure of a node to form a skip list based on red and blue links. We called it as a *red-blue lazy-list* or *rblazy-list*. This idea has been explored by Peri et al. in SVOSTMs [6]. *rblazy-list* maintains two-pointer corresponding to each node such as red link (RL) and blue link (BL). Where BL points to the live node and RL points to live node as well as a deleted node. Let us consider the same example as discussed above with this modification, key k_9 is directly searched from the head of the list with the help of BL as shown in Fig. 5. In this case, traversal time is efficient because any method of SF-KOSTM need not traverse the deleted nodes. To maintain the RL and BL in each node we modify the structure of *lazy-list* as $\langle key, lock, mark, vl, RL, BL, nNext \rangle$ and called it as *rblazy-list*.

3.3   Working of SF-KOSTM Algorithm

In this subsection, we describe the working of SF-KOSTM algorithm which includes the detail description of SF-KOSTM methods and challenges to make it starvation-free. This description can easily be extended to SF-MVOSTM and SF-MVOSTM-GC as well.

SF-KOSTM invokes *STM_begin(), STM_lookup(), STM_delete(), STM_insert()*, and *STM_tryC()* methods. *STM_lookup()* and *STM_delete()* work as rv_methods() which lookup the value of key k from shared memory and return it. Whereas *STM_insert()* and *STM_delete()* work as upd_methods() that modify the value of k in shared memory. We propose optimistic SF-KOSTM, so, upd_methods() first update the value of k in transaction local log *txLog* and the actual effect of upd_methods() will be visible after successful *STM_tryC()*. Now, we explain the functionality of each method as follows:

STM_begin(): When a thread Th_i invokes transaction T_i for the first time (or first incarnation), *STM_begin()* assigns a unique timestamp known as *current timestamp (cts)* using an atomic global counter *(gcounter)*. If T_i gets aborted then thread Th_i executes it again with the new incarnation of T_i, say T_j with the new *cts* until T_i commits but retains its initial *cts* as *initial timestamp (its)*. Th_i uses *its* to inform the SF-KOSTM system that whether T_i is a new invocation or an incarnation. If T_i is the first incarnation then *its* and *cts* are same as cts_i so, Th_i maintains $\langle its_i, cts_i \rangle$. If T_i gets aborted and retries with T_j then Th_i maintains $\langle it_i, ct_j \rangle$.

By assigning priority to the lowest *its* transaction (i.e. transaction have been in the system for a longer time) in *Single-Version OSTM*, *Starvation-Freedom* can easily be achieved as explained in Sect. 1. The detailed working of *Starvation-Free Single-Version OSTM (SF-SVOSTM)* is in accompanying technical report [24]. But achieving *Starvation-Freedom* in finite *K-versions OSTM (SF-KOSTM)* is challenging. Though the transaction T_i has lowest *its* but T_i may return abort because of finite versions T_i did not find a correct version to lookup from or overwrite a version. Table 1 shows the key insight to achieve the starvation-freedom in finite K-versions OSTM. Here, we considered two transaction T_{10} and T_{20} with *cts* 10 and 20 that performs *STM_lookup()* (or *l*) and *STM_insert()* (or *i*) on same key k. We assume that a version of k exists with *cts* 5, so, *STM_lookup()* of T_{10} and T_{20} find a previous version to lookup and never return abort. Due to the optimistic execution in SF-KOSTM, effect of *STM_insert()* comes after successful *STM_tryC()*, so *STM_lookup()* of a transaction comes before effect of its *STM_insert()*. Hence, a total of six permutations are possible as defined in Table 1. We can observe from Table 1 that in some cases T_{10} returns abort. But if T_{20} gets the lowest *its* then T_{20} never returns abort. This ensures that a transaction with lowest *its* and highest *cts* will never return abort. But achieving highest *cts* along with lowest *its* is a bit difficult because new transactions are keep on coming with higher *cts* using *gcounter*. So, to achieve the highest *cts*, we introduce a new timestamp as *working timestamp (wts)* which is significantly larger than *cts*.

STM_begin() maintains the *wts* for transaction T_i as wts_i, which is potentially higher timestamp as compare to cts_i. So, we derived,

$$wts_i = cts_i + C * (cts_i - its_i); \tag{1}$$

where C is any constant value greater than 0. When T_i is issued for the first time then wts_i, cts_i, and its_i are same. If T_i gets aborted again and again then drift between the cts_i and wts_i will increases. The advantage for maintaining wts_i is if any transaction keeps getting aborted then its wts_i will be high and its_i will be low. Eventually, T_i will get chance to commit in finite number of steps to achieve starvation-freedom. For simplicity, we use timestamp (*ts*) *i* of T_i as wts_i, i.e., $\langle wts_i = i \rangle$ for SF-KOSTM.

Observation 1. *Any transaction T_i with lowest its_i and highest wts_i will never abort.*

Table 1. Possible permutations of methods

S. No	Execution sequence	Possible actions by transactions
1	$l_{10}(k), i_{10}(k), l_{20}(k), i_{20}(k)$	$T_{20}(k)$ lookups the version inserted by T_{10}. No conflict
2	$l_{10}(k), l_{20}(k), i_{10}(k), i_{20}(k)$	Conflict detected at $i_{10}(k)$. Either abort T_{10} or T_{20}
3	$l_{10}(k), l_{20}(k), i_{20}(k), i_{10}(k)$	Conflict detected at $i_{10}(k)$. Hence, abort T_{10}
4	$l_{20}(k), l_{10}(k), i_{20}(k), i_{10}(k)$	Conflict detected at $i_{10}(k)$. Hence, abort T_{10}
5	$l_{20}(k), l_{10}(k), i_{10}(k), i_{20}(k)$	Conflict detected at $i_{10}(k)$. Either abort T_{10} or T_{20}
6	$l_{20}(k), i_{20}(k), l_{10}(k), i_{10}(k)$	Conflict detected at $i_{10}(k)$. Hence, abort T_{10}

Sometimes, the value of wts is significantly larger than cts. So, wts is unable to maintain *real-time order* between the transactions which violates the correctness of SF-KOSTM. To address this issue SF-KOSTM uses the idea of timestamp ranges [27–29] along with $\langle its_i, cts_i, wts_i \rangle$ for transaction T_i in $STM\_begin()$. It maintains the *transaction lower timestamp limit* ($tltl_i$) and *transaction upper timestamp limit* ($tutl_i$) for T_i. Initially, $\langle its_i, cts_i, wts_i, tltl_i \rangle$ are the same for T_i. $tutl_i$ would be set as a largest possible value denoted as $+\infty$ for T_i. After successful execution of $rv\_methods()$ or $STM\_tryC()$ of T_i, $tltl_i$ gets incremented and $tutl_i$ gets decremented[1] to respect the real-time order among the transactions. $STM\_begin()$ initializes the *transaction local log* ($txLog_i$) for each transaction T_i to store the information in it. Whenever a transaction starts it atomically sets its *status* to be *live* as a global variable. Transaction *status* can be $\langle live, commit, false \rangle$. After successful execution of $STM\_tryC()$, T_i sets its *status* to be *commit*. If the *status* of the transaction is *false* then it returns *abort*. For more details of $STM\_begin()$ please refer the accompanying technical report [24].

***STM_lookup*() and *STM_delete*() as rv_methods():** $rv\_methods(ht, k, val)$ return the value (val) corresponding to the key k from the shared memory as hash table (ht). We show the high-level overview of the rv_methods() in Algorithm 1. First, it identifies the key k in the transaction local log as $txLog_i$ for transaction T_i. If k exists then it updates the $txLog_i$ and returns the val at Line 3.

If key k does not exist in the $txLog_i$ then before identify the location in share memory rv_methods() check the *status* of T_i at Line 6. If *status* of T_i (or i) is *false* then T_i has to *abort* which says that T_i is not having the lowest *its* and highest *wts* among other concurrent conflicting transactions. So, to propose starvation-freedom in SF-KOSTM other conflicting transactions set the *status* of T_i as *false* and force it to *abort*.

If the *status* of T_i is not *false* and key k does not exist in the $txLog_i$ then it identifies the location of key k optimistically (without acquiring the locks similar to the *lazy-list* [25]) in the shared memory at Line 8. SF-KOSTM maintains the shared memory in the form of a hash table with M buckets as shown in SubSect. 3.2, where each bucket stores the keys in *rblazy-list*. Each

---

[1] Practically ∞ can not be decremented for $tutl_i$ so we assign the highest possible value to $tutl_i$ which gets decremented.

node contains two pointer $\langle RL, BL \rangle$. So, it identifies the two *predecessors (pred)* and two *current (curr)* with respect to each node. First, it identifies the pred and curr for key k in BL as $\langle preds[0], currs[1] \rangle$. After that it identifies the pred and curr for key k in RL as $\langle preds[1], currs[0] \rangle$. If $\langle preds[1], currs[0] \rangle$ are not marked then $\langle preds[0] = preds[1], currs[1] = currs[0] \rangle$. SF-KOSTM maintains the keys are in increasing order. So, the order among the nodes are $\langle preds[0].key \leq preds[1].key < k \leq currs[0].key \leq currs[1].key \rangle$.

rv_methods() acquire the lock in predefined order on all the identified preds and currs for key k to avoid the deadlock at Line 9 and do the *rv_Validation()* at Line 10. If $\langle preds[0] \vee currs[1] \rangle$ is marked or preds are not pointing to identified currs as $\langle (preds[0].BL \neq currs[1]) \vee (preds[1].RL \neq currs[0]) \rangle$ then it releases the locks from all the preds and currs and identify the new preds and currs for k in shared memory.

Algorithm 1 *rv_methods(ht, k, val):* It can either be $STM\_delete_i(ht, k, val)$ or $STM\_lookup_i(ht, k, val)$ on key k by transaction T_i.

```
1:  procedure rv_methods_i(ht, k, val)
2:      if (k ∈ txLog_i) then
3:          Update the local log of T_i and return val.
4:      else
5:          /*Atomically check the status of its own transac-
            tion T_i (or i).*/
6:          if (i.status == false) then return ⟨abort_i⟩.
7:          end if
8:          Identify the preds[] and currs[] for key k in
            bucket M_k of rblazy-list using BL and RL.
9:          Acquire locks on preds[] & currs[] in increasing
            order of keys to avoid the deadlock.
10:         if (!rv_Validation(preds[], currs[])) then
11:             Release the locks and goto Line 8.
12:         end if
13:         if (k ∉ M_k.rblazy-list) then
14:             Create a new node n with key k as:
                ⟨key=k, lock=false, mark=true, vl=ver,
                nNext=φ⟩./*n is marked*/
15:             Create version ver as:⟨ts=0, val=nil, rvl=i,
                vrt=0, vNext=φ⟩.
16:             Insert n into M_k.rblazy-list s.t. it is accessi-
                ble only via RLs. /*lock sets true*/
17:             Release locks; update the txLog_i with k.

18:             return ⟨val⟩. /*val as nil*/
19:         end if
20:         Identify the version ver_j with ts = j such that
            j is the largest timestamp smaller (lts) than i.
21:         if (ver_j == nil) then /*Finite Versions*/
22:             return ⟨abort_i⟩
23:         else if (ver_j.vNext != nil) then
24:             /*tutl_i should be less then vrt of next ver-
                sion ver_j*/
25:             Calculate tutl_i = min(tutl_i, ver_j.vNext
                .vrt − 1).
26:         end if
27:         /*tltl_i should be greater then vrt of ver_j*/
28:         Calculate tltl_i = max(tltl_i, ver_j.vrt + 1).
29:         /*If limit has crossed each other then abort T_i*/
30:         if (tltl_i > tutl_i) then return ⟨abort_i⟩.
31:         end if
32:         Add i into the rvl of ver_j.
33:         Release the locks; update the txLog_i with k
            and value.
34:     end if
35:     return ⟨ver_j.val⟩.
36: end procedure
```

If key k does not exist in the *rblazy-list* of corresponding bucket M_k at Line 13 then it creates a new node n with key k as $\langle key=k, lock=false, mark=true, vl=ver, nNext=\phi \rangle$ at Line 14 and creates a version (ver) for transaction T_0 as $\langle ts = 0, val = nil, rvl = i, vrt = 0, vNext = \phi \rangle$ at Line 15. Transaction T_i creates the version of T_0, so, other concurrent conflicting transaction (say T_p) with lower timestamp than T_i, i.e., $\langle p < i \rangle$ can lookup from T_0 version. Thus, T_i save T_p to abort while creating a T_0 version and ensures greater concurrency. After that T_i adds its wts_i in the *rvl* of T_0 and sets the *vrt* 0 as the timestamp of T_0 version. Finally, it inserts the node n into $M_k.rblazy-list$ such that it is accessible via RL only at Line 16. rv_methods() releases the locks and update the $txLog_i$ with key k and value as *nil* (Line 17). Eventually, it returns the *val* as *nil* at Line 18.

If key k exists in the $M_k.rblazy\text{-}list$ then it identifies the current version ver_j with $ts = j$ such that j is the *largest timestamp smaller (lts)* than i at Line 20 and there exists no other version with timestamp p by T_p on same key k such that $\langle j < p < i \rangle$. If ver_j is *nil* at Line 21 then SF-KOSTM returns *abort* for transaction T_i because it does not found a version to lookup otherwise it identifies the next version with the help of $ver_j.vNext$. If next version ($ver_j.vNext$ as ver_k) exist then T_i maintains the $tutl_i$ with the minimum of $\langle tutl_i \vee ver_k.vrt - 1 \rangle$ at Line 25 and $tltl_i$ with a maximum of $\langle tltl_i \vee ver_j.vrt + 1 \rangle$ at Line 28 to respect the *real-time order* among the transactions. If $tltl_i$ is greater than $tutl_i$ at Line 30 then transaction T_i returns *abort* (fail to maintains real-time order) otherwise it adds the ts of T_i (wts_i) in the rvl of ver_j at Line 32. Finally, it releases the lock and updates the $txLog_i$ with key k and value as the current version value ($ver_j.val$) at Line 33. Eventually, it returns the value as $ver_j.val$ at Line 35.

STM_insert() and STM_delete() as upd_methods(): Actual effect of *STM_insert()* and *STM_delete()* come after successful *STM_tryC()*. They create the version corresponding to the key in shared memory. We show the high level view of *STM_tryC()* in Algorithm 2. First, *STM_tryC()* checks the *status* of the transaction T_i at Line 39. If the *status* of T_i is *false* then T_i returns *abort* with similar reasoning explained above in rv_methods().

If the *status* is not false then *STM_tryC()* sort the keys (exist in $txLog_i$ of T_i) of *upd_methods()* in increasing order. It takes the method (m_{ij}) from $txLog_i$ one by one and identifies the location of the key k in $M_k.rblazy\text{-}list$ as explained above in rv_methods(). After identifying the preds and currs for k it acquire the locks in predefined order to avoid the deadlock at Line 46 and calls *tryC_Validation()* to validate the methods of T_i.

tryC_Validation() identifies whether the methods of invoking transaction T_i are able to insert or delete a version corresponding to the keys while ensuring the progress guarantee as *starvation-freedom* and maintaining the *real-time order* among the transactions. It does four steps for validation. Step 1: First, it does the *rv_Validation()* as explained in rv_methods() above. Step 2: If *rv_Validation()* is successful and key k is exist in the $M_k.rblazy\text{-}list$ then it identifies the current version ver_j with $ts = j$ such that j is the *largest timestamp smaller (lts)* than i. If ver_j is *not exist* then SF-KOSTM returns *abort* for transaction T_i because it does not found the version to replace. Step 3: If ver_j *exist* then T_i compares its_i with *its* of other *live* transactions present in $ver_j.rvl$. If its_i of T_i is less than the *its* of such transactions then T_i sets the *status* of all those transactions to be *false*, otherwise, T_i returns *abort*. Step 4: To maintain the *real-time order*, T_i update the $tltl_i$ and $tutl_i$ of it with the help of ver_j and its next version ($ver_j.vNext$) respectively (explained in rv_methods() above). Please find the detailed descriptions of *tryC_Validation()* in accompanying technical report [24].

If all the steps of the *tryC_Validation()* is successful then the actual effect of the *STM_insert()* and *STM_delete()* will be visible to the shared memory. At Line 53, *STM_tryC()* checks for *poValidation()*. When two subsequent methods $\langle m_{ij}, m_{ik} \rangle$ of the same transaction T_i identify the overlapping location of preds and currs in *rblazy-list*. Then *poValidation()* updates the current method m_{ik} preds and currs with the help of previous method m_{ij} preds and currs.

If m_{ij} is *STM_insert()* and key k is not exist in the $M_k.rblazy\text{-}list$ then it creates the new node n with key k as $\langle key=k,lock=false,mark=false, vl=ver,nNext=\phi \rangle$ at Line 55. Later, it creates a version (ver) for transaction T_0 and T_i as $\langle\ ts=0,\ val=nil,\ rvl=i,\ vrt=0,\ vNext=i\ \rangle$ and $\langle ts=i,\ val=v,\ rvl=\phi,\ vrt=i,\ vNext=\phi \rangle$ at Line 56. The T_0 version created by transaction T_i to helps other concurrent conflicting transactions (with lower timestamp than T_i) to lookup from T_0 version. Finally, it inserts the node n into $M_k.rblazy\text{-}list$ such that it is accessible via RL as well as BL at Line 57. If m_{ij} is *STM_insert()* and key k exists in the $M_k.rblazy\text{-}list$ then it creates the new version ver_i as $\langle ts=i,\ val=v,\ rvl=\phi,\ vrt=i,\ vNext=\phi \rangle$ corresponding to key k. If the limit of the version reaches to K then SF-KOSTM replaces the oldest version with $(K+1)^{th}$ version which is accessible via RL as well as BL at Line 60.

Algorithm 2 STM_tryC (T_i): Validate the upd_methods() of T_i and returns *commit*.

```
37: procedure STM_tryC(Ti)
38:    /*Atomically check the status of its own transaction
       Ti (or i)*/
39:    if (i.status == false) then return ⟨aborti⟩.
40:    end if
41:    /*Sort the keys of txLogi in increasing order.*/
42:    /*Method (m) will be either STM_insert or STM_-
       delete*/
43:    for all (mij ∈ txLogi) do
44:       if(mij==STM_insert ||mij==STM_delete)then
45:          Identify the preds[] & currs[] for key k in
             bucket Mk of rblazy-list using BL & RL.
46:          Acquire the locks on preds[] & currs[] in
             increasing order of keys to avoid deadlock.
47:          if (! tryC_Validation()) then
48:             return ⟨aborti⟩.
49:          end if
50:       end if
51:    end for
52:    for all (mij ∈ txLogi) do
53:       poValidation() modifies the preds[] & currs[] of
          current method which would have been updated
          by previous method of the same transaction.

54:       if ((mij==STM_insert)&&(k∉Mk.rblazy-list))
          then
55:          Create new node n with k as: ⟨key=k,
             lock=false, mark= false, vl=ver, nNext=φ⟩.
56:          Create first version ver for T0 and next for
             i: ⟨ts=i, val=v, rvl=φ, vrt=i, vNext=φ⟩.
57:          Insert node n into Mk.rblazy-list such that
             it is accessible via RL as well as BL.
58:          /*lock sets true*/
59:       else if (mij == STM_insert) then
60:          Add  ver:  ⟨ts=i,  val=v,  rvl=φ,  vrt=i,
             vNext=φ⟩ into Mk.rblazy-list & accessible
             via RL, BL. /*mark=false*/
61:       end if
62:       if (mij == STM_delete) then
63:          Add  ver:⟨ts=i,  val=nil,  rvl=φ,  vrt=i,
             vNext=φ⟩ into Mk.rblazy-list & accessible
             via RL only. /*mark=true*/
64:       end if
65:       Update preds[] & currs[] of mij in txLogi.
66:    end for
67:    Release the locks; return ⟨commiti⟩.
68: end procedure
```

If m_{ij} is *STM_delete()* and key k exists in the $M_k.rblazy\text{-}list$ then it creates the new version ver_i as $\langle ts=i,\ val=nil,\ rvl=\phi,\ vrt=i,\ vNext=\phi \rangle$ which is accessible via RL only at Line 63. At last it updates the preds and currs of each m_{ij} into its $txLog_i$ to help the upcoming methods of the same transactions in *poValidation()* at Line 65. Finally, it releases the locks on all the keys in a predefined order and returns *commit* at Line 67.

Theorem 1. *Any legal history H generated by SF-SVOSTM satisfies co-opacity.*

Theorem 2. *Any valid history H generated by SF-KOSTM satisfies local-opacity.*

Theorem 3. *SF-SVOSTM and SF-KOSTM ensure starvation-freedom in presence of a fair scheduler that satisfies Assumption 1(bounded-termination) and in the absence of parasitic transactions that satisfies Assumption 2.*

Please find the proof of theorems in accompanying technical report [24].

4 Experimental Evaluation

This section represents the experimental analysis of variants of the proposed Starvation-Free Object-based STMs (SF-SVOSTM, SF-MVOSTM, SF-MVOSTM-GC, and SF-KOSTM)[2] for two data structure *hash table* (HT-SF-SVOSTM, HT-SF-MVOSTM, HT-SF-MVOSTM-GC and HT-SF-KOSTM) and *linked-list* (list-SF-SVOSTM, list-SF-MVOSTM, list-SF-MVOSTM-GC and list-SF-KOSTM) implemented in C++. We analyzed that HT-SF-KOSTM and list-SF-KOSTM perform best among all the proposed algorithms. So, we compared our HT-SF-KOSTM with hash table based state-of-the-art STMs HT-KOSTM [20], HT-SVOSTM [6], ESTM [21], RWSTM [7, Chap. 4], HT-MVTO [16] and our list-SF-KOSTM with list based state-of-the-art STMs list-KOSTM [20], list-SVOSTM [6], Trans-list [22], Boosting-list [4], NOrec-list [23], list-MVTO [16], list-KSFTM [9].

Experimental Setup: The system configuration for experiments is 2 socket Intel(R) Xeon(R) CPU E5-2690 v4 @ 2.60 GHz with 14 cores per socket and 2 hyper-threads per core, a total of 56 threads. A private 32 KB L1 cache and 256 KB L2 cache is with each core. It has 32 GB RAM with Ubuntu 16.04.2 LTS running Operating System. Default scheduling algorithm of Linux with all threads have the same base priority is used in our experiments. This satisfies Assumption 1 (bounded-termination) of the scheduler and we ensure the absence of parasitic transactions for our setup to satisfy Assumption 2.

Methodology: We have considered three different types of workloads namely, W1 (Lookup Intensive - 5% insert, 5% delete, and 90% lookup), W2 (Mid Intensive - 25% insert, 25% delete, and 50% lookup), and W3 (Update Intensive - 45% insert, 45% delete, and 10% lookup). To analyze the absolute benefit of starvation-freedom, we used a customized application called as the *Counter Application* (refer the pseudo-code in the technical report [24]) which provides us the flexibility to create a high contention environment where the probability of transactions undergoing starvation on an average is very high. Our *high contention* environment includes only 30 shared data-items (or keys), number of threads ranging from 50 to 250, each thread spawns upon a transaction, where each transaction performs 10 operations depending upon the workload chosen. To study starvation-freedom of various algorithms, we have used *max-time* which

---

[2] Code is available here: https://github.com/PDCRL/SF-MVOSTM.

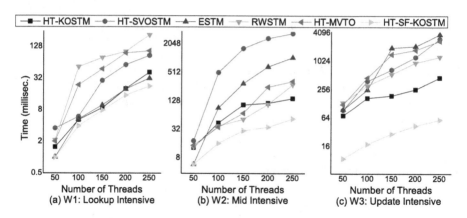

Fig. 6. Performance analysis of SF-KOSTM and state-of-the-art STMs on hash table

is the maximum time required by a transaction to finally commit from its first incarnation, which also involves time taken by all its aborted incarnations. We perform each of our experiments 10 times and consider the average of it to avoid the effect of outliers.

Results Analysis: All our results reflect the same ideology as proposed showcasing the benefits of Starvation-Freedom in Multi-Version OSTMs. We started our experiments with *hash table* data structure of bucket size 5 and compared *max-time* for a transaction to commit by proposed HT-SF-KOSTM (best among all the proposed algorithms shown in the technical report [24]) with hash table based state-of-the-art STMs. HT-SF-KOSTM achieved an average speedup of 3.9x, 32.18x, 22.67x, 10.8x and 17.1x over HT-KOSTM, HT-SVOSTM, ESTM, RWSTM and HT-MVTO respectively as shown in Fig. 6.

We further considered another data structure *linked-list* and compared *max-time* for a transaction to commit by proposed list-SF-KOSTM (best among all the proposed algorithms shown in the technical report [24]) with list based state-of-the-arts STMs. list-SF-KOSTM achieved an average speedup of 2.4x, 10.6x, 7.37x, 36.7x, 9.05x, 14.47x, and 1.43x over list-KOSTM, list-SVOSTM, Trans-list, Boosting-list, NOrec-list, list-MVTO, and list-KSFTM respectively as shown in Fig. 7. We consider the number of versions in the version list K as 5 and value of C as 0.1.

For additional experiments please refer the technical report [24] which shows the performance of HT-SF-KOSTM and list-SF-KOSTM under *low contention* is slightly lesser than non starvation-free HT-KOSTM and list-KOSTM. It also has plots of abort counts while varying the threads, best value of K and C, *stability* and *memory consumption*.

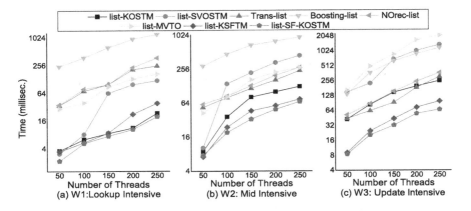

Fig. 7. Performance analysis of SF-KOSTM and state-of-the-art STMs on list

5 Conclusion

We proposed a novel *Starvation-Free K-Version Object-based STM (SF-KOSTM)* which ensure the *starvation-freedom* while maintaining the latest K-versions corresponding to each key and satisfies the correctness criteria as *local-opacity*. The value of K can vary from 1 to ∞. When K is equal to 1 then SF-KOSTM boils down to *Single-Version Starvation-Free OSTM (SF-SVOSTM)*. When K is ∞ then SF-KOSTM algorithm maintains unbounded versions corresponding to each key known as *Multi-Version Starvation-Free OSTM (SF-MVOSTM)*. To delete the unused version from the version list of SF-MVOSTM, we developed a separate Garbage Collection (GC) method and proposed SF-MVOSTM-GC. SF-KOSTM provides greater concurrency and higher throughput using higher-level methods. We implemented all the proposed algorithms for *hash table* and *linked-list* data structure but it is generic for other data structures as well. Results of SF-KOSTM shows significant performance gain over state-of-the-art STMs.

Acknowledgments. We are thankful to the anonymous reviewers for carefully reading the paper and providing us valuable suggestions.

References

1. Guerraoui, R., Kapalka, M.: On the correctness of transactional memory. In: PPoPP (2008)
2. Kuznetsov, P., Peri, S.: Non-interference and local correctness in transactional memory. Theor. Comput. Sci. **688**, 103–116 (2017)
3. Papadimitriou, C.H.: The serializability of concurrent database updates. J. ACM **26**(4), 631–653 (1979)
4. Herlihy, M., Koskinen, E.: Transactional boosting: a methodology for highly-concurrent transactional objects. In: PPOPP (2008)

5. Hassan, A., Palmieri, R., Ravindran, B.: Optimistic transactional boosting. In: PPoPP (2014)

6. Peri, S., Singh, A., Somani, A.: Efficient means of achieving composability using transactional memory. In: NETYS 2018 (2018)

7. Weikum, G., Vossen, G.: Transactional Information Systems: Theory, Algorithms, and the Practice of Concurrency Control and Recovery. Morgan Kaufmann, Burlington (2002)

8. Herlihy, M., Shavit, N.: The Art of Multiprocessor Programming, Revised Reprint, 1st edn. Morgan Kaufmann Publishers Inc., San Francisco (2012)

9. Chaudhary, V.P., Juyal, C., Kulkarni, S., Kumari, S., Peri, S.: Achieving starvation-freedom in multi-version transactional memory systems. In: Atig, M.F., Schwarzmann, A.A. (eds.) NETYS 2019. LNCS, vol. 11704, pp. 291–310. Springer, Cham (2019). https://doi.org/10.1007/978-3-030-31277-0_20

10. Bushkov, V., Guerraoui, R., Kapalka, M.: On the liveness of transactional memory. In: PODC 2012 (2012)

11. Herlihy, M., Shavit, N.: On the nature of progress. In: Fernàndez Anta, A., Lipari, G., Roy, M. (eds.) OPODIS 2011. LNCS, vol. 7109, pp. 313–328. Springer, Heidelberg (2011). https://doi.org/10.1007/978-3-642-25873-2_22

12. Guerraoui, R., Kapalka, M.: Principles of Transactional Memory, Synthesis Lectures on Distributed Computing Theory. Morgan and Claypool, San Rafael (2010)

13. Gramoli, V., Guerraoui, R., Trigonakis, V.: TM2C: a software transactional memory for many-cores. In: EuroSys 2012 (2012)

14. Waliullah, M.M., Stenström, P.: Schemes for avoiding starvation in transactional memory systems. Practice Exp. Concurr. Comput. **21**, 859–873 (2009)

15. Spear, M.F., Dalessandro, L., Marathe, V.J., Scott, M.L.: A comprehensive strategy for contention management in software transactional memory (2009)

16. Kumar, P., Peri, S., Vidyasankar, K.: A timestamp based multi-version STM algorithm. In: Chatterjee, M., Cao, J., Kothapalli, K., Rajsbaum, S. (eds.) ICDCN 2014. LNCS, vol. 8314, pp. 212–226. Springer, Heidelberg (2014). https://doi.org/10.1007/978-3-642-45249-9_14

17. Lu, L., Scott, M.L.: Generic multiversion STM. In: Afek, Y. (ed.) DISC 2013. LNCS, vol. 8205, pp. 134–148. Springer, Heidelberg (2013). https://doi.org/10.1007/978-3-642-41527-2_10

18. Fernandes, S.M., Cachopo, J.: Lock-free and scalable multi-version software transactional memory. In: PPoPP 2011, New York, NY, USA (2011)

19. Perelman, D., Byshevsky, A., Litmanovich, O., Keidar, I.: SMV: selective multiversioning STM. In: Peleg, D. (ed.) DISC 2011. LNCS, vol. 6950, pp. 125–140. Springer, Heidelberg (2011). https://doi.org/10.1007/978-3-642-24100-0_9

20. Juyal, C., Kulkarni, S., Kumari, S., Peri, S., Somani, A.: An innovative approach to achieve compositionality efficiently using multi-version object based transactional systems. In: Izumi, T., Kuznetsov, P. (eds.) SSS 2018. LNCS, vol. 11201, pp. 284–300. Springer, Cham (2018). https://doi.org/10.1007/978-3-030-03232-6_19

21. Felber, P., Gramoli, V., Guerraoui, R.: Elastic transactions. J. Parallel Distrib. Comput. **100**, 103–127 (2017)

22. Zhang, D., Dechev, D.: Lock-free transactions without rollbacks for linked data structures. In: SPAA 2016, New York, NY, USA (2016)

23. Dalessandro, L., Spear, M.F., Scott, M.L.: NOrec: streamlining STM by abolishing ownership records. In: Govindarajan, R., Padua, D.A., Hall, M.W. (eds.) PPOPP. ACM (2010)

24. Juyal, C., Kulkarni, S.S., Kumari, S., Peri, S., Somani, A.: Obtaining progress guarantee and greater concurrency in multi-version object semantics. CoRR abs/1904.03700 (2019)

25. Heller, S., Herlihy, M., Luchangco, V., Moir, M., III, W.N.S., Shavit, N.: A lazy concurrent list-based set algorithm. Parallel Process. Lett. **17**(4), 411–424 (2007)

26. Harris, T.L.: A pragmatic implementation of non-blocking linked-lists. In: Welch, J. (ed.) DISC 2001. LNCS, vol. 2180, pp. 300–314. Springer, Heidelberg (2001). https://doi.org/10.1007/3-540-45414-4_21

27. Riegel, T., Felber, P., Fetzer, C.: A lazy snapshot algorithm with eager validation. In: Dolev, S. (ed.) DISC 2006. LNCS, vol. 4167, pp. 284–298. Springer, Heidelberg (2006). https://doi.org/10.1007/11864219_20

28. Guerraoui, R., Henzinger, T.A., Singh, V.: Permissiveness in transactional memories. In: DISC, pp. 305–319 (2008)

29. Crain, T., Imbs, D., Raynal, M.: Read invisibility, virtual world consistency and probabilistic permissiveness are compatible. In: Xiang, Y., Cuzzocrea, A., Hobbs, M., Zhou, W. (eds.) ICA3PP 2011. LNCS, vol. 7016, pp. 244–257. Springer, Heidelberg (2011). https://doi.org/10.1007/978-3-642-24650-0_21

Improved-Zigzag: An Improved Local-Information-Based Self-optimizing Routing Algorithm in Virtual Grid Networks

Yonghwan Kim[1]([✉]), Masahiro Shibata[2], Yuichi Sudo[3], Junya Nakamura[4],
Yoshiaki Katayama[1], and Toshimitsu Masuzawa[3]

[1] Nagoya Institute of Technology,
Gokiso-cho, Showa-ku, Nagoya, Aichi 466-8555, Japan
kim@nitech.ac.jp
[2] Kyushu Institute of Technology, Fukuoka, Japan
[3] Osaka University, Osaka, Japan
[4] Toyohashi University of Technology, Toyohashi, Aichi, Japan

Abstract. A wireless network consisting of many wireless devices becomes popular and essential in distributed systems. In the wireless networks, each wireless device, also called *nodes*, can directly communicate with other devices located within its communication range. However, to communicate with the nodes outside the communication range, the message should be relayed to the target node via some other nodes. A *virtual grid network* is an overlay network on a wireless network which can be constructed by virtually dividing the area covered by the wireless network into geographical square regions of the same size, selecting a representative node at each region, and connecting the nodes of neighboring regions. A virtual grid network is utilized for realizing an energy-efficient wireless network because not all of the nodes in the system need to join the routing, moreover, a routing algorithm can be easily designed thanks to regularity of the grid topology. A local-information-based self-optimizing routing algorithm, *Zigzag*, in virtual grid networks was proposed. In this paper, we propose the locality-based model, named (α, β)-range model, based on the snapshot range α and communication range β to clearly specify the locality. Moreover, we propose a new self-optimizing routing algorithm *Improved-Zigzag* which improves *Zigzag* by reducing the snapshot range.

1 Introduction

Recently, various wireless devices, like smartphones or tablet computers, become popular and necessary to take advantages of many services. As the spread of wireless devices, a wireless network consisting of many wireless devices becomes also popular and important in the field of distributed systems [7–9]. In the wireless networks, each wireless device, also called a *node*, has limited wireless communication range, and can directly communicate, i.e., send or receive some messages,

© Springer Nature Switzerland AG 2019
M. Ghaffari et al. (Eds.): SSS 2019, LNCS 11914, pp. 228–242, 2019.
https://doi.org/10.1007/978-3-030-34992-9_18

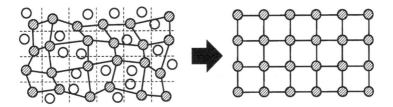

Fig. 1. A virtual grid network

with only the other devices located within its communication range. When the target node is outside the communication range, the message should be relayed to the target node via some other nodes, which is called *multi-hop communication* or *multi-hop routing*. A routing protocol is a protocol to determine the route (usually presented as the sequence of the relaying nodes) for sending messages to the target node. Many routing protocols in the wireless networks have been proposed [1–4,11,12] to realize effective multi-hop communication.

A *virtual grid network* is an overlay network for an energy-efficient routing. A virtual grid network can be constructed by virtually dividing the area covered by a wireless network into a grid of geographical square regions, called *cells*, of the same size. The size of each cell is determined by the communication range of each node such that any two nodes located in the adjacent two cells respectively can directly communicate with each other. Figure 1 represents an example of a virtual grid network. In each cell, a single node is selected as a *router* which is responsible for a relaying node, which means that every multi-hop communication is realized by the routers. And the non-routers, i.e., the nodes that are not selected as routers, need not relay messages, thus they can become inactive to save energy consumption. To increase the total lifetime of the wireless network, in each cell, the role of the router can be taken over among the nodes in the same cell, periodically or when needed.

In the virtual wireless network, a source node (the node initiates a message) sends a message to the router in the same cell, the router transfers, via routers, the message to the router in the cell containing the target node (the node delivers the message). As a result, the target node can receive the message from the router in the same cell. Figure 2(a) illustrates an example path from the router r_s which received a message from the source node to the router r_t sent a message to the target node. A wireless network consists of many (mobile) wireless devices. Hence, some wireless devices may move from the current cell to a (adjacent) different cell like Fig. 2(b). Consider that the source node or the target node in the virtual grid network moves to a different cell. In this case, the path (route) from the source node to the target node has to be extended to the next cell that the source or target node moves to like Fig. 2(c). However, if the source or target node repeatedly moves around the network, the path may become redundantly long (Fig. 2(c)). Since energy efficiency is one of the most important issues in

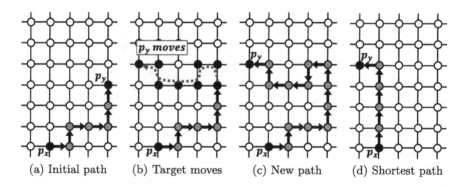

(a) Initial path   (b) Target moves   (c) New path   (d) Shortest path

Fig. 2. Redundant path constructed by movements of a node

wireless networks, a redundantly long path should be shortened to decrease the number of the relaying nodes like Fig. 2(d).

The shortest path between the source and the target node can be easily constructed from scratch using global information of the entire network, however, unacceptable communication cost and large memory are required in a large-scale wireless network. To solve this problem, Takatsu et al. proposed a local-information-based self-optimizing routing algorithm *Zigzag* [5] in virtual grid networks. In algorithm *Zigzag*, each router applies only local update rules based on its local information. Algorithm *Zigzag* requires no global information, e.g., a global coordination, locations of the source and target nodes, or the length of the path, but it can effectively optimize the path, i.e., construct the shortest path, regardless of the scale of the wireless network.

Algorithm *Zigzag* opened the possibility of local-information-based routing algorithms, which transforms any given path between two nodes into a shortest path between the nodes. Algorithm *Zigzag* uses only the local information of the path within a constant distance. This implies that each router uses the constant-range local information and communicates with the routers within a constant distance to execute the local updates.

In the field of the distributed computing, how to solve a problem using only local information, like algorithm *Zigzag*, is the one of the important challenges. Especially, to clarify the amount of information to solve the given problem is an important issue. However, there exist many types of the system models in the field of the distributed computing, and the evaluation of the amount of the information deeply depends on the assumed system model. In this paper, we introduce the locality-based model for the virtual grid network to evaluate the amount (range) of information to solve the problem.

Our Contribution: The contribution of this paper is twofold: (a) we propose a locality-based model, named (α, β)-range model, which is based on the snapshot range α (the range of the routing information that each router can refer to) and the communication range β (the range such that each router can send a message in one atomic step) in virtual grid networks to clearly specify the locality of local-information-based routing algorithms, and (b) we propose a new local-information-based self-optimizing routing algorithm which improves *Zigzag* by reducing the snapshot range α. This comparison can be realized by the proposed locality-based model.

2 Related Works

A virtual grid network is introduced in [10] which presents a routing protocol named GAF (Geographic Adaptive Fidelity). GAF reduces energy consumption of the wireless networks such as an ad-hoc network by turning off nodes that are unnecessary for routing messages with keeping a consistent level of routing fidelity. In [10], the entire area is divided into square regions called *cells* such that they form a grid. If each node has the communication range R, the length of each side of each cell r is set to satisfy $r \le \frac{R}{\sqrt{5}}$. Thus all nodes in the cell can communicate with all the nodes in its adjacent cells. A single node is selected as a router in every cell, and each router has responsibility for communication with the routers in its adjacent cells for routing. These routers compose a *virtual grid network*.

In a virtual grid network, each router communicates with the other routers in its adjacent cells for routing and non-router nodes communicate with the router in the same cell when they need to communicate with nodes outside of their communication range. When the non-router node does not need to communicate with any other nodes, it can sleep for saving energy.

Takatsu et al. proposed a local-information-based self-optimizing routing algorithm in virtual grid networks which is named *Zigzag* [5]. Algorithm *Zigzag* consists of only three simple rules for local updates to optimize the path shown in Fig. 3. In algorithm *Zigzag*, each node (or router in the virtual grid network) does not know the locations of the source and the target nodes. There is no coordination system, this means that each node does not have any location information. The detailed system model will be described in Sect. 3.

Algorithm *Zigzag* ensures convergence of any given path into a shortest path within reasonable convergence time $O(|P|)$, where $|P|$ is the length of the given path. *Zigzag* also preserves the connection between the source and the target nodes during the convergence.

Kim et al. [6] extended a virtual grid network from two-dimensional (2D) plane into three-dimensional (3D) space. Nodes are deployed in 3D space, and

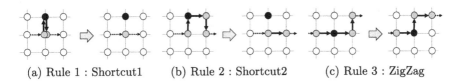

(a) Rule 1 : Shortcut1   (b) Rule 2 : Shortcut2   (c) Rule 3 : ZigZag

Fig. 3. Three rules of algorithm *Zigzag*

they have a spherical communication range instead of a circular one. Kim et al. proposed a local-information-based algorithm in the 3D space. Like algorithm *Zigzag*, the algorithm by Kim et al. uses only local information of the path within a constant distance and it consists of four local update rules. However, unlike algorithm *Zigzag*, the algorithm by Kim et al. requires the global coordination, i.e., every node agrees on the directions of three axes[1].

3   Preliminaries

3.1   System Model

A *virtual grid network* can be constructed by dividing the area covered by the wireless network into geographical square regions, called *cells*, of the same size, selecting a node, called a *router*, from each region, and connecting the routers of neighboring regions. In this paper, we assume that a virtual grid network is already constructed (Refer [10] for construction of a virtual grid network).

There are two special nodes, a source node v_s and a target node v_t. A message is firstly sent from v_s to the router in the same cell, denoted r_s. And the message is relayed by the routers from r_s to r_t which is the router in the cell where v_t exists, as a result, the message is forwarded to v_t by r_t. Note that v_s and r_s may be the same node, and v_t and r_t also may be the same. We say that two routers p_i and p_j are neighbors if the two corresponding cells are neighbors, i.e. the two cells share a common edge.

The virtual grid network is modeled as an undirected graph $G = (V, E)$ where V is a set of routers and E is a set of communication links between routers. The link set E represents all neighboring pairs of routers, that is, $(p_i, p_j) \in E$ holds if and only if two routers $p_i, p_j \in V$ are in neighboring cells. A path P from r_s to r_t is a sequence of routers: $P = (p_0, p_1, p_2, \cdots, p_n)$ satisfying $(p_i, p_{i+1}) \in E$ for each i $(0 \leq i < n)$.

Each router has no knowledge of its own location in a virtual grid network. However, all the nodes agree on the directions of two axes. This means each router does not have its own (x, y)-coordinate, instead, it has common sense of direction, e.g., every router knows the positive directions of both x-axis and y-axis. Each communication link incident to a router is locally labeled with U (Up), R (Right), L (Left), or D (Down) (Fig. 4).

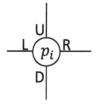

Fig. 4. Labels of links

Each router p_i in path P maintains the *routing information* consisting of the pair of (local) labels of the links: one is incoming link (r_s side) and another one is outgoing link (r_t side).

---

[1] In [5], algorithm *Zigzag* assumes the agreement of the directions of two axes, however the algorithm can be easily modified for the model so that every node agrees on only the chirality which means the orientation of the axes, e.g., clockwise.

3.2 Path Optimization Problem

The path optimization problem for transforming the given r_s-r_t path into the shortest r_s-r_t path in virtual grid networks is defined as follows.

Definition 1 (Path optimization problem). *Let r_s and r_t be any two distinct routers in a virtual grid network $G = (V, E)$ and let $P = (r_s = p_0, p_1, p_2, \cdots, p_n = r_t)$ be any path from r_s to r_t. The path optimization problem is to transform a given path P into a shortest path from r_s to r_t while r_s and r_t do not move.*

In this paper, we focus on a local-information-based self-optimizing algorithm to solve the path optimization problem which uses only the local information within a constant distance. This means that each router autonomously executes local updates (self-optimizing) with preserving the connection between r_s and r_t using the routing information of the routers within a constant distance (local-information-based).

We assume that a source or a target node moves slowly compared with the required amount of the time to computations or communications. This means that path extending by any node's move is occurred, the path can be instantly transformed into a shortest path by some computations and/or communications.

4 (α, β)-Range Model for Locality

We introduce a new locality-based model, (α, β)-range model, which has two parameters, the snapshot range α and the communication range β, to specify the locality of local-information-based routing algorithms.

4.1 Definition of (α, β)-Range Model

The Snapshot Range α. At first, we consider the range of the routing information that each router can refer to. Each router p_i in path P maintains its routing information, a pair (in, out) such that $in, out \in \{U, R, L, D, -\}$, which represents the directions of the previous (in) and next routers (out) in P. In the example illustrated in Fig. 5, router p_4 has (D, R) and router p_7 has (U, R), which mean that p_3 and p_5 are located downward and rightward from p_4, respectively and that p_6 and p_8 are located upward and rightward from p_7, respectively. Symbol '-' represents "no router", which is used in r_s and r_t. In the example of Fig. 5, the routing information of r_s is $(-, L)$.

A local-information-based routing algorithm allows nodes to locally change the path based on local information (or local shape of the path). If each router p_i in P has no information of other nodes, it can use only its routing information from which p_i can get the shape of the partial path of length 2 (p_i is in the middle). Otherwise, if each router p_i in path P can know the routing information of the other two nodes (or p_{i-1} and p_{i+1}) one hop away from itself in P, it can use the routing information of the two nodes. In the same way, if each router p_i

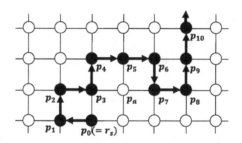

Fig. 5. An example path P

in path P can know the information of the other nodes within α hops (if exists) from itself, it can use the information of the partial path of length $2 * (\alpha + 1)$ (if exists) such that p_i is located in the middle. For example, if α is 2, router p_6 knows the routing information of p_4, p_5, p_7, p_8, and p_6 itself. We call α a *snapshot range*.

Recall the previous work *Zigzag* consisting of three local update rules. Figure 3 illustrates the three rules, and each local update is executed by the black router. In the first rule (Fig. 3(a)), the black router can detect whether the rule is applicable or not using only its own routing information ($\alpha = 0$). For the second or third rule, the black router requires the information of the node one hop away from itself ($\alpha = 1$) to detect whether the rules are applicable or not. However, as we will see later, information of further nodes is required to solve the conflicts between the applicable rules such that simultaneous application cannot preserve connection between r_s and r_t.

The Communication Range β. Now, we introduce the *communication range* β. This parameter represents the range such that each router can send a message in one atomic step to make the receiver node locally update its routing information. Specifically, each router p_i can send a message to all the routers within β hops, i.e., $p_{i-\beta}, p_{i-\beta+1}, \ldots, p_{i+\beta}$, in one atomic step.

Now we define the locality-based model of the virtual grid network as follows.

Definition 2. ((α, β)-Range Model). *(α, β)-Range Model is a virtual grid network model such that (a) each router in path P knows the information of the routers within α hops from itself along path P, (b) each router can communicate with the routers within β hops, in one atomic step.*

To help to understand, we show an example using Fig. 5. If the virtual grid network in Fig. 5 is the (3,2)-range model, p_5 knows the routing information of routers from p_2 to p_8 (or the partial path from p_1 to p_9). Note that the routing information of p_2 includes the routing direction from p_1, and this implies that p_5 can knows the (relative) location of p_1. If router p_5 detects some rules are applicable, it can send messages to p_3, p_4, p_6, or p_7. Not only these routers, but p_5 can also send a message to router p_a which is adjacent to its link labeled D if necessary (in Fig. 5).

4.2 Solvability of the Previous Work

In this subsection, we evaluate the previous work, algorithm *Zigzag* [5], using the (α, β)-range model. Algorithm *Zigzag* consists of three rules, and these rules may conflicted with each other. According to [5], there exist 12 patterns of conflicts. Table 1 shows the list of the local update rules that are conflicting with a local update rule at p_i and have higher priorities. For example, even if p_i detects rule 1 is applicable, it never executes its local update if p_{i-1} also detects rule 1 is applicable. Thus algorithm *Zigzag* uses information to resolve the conflicts.

Table 1. Conflicting rules with higher priority for local update at p_i

local update at p_i	Rule 1	Rule 2	Rule 3
local update at p_{i-2}	–	Rule 2	–
local update at p_{i-1}	Rule 1	Rule 1 or 2	Rule 2 or 3
local update at p_{i+1}	–	Rule 1	Rule 1 or 2
local update at p_{i+2}	–	Rule 1	Rule 1 or 2

Now we evaluate the actual value of parameter α so that every p_i can detect applicability of the rules and resolve the conflicts between local update rules.

- Rule 1: Router p_i can evaluate rule 1 using only its own routing information. However, p_i has to know the information of p_{i-1} to actually apply rule 1 since applicability of rule 1 at p_{i-1} prevents p_i from applying rule 1. Thus $\alpha \geq 1$ must hold.
- Rule 2: Router p_i can evaluate rule 2 if it knows p_{i+1}'s information. To check all conflicts, p_i has to know p_{i-2}'s information (for rule 2 at p_{i-2}) and p_{i+2}'s information (for rule 1 at p_{i+2}). Thus $\alpha \geq 2$ must hold.
- Rule 3: Router p_i can evaluate rule 3 if it knows p_{i+1}'s information. To check all conflicts, p_i has to know p_{i-1}'s information (for rule 2 or 3 at p_{i-1}) and p_{i+3}'s information (for rule 2 at p_{i+2}). Thus $\alpha \geq 3$ must hold.

As a result, to execute local updates correctly in algorithm *Zigzag*, each router has to know the information of the router within three hops from itself. Therefore, $\alpha \geq 3$ must hold.

In algorithm *Zigzag*, when p_i detects some local update rule can be applied without any conflict, it executes the appropriate local update with some other routers. Hence, some message communications are required to execute a local update. For rule 1, p_{i-1} and p_{i+1} have to update their routing information ($\beta = 1$). For rule 2, p_{i-1}, p_{i+1}, and p_{i+2} have to update their routing information ($\beta = 2$). Finally, for rule 3, p_{i+1}, p_{i+2}, and p_i's neighboring router (which is not in P) have to update their routing information ($\beta = 2$). Consequently, the following observation holds.

Observation 1. *Algorithm Zigzag [5] solves the path optimization problem in (3,2)-range model.*

Observation 1 considers the minimum snapshot range to detect applicability and conflict of update rules. However, if the snapshot range is larger, every router can know that the other nearby routers detect applicability and conflicts of some local updates, and thus it can detect whether its local rule can be actually applied. For instance, consider router p_i detect rule 2 is applicable without any conflict with p_{i-2} or p_{i-1}. If p_{i+2} know the routing information of p_{i-2}, p_{i+2} can know that p_i can actually apply local update for rule 2. Thus we consider an enough snapshot range α for $\beta = 0$.

To represent the range of the routers, we use the range notation: $[p_a : p_b]$ (where $a \le b$) which means all the routers from p_a to p_b in P. And we assume that two ranges can be merged. We can present the required range of the routing information to detect applicability (without considering any conflict) at p_i using the range notation: (a) p_i can evaluate rule 1 using only its routing information (denote $\alpha_1[p_i] = [p_i]$), (b) p_i can evaluate rule 2 using its routing information and p_{i+1}'s (denote $\alpha_2[p_i] = [p_i : p_{i+1}]$), and (c) p_i can evaluate rule 3 using its routing information and p_{i+1}'s (denote $\alpha_3[p_i] = [p_i : p_{i+1}]$).

Now we can calculate the snapshot range α for $\beta = 0$ of each rule, from the length of the range by merging two ranges as follows: (i) all the routers to detect applicability and conflict of update rules, and (ii) all the routers of which each rule causes the update of the routing information.

- Rule 1 at p_i: (i) Local update for Rule 1 causes the update of the routing information of p_{i-1}, p_i, and p_{i+1} (range $[p_{i-1}:p_{i+1}]$). (ii) To detect applicability of rule 1 at p_i, the evaluation of both rule 1 at p_{i-1} and rule 1 at p_i, and this range can be represented as $\alpha_1[p_{i-1}] + \alpha_1[p_i] = [p_{i-1} : p_i]$. The summation of these two ranges becomes $[p_{i-1}:p_{i+1}]$, and its length is 2. Thus $\alpha \ge 2$ must hold.
- Rule 2 at p_i: (i) Local update for Rule 2 causes the update of the routing information of the routers in $[p_{i-1}:p_{i+2}]$. (ii) To detect applicability of rule 2 at p_i, the routing information of the routers in $(\alpha_2[p_{i-2}] + \alpha_1[p_{i-1}] + \alpha_2[p_{i-1}] + \alpha_2[p_i] + \alpha_1[p_{i+1}] + \alpha_1[p_{i+2}]) = [p_{i-2} : p_{i+2}]$ are necessary. Thus $\alpha \ge 4$ must hold.
- Rule 3 at p_i: (i) Local update for Rule 3 causes the update of the routing information of the routers in $[p_{i-1}:p_{i+2}]$. (ii) To detect applicability of rule 3 at p_i, the routing information of the routers in $(\alpha_2[p_{i-1}] + \alpha_3[p_{i-1}] + \alpha_3[p_i] + \alpha_1[p_{i+1}] + \alpha_2[p_{i+1}] + \alpha_1[p_{i+2}] + \alpha_2[p_{i+2}]) = [p_{i-1} : p_{i+3}]$ are necessary. Thus $\alpha \ge 4$ must hold.

From the above discussion, the following observation holds.

Observation 2. *Algorithm Zigzag solves the path optimization problem in (4,0)-range model.* [2]

---

[2] Note that a router which is not in path P is joined into path P as the result of the local update for rule 3. Thus we assume that a router, which is not in path P but adjacent to the router in P, knows the routing information of its adjacent routers in P within the snapshot range.

5   The Proposed Algorithm *Improved-Zigzag*

In this section, we propose a new (local-information-based) self-optimizing routing algorithm *Improved-Zigzag* which is an improved version of *Zigzag* in virtual grid networks.

5.1   Algorithm *Improved-Zigzag*

By Observations 1 and 2, algorithm *Zigzag* solves the path optimization problem in (3,2)-range model or (4,0)-range model. In this section, we propose a new self-optimizing algorithm, *I-Zigzag (Improved Zigzag)* which can solves the path optimization problem in smaller (α, β)-range model, (2,1)-range model or (3,0) range-model.

Algorithm *I-Zigzag* consists of three local update rules which are similar to those in *Zigzag*. Figure 6 illustrates the three rules.

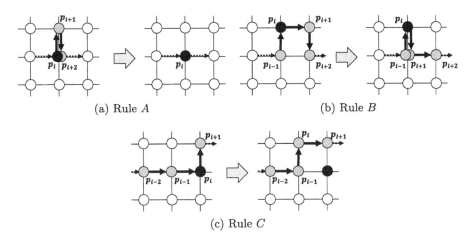

(a) Rule A (b) Rule B

(c) Rule C

Fig. 6. Three rules of algorithm *I-Zigzag*

Each router in P has its routing information consisting of two (local) labels of the links: one is the label of the link which is nearer to the source node, and another is the label of the link which is nearer to the target node. We denote these two link as $p_i.in$ and $p_i.out$ respectively. For example, in Fig. 6(b), $p_i.in$ remains D and $p_i.out$ becomes D from R.

Table 2 represents the predicates to evaluate local rules in algorithm *I-Zigzag*. We use the dot product to present the relationship between the directions of two links: if the directions of two links are the same, the result of the dot product becomes 1, e.g., $R \cdot R = 1$, and if they are orthogonal, the result becomes 0, e.g., $D \cdot L = 0$. And if they are in the opposite directions, the result becomes -1, e.g., $U \cdot D = -1$.

Table 2. Predicates to evaluate local rules

$isRuleA(p_i) \equiv (p_i.out \cdot p_{i+1}.out = -1)$
$isRuleB(p_i) \equiv (p_{i-1}.out \cdot p_i.out = 0) \wedge (p_{i-1}.out \cdot p_{i+1}.out = -1)$
$isRuleC(p_i) \equiv (p_{i-2}.out \cdot p_{i-1}.out = 1) \wedge (p_{i-1}.out \cdot p_i.out = 0)$

Table 3. Conflicting rules with higher priority in algorithm *I-Zigzag*

local update at p_i	Rule A	Rule B	Rule C
local update at p_{i-1}	Rule A or B	Rule B	–
local update at p_i	Rule C	Rule C	–

Table 3 shows the list of the local update rules that are conflicting with a local update rule at p_i and have higher priorities. Different from algorithm *Zigzag*, in algorithm *I-Zigzag*, p_i may evaluate two rules at the same time: (i) rules A and C, or (ii) rules B and C. Router p_i never executes rule A nor B if it detects rule C is applicable at the same time, i.e., rule C has a higher priority.

Algorithm *I-Zigzag* is basically based on algorithm *Zigzag*, but the second rule is changed and the other rules are evaluated by the different routers. These changes allow less conflicting rules compared with algorithm *Zigzag*'s (Table 1). Moreover, algorithm *I-Zigzag* can solve the path optimization problem using less information (smaller α). The evaluation of the proposed algorithm *I-Zigzag* will be presented in the next subsection.

Algorithm 1 presents the proposed algorithm *I-Zigzag* using the predicates in Table 2. Router p_i checks rule C at first because p_i may evaluate rules A or B at the same time and rule C has a higher priority. If rule C is evaluated by p_i (line 1), the local update for rule C is executed. Note that the message communications to p_{i+1} and p_{i+2} from p_i are required for this local update. Even if p_i detects rule A or rule B is applicable, to actually execute the local update for rule B or rule A, p_i has to verify no conflict exists (lines 7 and 12).

5.2   Proof Sketch of *I-Zigzag*

In this subsection, we show the correctness of the proposed algorithm. We do not give a complete proof, but give a proof sketch due to the lack of space.

Lemma 1. *Algorithm I-Zigzag never disconnects the path even if any set of updates is executed.*

Proof. Trivially, if there is no conflict among rules, the path is never disconnected by the definition of each rule. Therefore, we consider only the case two or more applicable rules are conflicted. If rule C is applicable at p_i, rule A or B can be applicable at the same time, but it is ignored by the priority among rules (refer Table 3) to preserve the disconnection of the path. Moreover, rules A and B are

Algorithm 1. Local-information-based path optimization algorithm *I-Zigzag*

1: **if** $isRuleC(p_i)$ **then**
2:    $p_{i-1}.out \leftarrow p_i.out$ ▷ Local update for rule C
3:    Let p_a be the router which is adjacent to p_{i-1}'s link labeled $p_i.out$
4:    $p_a.in \leftarrow p_{i+1}.in$, $p_a.out \leftarrow p_{i-1}.out$
5:    $p_{i+1}.in \leftarrow p_i.in$
6:    the part (p_{i-1}, p_i, p_{i+1}) of P is replaced with (p_{i-1}, p_a, p_{i+1})
7: **else if** $isRuleB(p_i) \wedge \neg(isRuleC(p_i) \vee isRuleB(p_{i-1}))$ **then**
8:    $p_i.out \leftarrow p_{i+1}.out$ ▷ Local update for rule B
9:    $p_a \leftarrow p_{i-1}$, $p_a.in \leftarrow p_{i+2}.in$, $p_a.out \leftarrow p_i.out$
10:    $p_{i+2}.in \leftarrow p_{i+1}.in$
11:    the part (p_i, p_{i+1}, p_{i+2}) of P is replaced with (p_i, p_a, p_{i+2})
12: **else if** $isRuleA(p_i) \wedge \neg(isRuleC(p_i) \vee isRuleA(p_{i-1}) \vee isRuleB(p_{i-1}))$ **then**
13:    $p_i.out \leftarrow p_{i+2}.out$ ▷ Local update for rule A
14:    the part (p_i, p_{i+1}, p_{i+2}) of P is replaced with (p_i)
15: **end if**

never applicable at p_i at the same time by the definitions of the rules. Hence, each router executes at most one local update.

Now we consider the conflict between two consecutive local updates by two different routers. If p_i executes the local update by rule A, two hops subpath after p_i is deleted. The local update for rule B changes two hops subpath after p_i, and the local update by rule C changes two hops subpath within one hop from p_i. This implies that even if rule A or B are applicable at routers p_i and p_{i-2} at the same time, these consecutive two local updates never cause the disconnection of the path. Moreover, the case such that rule A or B is applicable at p_{i-2} and rule C is applicable at p_i never occurs by the definition of the rules. Therefore, we need to consider only the conflict between the two concurrent local updates at p_{i-1} and p_i. Note that there are 9 patterns of the conflicts by the combination of the rules, and we denote each pattern of the conflict as (X,Y)-conflict where X and Y are the applicable rules at p_{i-1} and p_i respectively, for example, (B,C)-conflict means the conflict such that rule B is applicable at p_{i-1} and rule C is applicable at p_i at the same time.

By the priority among rules (Table 3), the three conflicts, (A,A)-conflict, (B,A)-conflict, and (B,B)-conflict, do not occur because p_i does not execute its local update. (C,A)-conflict and (C,B)-conflict update two disjoint subpaths, thus the path is not disconnected. The other four conflicts cannot occur by the definitions of the rules.   □

Lemma 2. *Algorithm I-Zigzag eventually transforms any given arbitrary path to a shortest one.*

Proof. If the current path is not a shortest path, there must be exist one or more routers which any local update rule is applicable at (we omit the proof). And we consider that the path is partitioned into the sequence of the line segments, each line segment consists of the consecutive same directions of the path. For example, in Fig. 2(a), the initial path is partitioned into the 4 line segments (p_x, p_1), (p_1, p_2), (p_2, p_3, p_4), and (p_4, p_5, p_y), where each p_i denotes $i-th$ router in the path. We define the potential function $\mathscr{F}$ of the path P using the line segments as follows: a sequence $\mathscr{F}(P) = (|P|, |s_1|, |s_2|, \ldots, |s_k|)$, where $|P|$ is the total length of the path P and each $|s_i|$ is the length of each line segment, which can be totally ordered by the lexicographical order. Note that $|P| = \sum_{i=1}^{k} |s_i|$ holds. Finally we show that whenever any local update is executed, the potential function $\mathscr{F}(P)$ decreases. The local update of rule A shorten the path, thus $\mathscr{F}(P)$ decreases. The local update of rule C bends the straight line of the path, which decreases $\mathscr{F}(P)$. However, the local update of rule B sometimes increases $\mathscr{F}(P)$, because if router p_i executes the local update of rule B, the line segment starting with p_{i+1} may lengthen by one. However, after the local update of rule B, the local update of rule A is eventually executed and this shorten the length of the path. Hence $\mathscr{F}(P)$ eventually decreases. □

5.3   Evaluation Using Range Model

In this subsection, we show the algorithm *I-Zigzag* works in (1,2)-range model and (3,0)-range model.

Lemma 3. *Algorithm I-Zigzag solves the path optimization problem in (1,2)-range model.*

Proof. (1) **Parameter** α: At first, we verify the snapshot range for the predicates in algorithm *I-Zigzag*. To judge the predicates in Table 2, the first predicate requires the routing information of p_{i+1}, the second one requires that of p_{i+1} (note that the routing information of p_{i-1} is not necessary for rule B because p_i can know $p_{i-1}.out$ from $p_i.in$), and the last one requires the information of p_{i-1} (the information of p_{i-2} is not necessary for the same reason above). Consequently, α should be 1 or more for predicates in algorithm *I-Zigzag*.

In Algorithm 1, router p_i should know the predicate of router p_{i-1} to check the conflicts. In line 7 of Algorithm 1, p_i checks $isRuleB(p_{i-1})$ which requires the routing information of p_{i-1} and p_i, hence $\alpha \geq 1$ must hold. In line 12 of Algorithm 1, p_i checks two predicates of $isRuleA(p_{i-1})$ and $isRuleB(p_{i-1})$, and both of them can also be determined by the information of p_{i-1} and p_i. Therefore, $\alpha \geq 1$ must hold.

(2) **Parameter** β: Router p_{i+2} is the farthest router from p_i when the local update for rule A or B is executed. And, router p_{i-2} is the farthest router from p_i when the local update for rule C. Therefore, $\beta \geq 2$ must hold. □

From Observation 1 and Lemma 3, the proposed algorithm *I-Zigzag* uses less information to solve the path optimization problem compared with algorithm *Zigzag*. We can consider algorithm *I-Zigzag* as the non-communication model ($\beta = 0$).

Lemma 4. *Algorithm I-Zigzag solves the path optimization problem in (3,0)-range model.*

Proof. **(1) Rule** A**:** Router p_{i+2} has to know the information of p_{i-1} to check the conflict of rule A of p_i without any message communication. Thus, $\alpha \geq 3$ must hold for rule A.

(2) Rule B**:** Router p_{i+2} has to know the information of p_{i-1} to check the conflict of rule B of p_i without any message communication. Thus, $\alpha \geq 3$ must hold for rule B.

(3) Rule C**:** Router p_{i-2} has to know the information of p_i to evaluate rule C of p_i without any message communication. Thus, $\alpha \geq 2$ must hold for rule C.

Therefore, $\alpha \geq 3$ holds for the non-communication model where $\beta = 0$.   □

From Lemmas 3 and 4, the following theorem holds.

Theorem 3. *Algorithm I-Zigzag solve the path optimization problem in (1,2)-range model or (3,0)-range model.*

By Theorem 3, Observations 1, and 2, the proposed algorithm *I-Zigzag* solves the path optimization problem using less information compared with algorithm *Zigzag* which solves the path optimization problem in (3,2)-range model or (4,0)-range model.

6 Conclusion

In this paper, to clarity locality for checking conditions (predicates of rules) and executing actions, we proposed a locality-based system model, (α, β)-range model, where α is the snapshot range α and β is the communication range. The (α, β)-range model can be used for clearly specifying path optimization algorithms from the viewpoint of the locality. Base on the (α, β)-range model, we showed that the local-information-based self-optimizing algorithm *Zigzag* [5] works in (3,2)-range model and (4,0)-range model. On the other hand, the proposed algorithm *I-Zigzag* can solve the path optimization problem in (1,2)-range model and (3,0)-range model, which improves the locality of *Zigzag*.

(α, β)-range model provides the measure for the locality of the local-information-based routing algorithm in virtual grid networks. As a result, we show that (1,2)-range and (3,0)-range are the upper bounds of the routing algorithms with and without communication respectively. It is clear that no algorithm solves the path optimization problem in (0,0)-range model because of the following reason: Each router can use its routing information only, therefore, only the rule 1 of *Zigzag* (or rule A of the proposed algorithm) can be used for the path optimization. Any other rule consisting of only its routing information cannot be used (we omit the proof). However, even this rule becomes applicable by any router, the router cannot inform some other router of the rule $(\beta = 0)$. Moreover, any other routers never knows the applicable rule of the other router $(\alpha = 0)$. On the other hand, we do not know whether the path optimization problem is solvable or not even in (1,0)-range model or (0,1)-range model yet. The future work is to find the lower bound of (α, β)-range model for the local-information-based routing algorithm.

Acknowledgements. This work was supported by JSPS KAKENHI Grant Numbers 18K18000, 18K18029, 18K18031, 19H04085, 19K11823, and JST SICORP Grant Numbers JPMJSC1606 and JPMJSC1806, Japan.

References

1. Dargie, W., Poellabauer, C.: Fundamentals of Wireless Sensor Networks: Theory and Practice. Wiley, Hoboken (2010)
2. Perkins, C., Royer, E.: Ad hoc on demand distance vector routing. In: Proceedings of the 2nd IEEE Workshop on Mobile Computing Systems and Applications (WMCSA 1999), pp. 90–100 (1999)
3. Al-Karaki, J.N., Kamal, A.E.: Routing techniques in wireless sensor networks: a survey. IEEE Wirel. Commun. **11**(6), 6–28 (2004)
4. Heinzelman, W., Chandrakasan, A., Balakrishnan, H.: Energy-efficient communication protocol for wireless microsensor networks. In: Proceedings of the 33rd Hawaii International Conference on System Sciences (HICSS 2000) (2000)
5. Takatsu, S., Ooshita, F., Kakugawa, H., Masuzawa, T.: Zigzag: local-information-based self-optimizing routing in virtual grid networks. In: Proceedings of the 33rd International Conference on Distributed Computing Systems (ICDCS), pp. 358–368, July 2013
6. Kim, Y., Katayama, Y.: A self-optimizing routing algorithm using local information in a 3-dimensional virtual grid network with theoretical and practical analysis. Int. J. Netw. Comput. **7**(2), 349–371 (2017)
7. Toh, C.-K.: Ad hoc mobile wireless networks: protocols and systems, 1st edn. Prentice Hall PTR, Upper Saddle River (2002)
8. de Morais Cordeiro, C., Agrawal, D.P.: Ad Hoc and Sensor Networks: Theory and Applications, 2nd edn. World Scientific, Singapore (2011)
9. Zheng, J., Jamalipour, A.: Wireless Sensor Networks: A Networking Perspective. Wiley-IEEE Press (2009)
10. Xu, Y., Heidemann, J., Estrin, D.: Geography-informed energy conservation for ad hoc routing. In: Proceedings of the 7th Annual International Conference on Mobile Computing and Networking (MobiCom 2001), pp. 70–84 (2001)
11. Braginsky, D., Estrin, D.: Rumor routing algorithm for sensor networks. In: The Proceedings of the First Workshop on Sensor Networks and Applications (WSNA) (2002)
12. Ye, F., Luo, H., Cheng, J., Lu, S., Zhang, L.: A Two-tier data dissemination model for large-scale wireless sensor networks. In: Proceedings of ACM/IEEE MOBICOM (2002)

Fault Tolerant Network Constructors

Othon Michail[1(✉)], Paul G. Spirakis[1,2(✉)], and Michail Theofilatos[1(✉)]

[1] Department of Computer Science, University of Liverpool, Liverpool, UK
{Othon.Michail,P.Spirakis,Michail.Theofilatos}@liverpool.ac.uk
[2] Computer Engineering and Informatics Department, University of Patras,
Patras, Greece

Abstract. In this work, we consider adversarial crash faults of nodes in the network constructors model [Michail and Spirakis, 2016]. We first show that, without further assumptions, the class of graph languages that can be (stably) constructed under crash faults is non-empty but small. When there is a finite upper bound f on the number of faults, we show that it is impossible to construct any *non-hereditary* graph language and leave as an interesting open problem the *hereditary case*. On the positive side, by relaxing our requirements we prove that: (i) permitting linear waste enables to construct on $n/(2f) - f$ nodes, any graph language that is constructible in the fault-free case, (ii) *partial constructibility* (i.e., not having to generate all graphs in the language) allows the construction of a large class of graph languages. We then extend the original model with a minimal form of *fault notifications*, and our main result here is a *fault-tolerant universal constructor* that requires linear waste in the population. Finally, we show that logarithmic local memories can be exploited for a no-waste fault-tolerant simulation of any such protocol.

Keywords: Network construction · Distributed protocol · Self stabilization · Fault tolerant protocol · Dynamic graph formation · Fairness · Self-organization

1 Introduction and Related Work

In this work, we address the issue of the dynamic formation of graphs under faults. We do this in a minimal setting, that is, a population of agents running *Population Protocols* that can additionally activate/deactivate links when they meet. This model, called *Network Constructors*, was introduced in [MS16], and is based on the *Population Protocol* (PP) model [AAD+06, AAER07] and the *Mediated Population Protocol* (MPP) model [MCS11]. We are interested in answering questions like the following: If one or more faults can affect the formation process, can we always re-stabilize to a correct graph, and if not, what

All authors were supported by the EEE/CS initiative NeST. The last author was also supported by the Leverhulme Research Centre for Functional Materials Design. This work was partially supported by the EPSRC Grant EP/P02002X/1 on Algorithmic Aspects of Temporal Graphs.

is the class of graph languages for which there exists a fault-tolerant protocol? What are the additional minimal assumptions that we need to make in order to find fault-tolerant protocols for a bigger class of languages?

Population Protocols run on networks that consist of computational entities called *agents* (or *nodes*). One of the challenging characteristics is that the agents have no control over the schedule of interactions with each other. In a population of n agents, repeatedly a pair of agents is chosen to interact. During an interaction their states are updated based on their previous states. In general, the interactions are scheduled by a *fair scheduler*. When the execution time of a protocol needs to be examined, a typical fair scheduler is the one that selects interactions uniformly at random.

Network Constructors (and its geometric variant [Mic18]) is a theoretical model that may be viewed as a minimal model for programmable matter operating in a dynamic environment [MS17]. Programmable matter refers to any type of matter that can *algorithmically* transform its physical properties, for example shape and connectivity. The transformation is the result of executing an underlying program, which can be either a centralized algorithm or a distributed protocol stored in the material itself. There is a wide range of applications, spanning from distributed robotic systems [GKR10], to smart materials, and many theoretical models (see, e.g., [DDG+14, DDG+18, MSS19, DLFS+19] and references therein), try to capture some aspects of them.

The main difference between PPs and Network Constructors is that in the PP (and the MPP) models, the focus is on computation of functions of some input values, while Network Constructors are mostly concerned with the stable formation of graphs that belong to some graph language. Fault tolerance must deal with the graph topology, thus, previous results on self-stabilizing PPs [AAFJ08, BBB13, DLFI+17, CLV+17] and MPPs [MOKY12] do not apply here.

In [MS16], Michail and Spirakis gave protocols for several basic network construction problems, and they proved several universality results by presenting generic protocols that are capable of simulating a Turing Machine and exploiting it in order to stably construct a large class of networks, in the absence of crash failures.

In this work, we examine the setting where *adversarial crash faults* may occur, and we address the question of which families of graph languages can be stably formed. Here, adversarial crash faults mean that an adversary knows the rules of the protocol and can select some node to be removed from the population at any time. For simplicity, we assume that the faults can only happen sequentially. This means that in every step at most one fault may occur, as opposed to the case where many faults can occur during each step. These cases are equivalent in the Network Constructors model w.l.o.g., but not in the extended version of this model (which allows fault notifications) that we consider later.

A main difference between our work and traditional self-stabilization approaches is that the nodes are supplied with constant local memory, while in principle they can form linear (in the population size) number of connections per node. Existing self-stabilization approaches that are based on restarting techniques cannot be directly applied here [DIM93, Dol00], as the nodes cannot distinguish whether they still have some activated connections with the remaining nodes, after a fault has occurred. This difficulty is the reason why it is not sufficient to just reset the state of a node in case of a fault. In addition, in contrast to previous self-stabilizing approaches [GK10, DT01] that are based on *shared memory* models, two adjacent nodes can only store 1 bit of memory in the edge joining them, which denotes the existence or not of a connection between them.

Angluin *et al.* [AAFJ08] incorporated the notion of self-stabilization into the population protocol model, giving self-stabilizing protocols for some fundamental tasks such as token passing and leader election. They focused on the goal of stably maintaining some property such as having a unique leader or a legal coloring of the communication graph.

Delporte-Gallet *et al.* [DGFGR06] studied the issue of correctly computing functions on the node inputs in the Population Protocol model [AAD+06], in the presence of crash faults and transient faults that can corrupt the states of the nodes. They construct a transformation which makes any protocol that works in the failure-free setting, tolerant in the presence of such failures, as long as modifying a small number of inputs does not change the output. Guerraoui and Ruppert [GR09] introduced an interesting model, called *Community Protocol*, which extends the Population Protocol model with unique identifiers and enough memory to store a constant number of other agents' identifiers. They show that this model can solve any decision problem in NSPACE($n \log n$) while tolerating a constant number of Byzantine failures.

Peleg [Pel09] studies logical structures, constructed over static graphs, that need to satisfy the same property on the resulting structure after node or edge failures. He distinguishes between the stronger type of fault-tolerance obtained for geometric graphs (termed *rigid fault-tolerance*) and the more flexible type required for handling general graphs (termed *competitive fault-tolerance*). It differs from our work, as we address the problem of constructing such structures over dynamic graphs.

1.1   Our Contribution

The goal of any Network Constructor (NET) protocol is to stabilize to a graph that belongs to (or satisfies) some graph language L, starting from an initial configuration where all nodes are in the same state and all connections are disabled. In [MS16], only the fault-free case was considered. In this work, we formally define the model that extends NETs allowing crash failures, and we examine protocols in the presence of such faults. Whenever a node crashes, it is removed from the population, along with all its activated edges. This leaves the remaining population in a state where some actions may need to be taken by the protocol in order to eventually stabilize to a correct network.

We first study the constructive power of the original NET model in the presence of crash faults. We show that the class of graph languages that is in principle constructible is non-empty but very small: for a potentially unbounded number of faults, we show that the only stably constructible language is the *Spanning Clique*. We also prove a strong impossibility result, which holds even if the size of graphs that the protocol outputs in populations of size n need only grow with n (the remaining nodes being *waste*). For a bounded number of faults, we show that any non-hereditary graph language is impossible to be constructed. However, we show that by relaxing our requirements we can extend the class of constructible graph languages. In particular, permitting linear waste enables to construct on $n/(2f) - f$ nodes, where f is a finite upper bound on the number of faults, any graph language that is constructible in a failure-free setting. Alternatively, by allowing our protocols to generate only a subset of all graphs in the language (called *partial constructibility*), a large class of graph languages becomes constructible (see Sect. 3).

In light of the impossibilities in the Network Constructors model, we introduce the minimal additional assumption of *fault notifications*. In particular, after a fault on some node u occurs, all nodes that maintain an active edge with u at that time (if any) are notified. If there are no such nodes, an arbitrary node in the population is notified. In that way, we guarantee that at least one node in the population will sense the removal of u. Nevertheless, some of our constructions work without notifications in the case of a crash fault on an isolated node (Sect. 4).

We obtain two fault-tolerant universal constructors. One of the main technical tools that we use in them, is a fault-tolerant construction of a stable path topology (i.e., a line). We show that this topology is capable of simulating a Turing Machine (abbreviated "TM" throughout this paper), and, in the event of a fault, is capable of always reinitializing its state correctly (Sect. 4.1). Our protocols use a subset of the population (called *waste*) in order to construct there a TM, while the graph which belongs to the required language L is constructed in the rest of the population (called *useful space*). Throughout this paper, we call waste all nodes that do not belong to the constructed graph $G \in L$ after stabilization, and remain either isolated nodes or part of a component such as the TM. The idea is based on [MS16], where they show several universality results by constructing on k nodes of the population a network G_1 capable of simulating a TM, and then repeatedly drawing a random network G_2 on the remaining $n - k$ nodes. The idea is to execute on G_1 the TM which decides the language L with input the network G_2. If the TM accepts, it outputs G_2, otherwise the TM constructs a new random graph.

This allows a fault-tolerant construction of any graph accepted by a TM in linear space, with waste $min\{n/2 + f(n), n\}$, where $f(n)$ is the number of faults in the execution. We finally prove that increasing the permissible waste to $min\{2n/3 + f(n),\ n\}$ allows the construction of graphs accepted by an $O(n^2)$-space Turing Machine, which is asymptotically the maximum simulation space that we can hope for in this model.

In the full version, we also provide a protocol Π' based on restarts such that, given any network constructor Π with notifications, Π' is a fault-tolerant version of Π without waste. However, the required memory per node in this protocol is $O(\log n)$ bits.

Finally, in Sect. 5 we conclude and discuss further research directions opened by this work.

The following table summarizes all results proved in this paper (Table 1).

Table 1. Summary of our results.

Constructible languages		
Without notifications		With notifications
Unbounded faults	Bounded faults	Unbounded faults
Only Spanning Clique	Non-hereditary impossibility	Fault-tolerant protocols: Spanning Star, Cycle Cover, Spanning Line
Strong impossibility even with linear waste	A representation of any finite graph (partial constructibility)	Universal Fault-tolerant Constructors (with waste)
	Any constructible graph language with linear waste	Universal Fault-tolerant Restart (without waste)

Due to space constraints, several technical details are omitted from this extended abstract. A full version with all proofs can be found at [MST19].

2 Model and Definitions

A Network Constructor (NET) is a distributed protocol defined by a 4-tuple $(Q, q_0, Q_{out}, \delta)$, where Q is a finite set of node-states, $q_0 \in Q$ is the initial node-state, $Q_{out} \subseteq Q$ is the set of output node-states, and $\delta : Q \times Q \times \{0,1\} \to Q \times Q \times \{0,1\}$ is the transition function, where $\{0,1\}$ is the set of edge states.

In the generic case, there is an *underlying interaction graph* $G_U = (V_U, E_U)$ specifying the permissible interactions between the nodes, and on top of G_U, there is a dynamic overlay graph $G_O = (V_O, E_O)$. A mapping function F maps every node in the overlay graph to a distinct underlay node. In this work, G_U is a *complete undirected interaction graph*, i.e., $E_U = \{uv : u, v \in V_U \text{ and } u \neq v\}$, while the overlay graph consists of a population of n initially *isolated* nodes (also called *agents*).

The NET protocol is stored in each node of the overlay network, thus, each node $u \in G_O$ is defined by a state $q \in Q$. Additionally, each edge $e \in E_O$ is defined by a binary state (*active/connected* or *inactive/disconnected*). Initially, all nodes are in the same state q_0 and all edges are inactive. The goal is for the nodes, after interacting and activating/deactivating edges for a while, to end up with a desired stable overlay graph, which belongs to some graph language L.

During a (pairwise) interaction, the nodes are allowed to access the state of their joining edge and either activate it (state $= 1$) or deactivate it (state $= 0$).

When the edge state between two nodes $u, v \in G_O$ is activated, we say that u and v are *connected*, or *adjacent* at that time t, and we write $u \underset{t}{\sim} v$.

In this work, we present a version of this model that allows *adversarial crash failures*. A crash (or *halting*) failure causes an agent to cease functioning and play no further role in the execution. This means that all the adjacent edges of $F(u) \in G_U$ are removed from E_U, and, at the same time, all the adjacent edges of $u \in G_O$ become inactive.

The execution of a protocol proceeds in discrete steps. In every step, an edge $e \in E_U$ between two nodes $F(u)$ and $F(v)$ is selected by an *adversary scheduler*, subject to some *fairness* guarantee. The corresponding nodes u and v interact with each other and update their states and the state of the edge $uv \in G_O$ between them, according to a joint transition function δ. If two nodes in states q_u and q_v with the edge joining them in state q_{uv} encounter each other, they can change into states q'_u, q'_v and q'_{uv}, where $(q'_u, q'_v, q'_{uv}) \in \delta(q_u, q_v, q_{uv})$. In the original model, G_U is the complete directed graph, which means that during an interaction, the interacting nodes have distinct roles. In our protocols, we consider a more restricted version, that is, *symmetric* transition functions, as we try to keep the model as minimal as possible. In particular, $\delta(q_u, q_v, q_{uv}) = (a, b, c)$ implies $\delta(q_v, q_u, q_{vu}) = (b, a, c)$.

A configuration is a mapping $C : V_I \cup E_I \rightarrow Q \cup \{0, 1\}$ specifying the state of each node and each edge of the interaction graph. An execution of the protocol on input I is a finite or infinite sequence of configurations, $C_0, C_1, C_2, \ldots$, each of which is a set of states drawn from $Q \cup \{0, 1\}$. In the initial configuration C_0, all nodes are in state q_0 and all edges are inactive. Let q_u and q_v be the states of the nodes u and v, and q_{uv} denote the state of the edge joining them. A configuration C_k is obtained from C_{k-1} by one of the following types of transitions:

1. **Ordinary transition:** $C_k = (C_{k-1} - \{q_u, q_v, q_{uv}\}) \cup \{q'_u, q'_v, q'_{uv}\}$ where $\{q_u, q_v, q_{uv}\} \subseteq C_{k-1}$ and $(q'_u, q'_v, q'_{uv}) \in \delta(q_u, q_v, q_{uv})$.
2. **Crash failure:** $C_k = C_{k-1} - \{q_u\} - \{q_{uv} : uv \in E_I\}$ where $\{q_u, q_{uv}\} \subseteq C_{k-1}$.

We say that C' is *reachable from* C and write $C \rightsquigarrow C'$, if there is a sequence of configurations $C = C_0, C_1, \ldots, C_t = C'$, such that $C_i \rightarrow C_{i+1}$ for all i, $0 \leq i < t$. The fairness condition that we impose on the scheduler is quite simple to state. Essentially, we do not allow the scheduler to avoid a possible step forever. More formally, if C is a configuration that appears infinitely often in an execution, and $C \rightarrow C'$, then C' must also appear infinitely often in the execution. Equivalently, we require that any configuration that is always reachable is eventually reached.

We define the output of a configuration C as the graph $G(C) = (V, E)$ where $V = \{u \in V_O : C(u) \in Q_{out}\}$ and $E = \{uv : u, v \in V, u \neq v, \text{ and } C(uv) = 1\}$. If there exists some step $t \geq 0$ such that $G(C_i) = G$ for all $i \geq t$, we say that the output of an execution $C_0, C_1, \ldots$ *stabilizes* (or *converges*) to graph G, every configuration C_i, for $i \geq t$, is called *output-stable*, and t is called the *running time* under our scheduler. We say that a protocol Π stabilizes eventually to a graph G of *type* L if and only if after a finite number of pairwise interactions, the graph defined by 'on' edges does not change and belongs to the graph language L.

Definition 1. *We say that a protocol Π constructs a graph language L if: (i) every execution of Π on n nodes stabilizes to a graph $G \in L$ s.t. $|V(G)| = n$ and (ii) $\forall G \in L$ there is an execution of Π on $|V(G)|$ nodes that stabilizes to G.*

Definition 2. *We say that a protocol Π partially constructs a graph language L, if: (i) requirement (i) from Definition 1 holds and (ii) $\exists G \in L$ s.t. no execution of Π on $|V(G)|$ nodes stabilizes to G.*

Definition 3 (Fault-tolerant protocol). *Let Π be a NET protocol that, in a failure-free setting, constructs a graph $G \in L$. Π is called f-fault-tolerant if for any population size $n > f$, any execution of Π constructs a graph $G \in L$, where $|V(G)| = n - f$. We also call Π fault-tolerant if the same holds for any number $f \leq n - 2$ of faults.*

Definition 4 (Constructible language). *A graph language L is called constructible (partially constructible) if there is a protocol that constructs (partially constructs) it. Similarly, we call L constructible under f faults, if there is an f-fault-tolerant protocol that constructs L, where f is an upper bound on the maximum number of faults during an execution.*

Definition 5 (Critical node). *Let G be a graph that belongs to a graph language L. Call u a critical node of G if by removing u and all its edges, the resulting graph $G' = G - \{u\} - \{uv : v \sim u\}$, does not belong to L. In other words, if there are no critical nodes in G, then any (induced) subgraph G' of G that can be obtained by removing nodes and all their edges, also belongs to L.*

Definition 6 (Hereditary Language). *A graph language L is called hereditary if for any graph $G \in L$, every induced subgraph of G also belongs to L. In other words, there is no graph $G \in L$ with critical nodes.*

This notion is known in the literature as *hereditary property* of a graph w.r.t. (with respect to) some graph language L. Observe that if there exists a graph G s.t. for any induced subgraph G' of G, $G' \in L$, does not imply that the same holds for any graph in L. Some examples of hereditary languages are "Bipartite graph", "Planar graph", "Forest of trees", "Clique", "Set of cliques", and "Maximum node degree $\leq \Delta$".

In this work, unless otherwise stated, a graph language L is an infinite set of graphs satisfying the following properties:

1. (*No gaps*): For all $n \geq c$, where $c \geq 2$ is a finite integer, $\exists G \in L$ of order n.
2. (*No Isolated Nodes*): $\forall G \in L$ and $\forall u \in V(G)$, it holds that $d(u) \geq 1$ (where $d(u)$ is the degree of u).

Even though graph languages are not allowed to contain isolated nodes, there are cases in which a protocol might be allowed to output one or more isolated nodes. In particular, if a protocol Π constructing L is allowed a waste of at most w, then whenever Π is executed on n nodes, it must output a graph $G \in L$ of order $|V(G)| \geq n - w$, leaving at most w nodes in one or more separate components (could be all isolated).

3   Network Constructors Without Fault Notifications

In this section, we study the constructive power of the original NET model in the presence of bounded and unbounded crash faults when no form of notification is available to the nodes.

3.1   Unbounded Number of Faults

We here consider the setting where the number of faults can be any number up to $n-2$. We prove that the only constructible graph language is *Spanning Clique* = $\{G : G$ is a spanning clique$\}$.

We first present a protocol which constructs the language *Spanning Clique* and we show that it can tolerate any number of faults. Let *Clique* be the following 2-state *symmetric* protocol.

Protocol Clique: $Q = \{b, r\}$, initial state b, and transition function δ : $(b,\ b,\ 0) \rightarrow (b,\ r,\ 0),\ (b,\ r,\ 0) \rightarrow (r,\ r,\ 0),\ (r,\ r,\ 0) \rightarrow (r,\ r,\ 1)$

Lemma 1. Clique *is a fault-tolerant protocol for* Spanning Clique.

In addition, we show that (due to the power of the adversary), no other graph language is constructible under unbounded faults.

Lemma 2. *Let Π be a protocol constructing a language L and $G \in L$ be a graph that Π outputs on $|V(G)|$ nodes. If G has an independent set $S \subseteq V$, s.t. $|S| \geq 2$, then there is an execution of Π on n nodes which stabilizes on $|S|$ isolated nodes (where $|S| = n - f$ and f is the number of faults in that execution).*

Theorem 1. *Let L be any graph language such that $L \neq$ Spanning Clique. Then, there is no protocol that constructs L if an unbounded number of crash failures may occur.*

The proof following theorem is a direct application of Lemma 2.

Theorem 2. *Let L be any graph language such that the graphs $G \in L$ have maximum independent sets whose size grows with $|V(G)|$. If the useful space of protocols is required to grow with n, then there is no protocol that constructs L in the unbounded-faults case.*

3.2   Bounded Number of Faults

The exact characterization established above, shows that under unbounded failures and without further assumptions, we cannot hope for non-trivial constructions. We now relax the power of the faults adversary, so that there is a *finite upper bound* f on the number of faults. In particular, fixing any such $0 \leq f \leq n$ in advance, it is guaranteed that $\forall n \geq 0$ and all executions of a protocol on n nodes, at most f nodes may fail during the execution. Then the class of constructible graph languages is naturally parameterized in f.

We first show that non-hereditary languages are not constructible under a single fault.

Theorem 3. *If there exists a critical node in G, there is no 1-fault-tolerant NET protocol that stabilizes to it.*

By Definition 6 and Theorem 3 it follows that.

Corollary 1. *If a graph language L is non-hereditary, it is impossible to be constructed under a single fault.*

Note that this does not imply that any hereditary language is constructible under a constant number of faults. We leave this as an interesting open problem.

On the positive side we show that in the case of a bounded number of faults, there is a non-trivial class of languages that is partially constructible. Consider the class of graph languages defined as follows. Any such language $L_{D,f}$ in the family is uniquely specified by a graph $D = ([k], H)$ and the finite upper bound $f < k$ on the number of faults. A graph $G = (V, E)$ belongs to $L_{D,f}$ iff there are k partitions $V_1, V_2, \ldots, V_k$ of V s.t. for all $1 \leq i, j \leq k$, $||V_i| - |V_j|| \leq f + 1$. In addition, E is constructed as follows. The graph $D = ([k], H)$, possibly containing self-loops, defines a neighboring relation between the k partitions. For every $(i, j) \in H$ (where possibly $i = j$), E contains all edges between partitions V_i and V_j, i.e., a complete bipartite graph between them (or a clique in case $i = j$). As no isolated nodes are allowed, every V_i must be fully connected to at least one V_j (possibly itself).

We first consider the case where $k = 2^\delta$, for some constant $\delta \geq 0$, and we provide a protocol that divides the population into k partitions. The protocol works as follows: initially, all nodes are in state c_0 (we call this the partition 0). When two nodes in states c_i, where $i \geq 0$ interact with each other, they update their states to c_{2i+1} and c_{2i+2}, moving to partitions $2i+1$ and $2i+2$ respectively. Interactions between nodes in different c-states (c_i, c_j, where $i \neq j$) do not affect the configuration. When $j = 2i + 1 \geq k - 1$ (or $j = 2i + 2 \geq k - 1$) for the first time, it means that the node has reached its final partition. It updates its state to P_m, where $m = j - k + 1$, thus, the final partitions are $\{P_0, P_1, \ldots, P_{k-1}\}$.

This process divides each partition into two partitions of equal size. However, in the case where the number of nodes is odd, a single node remains unmatched. For this reason, all nodes participate to the final formation of H regardless of whether they have reached their final partitions or not. There is a straightforward mapping of each internal partition to a distinct leaf of the binary tree, that is, each partition c_i behaves as if it were in partition P_i. In order to avoid false connections between the partitions, we also allow the nodes to disconnect from each other if they move to a different partition. This process guarantees that eventually all nodes end up in a single partition, and their connections are strictly described by H.

Lemma 3. *In the absence of faults, the above protocol divides the population into k partitions of at least $n/k - 1$ nodes each.*

Lemma 4. *In the case where up to f faults occur during the execution, each final partition has at least $n/k - f - 1$ nodes, where k is the number of partitions and $f < k$.*

Corollary 2. $||V_i| - |V_j|| \leq f + 1$, $\forall 1 \leq i, j \leq k$.

Theorem 4. *The language $L_{D,f}$, where k is a constant number, is partially constructible under f faults.*

Finally, we show that if we permit a waste linear in n, any graph language that is constructible in the fault-free NET model, becomes constructible under a bounded number of faults.

Theorem 5. *Take any NET protocol Π of the original fault-free model. There is a NET Π' such that when at most f faults may occur on any population of size n, Π' successfully simulates an execution of Π on at least $\frac{n}{2f} - f$ nodes.*

4   Notified Network Constructors

In light of the impossibility results of Sect. 3, we allow fault notifications when nodes crash, aiming at constructing a larger class of graph languages. In particular, we introduce a *fault flag* in each node, which is initially zero. When a node u crashes at time t, every node v which was adjacent to u at time t is notified, that is, the fault flag of all v becomes 1. In the case where u is an isolated node (i.e., it has no active edges), an arbitrary node w in the graph is notified, and its fault flag becomes 2. Then, the fault flag becomes immediately zero after applying a corresponding rule from the transition function.

More formally, the set of node-states is $Q \times \{0, 1, 2\}$, and for clarity in our descriptions and protocols, we define two types of transition functions. The first one determines the node and connection state updates of pairwise interactions ($\delta_1 : Q \times Q \times \{0, 1\} \rightarrow Q \times Q \times \{0, 1\}$), while the second transition function determines the node state updates due to fault notifications ($\delta_2 : Q \times \{0, 1, 2\} \rightarrow Q \times \{0\}$). This means that during a step t that a node u crashes, all its adjacent nodes are allowed to update their states based on δ_2 at that same step. If there are no any adjacent nodes to u, an arbitrary node is notified, thus, updating its state based on δ_2 at step t.

Proposition 1. *We provide fault-tolerant protocols for spanning star and cycle cover (see Protocol 3 and Protocol 4 in [MST19]).*

4.1   Universal Fault-Tolerant Constructors

In this section, we ask whether there is a generic fault-tolerant constructor capable of constructing a large class of graphs. We first give a fault-tolerant protocol that constructs a spanning line, and then we show that we can simulate a given TM on that line, tolerating any number of crash faults. The rules of the protocol and the proof of its correctness can be found in [MST19]. Finally, we exploit that in order to construct any graph language that can be decided by an $O(n^2)-$space TM, paying at most linear waste.

Lemma 5. *FT Spanning Line (see Protocol 5 in [MST19]) is fault-tolerant.*

Lemma 6. *There is a NET Π (with notifications) such that when Π is executed on n nodes and at most k faults can occur, where $0 \leq k < n$, Π will eventually simulate a given TM M of space $O(n - k)$ in a fault-tolerant way.*

Lemma 7. *There is a fault-tolerant NET Π (with notifications) which partitions the nodes into two groups U and D with waste at most $2f(n)$, where $f(n)$ is an upper bound on the number of faults that can occur. U is a spanning line with a unique leader in one endpoint and can eventually simulate a TM M. In addition, there is a perfect matching between U and D.*

Theorem 6. *For any graph language L that can be decided by a linear space TM, there is a fault-tolerant NET Π (with notifications) that constructs a graph in L with waste at most $min\{n/2 + f(n), n\}$, where $f(n)$ is an upper bound on the number of faults that can occur.*

We now show that if the constructed network is required to occupy $1/3$ instead of half of the nodes, then the available space of the TM-constructor dramatically increases from $O(n)$ to $O(n^2)$. We provide a protocol which partitions the population into three sets U, D and M of equal size $k = n/3$ (see Protocol 6 in [MST19]). The idea is to use the set M as a $\Theta(n^2)$ binary memory for the TM, where the information is stored in the $k(k-1)/2$ edges of M.

Theorem 7. *For any graph language L that can be decided by an $O(n^2)-$ space TM, there is a protocol that constructs L equiprobably with waste at most $min\{2n/3 + f(n), n\}$, where $f(n)$ is an upper bound on the number of faults.*

5 Conclusions and Open Problems

A number of interesting problems are left open for future work. Our only exact characterization was achieved in the case of unbounded faults and no notifications. If faults are bounded, non-hereditary languages were proved impossible to construct without notifications but we do not know whether hereditary languages are constructible. Relaxations, such as permitting waste or partial constructibility were shown to enable otherwise impossible transformations, but there is still work to be done to completely characterize these cases. In case of notifications, we managed to obtain fault-tolerant universal constructors, but it is not yet clear whether the assumptions of waste and local coin tossing that we employed are necessary and how they could be dropped. Apart from these immediate technical open problems, some more general related directions are the examination of different types of faults such as random, Byzantine, and communication/edge faults. Finally, a major open front is the examination of fault-tolerant protocols for stable dynamic networks in models stronger than NETs.

References

[AAD+06] Angluin, D., Aspnes, J., Diamadi, Z., Fischer, M.J., Peralta, R.: Computation in networks of passively mobile finite-state sensors. Distrib. Comput. **18**(4), 235–253 (2006)

[AAER07] Angluin, D., Aspnes, J., Eisenstat, D., Ruppert, E.: The computational power of population protocols. Distrib. Comput. **20**(4), 279–304 (2007)

[AAFJ08] Angluin, D., Aspnes, J., Fischer, M.J., Jiang, H.: Self-stabilizing population protocols. ACM Trans. Auton. Adapt. Syst. **3**(4), 1–28 (2008)

[BBB13] Beauquier, J., Blanchard, P., Burman, J.: Self-stabilizing leader election in population protocols over arbitrary communication graphs. In: Baldoni, R., Nisse, N., van Steen, M. (eds.) OPODIS 2013. LNCS, vol. 8304, pp. 38–52. Springer, Cham (2013). https://doi.org/10.1007/978-3-319-03850-6_4

[CLV+17] Cooper, C., Lamani, A., Viglietta, G., Yamashita, M., Yamauchi, Y.: Constructing self-stabilizing oscillators in population protocols. Inf. Comput. **255**, 336–351 (2017)

[DDG+14] Derakhshandeh, Z., Dolev, S., Gmyr, R., Richa, A.W., Scheideler, C., Strothmann, T.: Brief announcement: amoebot-a new model for programmable matter. In: Proceedings of the 26th ACM Symposium on Parallelism in Algorithms and Architectures, pp. 220–222. ACM (2014)

[DDG+18] Daymude, J.J., et al.: On the runtime of universal coating for programmable matter. Nat. Comput. **17**(1), 81–96 (2018)

[DGFGR06] Delporte-Gallet, C., Fauconnier, H., Guerraoui, R., Ruppert, E.: When birds die: making population protocols fault-tolerant. In: Gibbons, P.B., Abdelzaher, T., Aspnes, J., Rao, R. (eds.) DCOSS 2006. LNCS, vol. 4026, pp. 51–66. Springer, Heidelberg (2006). https://doi.org/10.1007/11776178_4

[DIM93] Dolev, S., Israeli, A., Moran, S.: Self-stabilization of dynamic systems assuming only read/write atomicity. Distrib. Comput. **7**(1), 3–16 (1993)

[DLFI+17] Di Luna, G.A., Flocchini, P., Izumi, T., Izumi, T., Santoro, N., Viglietta, G.: Population protocols with faulty interactions: the impact of a leader. In: Fotakis, D., Pagourtzis, A., Paschos, V.T. (eds.) CIAC 2017. LNCS, vol. 10236, pp. 454–466. Springer, Cham (2017). https://doi.org/10.1007/978-3-319-57586-5_38

[DLFS+19] Di Luna, G.A., Flocchini, P., Santoro, N., Viglietta, G., Yamauchi, Y.: Shape formation by programmable particles. Distrib. Comput. 1–33 (2019)

[Dol00] Dolev, S.: Self-stabilization. MIT Press, Cambridge (2000)

[DT01] Ducourthial, B., Tixeuil, S.: Self-stabilization with r-operators. Distrib. Comput. **14**(3), 147–162 (2001)

[GK10] Guellati, N., Kheddouci, H.: A survey on self-stabilizing algorithms for independence, domination, coloring, and matching in graphs. J. Parallel Distrib. Comput. **70**(4), 406–415 (2010)

[GKR10] Gilpin, K., Knaian, A., Rus, D.: Robot pebbles: one centimeter modules for programmable matter through self-disassembly. In: 2010 IEEE International Conference on Robotics and Automation (ICRA), pp. 2485–2492. IEEE (2010)

[GR09] Guerraoui, R., Ruppert, E.: Names trump malice: tiny mobile agents can tolerate byzantine failures. In: Albers, S., Marchetti-Spaccamela, A., Matias, Y., Nikoletseas, S., Thomas, W. (eds.) ICALP 2009. LNCS, vol. 5556, pp. 484–495. Springer, Heidelberg (2009). https://doi.org/10.1007/978-3-642-02930-1_40

[MCS11] Michail, O., Chatzigiannakis, I., Spirakis, P.G.: Mediated population protocols. Theoret. Comput. Sci. **412**(22), 2434–2450 (2011)

[Mic18]   Michail, O.: Terminating distributed construction of shapes and patterns in a fair solution of automata. Distrib. Comput. **31**(5), 343–365 (2018)

[MOKY12] Mizoguchi, R., Ono, H., Kijima, S., Yamashita, M.: On space complexity of self-stabilizing leader election in mediated population protocol. Distrib. Comput. **25**(6), 451–460 (2012)

[MS16]   Michail, O., Spirakis, P.G.: Simple and efficient local codes for distributed stable network construction. Distrib. Comput. **29**(3), 207–237 (2016)

[MS17]   Michail, O., Spirakis, P.G.: Network constructors: a model for programmable matter. In: Steffen, B., Baier, C., van den Brand, M., Eder, J., Hinchey, M., Margaria, T. (eds.) SOFSEM 2017. LNCS, vol. 10139, pp. 15–34. Springer, Cham (2017). https://doi.org/10.1007/978-3-319-51963-0_3

[MSS19]   Michail, O., Skretas, G., Spirakis, P.G.: On the transformation capability of feasible mechanisms for programmable matter. J. Comput. Syst. Sci. **102**, 18–39 (2019)

[MST19]   Michail, O., Spirakis, P.G., Theofilatos, M.: Fault tolerant network constructors. arXiv preprint arXiv:1903.05992 (2019)

[Pel09]   Peleg, D.: As good as it gets: competitive fault tolerance in network structures. In: Guerraoui, R., Petit, F. (eds.) SSS 2009. LNCS, vol. 5873, pp. 35–46. Springer, Heidelberg (2009). https://doi.org/10.1007/978-3-642-05118-0_3

Ring Exploration of Myopic Luminous Robots with Visibility More Than One

Shota Nagahama$^{(\boxtimes)}$, Fukuhito Ooshita, and Michiko Inoue

Nara Institute of Science and Technology, Takayama 8916-5, Ikoma, Nara, Japan
{nagahama.shota.nl1,f-oosita,kounoe}@is.naist.jp

Abstract. In this paper, we investigate ring exploration algorithms for autonomous myopic luminous robots. Myopic robots mean that they can observe nodes only within a certain fixed distance, and luminous robots mean that they have light devices that can emit a color from a set of constant number of colors. We consider the constraint that the visible distance is any constant of at least two and the number of colors of light devices is two. As a main contribution, in the fully synchronous, semi-synchronous, and asynchronous models, we prove that (1) two robots are necessary and sufficient to achieve perpetual exploration and (2) three robots are necessary and sufficient to achieve terminating exploration, where perpetual exploration requires every robot to visit every node infinitely many times and terminating exploration requires robots to terminate after every node is visited by a robot at least once. These results show the power of large visibility for luminous robots because, when the visible distance is one and the number of colors is two, three and four robots are necessary to achieve perpetual and terminating exploration, respectively, in the semi-synchronous and asynchronous models. We also show that the proposed perpetual exploration algorithm is universal, that is, the algorithm achieves perpetual exploration from any solvable initial configuration with two robots. On the other hand, we show that no universal algorithm exists for terminating exploration with three robots.

Keywords: Autonomous mobile robots · Exploration problem · Discrete environments

1 Introduction

1.1 Background and Motivation

Theoretical research on computing by autonomous mobile robots has attracted a lot of attention in the field of distributed computing. The research focuses on the minimum capabilities of robots that permit to achieve a given task, and clarifies the limitation of computing in such settings. As a model of robot operations,

This work was supported in part by JSPS KAKENHI Grant Number 18K11167 and JST SICORP Grant Number JPMJSC1806.

M. Ghaffari et al. (Eds.): SSS 2019, LNCS 11914, pp. 256–271, 2019.
https://doi.org/10.1007/978-3-030-34992-9_20

the Look-Compute-Move (LCM) model [17] is commonly used. In the LCM model, each robot repeats cycles of look, compute, and move phases. In the look phase, the robot observes positions of other robots. In the compute phase, the robot executes its algorithm using the observation as its input, and decides whether it moves somewhere or stays idle. In the move phase, it moves to a new position if the robot decided to move in the compute phase. To consider minimum capabilities, most studies assume that robots are identical (*i.e.*, robots execute the same algorithm and have no identifier), oblivious (*i.e.*, robots have no memory to record past history), and silent (*i.e.*, robots cannot communicate with other robots explicitly). Indeed, communication among robots is done only in the implicit way by observing positions of other robots and moving to a new position. Previous works considered task solvability of LCM robots in continuous environments (aka two- or three-dimensional Euclidean space) [14,17,18], while others considered discrete environments (aka graph networks) [5,13,15].

In this paper, we focus on graph networks. One of the most fundamental tasks in graph networks is exploration. Two types of exploration tasks have been well studied: perpetual exploration requires every robot to visit every node infinitely many times, and terminating exploration requires robots to terminate after every node is visited by a robot at least once. Perpetual exploration has been studied for rings [1] and grids [2]. Terminating exploration has been studied for rings [11,13], trees [12], grids [9], tori [10], and arbitrary networks [4].

All aforementioned works for exploration make the assumption that each robot has unlimited visibility, *i.e.*, it observes all other robots in the networks. However, this powerful ability somewhat contradicts the principle of very weak mobile entities. For this reason, recent studies consider the more realistic case of myopic robots [7,8]. A myopic robot has limited visibility, *i.e.*, it can see nodes (and robots on them) only within a certain fixed distance ϕ. Datta et al. studied terminating exploration of rings for $\phi = 1$ [7] and $\phi = 2, 3$ [8].

Since most results for $\phi = 1$ are negative, Ooshita and Tixeuil [16] consider a non-volatile visible light [6] to improve the task solvability. A robot endowed with such a light is called a luminous robot. Each luminous robot is equipped with a light device that can emit a constant number of colors to other robots, a single color at a time. The light color is non-volatile, so it can be used as a constant-space memory. Ooshita and Tixeuil [16] studied perpetual and terminating exploration of rings for $\phi = 1$ and showed that the number of robots required to achieve the tasks can be reduced compared to non-luminous robots. A remaining natural question is whether the number of required robots can be further reduced with $\phi \geq 2$. The question is the main topic in this paper.

Recently Bramas et al. studied exploration of an infinite grid in case of myopic luminous robots with $\phi = 1$ and in case of myopic non-luminous robots with $\phi = 2$ [3]. In their protocols, a few constant number of robots operate so that every node is visited by a robot at least once.

1.2   Our Contributions

We focus on ring exploration with myopic luminous robots in case of $\phi \geq 2$ (but $\phi \neq \infty$) and two colors. We give answers to the question: whether the number of required robots can be reduced in case of $\phi \geq 2$. Table 1 summarizes our contributions and related works. Note that robots with no light are equivalent to robots with a single color light.

Table 1. Ring exploration with myopic robots.

Reference	Exploration	Synchrony	ϕ	#colors	#robots	
					Necessary	Sufficient
[7]	Terminating	FSYNC	1	1	5	5
[7]	Terminating	SSYNC & ASYNC	1	1	Impossible	
[8]	Terminating	SSYNC & ASYNC	2	1	5	7
[8]	Terminating	SSYNC & ASYNC	3	1	5	5
[16]	Perpetual	FSYNC	1	2	2	2
[16]	Terminating	FSYNC	1	2	3	3
[16]	Perpetual	SSYNC & ASYNC	1	2	3	3
[16]	Terminating	SSYNC & ASYNC	1	2	4	4
This paper	Perpetual	FSYNC, SSYNC & ASYNC	≥ 2	2	2	2
This paper	Terminating	FSYNC, SSYNC & ASYNC	≥ 2	2	3	3

The answers to the question are that *(i)* in the fully synchronous model, the number of required robots cannot be reduced, and *(ii)* in the semi-synchronous and asynchronous models, the number of required robots can be reduced by one. In case of $\phi \geq 2$, we prove that, in the fully synchronous, semi-synchronous, and asynchronous models, two and three robots are necessary and sufficient to achieve perpetual and terminating exploration, respectively. Ooshita and Tixeuil [16] proved that, in case of $\phi = 1$, (1) in the fully synchronous model, two and three robots are necessary and sufficient to achieve perpetual and terminating exploration, respectively, and (2) in the semi-synchronous and asynchronous models, three and four robots are necessary and sufficient to achieve perpetual and terminating exploration, respectively. Therefore, the answers are true, and they characterize the power of large visibility for luminous robots.

Similarly to previous works for myopic robots, all algorithms proposed in this paper assume some specific initial configurations because most configurations are not solvable. For example, when myopic robots are deployed so that no robot can observe other robots, they cannot achieve exploration. However, our perpetual exploration algorithm achieves the best possible property, that is, they are universal. This means that, in the fully synchronous, semi-synchronous, and asynchronous models, the proposed algorithm solves perpetual exploration from any solvable initial configuration with two robots and two colors. As for terminating exploration, we show that no universal algorithm exists. That is, in the fully synchronous, semi-synchronous, and asynchronous models, no algorithm may solve terminating exploration from any solvable initial configuration with three robots. The approach used in this proof is also an important contribution of this paper. Although most works prove non-existence of universal algorithms by considering possible behaviors exhaustively, our proof does not require such consideration.

2 Preliminaries

In this section, we define the system model and terminologies used in this paper. They are almost identical to those in [16].

2.1 System Model

The system consists of n nodes and k mobile robots. The nodes $v_0, v_1, ..., v_{n-1}$ form an undirected and unoriented ring-shaped graph, where a link exists between v_i and v_{i+1}, for $i < n-1$, and between v_{n-1} and v_0. For simplicity we consider mathematical operations on node indices as operations modulo n. The indices $0, ..., n-1$ are used for notation purposes only and robots do not know them. Neither nodes nor links have identifiers or labels, and consequently robots cannot distinguish nodes and cannot distinguish links. Robots do not know n, the size of the ring. Each robot is on a node of the ring at each instant. When a robot r is on a node v, we say r *occupies* v. The distance between two nodes is the number of links in a shortest path between the nodes. The distance between two robots a and b is the distance between two nodes occupied by a and b. Two robots a and b are neighbors if the distance between a and b is one.

Robots we consider have the following characteristics and capabilities. Robots are *identical*, that is, robots execute the same deterministic algorithm and do *not* have unique identifiers. Robots are *luminous*, that is, each robot has a light (or state) that is visible to itself and other robots. A robot can choose the color of its light from a discrete set Col. When the set Col is finite, l denotes the number of available colors (*i.e.*, $l = |Col|$). Robots have no other persistent memory and cannot remember the history of past actions. Each robot can communicate by observing positions and colors of other robots (for collecting information), and by changing its color and moving (for sending information). Robots are *myopic*, that is, each robot r can observe positions and colors of robots within a fixed distance ϕ ($\phi > 0$) from its current position. Since robots are identical, they share the same ϕ.

Each robot executes an algorithm by repeating three-phase cycles: Look, Compute, and Move (L-C-M). During the *Look* phase, the robot takes a snapshot of positions and colors of robots within distance ϕ. During the *Compute* phase, the robot computes its next color and movement according to the observation in the Look phase. The robot may change its color at the end of the Compute phase. If the robot decides to move, it moves instantaneously to a neighboring node during the *Move* phase. To model asynchrony of executions, we introduce the notion of *scheduler* that decides when each robot executes phases. When the scheduler makes robot r execute some phase, we say the scheduler activates the phase of r or simply activates r. We consider three types of synchronicity: the FSYNC (full-synchronous) model, the SSYNC (semi-synchronous) model, and the ASYNC (asynchronous) model. In all models, time is represented by an infinite sequence of instants $0, 1, 2, \ldots$. No robot has access to this global time. In the FSYNC and SSYNC models, all the robots that are activated at an instant t execute a full cycle synchronously and concurrently between t and $t + 1$. In the FSYNC model, at every instant, the scheduler activates all robots. In the SSYNC model, at every instant, the scheduler selects a non-empty subset of robots and activates the selected robots. In the ASYNC model, the scheduler activates cycles of robots asynchronously: the time between Look, Compute, and Move phases is finite but unpredictable. Note that in the ASYNC model, a robot r can move based on the outdated view obtained during the previous Look phase. Throughout the paper we assume that the scheduler is *fair*, that is, each robot is activated infinitely often.

In the sequel, $M_i(t)$ denotes the multiset of colors of robots located in node v_i at an instant t. If v_i is not occupied by any robot at t, then $M_i(t) = \emptyset$ holds, and v_i is *free* at t. Then, v_i is a *tower* at t if $|M_i(t)| \geq 2$. A *configuration* $C(t)$ of the system at an instant t is defined as $C(t) = (M_0(t), M_1(t), \ldots, M_{n-1}(t))$. If t is clear from the context, we simply write $C = (M_0, M_1, \ldots, M_{n-1})$. If there exists an index x such that $M_{x+i} = M_{x-i}$ holds for any i, or if $M_{x+i} = M_{x-(i+1)}$ holds for any i (*i.e.*, there exists at least one axis of symmetry in the configuration), configuration C is called *symmetric*.

Let S be a set of robots. The visibility graph of S in configuration C is defined as graph $G_C(S) = (S, L_C)$, where $\{r_1, r_2\} \in L_C$ holds iff robots r_1 and r_2 can see each other (*i.e.*, on the ring, the distance between r_1 and r_2 is at most ϕ) in configuration C. We say S is *connected* in C if $G_C(S)$ is connected; otherwise, S is disconnected in C.

When a robot takes a snapshot of its environment, it gets a *view* up to distance ϕ. Consider a robot r on node v_i; then, r obtains two views: the forward view and the backward view. The forward and backward views of r are defined as $V_f = (c_r, M_{i-\phi}, \ldots, M_{i-1}, M_i, M_{i+1}, \ldots, M_{i+\phi})$, and $V_b = (c_r, M_{i+\phi}, \ldots, M_{i+1}, M_i, M_{i-1}, \ldots, M_{i-\phi})$, respectively, where c_r denotes r's color. If the forward view and the backward view of r are identical, then r's view is *symmetric*. In this case, since we assume unoriented rings, r cannot distinguish between the two directions when it moves, and the scheduler decides which direction r moves to. If r observes no other robot in its view, r is *isolated*.

2.2 Algorithm

An algorithm is described as a set of rules. Each rule is represented in the following manner $< Label >:< Guard >::< Action >$. The guard $< Guard >$ represents possible views obtained by a robot. If a forward or backward view of robot r matches a guard in some rule, we say r is *enabled*. We also say the corresponding rule $<Label>$ is enabled. If a robot is enabled, the robot may change its color and move based on the corresponding action $<Action>$ during the Compute and Move phases. If a forward or backward view of r matches several guards, one of the corresponding rules is enabled (it is decided by the scheduler). However, every algorithm proposed in this paper is described as a set of rules such that no two guards include the same view.

2.3 Execution, Problem, and Exploration Problem

An execution from initial configuration C_0 is a maximal sequence of configurations $E = C_0, C_1, ..., C_i, ...$ such that, for any $j > 0$, we have *(i)* $C_{j-1} \neq C_j$, *(ii)* C_j is obtained from C_{j-1} after some robots move or change their colors, and *(iii)* for every robot r that moves or changes its color between C_{j-1} and C_j, there exists $0 \leq j' \leq j$ such that r takes its decision to move or change its color according to its algorithm and its view in $C_{j'}$. The term "*maximal*" means that the execution is either infinite or ends in a *terminal configuration*, i.e., a configuration in which no robot is enabled.

A problem $\mathcal{P}$ is defined as a set of executions: An execution E solves $\mathcal{P}$ if $E \in \mathcal{P}$ holds. An algorithm $\mathcal{A}$ solves problem $\mathcal{P}$ from initial configuration C_0 if any execution from C_0 solves $\mathcal{P}$. We simply say an algorithm $\mathcal{A}$ solves problem $\mathcal{P}$ if there exists an initial configuration C_0 such that $\mathcal{A}$ solves $\mathcal{P}$ from C_0. For configuration C and problem $\mathcal{P}$, C is solvable for $\mathcal{P}$ if there exists an algorithm (specific to C) that solves $\mathcal{P}$ from initial configuration C. Let $C_s(\mathcal{P})$ be a set of all configurations solvable for $\mathcal{P}$. We say algorithm $\mathcal{A}$ is universal with respect to problem $\mathcal{P}$ if $\mathcal{A}$ solves $\mathcal{P}$ from any initial configuration in $C_s(\mathcal{P})$. That is, a universal algorithm solves $\mathcal{P}$ from any solvable initial configuration.

In this paper, we consider the perpetual exploration problem and terminating exploration problem in case of $\phi \geq 2$.

Definition 1 (Perpetual exploration problem). *Perpetual exploration is defined as a set of executions E such that every robot visits every node infinitely many times in E.*

Definition 2 (Terminating exploration problem). *Terminating exploration is defined as a set of executions E such that (1) every node is visited by at least one robot in E and (2) there exists a suffix of E such that no robots are enabled.*

2.4 Descriptions

Let $C = (M_0, ..., M_{n-1})$ be a configuration. We say $C' = (M'_0, ..., M'_{n'-1})$ is a sub-configuration of C if there exists x such that $M_{x+i} = M'_i$ holds for any i ($0 \leq i \leq n' - 1$). In this case, we say n' is the length of sub-configuration C'. We sometimes describe a sub-configuration $C' = (M'_0, ..., M'_{n'-1})$ by listing all colors in M_i' as the i-th column. That is, when $M_i' = \{c_1^i, ..., c_{|M_i'|}^i\}$ holds for each i $(0 \le i \le n' - 1)$, we describe C' as follows:

$$
\begin{array}{ccccc}
c_{|M_0'|}^0 & & c_{|M_i'|}^i & & \\
c_{|M_0'|-1}^0 & c_{|M_1'|}^1 & c_{|M_i'|-1}^i & & c_{|M_{n'-1}'|}^{n'-1} \\
\vdots & \vdots & \vdots & \cdots & \vdots \\
c_1^0 & c_1^1 & c_1^i & & c_1^{n'-1}
\end{array}
$$

When $M_i' = \emptyset$ holds, we write $\emptyset$ as the i-th column. If h free nodes exist successively, we sometimes write $\emptyset^h$ instead of writing h columns with $\emptyset$. For simplicity, when C' is a sub-configuration of C and all robots appear in C', we use C' instead of C to represent configuration C. We also use this description to represent views of robots.

Throughout the paper, we consider the case of $\phi \ge 2$. We describe a rule in an algorithm in the following manner:

$$
\mathcal{R}_{rule} : \quad
\begin{array}{ccccc}
 & c_{0,m_0} & & c_{\phi,m_\phi} & \\
c_{-\phi,m_{-\phi}} & c_{0,m_0-1} & & c_{\phi,m_\phi-1} & \\
\vdots & \cdots & \vdots & \cdots & \vdots \\
c_{-\phi,1} & (c_{0,1}) & & c_{\phi,1} &
\end{array}
\quad :: c_{new}, Movement
$$

Notation $\mathcal{R}_{rule}$ is a label of the rule. The middle part represents a guard. This represents a view $V = (c_{0,1}, M_{-\phi}, ..., M_{-1}, M_0, M_1, ..., M_\phi)$, where $M_i = \{c_{i,1}, ..., c_{i,m_i}\}$ holds for $i \in \{-\phi, ..., -1, 0, 1, ..., \phi\}$. Intuitively, each column represents colors of robots on a single node and a color within parentheses represents its current color. To simplify the description, we will use a wild card X to represent the guard. When we write X as the i-th column, M_i is arbitrary. If h columns with X exist successively, we sometimes write X^h instead of writing h columns with X. Similarly, if h columns with $\emptyset$ exist successively, we sometimes write $\emptyset^h$ instead of writing h columns with $\emptyset$. The notation h sometimes includes a variable i. In this case, we describe the range of i on the right side of the guard. If a forward or backward view of robot r is equal to V, r is enabled. In this case, if the scheduler activates r, it executes an action represented by $c_{new}, Movement$. Notation c_{new} represents a new color of the robot. Notation $Movement$ can be $\perp$, $\leftarrow$, $\rightarrow$, or $\leftarrow \vee \rightarrow$ and represents the movement: (1) $\perp$ implies a robot does not move, (2) $\leftarrow$ (resp., $\rightarrow$) implies a robot moves toward the node such that a set of robot colors is M_{-1} (resp., M_1), and (3) $\leftarrow \vee \rightarrow$ implies a robot moves toward one of two directions (the scheduler decides the direction). When the view V described in a guard is symmetric, $Movement$ should be either $\perp$ or $\leftarrow \vee \rightarrow$. As an example, consider the following rule.

$$
\mathcal{R}_{ex} : \quad \emptyset^{\phi-i-1} \; \mathsf{W} \; \emptyset^i \; \overset{\mathsf{G}}{(\mathsf{W})} \; X^\phi \; (1 \le i \le \phi - 1) :: \mathsf{G}, \leftarrow
$$

Robot r is enabled by $\mathcal{R}_{ex}$ if (1) the color of r is W, (2) the current node is occupied by two robots with colors G and W, (3) in one direction, there exists a node occupied by exactly one robot with color W and the distance to the node is $i+1$ $(1 \leq i \leq \phi - 1)$, and (4) in the same direction as the third condition, the other nodes are free. If r is enabled by $\mathcal{R}_{ex}$, r changes its color to G and moves in the same direction as the third and fourth conditions.

3 Perpetual Exploration

In this section, we provide a perpetual exploration algorithm for $n \geq 2\phi + 1$ in case of $\phi \geq 2$, $l = 2$, and $k = 2$. Note that one robot cannot achieve perpetual exploration clearly because the direction of its movement is decided by the scheduler. A set of colors is $Col = \{G, W\}$. The algorithm is given in Algorithm 1. In the initial configurations, two robots with colors G and W are connected. Each robot can see the other robot only in one direction because of $n \geq 2\phi + 1$. When the distance between two robots is one or zero, the robot with color W moves against the other robot by rule $R1$ or $R2$. When the distance between two robots is two or greater, the robot with color G moves toward the other robot by rule $R3$. This implies two robots move in the same direction. Hence, two robots continue to move and achieve perpetual exploration. Clearly we have the following theorem.

Theorem 1. *In case of $\phi \geq 2$, $Col = \{G, W\}$, and $k = 2$, Algorithm 1 solves perpetual exploration from initial configurations* $\genfrac{}{}{0pt}{}{W}{G}$, $G \emptyset^i W$, $W \emptyset^i G$ $(0 \leq i \leq \phi - 1)$ *for $n \geq 2\phi + 1$ in the ASYNC model.*

Algorithm 1. Asynchronous Perpetual Exploration for $\phi \geq 2, l = 2, k = 2$

Initial configurations

  $\genfrac{}{}{0pt}{}{W}{G}$, $G \emptyset^i W$, and $W \emptyset^i G$ $(0 \leq i \leq \phi - 1)$

Rules

  $R1$: $\emptyset^\phi \genfrac{}{}{0pt}{}{G}{(W)} \emptyset^\phi$ :: $W, \leftarrow \vee \rightarrow$

  $R2$: $\emptyset^\phi$ (W) G $\emptyset^{\phi-1}$ :: $W, \leftarrow$

  $R3$: $\emptyset^{\phi-i-1}$ W $\emptyset^i$ (G) $\emptyset^\phi$ $(1 \leq i \leq \phi - 1)$ :: $G, \leftarrow$

In addition, we can prove that Algorithm 1 is universal with respect to perpetual exploration for $n \geq 2\phi + 4$ in case of $\phi \geq 2$, $l = 2$, and $k = 2$. This is because other initial configurations are unsolvable for $n \geq 2\phi + 4$. To prove the impossibilities, we first introduce Lemma 1 (due to the lack of space, we omit the proof of Lemma 1). For configuration C, we define $V_r(C)$ as a set of nodes occupied by at least one robot. We say a set of two neighboring nodes $T = \{v_i, v_{i+1}\}$ is a territory of robots on node v if $v \in T$ holds. We say a territory set $\mathcal{T}$ is independent if, for every pair of territories $T_1, T_2 \in \mathcal{T}$, the distance between any node in T_1 and any node in T_2 is at least $\phi + 1$.

Lemma 1. *Consider a configuration C such that every distance between two nodes in $V_r(C)$ is at least $\phi + 1$ and, for every node $v \in V_r(C)$, robots on v have the same color. If there exists a territory set T such that T is independent and every node in $V_r(C)$ belongs to some territory in T, there is an execution starting with C where robots on $v \in V_r(C)$ cannot go out of their territory in T in the FSYNC, SSYNC, and ASYNC models.*

Now we can prove that Algorithm 1 is universal with respect to perpetual exploration for $n \geq 2\phi + 4$ in case of $\phi \geq 2$, $l = 2$, and $k = 2$.

Theorem 2. *In case of $\phi \geq 2$, $Col = \{G, W\}$, and $k = 2$, Algorithm 1 is universal with respect to perpetual exploration for $n \geq 2\phi + 4$ in the FSYNC, SSYNC, and ASYNC models.*

Proof. (Sketch) Initial configurations other than ones described in Algorithm 1 are divided into two cases: (1) two robots are disconnected or (2) two robots have the same color. In Case 1, we can define an independent territory set. In Case 2, robots move in symmetric manners, and hence, to achieve exploration, eventually each robot moves against the other robot. Consequently they become disconnected and we can define an independent territory set. Therefore, in both cases, there is an execution where two robots cannot achieve exploration from Lemma 1.

4 Terminating Exploration

4.1 Impossibility of Two Robots

In this subsection, we prove that no algorithm solves terminating exploration for $k = 2$.

Theorem 3. *In case of $k = 2$, no algorithm solves terminating exploration in the FSYNC model. This holds even if robots can use an infinite number of colors.*

Proof. Assume that such algorithm $\mathcal{A}$ exists. Consider an execution $E = C_0, C_1, \ldots$ of $\mathcal{A}$ in a n_1-node ring R_1 ($n_1 \geq 2\phi + 4$). Let i be the minimum index such that two robots terminate or become disconnected at C_i. Next, for some $n_2 > 2(i+1)$, let us consider an execution $E' = C'_0, C'_1, \ldots$ of $\mathcal{A}$ in a n_2-node ring R_2. Clearly, as long as two robots keep connected, they do not recognize the difference between R_1 and R_2. Hence, in E', two robots move similarly to E until C'_i. If two robots terminate at C'_i, they have visited at most $2(i+1)$ nodes and thus they do not achieve exploration. If two robots become disconnected at C'_i, we can define an independent territory set at C'_i. From Lemma 1, two robots cannot visit the remaining nodes and thus they achieve exploration. This is a contradiction.

Algorithm 2. Asynchronous Terminating Exploration for $\phi \geq 2, l = 2, k = 3$

Initial configurations
   $W \emptyset^i W \emptyset^j G$ and $G \emptyset^i W \emptyset^j W$ $(0 \leq i \leq \phi - 1, 0 \leq j \leq \phi - 1)$

Rules
   $R1:$ $X^{\phi-i-1} W \emptyset^i (G) X^\phi (1 \leq i \leq \phi - 1) :: G, \leftarrow$
   $R2:$ $X^{\phi-i-1} W \emptyset^i (W) G X^{\phi-1} (1 \leq i \leq \phi - 1) :: W, \leftarrow$
   $R3:$ $X^{\phi-1} W (W) G X^{\phi-1} :: G, \perp$
   $R4:$ $\emptyset^\phi (W) G X^{\phi-1} :: W, \leftarrow$

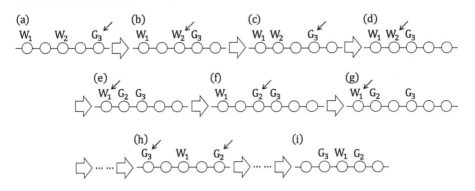

Fig. 1. An execution of Algorithm 2 ($\phi = 2$)

4.2   A Terminating Exploration Algorithm for Three Robots

In this subsection, we give a terminating exploration algorithm for $n \geq 3\phi + 1$ in case of $\phi \geq 2$, $l = 2$, and $k = 3$. A set of colors is $Col = \{G, W\}$. The algorithm is given in Algorithm 2.

  An execution of Algorithm 2 for $\phi = 2$ and $n \geq 3\phi + 1$ is given in Fig. 1. In the figure, W_i (resp., G_i) represents robot r_i with color W (resp., G). Arrows represent that indicated robots are enabled. Let us consider a configuration $W \emptyset^i W \emptyset^j G$ $(0 \leq i \leq \phi - 1, 0 \leq j \leq \phi - 1)$, and assume that r_1, r_2, and r_3 compose the configuration in this order (Fig. 1(a)). Here if the distance between r_2 and r_3 is two or greater, r_3 is enabled with rule $R1$ and moves toward r_2. Since a robot with color W is enabled only in configurations where it finds a robot with color G at its neighboring node, only r_3 is enabled until the configuration becomes $W \emptyset^i W G$ (Fig. 1(b)). At configuration $W \emptyset^i W G$, r_2 moves toward r_1 by rule $R2$, and then the configuration becomes $W \emptyset^{i-1} W \emptyset G$ (Fig. 1(c)). From this configuration, r_3 moves toward r_2 by rule $R1$. After that, r_2 and r_3 continue to move by rules $R2$ and $R1$ until the configuration becomes $W W G$ (Fig. 1(d)). At configuration $W W G$, r_2 changes its color to G by rule $R3$. At configuration $W G G$ (Fig. 1(e)), r_1 moves against r_2 by rule $R4$. At configuration $W \emptyset G G$ (Fig. 1(f)), r_2 moves toward r_1 by rule $R1$, and then the configuration becomes $W G \emptyset G$ (Fig. 1(g)). After that, r_1 and r_2 continue to move to the same direction by rule $R4$ and $R1$. After r_1 and r_2 explore the ring, the configuration becomes $G \emptyset^{\phi-1} W \emptyset G$

(Fig. 1(h)). From this configuration, r_2 and r_3 move toward r_1 by rule $R1$. Since a robot is not enabled when it sees other robots with color G in both directions, r_1 is not enabled while r_2 and r_3 move. Hence, the configuration becomes G W G (Fig. 1(i)). In this configuration, no robots are enabled.

Thus Algorithm 2 solves terminating exploration. Hence we have the following theorem.

Theorem 4. *In case of $\phi \geq 2$, $Col = \{G, W\}$, and $k = 3$, Algorithm 2 solves terminating exploration from initial configurations $W \emptyset^i W \emptyset^j G$ and $G \emptyset^i W \emptyset^j W$ $(0 \leq i \leq \phi - 1, 0 \leq j \leq \phi - 1)$ for $n \geq 3\phi + 1$ in the ASYNC model.*

Note that we can construct another algorithm by swapping the roles of colors G and W in Algorithm 2. Clearly this algorithm solves terminating exploration from configurations such that colors G and W are swapped from solvable configurations for Algorithm 2. This implies configurations $G \emptyset^i G \emptyset^j W$ and $W \emptyset^i G \emptyset^j G$ $(0 \leq i \leq \phi - 1, 0 \leq j \leq \phi - 1)$ are also solvable. Hence, we have the following lemma.

Lemma 2. *In case of $\phi \geq 2$, $Col = \{G, W\}$, $k = 3$, and $n \geq 3\phi + 1$, configurations $W \emptyset^i W \emptyset^j G$, $G \emptyset^i W \emptyset^j W$, $G \emptyset^i G \emptyset^j W$, and $W \emptyset^i G \emptyset^j G$ $(0 \leq i \leq \phi - 1, 0 \leq j \leq \phi - 1)$ are solvable for terminating exploration in the ASYNC model.*

4.3 Nonexistence of Universal Algorithm

In this subsection, we prove that there exists no universal algorithm with respect to terminating exploration for in case of $\phi \geq 2$, $l = 2$, and $k = 3$. This validates the assumption that Algorithm 2 starts from some designated initial configuration.

Theorem 5. *In case of $\phi \geq 2$, $l = 2$, and $k = 3$, no universal algorithm exists with respect to terminating exploration in the FSYNC, SSYNC, and ASYNC models.*

The outline of the proof is as follows. Let C_S be the set of configurations where all robots are connected and no tower exists in case of $\phi \geq 2$, $Col = \{G, W\}$, and $k = 3$. We first prove that all initial configurations in C_S are solvable in the ASYNC model. Then, we prove that, in the FSYNC model, there exists no algorithm that solves terminating exploration from any initial configuration in C_S. Consequently, no universal algorithm exists with respect to terminating exploration in the FSYNC, SSYNC, ASYNC models.

Now we prove that all initial configurations in C_S are solvable in the ASYNC model. We describe C_S as follows:

$$C_S = C_1 \cup C_2 \cup C_3$$
$$C_1 = \cup_{0 \leq i \leq \phi - 1, 0 \leq j \leq \phi - 1}\{\, W \emptyset^i W \emptyset^j G, G \emptyset^i W \emptyset^j W,$$
$$G \emptyset^i G \emptyset^j W, W \emptyset^i G \emptyset^j G \,\}$$
$$C_2 = \cup_{0 \leq i \leq \phi - 1, 0 \leq j \leq \phi - 1}\{\, G \emptyset^i W \emptyset^j G, W \emptyset^i G \emptyset^j W \,\}$$
$$C_3 = \cup_{0 \leq i \leq \phi - 1, 0 \leq j \leq \phi - 1}\{\, G \emptyset^i G \emptyset^j G, W \emptyset^i W \emptyset^j W \,\}$$

From Lemma 2, initial configurations in C_1 are solvable. In addition, we can construct algorithms that solve terminating exploration from initial configurations in C_2 and C_3 (due to the lack of space, we omit algorithms for C_2 and C_3). From the above discussion, initial configurations in C_1, C_2, and C_3 are solvable. Therefore, we have the following lemma.

Lemma 3. *In case of $\phi \geq 2$, $Col = \{G, W\}$, $k = 3$, and $n \geq 3\phi + 1$, configurations in C_S, namely, configurations where all robots are connected and no tower exists are solvable for terminating exploration in the ASYNC model.*

In the following, we prove that, even in the FSYNC model, no *single* algorithm solves terminating algorithm from any configuration in C_S.

Lemma 4. *In the FSYNC model, there exists no algorithm that solves terminating exploration from any initial configuration in C_S.*

To prove Lemma 4 by contradiction, we assume that such an algorithm $\mathcal{A}$ exists. Consider an execution $E = C_0, C_1, \ldots$ of $\mathcal{A}$ in a n-node ring R_1 ($n \geq 3\phi + 3$) starting from an initial configuration $C_0 \in C_S$. Without loss of generality, we assume that, in C_0, robots r_1, r_2, and r_3 occupy nodes v_0, v_x ($0 < x \leq \phi$), and v_y ($x < y \leq x + \phi$), respectively. We define the left and right directions as directions from v_i to v_{i-1} and v_i to v_{i+1}, respectively. Next we define the leftmost and rightmost nodes. In C_0, since all robots stay at nodes from v_0 to v_y, we regard v_0 and v_y as the leftmost and rightmost nodes, respectively. Intuitively, the leftmost node moves to the left (resp., right) direction if a robot on the leftmost node moves to the left (resp., right) direction. The rightmost node is also defined similarly. Formally the leftmost node $L(i)$ and the rightmost node $R(i)$ in C_i are defined as follows.

- In C_0, the leftmost node $L(0)$ is v_0 and the rightmost node $R(0)$ is v_y.
- In C_i ($i > 0$), the leftmost node $L(i)$ is defined from the leftmost node $v_L = L(i-1)$ in C_{i-1}. If at least one robot moves from v_L to v_{L-1} during C_{i-1} to C_i, $L(i)$ is v_{L-1}. Otherwise, if some robot still exists on v_L in C_i, $L(i)$ is v_L. Otherwise (i.e., during C_{i-1} to C_i, all robots on v_L move to v_{L+1} and no robot moves from v_{L+1} to v_L), $L(i)$ is v_{L+1}.
- In C_i ($i > 0$), the rightmost node $R(i)$ is defined from the rightmost node $v_R = R(i-1)$ in C_{i-1}. If at least one robot moves from v_R to v_{R+1} during C_{i-1} to C_i, $R(i)$ is v_{R+1}. Otherwise, if some robot still exists on v_R in C_i, $R(i)$ is v_R. Otherwise, $R(i)$ is v_{R-1}.

At each configuration C_i, we define a leftmost (resp., rightmost) robot as a robot on the leftmost (resp., rightmost) node. In the following, we prove that, for any configuration in C_S, there exists an execution that reaches a configuration in C_S again. This implies that $\mathcal{A}$ cannot terminate and hence cannot solve the terminating exploration. To prove the proposition, we focus on an execution $E = C_0, C_1, \ldots$ starting from $C_0 \in C_S$ that satisfies the following conditions:

- If leftmost robots have symmetric views and decide to move, the scheduler make them move to the right direction.

– If rightmost robots have symmetric views and decide to move, the scheduler make them move to the left direction.

Intuitively, the scheduler tries to make leftmost robots and rightmost robots move so that leftmost (resp., rightmost) robots do not observe rightmost (resp., leftmost) robots in the left (resp., right) direction. Despite this behavior, we have the following lemma.

Lemma 5. *There exists an index i_c such that a leftmost robot sees a rightmost robot in the left direction at C_{i_c}.*

Proof. For contradiction, assume that such instant does not exist. Let i be the index such that robots terminate at C_i in E. Next, in a n'-node ring R_2 ($n' > 3(i+1)$), let us consider an execution $E' = C'_0, C'_1, \ldots$ of $\mathcal{A}$ such that three robots form the same sub-configuration as C_0 at C'_0. From the assumption, since leftmost robots do not see rightmost robots in the left direction at each configuration, three robots do not recognize the difference between R_1 and R_2. Hence, in E', they move similarly to E and terminate at C'_i. Three robots have visited at most $3(i+1)$ nodes until the configuration becomes C'_i and thus they do not achieve exploration. This is a contradiction.

Let i'_c be the minimum index such that a leftmost robot sees a rightmost robot in the left direction at $C_{i'_c}$. By proving the following two lemmas, we show that $C_{i'_c}$ is in C_S.

Lemma 6. *All robots are connected at $C_{i'_c}$.*

Proof. For contradiction, assume that not all robots are connected at $C_{i'_c}$. Since a leftmost robot and a rightmost robot are connected at $C_{i'_c}$, the other robot is isolated at $C_{i'_c}$. In a n'-node ring R_2 ($n' \geq n + 3$), let us consider an execution $E' = C'_0, C'_1, \ldots$ of $\mathcal{A}$ such that three robots form the same sub-configuration as C_0 at C'_0. Since leftmost robots do not see rightmost robots in the left direction until instant $i'_c - 1$, three robots do not recognize the difference between R_1 and R_2 until instant $i'_c - 1$. Hence, in E', they move similarly to E until $C'_{i'_c-1}$. Since $n' \geq n + 3$, the distance between the leftmost robot and the rightmost robot is $\phi + 2$ or greater at $C'_{i'_c}$. In addition, the other robot is isolated at $C'_{i'_c}$. Thus we can define an independent territory set at $C'_{i'_c}$. From Lemma 1, three robots cannot visit the remaining nodes and thus they cannot achieve exploration. This is a contradiction.

Lemma 7. *No tower exists at $C_{i'_c}$.*

Proof. Consider configuration $C_{i'_c-1}$, that is, a configuration immediately before a leftmost robot sees a rightmost robot in the left direction. Since there exist only three robots and $n \geq 3\phi + 3$ holds, either a leftmost robot or a rightmost robot is isolated in $C_{i'_c-1}$.

We consider the case that the rightmost robot, say r, is isolated. We can prove the other case in the same manner. Since the view of r is symmetric in

$C_{i'_c-1}$, the scheduler decides the direction of the movement from the construction of E. That is, the scheduler makes r move to the left direction if r decides to move. However, if r moves to the left direction, leftmost robots cannot see r in the left direction in $C_{i'_c}$. Hence, r does not move in $C_{i'_c-1}$. Thus a leftmost robot must move in the left direction during $C_{i'_c-1}$ to $C_{i'_c}$. Otherwise, leftmost robots cannot see r in the left direction in $C_{i'_c}$. If leftmost robots make a tower in $C_{i'_c-1}$, since the views of them are symmetric in $C_{i'_c-1}$, the scheduler decides the direction of the movement from the construction of E. That is, the scheduler makes leftmost robots move to the right direction if they decide to move, and thus they cannot see r in the left direction in $C_{i'_c}$. Hence, no tower exists at $C_{i'_c-1}$. Since the leftmost robot moves in the left direction during $C_{i'_c-1}$ to $C_{i'_c}$, it does not make a tower in $C_{i'_c}$. In addition, since r is isolated at $C_{i'_c-1}$, r does not make a tower in $C_{i'_c}$. Therefore, no tower exists at $C'_{i'_c}$.

From Lemmas 6 and 7, in the FSYNC model, for any configuration in C_S, there exists an execution that reaches a configuration in C_S again. This implies that $\mathcal{A}$ cannot terminate and hence cannot solve the terminating exploration. Therefore, we have Lemma 4.

Lemma 4 implies, also in the SSYNC and ASYNC models, there exists no algorithm that solves terminating exploration from any initial configuration in C_S. In addition, from Lemma 3, all configurations in C_S are solvable in the FSYNC, SSYNC, and ASYNC models. This implies, in the FSYNC, SSYNC, and ASYNC models, there exists no algorithm that solves terminating exploration from any solvable initial configuration. Therefore, we have Theorem 5.

5   Conclusions

In this paper, we investigated ring exploration algorithms for myopic luminous robots. Previous work considered the case where the visible distance ϕ is one and the number of colors of lights is two, and revealed the number of required robots to achieve perpetual and terminating exploration. A remaining natural question is whether the number of required robots can be further reduced in case of $\phi \geq 2$. In this case, we proved that, in the fully synchronous, semi-synchronous, and asynchronous models, two and three robots are necessary and sufficient to achieve perpetual and terminating exploration, respectively. These results revealed the answers to the question: *(i)* in the fully synchronous model, the number of required robots cannot be reduced, and *(ii)* in the semi-synchronous and asynchronous models, the number of required robots can be reduced by one. We also showed that our perpetual exploration algorithm is universal, and that no universal algorithm exists for terminating exploration with three robots.

For the future work, it is interesting to consider exploration algorithms that handle continuous ring environments or tolerate some faults. It is also interesting to consider other tasks and topologies with myopic luminous robots.

References

1. Blin, L., Milani, A., Potop-Butucaru, M., Tixeuil, S.: Exclusive perpetual ring exploration without chirality. In: Lynch, N.A., Shvartsman, A.A. (eds.) DISC 2010. LNCS, vol. 6343, pp. 312–327. Springer, Heidelberg (2010). https://doi.org/10.1007/978-3-642-15763-9_29

2. Bonnet, F., Milani, A., Potop-Butucaru, M., Tixeuil, S.: Asynchronous exclusive perpetual grid exploration without sense of direction. In: Fernàndez Anta, A., Lipari, G., Roy, M. (eds.) OPODIS 2011. LNCS, vol. 7109, pp. 251–265. Springer, Heidelberg (2011). https://doi.org/10.1007/978-3-642-25873-2_18

3. Bramas, Q., Devismes, S., Lafourcade, P.: Brief announcement: infinite grid exploration by disoriented robots. In: 26th International Colloquium onStructural Information and Communication Complexity SIROCCO 2019. L'Aquila, Italy, Jul 2019. https://hal.archives-ouvertes.fr/hal-02145822

4. Chalopin, J., Flocchini, P., Mans, B., Santoro, N.: Network exploration by silent and oblivious robots. In: Thilikos, D.M. (ed.) WG 2010. LNCS, vol. 6410, pp. 208–219. Springer, Heidelberg (2010). https://doi.org/10.1007/978-3-642-16926-7_20

5. D'Angelo, G., Navarra, A., Nisse, N.: A unified approach for gathering andexclusive searching on rings under weak assumptions. Distrib. Comput. 30(1), 17–48 (2017). https://doi.org/10.1007/s00446-016-0274-y

6. Das, S., Flocchini, P., Prencipe, G., Santoro, N., Yamashita, M.: Autonomous mobile robots with lights. Theoret. Comput. Sci. 609, 171–184 (2016). https://doi.org/10.1016/j.tcs.2015.09.018

7. Datta, A.K., Lamani, A., Larmore, L.L., Petit, F.: Ring exploration by oblivious agents with local vision. In: IEEE 33rd International Conference on Distributed Computing Systems, pp. 347–356 (2013). https://doi.org/10.1109/ICDCS.2013.55

8. Datta, A.K., Lamani, A., Larmore, L.L., Petit, F.: Ring exploration by oblivious robots with vision limited to 2 or 3. In: Higashino, T., Katayama, Y., Masuzawa, T., Potop-Butucaru, M., Yamashita, M. (eds.) SSS 2013. LNCS, vol. 8255, pp. 363–366. Springer, Cham (2013). https://doi.org/10.1007/978-3-319-03089-0_31

9. Devismes, S., Lamani, A., Petit, F., Raymond, P., Tixeuil, S.: Optimal grid exploration by asynchronous oblivious robots. In: Richa, A.W., Scheideler, C. (eds.) SSS 2012. LNCS, vol. 7596, pp. 64–76. Springer, Heidelberg (2012). https://doi.org/10.1007/978-3-642-33536-5_7

10. Devismes, S., Lamani, A., Petit, F., Tixeuil, S.: Optimal torus exploration by oblivious robots. In: Bouajjani, A., Fauconnier, H. (eds.) NETYS 2015. LNCS, vol. 9466, pp. 183–199. Springer, Cham (2015). https://doi.org/10.1007/978-3-319-26850-7_13

11. Devismes, S., Petit, F., Tixeuil, S.: Optimal probabilistic ring exploration by semi-synchronous oblivious robots. Theoret. Comput. Sci. 498, 10–27 (2013). https://doi.org/10.1016/j.tcs.2013.05.031

12. Flocchini, P., Ilcinkas, D., Pelc, A., Santoro, N.: Remembering without memory: tree exploration by asynchronous oblivious robots. Theoret. Comput. Sci. 411(14–15), 1583–1598 (2010). https://doi.org/10.1016/j.tcs.2010.01.007

13. Flocchini, P., Ilcinkas, D., Pelc, A., Santoro, N.: Computing without communicating: Ring exploration by asynchronous oblivious robots. Algorithmica 65(3), 562–583 (2013). https://doi.org/10.1007/s00453-011-9611-5

14. Flocchini, P., Prencipe, G., Santoro, N., Widmayer, P.: Gathering of asynchronous robots with limited visibility. Theoret. Comput. Sci. 337(1–3), 147–168 (2005). https://doi.org/10.1016/j.tcs.2005.01.001

15. Klasing, R., Kosowski, A., Navarra, A.: Taking advantage of symmetries: gathering of many asynchronous oblivious robots on a ring. Theoret. Comput. Sci. **411**(34–36), 3235–3246 (2010). https://doi.org/10.1016/j.tcs.2010.05.020

16. Ooshita, F., Tixeuil, S.: Ring exploration with myopic luminous robots. In: Izumi, T., Kuznetsov, P. (eds.) SSS 2018. LNCS, vol. 11201, pp. 301–316. Springer, Cham (2018). https://doi.org/10.1007/978-3-030-03232-6_20

17. Suzuki, I., Yamashita, M.: Distributed anonymous mobile robots: formation of geometric patterns. SIAM J. Comput. **28**(4), 1347–1363 (1999). https://doi.org/10.1137/S009753979628292X

18. Yamauchi, Y., Uehara, T., Kijima, S., Yamashita, M.: Plane formation by synchronous mobile robots in the three-dimensional Euclidean space. J. ACM **64**(3), 16 (2017). https://doi.org/10.1145/3060272

Brief Announcement: Self-stabilizing Construction of a Minimal Weakly $\mathcal{ST}$-Reachable Directed Acyclic Graph

Junya Nakamura[1(✉)], Masahiro Shibata[2], Yuichi Sudo[3], and Yonghwan Kim[4]

[1] Toyohashi University of Technology, 1–1 Tempaku, Hibarigaoka,
Toyohashi, Aichi 441–8580, Japan
`junya@imc.tut.ac.jp`
[2] Kyushu Institute of Technology, Iizuka, Fukuoka, Japan
[3] Osaka University, Suita, Osaka, Japan
[4] Nagoya Institute of Technology, Nagoya, Aichi, Japan

Abstract. In this paper, we propose a self-stabilizing algorithm to construct a minimal weakly $\mathcal{ST}$-reachable directed acyclic graph (DAG). Given an arbitrary simple, connected, and undirected graph $G = (V, E)$ and two sets of vertices, senders $\mathcal{S}(\subset V)$ and targets $\mathcal{T}(\subset V)$, a directed subgraph $\overrightarrow{G}$ of G is a weakly $\mathcal{ST}$-reachable DAG on G if $\overrightarrow{G}$ is a DAG and every sender can reach at least one target, and every target is reachable from at least one sender in $\overrightarrow{G}$. We say that a weakly $\mathcal{ST}$-reachable DAG $\overrightarrow{G}$ on G is minimal if any proper subgraph of $\overrightarrow{G}$ is no longer a weakly $\mathcal{ST}$-reachable DAG. The weakly $\mathcal{ST}$-reachable DAG on G, which we consider here, is a relaxed version of the original (or *strongly*) $\mathcal{ST}$-reachable DAG on G where all targets are reachable from all senders. A strongly $\mathcal{ST}$-reachable DAG G does not always exist; even if we focus on the case $|\mathcal{S}| = |\mathcal{T}| = 2$, some G has no strongly $\mathcal{ST}$-reachable DAG. On the other hand, the proposed algorithm always construct a weakly $\mathcal{ST}$-reachable DAG for any given graph $G = (V, E)$ and any $\mathcal{S}, \mathcal{T} \subset V$.

Keywords: Directed acyclic graph · ST-reachable DAG · Self-stabilization

1 Introduction

Nowadays, wireless networks, e.g., Wireless Sensor Networks (WSN) or Internet of Things (IoT), attract lots of attention in the area of distributed computing. In a wireless network, generally, each node can communicate with only other nodes within a limited range; thus, routing a message from a sender node to a destination node via other nodes plays an important role. In the literature, many routing algorithms for wireless networks were proposed. In the routing task for wireless networks, the following properties are important due to the instability of nodes and their limited power source: (1) reachability between sender and target

© Springer Nature Switzerland AG 2019
M. Ghaffari et al. (Eds.): SSS 2019, LNCS 11914, pp. 272–276, 2019.
https://doi.org/10.1007/978-3-030-34992-9_21

Table 1. Summary of the related DAG construction algorithms. FT and SS are abbreviations of Fault-Tolerance and Self-stabilizing, respectively.

	Alg. Type	$|\mathcal{S}|$	$|\mathcal{T}|$	Topology	Edge direction	Reachability	FT
[1]	Distributed	1	1	Biconnected	All	Strong	SS
[2]	Distributed	1	1	Biconnected	All	Strong	SS
[4]	Distributed	1	1	Connected	Maximal	Strong	SS
[3]	Distributed	1	2	Connected	Maximal	Strong	SS
[3]	Distributed	2	2	Connected	Maximal	Weak	SS
[5]	Distributed	2	2	Connected*	Minimal	Strong	SS
Our result	Distributed	Any	Any	Connected	Minimal	Weak	SS

*Each node detects an error if a graph does not satisfy a necessary condition to construct an $\mathcal{ST}$-reachable DAG.

nodes guaranteed by a routing algorithm, (2) the number of nodes necessary to participate in the task, and (3) fault-tolerance.

For this routing task, Kim et al. proposed an $\mathcal{ST}$-*reachable directed acyclic graph (DAG)* [5] on a wireless network, which provides reachability from every sender node $s \in \mathcal{S}$ to every target node $t \in \mathcal{T}$. However, they also proved in [5] that this construction problem is not always solvable. A graph G, and sets $\mathcal{S}$ and $\mathcal{T}$ must satisfy a certain condition to have an $\mathcal{ST}$-reachable DAG even if $|\mathcal{S}| = |\mathcal{T}| = 2$.

To circumvent this impossibility, in this paper, we consider a weaker version of $\mathcal{ST}$-reachable DAG called *weakly $\mathcal{ST}$-reachable DAG*. A subgraph $\overrightarrow{G}$ of G is a weakly $\mathcal{ST}$-reachable DAG if (1) every sender node $s \in \mathcal{S}$ can reach at least one target node $t \in \mathcal{T}$, (2) every target node $t \in \mathcal{T}$ is reachable from at least one sender node $s \in \mathcal{S}$, and (3) $\overrightarrow{G}$ has no cycle. This DAG can be constructed on any connected network since this DAG weakens the requirements of the original (or *strong*) $\mathcal{ST}$-reachable DAG on reachability from the senders to the targets.

We propose a distributed algorithm that constructs a minimal weakly $\mathcal{ST}$-reachable DAG on given a simple, connected, and undirected graph $G = (V, E)$ and two sets $\mathcal{S}, \mathcal{T} \subset V$. The algorithm is self-stabilizing; thus, it tolerates any number of transient failures. The convergence time of the proposed algorithm is $O(D)$ (asynchronous) rounds where D is the diameter of a given graph G.

2   Related Work

Table 1 summarizes the related algorithms that construct some kinds of DAG from sender nodes to target nodes on a given graph. The most important aspect of the table is reachability. A DAG with strong reachability ensures that every target node is reachable from every sender node. On the other hand, a DAG with weak reachability guarantees that every sender can reach at least one target node, and every target node is reachable from at least one sender node; thus, a sender node may not be able to reach some target node.

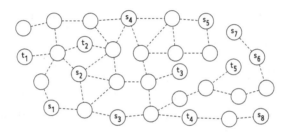

Fig. 1. A given example graph $G = (V, E, \mathcal{S}, \mathcal{T})$ where $\mathcal{S} = \{s_1, s_2, \ldots, s_8\}$ and $\mathcal{T} = \{t_1, t_2, \ldots, t_5\}$. A dashed line represents an edge $e \in E$.

3   Problem Specification

Let $G' = (V', E')$ be any (undirected) subgraph of G (note that $G = G'$ may hold). We say that a directed graph or digraph $\overrightarrow{G} = (V'', A)$ is a directed subgraph of G' if a set V'' of nodes satisfies $V'' \subseteq V'$ and a set A of arcs satisfies $(u, v) \in A \Rightarrow \{u, v\} \in E'$ where (u, v) denotes an arc from u to v. We say that a directed subgraph $\overrightarrow{G}$ of $G = (V, E, \mathcal{S}, \mathcal{T})$ is a *weakly $\mathcal{ST}$-reachable DAG* of G if all of the following conditions hold:

C1 every sender in $\mathcal{S}$ can reach at least one target in $\mathcal{T}$ in $\overrightarrow{G}$,
C2 every target in $\mathcal{T}$ is reachable from a sender in $\mathcal{S}$ in $\overrightarrow{G}$, and
C3 there is no directed cycle in $\overrightarrow{G}$.

Here, for any two nodes u and v, we say that v is *reachable* from u, or u *can reach* v in $\overrightarrow{G}$ if there exists a directed path from u to v in $\overrightarrow{G}$. In addition, we say that a weakly $\mathcal{ST}$-reachable DAG $\overrightarrow{G}$ of G is *minimal* if condition C1 or C2 becomes unsatisfied when any arc of $\overrightarrow{G}$ is removed.

4   Proposed Algorithm

In this section, we propose a self-stabilizing algorithm called MWSTDAG that constructs a minimal weakly $\mathcal{ST}$-reachable DAG on a given graph $G = (V, E, \mathcal{S}, \mathcal{T})$. This algorithm is built by fair-composition and has four layers. Each node may have *red* or *blue* color. The red color assigned in layer 1 indicates that the node can reach a target node by tracing an L1 tree. The blue color assigned in layer 2 indicates that the node can reach a red node; thus, a blue node can reach a target node through the red node (if the configuration is legitimate for layers 1 and 2). These colors are propagated from a lower node to a higher node in their L1 and L2 trees.

We describe how each layer works with an example graph depicted in Fig. 1. The first layer builds Breadth-First-Search (BFS) trees on a given network G and checks reachability from sender nodes to target nodes on the trees (Fig. 2). Also, the second layer builds another kind of BFS trees to ensure reachability to target

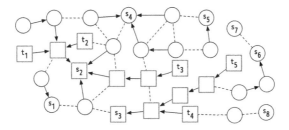

Fig. 2. A legitimate configuration of Layer 1. Solid arrows represent the generated L1 BFS trees rooted at sender nodes, and a red square is a red node. (Color figure online)

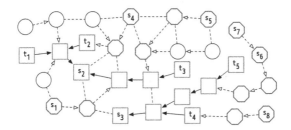

Fig. 3. A legitimate configuration of layer 2. Dashed arrows represent the constructed L2 BFS trees rooted at red nodes, and a blue octagon is a blue node. For simplicity, an arrow of an L1 tree rooted at a colorless sender node is omitted. (Color figure online)

nodes from the sender nodes that cannot reach to any target node in the layer 1 trees (Fig. 3). The third layer constructs a (possibly non-minimal) weakly $\mathcal{ST}$-reachable DAG based on the trees constructed in layers 1 and 2 (Fig. 4). The final layer detects and removes redundant arcs in the DAG to guarantee the minimality of the generated weakly $\mathcal{ST}$-reachable DAG (Fig. 5).

The removal in layer 4 plays an important role to guarantee the minimality of a constructed weakly $\mathcal{ST}$-reachable DAG and is conducted with the following idea: Basically, if a red node v has at least one incoming arc from a blue node, we can remove all but one arc from red nodes to v without violating the reachability requirement of a weakly $\mathcal{ST}$-reachable DAG. However, there is an exception. If a sender node becomes unreachable to any target node by the removal, the algorithm must not remove such an arc. Because only a red node u having two or more outgoing arcs to red nodes can detect such an arc, we propagate information whether there is a red node having an incoming arc from a blue node, from u's descendant nodes to u in an L1 tree.

We can prove the following theorem for the proposed algorithm MWSTDAG although its proof is omitted due to the space limitation.

Theorem 1. *Algorithm MWSTDAG is a silent self-stabilizing algorithm for the minimal weakly $\mathcal{ST}$-reachable DAG construction problem. Starting from any configuration, every fair execution of MWSTDAG reaches a final configuration within $O(D)$ asynchronous rounds.*

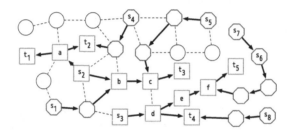

Fig. 4. A legitimate configuration of layer 3. Black bold arrows represent the constructed weakly $\mathcal{ST}$-reachable DAG. For simplicity, a dashed arrow of an L2 tree from a colorless node is omitted. (Color figure online)

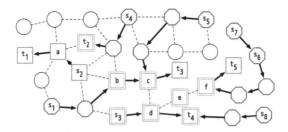

Fig. 5. A legitimate configuration of layer 4. A double red square is a branch node. (Color figure online)

Acknowledgments. This work was supported by JSPS KAKENHI Grant Numbers 18K18000, 18K18029, and 18K18031.

References

1. Chaudhuri, P., Thompson, H.: A self-stabilizing algorithm for the st-order problem. Int. J. Parallel Emergent Distrib. Syst. **23**(3), 219–234 (2008)
2. Karaata, M.H., Chaudhuri, P.: A dynamic self-stabilizing algorithm for constructing a transport net. Computing **68**(2), 143–161 (2002)
3. Kim, Y., Aono, H., Katayama, Y., Masuzawa, T.: A self-stabilizing algorithm for constructing a maximal (2,2)-directed acyclic mixed graph. In: the 6th International Symposium on Computing and Networking (CANDAR) (2018)
4. Kim, Y., Ohno, H., Katayama, Y., Masuzawa, T.: A self-stabilizing algorithm for constructing a maximal (1, 1)-directed acyclic mixed graph. Int. J. Netw. Comput. **8**(1), 53–72 (2018)
5. Kim, Y., Shibata, M., Sudo, Y., Nakamura, J., Katayama, Y., Masuzawa, T.: A self-stabilizing algorithm for constructing an $\mathcal{ST}$-reachable directed acyclic graph when $|\mathcal{S}| \leq 2$ and $|\mathcal{T}| \leq 2$. In: Proceedings of the 39th IEEE International Conference on Distributed Computing Systems (ICDCS), pp. 2228–2237 (2019)

Adaptive Versioning in Transactional Memories

Pavan Poudel and Gokarna Sharma[✉]

Department of Computer Science, Kent State University, Kent, OH 44242, USA
{ppoudel,sharma}@cs.kent.edu

Abstract. Transactional memory has been receiving much attention from both academia and industry. In transactional memory, program code is split into *transactions*, blocks of code that appear to execute atomically. Transactions are executed speculatively and the speculative execution is supported through data versioning and conflict detection and resolution mechanisms. *Lazy* versioning makes aborts fast but penalizes commits, whereas *eager* versioning makes commits fast but penalizes aborts. In this paper, we present an *adaptive* versioning approach that dynamically switches between eager and lazy versioning at runtime based on appropriate system parameters so that the performance of a transactional memory system is always better than that is obtained using either eager or lazy versioning individually. We implemented our adaptive versioning approach in the latest TinySTM distribution and extensively evaluated it through 5 micro-benchmarks and 8 complex benchmarks from STAMP and STAMPEDE suites. The results show significant benefits of our approach, giving performance improvements as much as 6.3x for execution time and as much as 170x for number of aborts.

1 Introduction

Concurrent processes (threads) need to synchronize to avoid introducing inconsistencies while accessing shared data objects. Traditional synchronization mechanisms such as locks and barriers have well-known limitations and pitfalls, including deadlock, priority inversion, reliance on programmer conventions, and vulnerability to failure or delay. *Transactional memory* (TM) [16,27] has emerged as an attractive alternative. Several commercial processors provide direct hardware support for TM, including Intel's Haswell [17] and IBM's Blue Gene/Q [14], zEnterprise EC12 [23], and Power8 [6]. There are proposals for adapting TM to clusters of GPUs [5,12,20].

Using TM, program code is split into *transactions*, blocks of code that appear to execute atomically. Transactions are executed *speculatively*: synchronization conflicts or failures may cause an executing transaction to *abort*: its effects are rolled back and the transaction is restarted. In the absence of conflicts or failures, a transaction typically *commits*, causing its effects to become visible. Supporting

This work is supported by the National Science Foundation grant CCF-1936450.

M. Ghaffari et al. (Eds.): SSS 2019, LNCS 11914, pp. 277–295, 2019.
https://doi.org/10.1007/978-3-030-34992-9_22

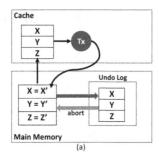 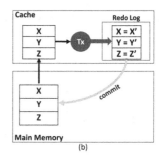

(a)   (b)

Fig. 1. An illustration of how a transaction T_x is executed using (a) eager versioning and (b) lazy versioning. Figure (a) depicts two kinds of operations in eager versioning, the first to copy the data from original memory locations to a log area (called *undo* log) in main memory and the second to copy the data back from the log area to the original memory locations, in case T_x aborts. If T_x commits, the data in the log area is simply discarded. Figure (b) depicts two kinds of operations in lazy versioning, the first to copy the updated values in cache to a log area (called *redo* log) in cache and the second to copy data from the log area to the original memory locations, in case T_x commits. If T_x aborts, the data in the log area is simply discarded. (Color figure online)

this speculative execution requires *data version management* and *conflict detection and resolution* mechanisms. The majority of the existing TM systems can be distinguished on how they implement these concepts. This is true for TM systems in hardware, called hardware TMs (HTMs) [4,13,22,26], as well as in software, called software TMs (STMs) [2,8,9].

Versioning handles the simultaneous storage of both *new* data (to be visible if transaction commits) and *old* data (retained if transaction aborts). At most one of these values can be stored "in place" (the original memory location), while the other value must be stored "on the side" (e.g., in cache or main memory). On a store, a TM system can either use *eager* versioning and put the new value in place or use *lazy* versioning to (temporarily) leave the old value in place. Figure 1 depicts how a transaction T_x is executed using eager and lazy versioning. Due to the working principle, lazy versioning makes aborts fast, but penalizes (the most frequent) commits, whereas eager versioning makes commits fast, but penalizes (the most frequent) aborts [22].

Conflict detection signals an overlap between the *write set* (data written) of one transaction and the write set or *read set* (data read) of other concurrent transactions. Conflict detection is called *eager* if it detects offending loads or stores immediately and *lazy* if it defers detection until later when transactions commit. Table 1 illustrates some existing TM systems that use lazy versus eager versioning and lazy versus eager conflict detection. *Conflict resolution* (or management) strategies are then used to decide on which conflicting transaction(s) to continue and which transaction(s) to wait (or abort and restart) the execution.

Table 1. Versioning and conflict detection mechanisms used in some TM systems.

		Versioning	
		Lazy	Eager
Conflict	Lazy	TCC [13], Norec [8], RSTM [2], SwissTM [9]	None
	Eager	LTM [4], VTM [26], RSTM [2], SwissTM [9]	UTM [4], LogTM [22], RSTM [2]

Both eager and lazy versioning along with both eager and lazy conflict detection and resolution have been studied heavily in the past for TM systems [2,4,8,9,13,22,26]. However, which versioning is better is still not clear and the studies provide contradictory conclusions. For example, consider two widely popular HTM implementations LOGTM [22] and UTM [4]. They advocate that TM should ideally use eager versioning and eager conflict detection since in eager versioning transaction commits are faster than transactions aborts. Moreover, commits are much more common than aborts in practical applications. In addition, eager conflict detection finds conflicts early and reduces the wasted work by conflicting transactions. On the other hand, consider again widely popular HTM implementation TCC [13]. They use lazy versioning and lazy conflict detection. Other HTMs such as VTM [26] and LTM [4] advocate lazy versioning with eager conflict detection. This is also the case in STMs as some use eager, some use lazy, and some use the combination of eager and lazy approaches [2,8,9].

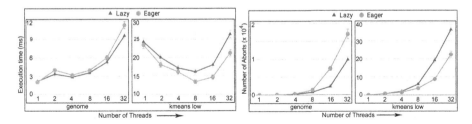

Fig. 2. An illustration of performance discrepancies in execution time (left) and number of aborts (right) in *genome* and *kmeans* benchmarks using eager and lazy versioning.

There is no study that elaborates the performance gap between eager and lazy versioning for TM systems. Figure 2 illustrates the performance discrepancies using eager and lazy versioning while executing *genome* and *kmeans* benchmarks from STAMP benchmark suite [21]. Lazy versioning performs well for *genome* where as for *kmeans* the opposite is true. This is mainly because of the fact that the versioning used is not appropriate for the workload and caused more number of aborts, subsequently increasing the execution time. Nevertheless, there are two major issues in selecting an appropriate versioning for TM systems. First, to select an appropriate versioning, a priori knowledge on the workload (write-dominated or read-dominated) and contention scenario (low or high) is needed. Second, such knowledge is difficult to obtain prior to runtime.

Contributions. In this paper, we demonstrate that we can obtain the best of both worlds without any a priori knowledge on the workload and contention scenario. Particularly, we present an *adaptive* versioning for TM systems, which we call ADAPTIVE, that dynamically switches the execution using either lazy or eager versioning at runtime, always achieving performance on any workload and contention scenario better than that is obtained using either lazy or eager versioning individually. For the experimental evaluation, we incorporated ADAPTIVE in the latest TinySTM implementation [10,11] and ran experiments against a diverse set of TM benchmarks [10,11,15]. Specifically, we used 5 microbenchmarks (bank, red black tree, hash set, linked list, and skip list) and 8 complex benchmarks (yada, vacation, ssca2, labyrinth, kmeans, intruder, genome, and bayes) from STAMP and STAMPEDE benchmarks [21,24]). We measured the performance of ADAPTIVE w.r.t. two crucial performance metrics.

- **execution time:** the total time to complete executing a set of transactions. This is the time interval from the beginning of the first transaction executed until the last transaction finishes and commits. In a dynamic setting, the execution time translates to *throughput*, the number of committed transactions per time step.
- **number of aborts:** the total number of transaction aborts until the current time. If compared with the total number of transaction commits until the current time, it provides *abort-to-commit ratio* (ACR), a useful metric.

Both metrics are fundamental and used extensively in evaluating TM systems. The number of aborts directly affect execution time since it is likely that the execution time increases with the increasing number of aborts requiring more number of transaction restarts.

The results show that, when using lazy versioning with eager conflict detection, ADAPTIVE achieves up to 6.3× better performance than lazy versioning and up to 5.5× better performance than eager versioning. When using lazy versioning with lazy conflict detection, ADAPTIVE achieves up to 3.7× better performance than lazy versioning and up to 5× better performance than eager versioning. The minimum performance gain for ADAPTIVE is 1.12. These results suggest that switching between eager and lazy versioning dynamically at runtime provides a way to exploit the positive aspects of both versioning methods for TM systems. ADAPTIVE is general enough to be applied to both HTMs and STMs, although we only report results from a STM implementation. In summary, we have the following three contributions.

- (**Section** 4) We introduce a novel versioning approach, ADAPTIVE, that switches between eager and lazy versioning dynamically at runtime.
- (**Section** 5) We discuss implementation issues related to ADAPTIVE and present two optimizations.

- (**Section** 6) We evaluate experimentally the performance of ADAPTIVE using five micro-benchmarks and 8 complex benchmarks from STAMP and STAM-PEDE, report the results obtained, and provide observations on the obtained results.

2   Related Work

The previous studies on TM mostly supported speculative execution of transactions using either eager or lazy versioning. There is no work that elaborates on the impact of using eager and lazy versioning on the performance of TM systems. In fact, as outlined in Table 1, the majority of well-known TM systems make contradictory conclusions on whether to use eager or lazy versioning. We focus in this paper on the impact of the eager and lazy versioning on the performance of TM systems. Particularly, we propose an adaptive versioning (switching between eager and lazy versioning at runtime without needing any a priori knowledge on workload and contention scenarios) and achieve significant improvements in execution time and number of aborts (two crucial performance metrics for evaluating a TM system) compared to that of using either eager or lazy versioning individually. Our approach is simple and may provide insights into future TM system designs and implementations.

The performance gap of using eager and lazy versioning is relatively well-studied for crash consistency in non-volatile memories. One recent work is [25] where they presented an adaptive versioning approach like the one presented here but specifically tailored to non-volatile memories. In particular, they focused on minimizing *the number of data movements* while running workloads through these versioning methods. However, their approach increased the execution time in several benchmarks. The approach we study here is tailored for TM systems in volatile memories.

The other closely related works are as follows. Wan *et al.* [29] empirically evaluated eager and lazy versioning on the open source non-volatile memory library (NVML) [1] for some constrained workloads, and suggested that *"one logging method does not fit all workloads"*. Particularly, they reported that (i) lazy versioning significantly outperforms eager versioning for workloads in which a transaction updates large number of different objects, while it underperforms eager versioning for read-dominated workloads, and (ii) eager versioning is more sensitive to *read-to-write* ratios whereas lazy versioning is less sensitive to those ratios [29]. The other works mostly proposed methods to provide crash consistency through either eager or lazy versioning, and there is no work that elaborates the performance gap between eager and lazy versioning. Coburn *et al.* [7] suggested a STM implementation for persistent memory NV-HEAPS using eager versioning. Volos *et al.* [28] suggested a TinySTM [10,11] variation MNEMOSYNE for persistent memory using lazy versioning. NV-HEAPS [7] and MNEMOSYNE [28] drew absolutely opposite conclusions on whether eager or lazy versioning is better for persistent memory. The former prefers to use eager versioning, and the latter opts to use lazy versioning. Recently, Alistarh *et al.* [3] studied two variants of the transactional conflict problem and provided optimal solutions for both the variants.

3   Preliminaries

Model. We consider a computer system with volatile shared main memory, many processing cores, and hard disk drive. All shared main memory is cacheable and caches are volatile and coherent. We assume that all the writes of a committed transaction can be accommodated in the cache, i.e., once a transaction commits but before the commit is reflected in original memory locations in main memory, all its newly modified data is in volatile cache. We run workloads using the TinySTM execution model [10,11]. We assume that the execution starts at time $t_0 = 0$. We measure in execution time the time for all the transactions within a benchmark to finish execution and commit, except for micro-benchmarks where we consider time to execute and commit 10,000 transactions. We also assume that only a single-version of data is stored in each eager, lazy, and adaptive versioning, which is essentially different from techniques, such as those given in [18], of storing multiple versions.

Eager Versioning. Eager versioning is supported through so-called *undo logs*. Undo logs are stored in cacheable main memory. In this method, a transaction works by first copying the data in original memory locations to a undo log area and then performs updates in-place in the original data locations (in main memory). In the event the transaction aborts, any modifications to the original memory locations are *rolled back* using the old data stored in the undo log. The left of Fig. 1 illustrates eager versioning.

Lazy Versioning. Lazy versioning is supported through so-called *redo logs*. Redo logs are stored in cache. In this method, a transaction copies data in each memory location that it is going to read/write to a redo log area, appends all its data updates to that log area, and then writes the data back to original memory locations when transaction commits. If the transaction fails, the updates in log area are simply discarded. Therefore, the writing of data in redo log back to the original memory locations happens only when transaction commits. The right of Fig. 1 illustrates lazy versioning.

4   Basic Adaptive Versioning

We now describe our approach, ADAPTIVE, that runs transactions using either eager or lazy versioning, switching between them dynamically at runtime. Figure 3 compares ADAPTIVE with eager and lazy versioning. We will introduce techniques to improve ADAPTIVE in Sect. 5.

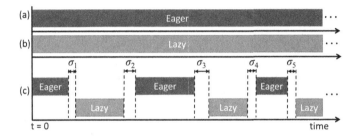

Fig. 3. An illustration of (a) eager, (b) lazy, and (c) basic adaptive versioning. The time gap σ_* while switching from eager (lazy) to lazy (eager) is to let finish executing in-flight transactions. This helps in avoiding potential data versioning inconsistencies.

High Level Overview. The high level idea in ADAPTIVE is to switch the versioning method depending on performance. That is, if the versioning currently used is hampering the performance, then we switch the versioning to improve the performance.

Now a fundamental question is how to identify and measure an indicator that reflects appropriately the effect of the versioning method on performance. Fortunately, in TM, if the number of aborts are increasing compared to the number of commits, then it might be a valid indicator of performance degradation due to the versioning method currently used. Therefore, we pick abort to commit ratio (ACR) as a performance indicator for any versioning method. ACR has also been used quite heavily in the TM literature as a vital indicator of performance, for example, see [19]. Ideally, the goal is to have no aborts, i.e., $ACR = 0$. However, in practice, this may not be feasible and the goal is to minimize ACR as much as possible.

Formally, ACR can be defined as follows: $ACR = \frac{N_{abort}}{N_{commit}}$, where N_{abort} is the total number of aborted transactions and N_{commit} is the total number of committed transactions from time 0 up to t. For eager (and lazy) versioning, we can compute ACR_{Eager} (and ACR_{Lazy}) based on the number of transactions committed and aborted using eager (lazy) versioning. To facilitate when to switch from one to another, we identify a *threshold* on ACR for both eager and lazy. We denote them by $Threshold_{Eager}$ and $Threshold_{Lazy}$, respectively. Let a transaction T be running at current time t using lazy versioning. If $ACR_{Lazy} < Threshold_{Lazy}$, then the versioning method is switched to Eager for transactions that start (or restart) execution after time $t' > t$. An analogous approach is used if currently T is executing using eager versioning.

Detailed Description. Let $N_{Ecommit}(N_{Lcommit})$ be the number of transaction commits in ADAPTIVE from time $t_0 = 0$ until the current time $t > t_0$ executed using eager (lazy) versioning. Similarly, let $N_{Eabort}(N_{Labort})$ be the total number of transaction aborts in ADAPTIVE from time $t_0 = 0$ until time $t > t_0$ executed using eager (lazy) versioning. Furthermore, let N_{commit} and N_{abort} be the total number of commits and aborts in ADAPTIVE from $t_0 = 0$ until time $t > t_0$. Notice that $N_{commit} = N_{Ecommit} + N_{Lcommit}$ and $N_{abort} = N_{Eabort} + N_{Labort}$.

Based on $N_{Ecommit}, N_{Lcommit}, N_{Eabort}$, and N_{Labort}, we compute ACR_{Eager} and ACR_{Lazy} at each time step $t > t_0$. These ratios ACR_{Eager} and ACR_{Lazy} are then compared with $Threshold_{Eager}$ and $Threshold_{Lazy}$ parameters (computed in the next section). Therefore, at any time $t > t_0$, the transaction T that is ready-to-execute will be executed as follows.

- Suppose the versioning currently used is $L_{cur} = Eager$. If $ACR_{Eager} > Threshold_{Eager}$, then L_{cur} is switched to $Lazy$ (i.e., $L_{cur} \leftarrow Lazy$) and T will be executed using lazy versioning.
- Suppose the versioning method currently used is $L_{cur} = Lazy$. If $ACR_{Lazy} < Threshold_{Lazy}$, then L_{cur} is switched to $Eager$ (i.e., $L_{cur} \leftarrow Eager$) and T will be executed using eager versioning.

In the special case of $t_0 = 0$, $N_{Ecommit}, N_{Lcommit}, N_{Eabort}$, and N_{Labort} are all zero. Therefore, a simple approach is to execute T using either lazy or eager versioning. However, if some information regarding the workload is available, then we can decide on which versioning method to use. Suppose, the read and write sets of T are available. Let $Wset(T)$ be the *write set* of T which is essentially the memory locations that T would modify while executing. Similarly, let $Rset(T)$ be the *read set* of T which is essentially the memory locations that T would read (but not modify) while executing. $RW(T) = Rset(T) + Wset(T)$, where $RW(T)$ denotes the total number of memory locations that T reads and modifies while executing. If $|Wset(T)| > |Rset(T)|$, then T is executed using lazy versioning, otherwise using eager versioning.

Computing Switching Thresholds $Threshold_{Eager}$ **and** $Threshold_{Lazy}$. Let N be the total number of transactions in any workload. When the workload finishes execution and all transactions commit, we have that $N_{commit} = N$ and $N_{abort} \geq 0$ (if each transaction commits without aborting, then $N_{abort} = 0$, otherwise $N_{abort} > 0$).

Suppose, each transaction T spends α amount of time while moving data from one memory location to other. Consider the case of executing T using eager versioning. Let τ_{Eager} be the total amount of time spent while (i) versioning data from the original memory locations to the undo log area and (ii) updating data from the undo log area back to the original memory locations. The first kind of operations are shown as a blue arrow in Fig. 1(a) and the second kind of operations are shown as a red arrow in Fig. 1(a). The first kind of operations are always done in eager versioning and the second kind of operations are done only when the transaction aborts. That means, for an aborted transaction, data movement is performed two times, one for versioning, other for rollback. Therefore, for eager versioning, $\tau_{Eager} = (N_{commit} + 2N_{abort}) \cdot \alpha$.

Similarly, for lazy versioning, $\tau_{Lazy} = (2N_{commit} + N_{abort}) \cdot \alpha$.

Based on 3 different cases below, we can see 3 scenarios for τ_{Eager} and τ_{Lazy}:

- Case 1: **If** $N_{commit} = N_{abort}$, **then** $\tau_{Eager} = \tau_{Lazy}$

- Case 2: If $N_{commit} > N_{abort}$, then $\tau_{Eager} < \tau_{Lazy}$
- Case 3: If $N_{commit} < N_{abort}$, then $\tau_{Eager} > \tau_{Lazy}$

Moreover, equation for τ_{Eager} suggests that in eager versioning, total time spent for an aborted transaction is twice as much as the time spent for a committed transaction. Then it is immediate that the eager versioning performs better until $N_{commit} \geq 2N_{abort}$; i.e. $\frac{N_{abort}}{N_{commit}} \leq \frac{1}{2}$. Thus, we get $Threshold_{Eager} = \frac{1}{2}$ and switch to lazy versioning when $ACR_{Eager} > \frac{1}{2}$. Similarly, equation for τ_{Lazy} suggests that the lazy versioning performs better until $2N_{commit} \leq N_{abort}$; i.e. $\frac{N_{abort}}{N_{commit}} \geq 2$. Then, we get $Threshold_{Lazy} = 2$ and switch to eager versioning when $ACR_{Lazy} < 2$.

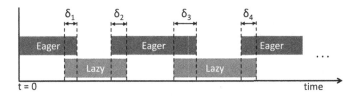

Fig. 4. An illustration of the better time barrier design. The interval δ_* between $Eager$ and $Lazy$ represents the time taken by in-flight transactions to finish their executions after versioning method is switched. The new transaction that do not conflict with transactions using previous versioning can execute concurrently with in-flight transactions.

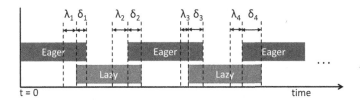

Fig. 5. An illustration of the better switching mechanism. λ_* represents the time interval in which versioning is not switched. δ_* resembles better time barrier of Fig. 4.

Time Barrier Requirement and Design. The ideal scenario in ADAPTIVE is to let each transaction T run Algorithm and decide which versioning (eager or lazy) to use for it to execute individually based on the parameters obtained at runtime. Let S_j be a set of transactions arrived before T. Suppose $L_{cur} = Eager$, which means that $L_{prev} = Lazy$. Suppose the versioning changed to $Eager$ from $Lazy$ after the transactions in S_j started execution but before T. If we run T using $Eager$ immediately and T conflicts with any of the transaction $T_j \in S_j$, then the conflict detection and resolution mechanisms interfere, hampering consistency. A simple approach to handle this situation is to ask T to wait until all transactions in S_j finish execution, which we call a *basic time barrier* (as shown in Fig. 3). The barrier reduces total number of aborts but due to a time delay before switching, it increases total execution time [25]. We provide a *better time barrier* design (described in Sect. 5) that will minimize this overhead.

5   Optimizations on Basic Adaptive Versioning

We provide two optimizations to basic ADAPTIVE. The first optimization is on time barrier design. The second optimization is on switching mechanism.

Better Time Barrier Design. Figure 4 illustrates the idea of better time barrier design. Consider a transaction T. Let S_j be a set of transactions arrived before T. Suppose $L_{cur} = Eager$, which means $L_{prev} = Lazy$. Suppose the versioning changed to $Eager$ from $Lazy$ after the transactions in S_j started execution but before T starts execution. In the basic time barrier design, T has to wait until all transactions in S_j finish execution. In this design, we ask T to start execution as soon as it is ready. If T does not conflict with transactions in S_j, we are done, otherwise, T aborts. If T conflicts with $T' \notin S_j$, it is handled as per the conflict resolution strategy used.

Better Switching Mechanism. Let $L_{cur} = Eager$. Suppose at time t, ADAPTIVE decides to switch to $Lazy$. We discuss here a mechanism so that ADAPTIVE does not switch to $Lazy$ at t but waits until a switching interval threshold SW_INT. We define SW_INT as the number of transactions after t for which the decision is to execute using $Lazy$. Let λ be the execution time interval during which all transactions in the interval SW_INT finish execution. Execution switches from $Eager$ to $Lazy$ at time $t + \lambda$. Figure 5 illustrates the design of better switching mechanism.

6   Experimental Evaluation

In this section, we evaluate the performance of optimized[1] ADAPTIVE (better time barrier and switching mechanism). The evaluation is performed in a STM implementation using TinySTM [10,11] modified appropriately to incorporate ADAPTIVE. The tests were executed on an Intel Xeon(R) E5-2620 v4 @ 4.20 GHz, 64-bit processor with 32 cores. Each core has private L1 and L2 caches, whose sizes are 64 KB and 256 KB, respectively. There is also an 20 MB L3 cache shared by all 32 cores and 32 GB main memory. The results are the average of 10 experimental runs. The results are for varying number of threads from 1 to

---

[1] The experimental results conducted on basic ADAPTIVE showed that the number of aborts always decrease in all the benchmarks but execution time for some benchmarks increase compared to the execution times obtained using eager and lazy versioning individually.

32. First, we present the experimental results for optimized ADAPTIVE with better time barrier using *suicide* conflict resolution strategy. Later, we extend the results using both better time barrier and switching mechanism. We also compare the performance of optimized ADAPTIVE against four different conflict resolution strategies.

Compared Versioning Methods. We developed a STM-based implementation using TinySTM [10,11]. TinySTM has implemented separately both lazy and eager versioning (called *Lazy* and *Eager*) through *Write_Back* and *Write_Through* designs, respectively. With *Write_Through* design, transactions directly write to original memory locations and revert their updates in case the transactions abort. However, with *Write_Back* design, transactions work on a copy of data and delay their updates to the original memory locations until commit [10,11]. Furthermore, *Write_Back* design has two different implementations: *Write_Back_ETL* and *Write_Back_CTL*. *Encounter-time locking* (ETL) detects conflicts early at the time of write and acquires the lock on the memory address before it is written. *Commit-time locking* (CTL) defers conflict detection on memory address until commit, i.e., the lock is acquired on the memory address at the commit time. Therefore, there are two different implementations of *Lazy* in TinySTM: one based on *ETL* called *Lazy_ETL* and another based on *CTL* called *Lazy_CTL*. We obtain adaptive design *Adaptive_ETL* using *Lazy_ETL* and *Eager* versioning. Similarly, we obtain adaptive design *Adaptive_CTL* using *Lazy_CTL* and *Eager* versioning. We run experiments with five different designs *Lazy_ETL*, *Lazy_CTL*, *Eager*, *Adaptive_ETL*, and *Adaptive_CTL*.

Results on Micro-benchmarks. The execution time results in 5 different micro-benchmarks are provided in Fig. 6. Figure 7 provides the result for the number of aborts. The results are for 10,000 transactions, each executed with *update rate* of 20%. Figure 6 shows that the execution time decreases notably in ADAPTIVE as compared to the other versioning methods with the increase in number of threads for all the micro-benchmarks. Specifically, *Adaptive_ETL* achieved up to 6.3× better execution time than *Lazy_ETL* and *Adaptive_CTL* achieved up to 3.7× better execution time than *Lazy_CTL*. Compared to *Eager*, *Adaptive_ETL* achieved up to 5.5× better execution time and *Adaptive_CTL* achieved up to 5× better execution time. The minimum execution gain for *Adaptive_ETL* beyond 4 number of threads is 1.23 and for *Adaptive_CTL* is 1.20. Due to high contention for memory access when transactions are executed with more number of threads, the number of aborts increases with the increasing number of threads. Figure 7 shows that ADAPTIVE minimizes number of aborts. Specifically, *Adaptive_ETL* achieved up to 2.6× less number of aborts than *Lazy_ETL* and up to 5.8× less number of aborts than *Eager*. *Adaptive_CTL* achieved up to 2.2× less number of aborts than *Lazy_CTL* and up to 8× less number of aborts than *Eager*.

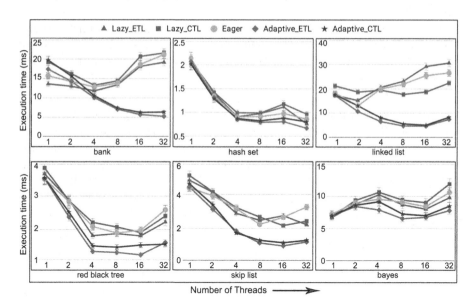

Fig. 6. Execution time in micro and bayes benchmark using better time barrier.

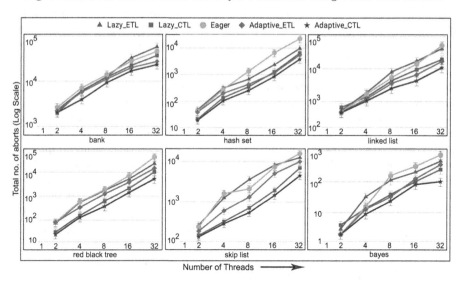

Fig. 7. Number of aborts in micro and bayes benchmark using better time barrier.

Results on STAMP Benchmarks. Figures 8 and 9, respectively, provide the execution time and number of aborts results for STAMP benchmarks. Regarding execution time, *Adaptive_ETL* has up to 1.78× better time than *Lazy_ETL* and *Adaptive_CTL* has up to 1.74× better time than *Lazy_CTL*. Compared to *Eager*, the execution time improvement in *Adaptive_ETL* and *Adaptive_CTL*

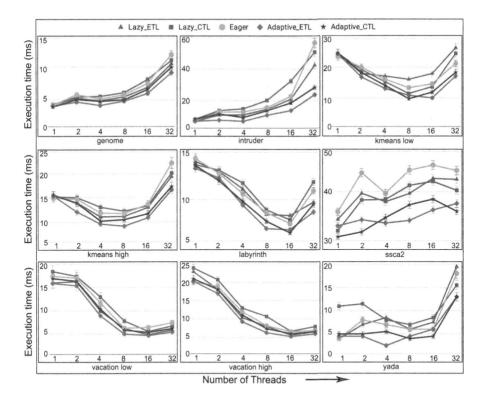

Fig. 8. Execution time in STAMP benchmarks using better time barrier.

is up to 2.36× and 2×, respectively. The minimum execution gain obtained in *Adaptive_ETL* is 1.13 and in *Adaptive_CTL* is 1.12 with the threads grater than 4. From Fig. 9, we observed that the number of aborts significantly increases in all the applications of STAMP benchmark when transactions are executed in more than 8 number of threads. Still, ADAPTIVE has significantly less aborts compared to *Lazy* and *Eager*. *Adaptive_ETL* has up to 16× less aborts than *Lazy_ETL* and up to 13× less aborts than *Eager*. Similarly, *Adaptive_CTL* has up to 2.5× less aborts than *Lazy_CTL* and up to 170× less aborts than *Eager*.

Results on STAMPEDE Benchmarks. Similar to micro and STAMP benchmarks, ADAPTIVE has better performance compared to *Lazy* and *Eager* in STAMPEDE benchmarks, for both execution time and number of aborts (Fig. 10). For execution time, *Adaptive_ETL* performed up to 1.72× better than *Lazy_ETL* and *Adaptive_CTL* performed up to 1.54× better than *Lazy_CTL*. Compared to *Eager*, *Adaptive_ETL* performed up to 1.68× better and *Adaptive_CTL* performed up to 1.91× better. The minimum execution gain obtained in *Adaptive_ETL* is 1.14 and in *Adaptive_CTL* is 1.12 with the threads greater than 4. For number of aborts, *Adaptive_ETL* performed up to

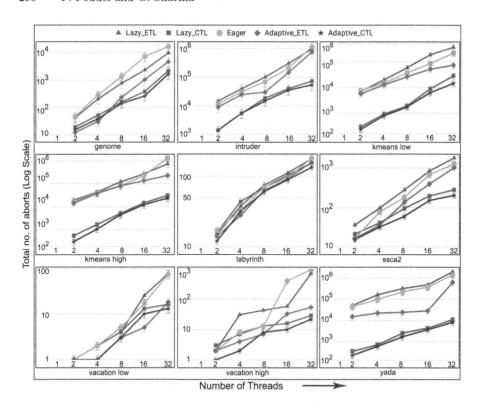

Fig. 9. Number of aborts in STAMP benchmarks using better time barrier.

4.1× better than *Lazy_ETL* and *Adaptive_CTL* performed up to 72× better than *Lazy_CTL*. Compared to *Eager*, *Adaptive_ETL* performed up to 10× better and *Adaptive_CTL* performed up to 124× better.

In all the benchmarks, the minimum execution gain for ADAPTIVE ranges between 1 and 1.16 when running with threads up to 4 numbers.

Further Results. The results in Figs. 6, 7, 8, 9 and 10 only considered optimized ADAPTIVE w.r.t. better time barrier. We also performed experiments for ADAPTIVE using both, better time barrier and better switching mechanism. We varied the switching interval threshold (*SW_INT*) from 2 up to 10. The results indicate that instead of switching versioning immediately, using the better switch mechanism increases the performance. However, for *SW_INT* > 2, the performance gradually reduces and becomes worse while reaching *SW_INT* = 10. Figure 11 shows the execution time for STAMP benchmarks when executed with both better time barrier and better switch mechanism (*SW_INT* = 2). The improvement is up to 1.09× compared to ADAPTIVE with better time barrier. Alongwith decreasing the total number of aborts, the better switch mechanism decreases the total number of switches between the versioning methods which helps to get

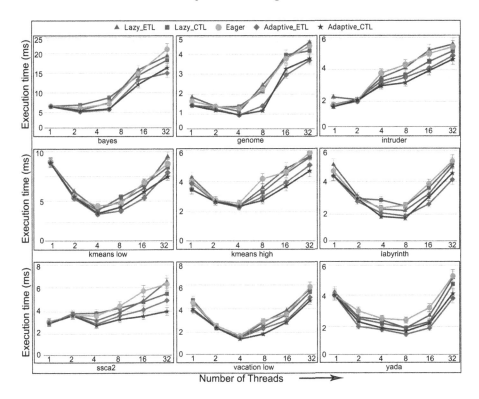

Fig. 10. Execution time in STAMPEDE benchmarks using better time barrier.

the improvement on execution time. Figure 12 illustrates the reduction of total number of switches using better switch mechanism for STAMP benchmarks. The experiments on micro-benchmarks and STAMPEDE showed similar results.

The experiments so far use $Threshold_{Eager} = \frac{1}{2}$ and $Threshold_{Lazy} = 2$ as computed in Sect. 4. It is natural to ask whether these are the ideal threshold values. Therefore, for $Threshold_{Eager}$, we used $\frac{1}{4}$ and $\frac{3}{4}$, whereas for $Threshold_{Lazy}$, we used 1 and 3. We performed experiments by using two different combinations of $Threshold_{Eager}$ and $Threshold_{Lazy}$, $(\frac{1}{4}, 1)$ and $(\frac{3}{4}, 3)$. We noticed the increase in both execution time and number of aborts in all the benchmarks for both the combinations. This suggests that the threshold values computed in Sect. 4 are appropriate.

The results reported in Figs. 6, 7, 8, 9, 10, 11 and 12 use *suicide* as a conflict resolution strategy. We were interested to see whether other strategies perform better than *suicide*. Therefore, we performed experiments using 4 different conflict resolution strategies *suicide, delay, back-off,* and *kill*. The results showed not significant change on performance in some of the benchmarks, while in the rest, the selection of conflict resolution strategy affected the performance. For example, *genome* and *intruder* performed better with *suicide* whereas, *kmeans*

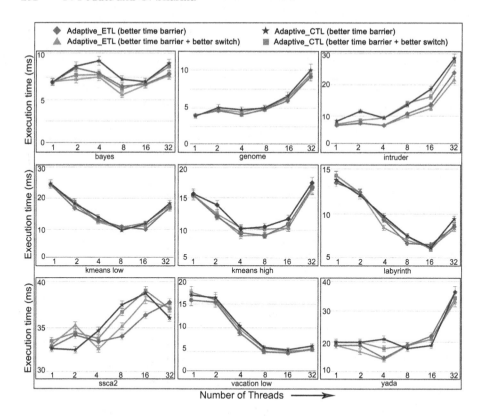

Fig. 11. Execution time in STAMP benchmarks using better barrier and better switch.

performed better with *back-off*. In overall, *suicide* performed better than the rest in most of the benchmarks.

Finally, we performed experiments starting the execution initially using eager and lazy versioning. We observed that the initial selection of versioning does not affect performance significantly in both micro and complex benchmarks except *intruder* and *kmeans* from STAMP in which ADAPTIVE performed better when starting with *Eager* than *Lazy* for up to 4 threads. This is mainly because transactions have almost constant abort rate and versioning method change is not necessary.

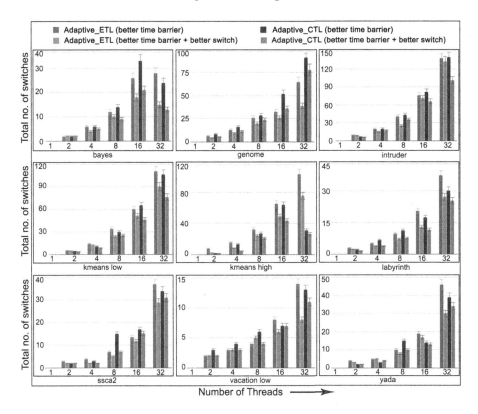

Fig. 12. Illustration of decrease in total number of switches between versioning methods using better switch mechanism.

7   Concluding Remarks

TM has been receiving much attention from both academia and industry. One of the most challenging issues in TM is how to ensure consistency of the shared data through speculative execution. Eager and lazy versioning have been used individually to support speculative execution in existing TM systems. However, whether to use eager or lazy versioning is better is not clear and previous studies contradict on the recommendations. In this paper, we have presented an adaptive framework that dynamically switches between eager and lazy versioning at runtime. Our framework is quite simple and achieves significantly better performance for execution time and number of aborts compared to eager and lazy versioning running individually in 5 micro-benchmarks and 8 applications from STAMP and STAMPEDE benchmarks. We believe that our results and techniques will be helpful in choosing proper versioning for TM systems.

References

1. The Persistent Memory Development Kit (PMDK). https://github.com/pmem/pmdk/. Accessed 14 Feb 2019
2. RSTM. http://www.cs.rochester.edu/research/synchronization/rstm/index.shtml. Accessed 14 Feb 2019
3. Alistarh, D., Haider, S.K., Kübler, R., Nadiradze, G.: The transactional conflict problem. In: SPAA, pp. 383–392 (2018)
4. Ananian, C.S., Asanovic, K., Kuszmaul, B.C., Leiserson, C.E., Lie, S.: Unbounded transactional memory. In: HPCA, pp. 316–327 (2005)
5. Bocchino, R.L., Adve, V.S., Chamberlain, B.L.: Software transactional memory for large scale clusters. In: PPoPP, pp. 247–258 (2008)
6. Cain, H.W., Michael, M.M., Frey, B., May, C., Williams, D., Le, H.Q.: Robust architectural support for transactional memory in the power architecture. In: ISCA, pp. 225–236 (2013)
7. Coburn, J., et al.: NV-Heaps: making persistent objects fast and safe with next-generation, non-volatile memories. In: ASPLOS, pp. 105–118 (2011)
8. Dalessandro, L., Spear, M.F., Scott, M.L.: NOrec: streamlining STM by abolishing ownership records. In: PPOPP, pp. 67–78 (2010)
9. Dragojevic, A., Guerraoui, R., Kapalka, M.: Stretching transactional memory. In: PLDI, pp. 155–165 (2009)
10. Felber, P., Fetzer, C., Marlier, P., Riegel, T.: Time-based software transactional memory. IEEE Trans. Parallel Distrib. Syst. **21**(12), 1793–1807 (2010)
11. Felber, P., Fetzer, C., Riegel, T.: Dynamic performance tuning of word-based software transactional memory. In: PPOPP, pp. 237–246 (2008)
12. Fung, W.W.L., Singh, I., Brownsword, A., Aamodt, T.M.: Hardware transactional memory for GPU architectures. In: MICRO, pp. 296–307 (2011)
13. Hammond, L., et al.: Transactional memory coherence and consistency. SIGARCH Comput. Archit. News **32**(2), 102 (2004)
14. Haring, R., et al.: The IBM Blue Gene/Q compute chip. IEEE Micro **32**(2), 48–60 (2012)
15. Herlihy, M., Luchangco, V., Moir, M., Scherer III., W.N.: Software transactional memory for dynamic-sized data structures. In: PODC, pp. 92–101 (2003)
16. Herlihy, M., Moss, J.E.B.: Transactional memory: architectural support for lock-free data structures. In: ISCA, pp. 289–300 (1993)
17. Intel (2012). http://software.intel.com/en-us/blogs/2012/02/07/transactional-synchronization-in-haswell
18. Keidar, I., Perelman, D.: Multi-versioning in transactional memory. In: Guerraoui, R., Romano, P. (eds.) Transactional Memory. Foundations, Algorithms, Tools, and Applications. LNCS, vol. 8913, pp. 150–165. Springer, Cham (2015). https://doi.org/10.1007/978-3-319-14720-8_7
19. Keidar, I., Perelman, D.: On avoiding spare aborts in transactional memory. Theory Comput. Syst. **57**(1), 261–285 (2015)
20. Manassiev, K., Mihailescu, M., Amza, C.: Exploiting distributed version concurrency in a transactional memory cluster. In: PPoPP, pp. 198–208 (2006)
21. Minh, C.C., Chung, J., Kozyrakis, C., Olukotun, K.: STAMP: Stanford transactional applications for multi-processing. In: IISWC, pp. 35–46 (2008)
22. Moore, K.E.: LogTM: log-based transactional memory. In: HPCA, pp. 258–269 (2006)

23. Nakaike, T., Odaira, R., Gaudet, M., Michael, M.M., Tomari, H.: Quantitative comparison of hardware transactional memory for Blue Gene/Q, zEnterprise EC12, Intel Core, and POWER8. In: ISCA, pp. 144–157 (2015)

24. Nguyen, D., Pingali, K.: What scalable programs need from transactional memory. In: ASPLOS, pp. 105–118 (2017)

25. Poudel, P., Sharma, G.: An adaptive logging framework for persistent memories. In: Izumi, T., Kuznetsov, P. (eds.) SSS 2018. LNCS, vol. 11201, pp. 32–49. Springer, Cham (2018). https://doi.org/10.1007/978-3-030-03232-6_3

26. Rajwar, R., Herlihy, M., Lai, K.: Virtualizing transactional memory. In: ISCA, pp. 494–505. IEEE Computer Society, Washington, DC (2005)

27. Shavit, N., Touitou, D.: Software transactional memory. Distrib. Comput. **10**(2), 99–116 (1997)

28. Volos, H., Tack, A.J., Swift, M.M.: Mnemosyne: lightweight persistent memory. In: ASPLOS, pp. 91–104 (2011)

29. Wan, H., Lu, Y., Xu, Y., Shu, J.: Empirical study of redo and undo logging in persistent memory. In: NVMSA, pp. 1–6 (2016)

Brief Announcement Blockguard: Adaptive Blockchain Security

Shishir Rai, Kendric Hood, Mikhail Nesterenko[✉], and Gokarna Sharma

Department of Computer Science, Kent State University, Kent, OH 44242, USA
{srai,khood5}@kent.edu, {mikhail,sharma}@cs.kent.edu

Abstract. We change the security of blockchain transactions by varying the size of consensus committees. To improve performance, such committees operate concurrently. We present two algorithms that allow adaptive security by forming concurrent variable size consensus committees on demand. One is based on a single joint blockchain, the other is based on separate sharded blockchains. For in-committee consensus, we implement synchronous Byzantine fault tolerance algorithm (BFT), asynchronous BFT and proof-of-work consensus. We evaluate the performance of our adaptive security algorithms.

1 Definitions and Committee Consensus Algorithms

A set of n *peer processes* (or *peers*) forms a network to maintain the blockchain. The *blockchain* is a sequence of blocks or transactions. We use the terms interchangeably, i.e. we assume that a block contains a single transaction. A *transaction* is a unit of blockchain recording. Each subsequent transaction is cryptographically linked to the previous one. The first transaction in the blockchain is the *genesis* transaction. Peers communicate through broadcasts. Message delivery is FIFO. There is no message loss. Messages cannot be forged. Peers are either *honest* or *Byzantine*. A set of peers that cooperate to approve a transaction despite actions of Byzantine peers is a *consensus committee*.

Sharding. A *(recording) group* is a set of processes that maintain a single blockchain. There are as many groups as there are separate blockchains. In case of sharding, a peer in the consensus committee that approves a certain transaction in a blockchain does not necessarily belong to the group that records it. However, a peer may belong to only one recording group and only one consensus committee at a time.

PBFT and SBFT. In PBFT [2] The committee of peers elect the *leader*. The leader runs consensus on every transaction. It initiates several message exchanges with other committee peers. A non-leader Byzantine peer may delay messages or send incorrect messages. A Byzantine leader may temporarily block the consensus by sending different messages to different peers or not sending

---

See [4] for full text of the paper.

© Springer Nature Switzerland AG 2019
M. Ghaffari et al. (Eds.): SSS 2019, LNCS 11914, pp. 296–300, 2019.
https://doi.org/10.1007/978-3-030-34992-9_23

messages altogether. In either case, the honest peers discover the Byzantine leader and replace it by forcing a *view change*. PBFT is guaranteed to withstand up to $f < n/3$ Byzantine peers regardless of the message propagation delay. The operation of SBFT [1] is similar to PBFT. This algorithm relies on at least one honest peer confirming the transaction. However, it assumes that there is a bound on communication delay between honest peers. If a message is not received after a certain delay, it is guaranteed never to arrive. Thus, the algorithm has to delay to ascertain this lack of message receipt. In practice this may make SBFT slower. However, it has higher resilience threshold. It can tolerate up to $f < n/2$ Byzantine peers.

PoW. We implement proof-of-work consensus similar to Nakamoto [3]. To attach a new transaction to the blockchain, a peer *mines* the transaction by solving a computationally intensive task that links the new and previous transaction. Several peers may mine transactions concurrently. This is a *fork* in the blockchain. A branch of a fork may be extended by the addition of mined transactions on top of the current block. The shorter branch is discarded. PoW consensus operates correctly provided that the computational power of honest peers exceeds that of Byzantine peers. If peers have the same computational power, PoW consensus tolerates up to $f < n/2$ Byzantine peers.

2   The Adaptive Security Problem and Solutions

The Problem. *The* Adaptive Security Problem *requires, as a solution, an adaptive security algorithm, to assign committees to the transactions such that each committee satisfies the transaction security level.* We consider an adaptive security algorithm that selects appropriate size committees and processes transactions with as much parallelism as possible. We present two such algorithms: *Composite Blockguard* and *Dynamic Blockguard*.

Composite Blockguard Adaptive Security Algorithm. In this algorithm, peers are divided into storage groups maintaining independent blockchains. The algorithm maintains a list of idle groups and pending transactions. Once a new transaction arrives or a consensus committee is done, Composite Blockguard finds appropriate number of available groups, forms a consensus committee to process the next pending transaction and dispatches the transaction. If not enough idle groups are available, the pending transactions wait.

Dynamic Blockguard Adaptive Security Algorithm. This algorithm has a single blockchain and thus a single recording group. A consensus committee is selected out of this group of peers. Multiple consensus committees may operate concurrently if their members do not intersect. This means that the committees have to concurrently write to the same blockchain. To ensure the integrity of the blockchain, the computation proceeds by alternating two stages: consensus stage and recording stage. In the *consensus* stage, committees agree on blocks to be written to the blockchain. Every committee must reach consensus before any committee may proceed to the next stage. In the *recording* stage, each committee broadcasts the transaction to the group maintaining the blockchain. That is, they broadcast it to the whole network. Each written transaction is cryptographically linked to all the written transaction in the previous recording stage. This way, the resultant blockchain is a series-parallel graph. *Committee selection window* is the set of unique peers that published in the blockchain most recently. Committee peers are picked at random from the committee selection window.

3   Performance Evaluation

Setup. We evaluate the performance of Composite and Dynamic Blockguard using abstract simulation. The behavior of each algorithm is represented as a sequence of rounds. In every round, each peer may receive a single new message, do local computation and send messages to other peers.

Byzantine peers' goal is to successfully commit a fraudulent transaction to the blockchain, we model this as follows. A committee is *reliable* if the number of Byzantine peers in it does not exceed its tolerance threshold, *defeated* otherwise. The tolerance threshold is 1/3 for PBFT and 1/2 for SBFT and PoW. Defeated committees commit only fraudulent transactions to the blockchain, and reliable committees never commit fraudulent transactions. Byzantine leaders propose only fraudulent transactions. If a fraudulent transaction is proposed in a reliable committee then a view change occurs. This repeats until a non-byzantine leader is found. In PoW, if a Byzantine peer is the first to mine in a reliable committee then nothing is recorded and mining restarts.

Experiment Parameters and Evaluation Metrics. Unless stated otherwise, in the below experiments, the parameters are set as follows. The fraction of Byzantine faults is $n/10$. The number of peers in the network is 1024. There are 1000 rounds in a computation. Each data point is the average of 10 computations. A new transaction is generated every two rounds. We have 5 security levels. The highest security level is the 5-th level which contains the whole network. Each lower level contains half of the peers of the higher level. We use geometric distribution to select the security level of newly generated transaction. In PoW, we use binomial distribution to determine the number of rounds it takes the peers to mine a transaction. The mode, i.e. most frequently occurring value, is 5 and variance 2.5. We vary maximum message delay and the fraction of Byzantine peers in the network. We consider a transaction approval as a consensus. We compute the following metrics. *Throughput* is the number of consensuses per round. Consensuses of defeated committees are not counted. *(Transaction) waiting time* is computed as follows. For coordinated consensus algorithms, i.e. PBFT and SBFT, it is the number of rounds from the moment the transaction is generated till the first peer determines that the transaction is committed. For PoW, it is the time for this transaction to be mined. The waiting time for transactions of defeated committees is counted.

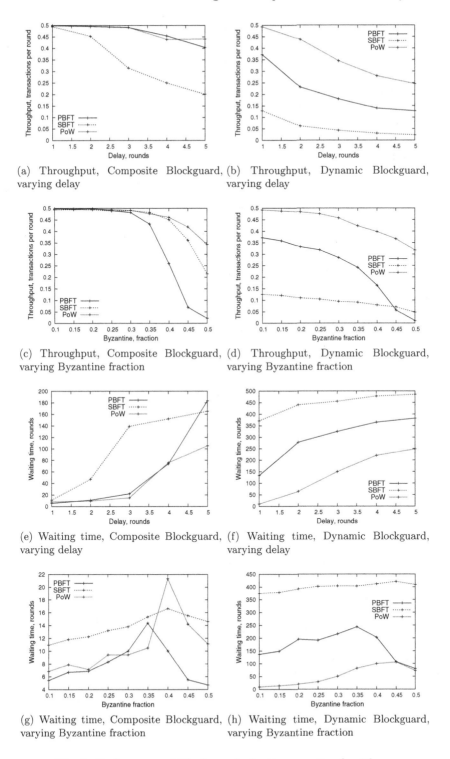

(a) Throughput, Composite Blockguard, varying delay

(b) Throughput, Dynamic Blockguard, varying delay

(c) Throughput, Composite Blockguard, varying Byzantine fraction

(d) Throughput, Dynamic Blockguard, varying Byzantine fraction

(e) Waiting time, Composite Blockguard, varying delay

(f) Waiting time, Dynamic Blockguard, varying delay

(g) Waiting time, Composite Blockguard, varying Byzantine fraction

(h) Waiting time, Dynamic Blockguard, varying Byzantine fraction

Fig. 1. Performance of Blockguard adaptive security algorithms.

Algorithm Performance Experiments. The results of the performance evaluation of the adaptive security algorithms are shown in Fig. 1. Figures 1a and b demonstrate how throughput depends on the network delay for Composite and Dynamic Blockguard respectively. As network delay increases, the throughput declines. However, different consensus committees react to this increase differently. PBFT has the best performance and lowest decline since the committees just wait for the actual messages to arrive. SBFT exhibits the most sensitivity to the network delay. The reason is that SBFT has to wait for the maximum delay to determine that the message is not coming. Let us discuss Figs. 1c and d. It shows that the performance of Composite and Dynamic Blockguard decreases as the fraction of Byzantine peers in the network increase. This is due to Byzantine peers slowing down the consensus algorithms. PBFT suffers the most since its tolerance threshold is only a third of the peers.

Figures 1e and f show the dependency of transaction waiting time on network delay. As expected, the waiting time increases with delay. SBFT is the most vulnerable to this increase since it has to wait for maximum delay time. Figures 1g and h show how waiting time varies with the fraction of Byzantine peers. Let us explain the trends in the data. As the consensus committee approaches its resiliency threshold, the number of view changes or repeated transaction mining increases which increases the transaction waiting time. If the fraction is away from this threshold, the committees are either reliable or defeated. In either case the waiting time is relatively low. Thus, there is a peak near $n/3$ for PBFT and near $n/5$ for SBFT and PoW. This trend is less pronounced in Dynamic Blockguard since it is masked by synchronization across consensus committees in the same stage.

The results of our experiments indicate that both Composite and Dynamic blackguard algorithm provide adaptive security with a trade-off between performance and security parameters.

References

1. Abraham, I., Devadas, S., Nayak, K., Ren, L.: Brief announcement: practical synchronous Byzantine consensus. In: DISC, pp. 41:1–41:4 (2017)
2. Castro, M., Liskov, B.: Practical Byzantine fault tolerance and proactive recovery. ACM Trans. Comput. Syst. **20**(4), 398–461 (2002)
3. Nakamoto, S.: Bitcoin: a peer-to-peer electronic cash system (2008). http://bitcoin.org/bitcoin.pdf
4. Rai, S., Hood, K., Nesterenko, M., Sharma, G.: Blockguard: adaptive blockchain security. arXiv e-prints arXiv:1907.13232, July 2019

Brief Announcement: Fully Anonymous Shared Memory Algorithms

Michel Raynal[1,2] and Gadi Taubenfeld[3(✉)]

[1] Univ Rennes IRISA, Rennes, France
[2] Department of Computing, Polytechnic University, Kowloon, Hong Kong
[3] The Interdisciplinary Center, 46150 Herzliya, Israel
tgadi@idc.ac.il

Abstract. Process anonymity has been studied for a long time. Memory anonymity is more recent. In an anonymous memory system, there is no a priori agreement among the processes on the names of the shared registers they access. This article introduces the *fully anonymous* model, namely a model in which both the processes and the memory are anonymous. It is shown that fundamental problems such as mutual exclusion, consensus, and its weak version called set agreement, can be solved despite full anonymity, the first in a failure-free system, the others in the presence of any number of process crashes.

1 Introduction

Process Anonymity. The notion of *process anonymity* has been studied for a long time from an algorithmic and computability point of view, both in message-passing systems and shared memory systems. Process anonymity means that processes have no identity, have the same code and the same initialization of their local variables (otherwise they could be distinguished). We assume a system that is composed of a finite set of $n \geq 2$ asynchronous, anonymous processes denoted $p_1, \ldots, p_n$. Each process knows the number of processes and the number of registers. The subscript i in p_i is only a notational convenience, which is not known by the processes.

Memory Anonymity. The notion of *memory anonymity* has been recently introduced in [8]. Let us consider a shared memory R made up of m atomic registers. Such a memory can be seen as an array with m entries, namely $R[1 \ldots m]$. In a non-anonymous memory system, for each index x, the name $R[x]$ denotes the same register for each process that accesses the address $R[x]$. Hence in a non-anonymous memory, there is an a priori agreement on the names of the shared registers.

The situation is different in an anonymous memory, where there is no a priori agreement on the names of the registers. Moreover, all the registers of an anonymous memory are assumed to be initialized to the same value (otherwise, their initial values could provide information allowing processes to distinguish

© Springer Nature Switzerland AG 2019
M. Ghaffari et al. (Eds.): SSS 2019, LNCS 11914, pp. 301–306, 2019.
https://doi.org/10.1007/978-3-030-34992-9_24

them). The interested reader will find an introductory survey on both process and memory anonymity in [4].

Anonymous Shared Memory. The shared memory is made up of $m \geq 1$ atomic anonymous registers denoted $R[1 \ldots m]$. Hence, *all* the registers are anonymous. As already indicated, due to its anonymity, $R[x]$ does not necessarily indicate the same object for different processes. More precisely, a memory-anonymous system is such that:

- For each process p_i an adversary is defining a permutation $f_i()$ over $\{1, 2, \cdots, m\}$, such that when p_i uses the address $R[x]$, it actually accesses $R[f_i(x)]$,
- No process knows the permutations, and
- All the registers are initialized to the same default value denoted $\perp$.

Table 1. Illustration of an anonymous memory model

Identifiers for an external observer	Local identifiers for process p_i	Local identifiers for process p_j
$R[1]$	$R_i[2]$	$R_j[3]$
$R[2]$	$R_i[3]$	$R_j[1]$
$R[3]$	$R_i[1]$	$R_j[2]$
Permutation	$f_i() : [2, 3, 1]$	$f_j() : [3, 1, 2]$

An example is presented in Table 1. To make apparent the fact that $R[x]$ can have a different meaning for different processes, we write $R_i[x]$ when p_i invokes $R[x]$.

Anonymous Register Models. We consider two types of anonymous register models.

- RW (read/write) model. In this model all, the registers can be atomically read or written by any process.
- RMW (read/modify/write) model. In this model, each register can be atomically read, written or accessed by an operation that atomically reads the register and (according to the value read) possibly modifies it.

Practical Motivation. It was recently shown that epigenetic cell modifications can be modeled by anonymous entities cooperating through anonymous communication media [6]. Hence, fully anonymous distributed systems could inspire bio-informatics (and be inspired by it) [2].

This article considers the following problems.

Mutual Exclusion. Mutual exclusion is the oldest and one of the most important synchronization problems. Formalized by E.W. Dijkstra in the mid-sixties, it consists in building what is called a lock (or mutex) object, defined by two operations, denoted acquire() and release(). For a formal definition, we refer the reader to [3,7].

Consensus. The consensus problem consists in building a one-shot operation, denoted propose(), which takes an input parameter (called *proposed* value) and returns a result (called *decided* value). A process may invoke the operation at most once. The meaning of this operation is defined as follows: (1) Validity: A decided value must be a proposed value; and (2) Agreement: No two processes decide on different values.

The problem must also satisfy one of the following progress conditions: (1) Wait-freedom: If a process does not crash, it must decide; or the weaker (2) Obstruction-freedom: If a process does not crash, and executes alone during a long enough period, it must decide. While wait-free consensus can be solved from registers in a non-anonymous RMW system, it cannot be solved in a non-anonymous RW system. It is possible to solve (the weaker) obstruction-free consensus in non-anonymous RW system.

Set Agreement. Set agreement captures a weaker form of consensus in which the agreement property is weakened as follows: At most $n - 1$ different values are decided upon. That is, in any given run, the size of the set of the decision values is at most $n - 1$.

The set agreement problem as defined above is also called the $(n - 1)$-set agreement problem. While much weaker than consensus, as consensus, wait-free set agreement cannot be solved in non-anonymous RW memory systems.

Content of the Paper. Table 2 describes the technical content of the paper. As an example, the second line states that Sect. 3 presents a consensus algorithm for an anonymous RMW system for $n > 1$ and $m \geq 1$. As far as the mutex algorithm is concerned, it is also shown that $m \in M(n)$, where $M(n) = \{m$ such that $\forall\ \ell :\ 1 < \ell \leq n: \gcd(\ell, m) = 1\}$ is a *necessary and sufficient* condition on the size of the memory.

Table 2. Results and the structure of the paper

Problem	Section	Tolerate failures	Register type	Progress condition	Number of processes n	Number of registers m
Mutual exclusion	2	No	RMW	Deadlock-freedom	$n > 1$	$m \in M(n)$
Consensus	3	Yes	RMW	Wait-freedom	$n > 1$	$m \geq 1$
Set agreement	4	Yes	RW	Obstruction-freedom	$n > 1$	$m \geq 3$
Consensus	5	Yes	RW	Obstruction-freedom	$n = 2$	$m \geq 3$

2   Fully Anonymous Mutex Using RMW Registers

The mutual exclusion problem can be solved for non-anonymous processes in both the anonymous RW register model and the anonymous RMW register model [1,8]. However, there is no mutual exclusion algorithm when the processes are anonymous, even when using non-anonymous RW registers. To see that, simply consider an execution in which the anonymous processes run in lock-steps (i.e., one after the other) and access the RW registers in the same order. In such a run it is not possible to break symmetry as the local states of the processes will be exactly the same after each such lock-step.

Let us recall that two integers x and y are said to be *relatively prime* if their greatest common divisor is 1, notice that a number is *not* relatively prime to itself. Let $M(n) = \{m$ such that $\forall \ell : 1 < \ell \leq n : \gcd(\ell, m) = 1\}$.

Theorem 1. *There is a deadlock-free mutual exclusion algorithm for $n \geq 2$ anonymous processes using $m \geq 1$ anonymous RMW registers if and only if $m \in M(n)$.*

The proof of the *if direction*, follows from the very existence of the deadlock-free mutual exclusion algorithm for n anonymous processes using m anonymous RMW registers, where $m \in M(n)$, presented in the full version of the paper [5]. The proof of *only if direction*, is a consequence of the lower bound result from [1], which states that $m \in M(n)$ is a necessary and sufficient condition for symmetric deadlock-free mutual exclusion for n non-anonymous processes and m anonymous RMW registers.

3   Fully Anonymous Wait-Free Consensus Using RMW Registers

We describe below a straightforward wait-free consensus algorithm for any number $m \geq 1$ of anonymous registers. This algorithm assumes that the set of proposed values is totally ordered. Each process tries to write the value it proposes into each anonymous register. Assuming that at least one process that does not crash invokes propose(), there is a finite time after which (whatever the concurrency/failure pattern is) each anonymous register contains a proposed value. Then, using the same deterministic rule (for example by choosing the maximum value) the processes decide on the same value.

4   Fully Anonymous Obstruction-Free Set Agreement Using RW Registers

We describe below an obstruction-free set agreement algorithm for $n \geq 2$ anonymous processes using $m \geq 3$ anonymous RW registers. Each anonymous RW register can store the preference of a process. Each participating process scans the m RW registers trying to write its preference into each one of the m registers. Before each write, the process scans the shared array and operates as follows:

- If its preference appears in all the m registers, it reads the array again, and if, for the second time, its preference appears in all the m registers, it decides on its preference and terminates.
- Otherwise, if some preference appears in more than half of the registers, the process adopts this preference as its new preference.

Afterward, the process finds some arbitrary entry in the shared array that does not contain its current preference and writes its current preference into that entry. Once the process finishes writing it repeats the above steps. The exact code and a detailed subtle proof of the algorithm can be found in the full version of this paper [5].

5   Fully Anonymous 2-Process Obstruction-Free Consensus Using RW Registers

As the reader can easily check, instantiating of the obstruction-free set agreement algorithm from the previous section with $n = 2$ provides us with 2-process obstruction-free consensus built using $m \geq 3$ RW registers.

A Conjecture. Let us consider the set agreement algorithm from the previous section, in which it is assumed that $n \geq 2$. We conjecture that when the requirement $m \geq 3$ in this algorithm is strengthened to $m \geq 2n - 1$ the resulting algorithm solves obstruction-free consensus for n processes.

Finally, it was recently proved in [9] that there is no obstruction-free consensus algorithm for two non-anonymous processes using only anonymous bits. Thus, as was shown in [9], anonymous bits are strictly weaker than anonymous (and hence also non-anonymous) multi-valued registers.

6   Discussion

This article has several contributions. The first is the introduction of the notion of *fully anonymous* shared memory systems, namely, systems where the processes are anonymous and there is no global agreement on the names of the shared registers. The article has then addressed the design of mutual exclusion, consensus and set agreement algorithms in specific contexts where the anonymous registers are read/write (RW) registers or more powerful read/modify/write (RMW) registers. It has been shown that, for fully anonymous mutual exclusion using RMW registers, the condition on the number m of registers, namely $m \in M(n)$, is both necessary and sufficient, extending thereby a result of [1] (which was for non-anonymous processes and anonymous registers).
A full version of this paper is available at [5].

Acknowledgments. M. Raynal was partially supported by the French ANR project DESCARTES (16-CE40-0023-03) devoted to layered and modular structures in distributed computing.

References

1. Aghazadeh, Z., Imbs, D., Raynal, M., Taubenfeld, G., Woelfel, Ph.: Optimal memory-anonymous symmetric deadlock-free mutual exclusion. In: Proceedings of 38th ACM Symposium on Principles of Distributed Computing, PODC 2019, 10 p. ACM Press (2019)
2. Navlakha, S., Bar-Joseph, Z.: Distributed information processing in biological and computational systems. Commun. ACM **58**(1), 94–102 (2015)
3. Raynal, M.: Concurrent Programming: Algorithms, Principles and Foundations, 515 p. Springer, Heidelberg (2013). https://doi.org/10.1007/978-3-642-32027-9. ISBN 978-3-642-32026-2
4. Raynal, M., Cao, J.: Anonymity in distributed read/write systems: an introductory survey. In: Podelski, A., Taïani, F. (eds.) NETYS 2018. LNCS, vol. 11028, pp. 122–140. Springer, Cham (2019). https://doi.org/10.1007/978-3-030-05529-5_9
5. Raynal M., Taubenfeld G.: Fully anonymous shared memory algorithms, 16 p. ArXiv-1909.05576 (2019)
6. Rashid S., Taubenfeld G., Bar-Joseph Z.: Genome wide epigenetic modifications as a shared memory consensus problem. In: 6th Workshop on Biological Distributed Algorithms, BDA 2018, London (2018)
7. Taubenfeld, G.: Synchronization Algorithms and Concurrent Programming. Pearson Education/Prentice Hall, London/Upper Saddle River, 423 p. (2006). ISBN 0-131-97259-6
8. Taubenfeld, G.: Coordination without prior agreement. In: Proceedings of 36th ACM Symposium on Principles of Distributed Computing, PODC 2017, pp. 325–334. ACM Press (2017)
9. Taubenfeld, G.: Set agreement power is not a precise characterization for oblivious deterministic anonymous objects. In: Censor-Hillel, K., Flammini, M. (eds.) SIROCCO 2019. LNCS, vol. 11639, pp. 293–308. Springer, Cham (2019). https://doi.org/10.1007/978-3-030-24922-9_20

A Topological View of Partitioning Arguments: Reducing k-Set Agreement to Consensus

Hugo Rincon Galeana[1] , Kyrill Winkler[2] , Ulrich Schmid[2($\boxtimes$)] ,
and Sergio Rajsbaum[1]

[1] Instituto de Matemáticas, UNAM, CDMX, 04510 Mexico, D.F., Mexico
[2] TU Wien, ECS Group (E191-02), Treitlstrasse 1–3, 1040 Vienna, Austria
s@ecs.tuwien.ac.at

Abstract. The objective of this paper is to understand the effect of partitioning in distributed computing models. In spite of being quite similar agreement problems, (deterministic) consensus (1-set agreement) and k-set agreement (for $k > 1$) require surprisingly different techniques for proving impossibilities. There is a widely applicable generic theorem, however, which allows to reduce the impossibility of k-set agreement to consensus in message-passing models that allow some partitioning. In this paper, we provide the topological representation of this theorem, which reveals how partitioning is reflected in the protocol complex: It turns out that this leads to a "color splitting" of the algorithm's decision map, which separates the sub-complexes representing the partitioned processes. We also harvest a general advantage of topological results, which allowed us to carry over our findings to shared memory systems. We first demonstrate the utility of our reduction theorem by proving that d-set agreement cannot be solved in the d-solo asynchronous read-write model even when a single process may crash, not just in the wait-free case. Moreover, our new insights into the structure of protocol complexes gave us the idea for a simple proof of the fact that no partitioning argument can provide a valid impossibility proof for wait-free set agreement in the iterated immediate snapshot model: For any set of partition-compatible runs (which do not contain runs where all processes always have a complete view), we provide a way to construct a simple algorithm that solves set agreement.

Keywords: Algebraic topology · Consensus · Set agreement · Partitioning arguments · Shared memory

1 Introduction

Partitions, i.e., sets of processes that cannot always communicate with each other, are a fundamental combinatorial notion in distributed computability

This work has been supported by the PAPIIT-UNAM grant IN109917 and the Austrian Science Fund (FWF) projects RiSE/SHiNE (S11405) and ADynNet (P28182).

© Springer Nature Switzerland AG 2019
M. Ghaffari et al. (Eds.): SSS 2019, LNCS 11914, pp. 307–322, 2019.
https://doi.org/10.1007/978-3-030-34992-9_25

[13,16]. Studying the effect of partitioning on the (im)possibility of agreement problems has been a focal point at least since the realization that asynchronous consensus is solvable with a minority of initially crashed processes [15] and the observation that a shared register cannot be implemented on top of a message passing system with a majority of faulty processes [5]. Since then, partitioning arguments have been applied successfully to many distributed computing problems [14]. Essentially, these arguments exploit the fact that one cannot guarantee agreement among processes of a distributed system that never, neither directly nor indirectly, communicate with each other.

The objective of this paper is to understand the computational power of models where partitioning can occur. For this purpose, we focus on the k-set agreement problem and its special cases consensus and set agreement. In these problems, every process owns a local input value taken from a finite domain $\mathcal{V}$ (to rule out trivial solutions, we assume that $|\mathcal{V}| \geq k$), and must irrevocably assign a local output value (also called decision value) that must be the input value of some process and satisfy certain properties. For consensus, no two processes may decide on, i.e., assign, different values. For set agreement among n processes, the number of different decision values must be at most $n - 1$ system-wide. The general case is k-set agreement, which requires that the number of different decision values is at most k. We will focus solely on deterministic algorithms.

Due to the landmark FLP impossibility result [15], which employs (now classic) bivalence proofs, it is well-known that consensus is impossible to solve in asynchronous systems if a single process may crash. The corresponding result for general k-set agreement is the impossibility of solving this problem in asynchronous systems where $f \geq k$ processes may crash. Surprisingly, establishing the latter result requires quite involved techniques based on algebraic topology resp. a variant of Sperner's lemma [11,20,23], which have not been matched by combinatorial proofs so far in their full generality (see the related work for some exceptions for special cases).

Despite this apparent "proof-incompatibility" of consensus and k-set agreement for $k > 1$, Biely, Robinson, and Schmid showed in [9] that, in message-passing models that allow partitioning, impossibility proofs for k-set agreement can be reduced to impossibility proofs for consensus: They provided a theorem (called BRS theorem in the sequel), which uses a partitioning argument as a means for reduction and is generic w.r.t. the underlying system model: Essentially, if failures and asynchrony allow for runs where the system partitions into k parts, the processes must decide on their own in every partition. By choosing distinct proposal values, solving k-set agreement in such runs requires solving consensus in every partition. Consequently, the impossibility of k-set agreement can be proved by showing that it is impossible to reach consensus in at least one of these partitions. Note that the BRS theorem actually works for fairly weak forms of partitioning, where some communication between partitions is still possible.

In [9], the authors applied the BRS theorem to various message-passing models, including purely asynchronous systems with crash failures, synchronous systems with omission failures, dynamic networks with omission failures, and even asynchronous systems with failure detectors.

Main Contributions: In order to understand the impact of partitioning in distributed computing models, we present a topological version/interpretation of the BRS theorem and, on the positive side, show that it can even be applied in the shared memory setting. On the negative side, we prove that one cannot show the impossibility of set agreement in the iterated immediate snapshot model based on a partitioning argument. In more detail:

(1) We provide a topological variant of a slightly generalized version of the BRS theorem and its proof for message-passing systems, which reveals that partitioning is reflected in the protocol complex by a "color splitting" of the algorithm's decision map that separates the sub-complexes representing the partitioned processes. This insight into the structure of the protocol complexes of partitionable models is of independent interest, as the second major contribution of our paper reveals.

(2) We exploit the natural genericity of topological results to translate the BRS theorem to the shared memory model. First, we apply it to d-set agreement in the d-solo asynchronous read-write model, where up to d processes may run solo: We provide a simple and illustrative proof that this problem is not just wait-free impossible, as has been shown in [18] already, but even impossible if just a single process may crash. Second, we used our new insights obtained in (1) to prove that one cannot hope to show the impossibility of wait-free set agreement in the iterated immediate snapshot model using any form of partitioning arguments: For any set of partition-compatible runs (which do not contain runs where all processes always have a complete view), we provide a way to construct a simple algorithm that solves set agreement.

In a nutshell, the results of our paper reveal how the partitioning argument implemented by means of the BRS theorem actually works: It is the "color-split" structure of the resulting protocol complex, which effectively allows to avoid a complex global topological analysis and to perform a simple reduction to the solvability of consensus in some partition instead. In the case of set agreement, the existence of any such splitting already guarantees a solution algorithm. We conjecture that we will be able to come up with similar statements also for other problems, in particular, for general k-set agreement. It is important to note, though, that our results apply only to systems where some partitioning can occur. They cannot hence replace results like [11,20,23] in general.

Related Work: We are not aware of much research that considers non-trivial reductions of k-set agreement to consensus. However, quite some papers, like [21], prove the impossibility of k-set agreement by partitioning the system into more than k sets of processes that decide independently.

For message-passing systems, besides [9], Biely et al. have employed reduction already in [8] to show that consensus is impossible in certain partially synchronous models, and to prove the tightness of the generalized loneliness failure detector $L(k)$ for k-set agreement. Similar reduction arguments were employed in [12] and, in particular, in [10], where certain k-set agreement runs with disjoint participants are pasted together in order to prove the necessity of the generalized quorum failure detector Σ_k for solving k-set agreement. In [4], a reduction to asynchronous set agreement is used to derive a lower bound on the minimum size of a "synchronous window" that is necessary for k-set agreement.

For shared memory systems, [1] shows the wait-free equivalence of k-set agreement and k-simultaneous consensus using read/write atomic registers. The latter problem allows processes to participate, with the same input, in k independent consensus instances that run simultaneously, but requires a decision only in one of those. Whereas this can be seen as an explicit form of partitioning, it is obviously much less general than the partitioning allowed by the BRS theorem.

Regarding different proof techniques for consensus and k-set agreement, the only alternatives to the celebrated impossibility proofs for k-set agreement [11,20,23] known to us, which are all based on algebraic topology resp. different proofs of Sperner's lemma, are the combinatorial impossibility proof for k-set agreement in wait-free environments provided in [6] and the counting-based impossibility of general wait-free colored and colorless tasks in [7]. The latter two results do not generalize to $k = f < n - 1$ crash failures, however.

In [2,3], Alistarh et al. described an approach for proving general impossibilities for wait-free tasks in the iterated immediate snapshot model by introducing *extension-based proofs*. The idea is to consider a game between a prover (constructing a schedule) and the protocol (specified by its decision map Δ, unknown to the prover). By constructing a protocol on the fly, via an adversarial strategy w.r.t. the prover, the authors could prove that the wait-free k-set agreement impossibility cannot be established by an extension-based proof. Part (2) of our work differs from [2] in that we restrict our attention to partitioning arguments, i.e., do not aim at general combinatorial proofs (not to speak of bivalence arguments) as does the latter. This restriction greatly reduces the effort needed to show that partitioning arguments are not sufficient for showing the wait-free set agreement impossibility: Rather than adversarially constructing a protocol following the prover's strategy, the construction of our set agreement protocol just instantiates a simple generic algorithm and is hence relatively easy.

Paper Organization: After a short introduction to topological modeling of distributed systems in Sect. 2, we translate the definitions and concepts underlying the original BRS theorem and prove[1] some basic lemmas in Sect. 3. In Sect. 4, we develop our topological version of the BRS theorem, in Sect. 5, we prove that set agreement can be solved for any set of partition-compatible runs. We conclude in Sect. 6 with some open questions.

2 Topological Modeling of Distributed Systems

We now briefly describe the basics of modeling distributed systems using algebraic topology, see e.g. [17] for a comprehensive introduction. Whereas this powerful approach became particularly popular for asynchronous shared memory systems [20], it is well-suited for message passing systems as well [19].

---

[1] Lacking space forced us to relegate all proofs into the full version [22] of our paper.

We consider a set $\Pi = \{p_1, \ldots, p_n\}$ of processes, each with its own unique identifier, which may suffer from crash or omission failures, i.e., may also lose messages. Except in Sect. 5, where we use the standard iterated snapshot asynchronous shared memory model, we will primarily deal with message passing protocols, where each process $p_i \in \Pi$ has an individual *message buffer* m_j for received messages from every potential sender process $p_j \in \Pi$ in its local state $(p_i, m_1, \ldots, m_n)$, where $(m_1, \ldots, m_n)$ is called p_i's *view*. Note that m_i is assumed to contain p_i's local variables. Valid messages are taken from a possibly infinite set M. Typically, we will consider deterministic full information protocols, where processes send messages that consist of the entire history of local states. We represent "p_2 receives message $m \in M$ from p_1" by appending m to the message buffer of p_2 that corresponds to p_1. We will assume that the initial input is given as a message from a process p_i to itself, and that the decision value is also appended to the message buffer reserved for itself. A global state of the system is a vector of local states, one for each $p_i \in \Pi$.

We define a run of a given protocol as a valid infinite sequence of global states, in which eventually each non-crashing process reaches a final decision state, i.e., a state where it has decided. Note carefully that, since we consider full-history protocols, the final decision state of a process contains its view of the entire run. Clearly, the decision of the process is determined by its local view at the first time it reached a decision state. We call such local views minimal final views. Note that we will assume that processes may still send messages after they have decided; messages that reach a process after a final decision state do not change the process' decision, however. Observe that if run α and run β have the same minimal final views for each process, then they have the same decisions. Therefore, we can restrict our attention to the equivalence classes of runs where two runs are equivalent if all processes have the same minimal final view.

We define a task $T_\Pi = \langle \mathcal{I}, \mathcal{O}, \Delta \rangle$ as a tuple, where $\mathcal{I}$ and $\mathcal{O}$ are chromatic simplicial complexes, which model the valid inputs and outputs for a set Π of processes, and $\Delta : \mathcal{I} \to 2^{\mathcal{O}}$ is a valid decision function that maps valid input configurations to sets of valid output configurations. Both complexes are chromatic, with coloring χ (formally, a simplicial map from the complex to a simplex of matching dimension, i.e., one that maps simplices to simplices) that attaches a unique label (in fact, a process id from Π) to every vertex such that no two neighbors in any 1-simplex have the same label. Note that we will usually write T instead of T_Π for brevity.

More formally, the input complex $\mathcal{I} = \langle V(\mathcal{I}), F(\mathcal{I}) \rangle$ is given by its set of vertices $V(\mathcal{I})$ and its set of faces $F(\mathcal{I})$: $V(\mathcal{I}) = \{(p_i, v_i) \mid p_i \in \Pi, \ v_i \in V_i\}$ where V_i is the set of valid inputs for p_i, and $F(\mathcal{I}) = \{\sigma \subseteq V(\mathcal{I}) \mid \sigma$ is part of a valid input cfg.$\}$.

The output complex $\mathcal{O} = \langle V(\mathcal{O}), F(\mathcal{O}) \rangle$ is given by its set of vertices $V(\mathcal{O})$ and its set of faces $F(\mathcal{O})$: $V(\mathcal{O}) = \{(p_i, v_i) \mid p_i \in \Pi, \ v_i \in \hat{V}_i\}$ where $\hat{V}_i$ is the set of valid outputs for p_i, $F(\mathcal{O}) = \{\sigma \subseteq V(\mathcal{O}) \mid \sigma$ is part of a valid output cfg.$\}$.

The decision function $\Delta : F(\mathcal{I}) \to 2^{F(\mathcal{O})}$ is a function with the property that, for every $\sigma \subseteq \rho$ and $\rho' \in \Delta(\rho)$, there is some $\sigma' \subseteq \rho'$ with $\sigma' \in \Delta(\sigma)$. Moreover,

Δ is a chromatic map, i.e., $\chi(\Delta(\sigma)) \subseteq \chi(\sigma)$, where $\chi(\sigma)$ gives the set of colors of the vertices in σ and $\chi(S) = \bigcup_{\sigma \in S} \chi(\sigma)$ for every set S of simplices.

We define the protocol complex $\mathcal{P}_\mathcal{M} = \langle V(\mathcal{P}_\mathcal{M}), F(\mathcal{P}_\mathcal{M}) \rangle$ for a given protocol P and model $\mathcal{M}$ as $V(\mathcal{P}_\mathcal{M}) = \{(p_i, v_i) \mid p_i \in \Pi, \ v_i \in \overline{V}_i\}$, where $\overline{V}_i$ is the set of valid minimal final views for p_i in protocol P under a given model $\mathcal{M}$. The set of faces is $F(\mathcal{P}_\mathcal{M}) = \{\sigma \subseteq V(\mathcal{P}_\mathcal{M})\}$, where σ corresponds to a valid configuration of minimal final views of a run. The chromatic function $\chi : V(\mathcal{P}_\mathcal{M}) \to \Pi$ is given by the id of each process, that is $\chi(p_i, v_i) = p_i$. The decision map for a protocol $\mu : V(\mathcal{P}_\mathcal{M}) \to 2^{V(\mathcal{O})}$ is a chromatic vertex map that maps final views of a process to valid outputs for a task.

Since the initial input values are self-messages in our model, each run is produced by a unique configuration of initial input values. Therefore, for each task T, there exists a chromatic simplicial map $i_T : F(\mathcal{P}_\mathcal{M}) \to F(\mathcal{I})$ that maps simplices to simplices of matching dimension, such that $i_T(\sigma)$ is the initial input configuration for each process in σ. A protocol P solves a task T in model $\mathcal{M}$ if and only if the decision map for the protocol is a simplicial map that carries Δ, i.e., ensures $\mu(\sigma) \subseteq \Delta(i_T(\sigma))$. Note that, since the decision map μ needs to be chromatic, it is determined by the mapping values at the facets, i.e., the maximal faces. Therefore, the facet decision map $\hat{\mu} : \hat{F}(\mathcal{P}_\mathcal{M}) \to 2^{V(\mathcal{O})}$, which is just the restriction of μ to the facets $\hat{F}(\mathcal{P}_\mathcal{M})$ in $F(\mathcal{P}_\mathcal{M})$, fully determines μ.

3 BRS Basic Definitions

In this section, we recast the foundations of the BRS theorem introduced in [9] in our topological framework. Most of the concepts introduced here allow to relate the runs of different algorithms in different models. After all, our purpose is to reduce a run of a k-set agreement algorithm A in some model $\mathcal{M}$ to a run of a consensus algorithm B in some model $\mathcal{M}'$, with a different set of processes and possibly different synchrony assumptions and failure models. The restricted model only requires that algorithm A is computationally compatible with $\mathcal{M}'$, i.e, that A can be executed in $\mathcal{M}'$. Since it is primarily the number of processes in $\mathcal{M}'$ and $\mathcal{M}'$ that matter here, we will follow [9]and sloppyly write $\mathcal{M} = \langle \Pi \rangle$ and $\mathcal{M}' = \langle D \rangle$ in the sequel. Note carefully, however, that the runs in $\mathcal{M}'$ do not necessarily correspond to runs in $\mathcal{M}$ and vice versa, due to $|\langle D \rangle| \neq |\langle \Pi \rangle|$ and the usually different synchrony assumptions and failure models.

A pivotal concept here is a restricted algorithm.

Definition 1 (Restricted algorithm). *Let A be an algorithm for a model $\mathcal{M} = \langle \Pi \rangle$ that consists of the set of processes Π, and $D \subseteq \Pi$ a nonempty set of processes. Consider a restricted model $\mathcal{M}' = \langle D \rangle$. To restrict algorithm A for model $\mathcal{M}$ to an algorithm for model $\mathcal{M}'$, we just drop all messages sent from D to the outside. We call the restricted algorithm $A_{|D} = B$.*

Restricted algorithms induce protocol complexes with specific properties. However, even if protocol B corresponds to a restriction of A to D, since message buffers for processes not in D are not present in $\mathcal{B}$ (the protocol complex of B),

$\mathcal{B}$ is strictly different from any protocol complex that includes Π in its set of processes. Therefore, we need to define a way to extend the views in $\mathcal{B}$ in a way that could possibly match a protocol complex with processes Π executed in $\mathcal{M}$. The natural way of doing this is by adding empty message buffers denoted by $\perp$ for any process not in D.

Definition 2 (Protocol complexes of restricted algorithms). *Given an algorithm A for $\mathcal{M}$ and a restricted algorithm B for $\mathcal{M}'$, let $\mathcal{B}$ be the protocol complex for B executed in $\mathcal{M}'$. We define the extended complex of B with respect to Π, $\mathcal{A}_D$, as follows: $V(\mathcal{A}_D) = \{(p, w, \perp, \ldots, \perp) \mid (p, w) \in V(\mathcal{B})\}$ and $\perp$ represents empty message buffers for processes in $\Pi \backslash D$, $F(\mathcal{A}_D) = \{\sigma \subseteq V(\mathcal{A}_D) \mid \exists \hat{\sigma} \in F(\mathcal{B}), (p, w, \perp, \ldots, \perp) \in \sigma \Rightarrow (p, w) \in \hat{\sigma}\}$.*

The following Lemma 1 shows that $\mathcal{A}_D$ is isomorphic to $\mathcal{B}$, i.e., $\mathcal{A}_D$ is an "extended view copy" of $\mathcal{B}$. This isomorphic copy $\mathcal{A}_D$ of the protocol complex $\mathcal{B}$ will turn out to be essential, since it corresponds to a subcomplex of $\mathcal{A}$ under certain conditions (Definition 4). Note carefully that, per se, this is not necessarily the case as, e.g., the synchrony model for $\mathcal{M}$ may forbid empty message buffers.

Lemma 1 (Isomorphic complex). *Let $\mathcal{A}_D$ and $\mathcal{B}$ be as defined above. Then there exists a chromatic bijective simplicial map $\mu : \mathcal{A}_D \to \mathcal{B}$.*

The following definition captures the notion of indistinguishability of runs:

Definition 3 (Indistinguishability of runs). *Runs α and β are indistinguishable for a process p if p has the same sequence of states in α and β until p decides. For a non-empty set D of processes, we say that $\alpha \overset{D}{\sim} \beta$ if α is indistinguishable from β until decision for all $p \in D$.*

Note that since p has the same sequence of states until decision for runs α and β, then the minimal final view for p (that contains the full history) is the same for both runs α and β. This means that the simplices σ_α and σ_β that correspond to runs α and β share vertex $(p, s_w, m_1, \ldots, m_k)$, where $(s_w, m_1, \ldots, m_k)$ corresponds to the minimal final view of p at both runs α and β. This translates naturally to $D\text{-skel}(\sigma_\alpha) = D\text{-skel}(\sigma_\beta)$, where $D\text{-skel}(\sigma_\alpha) = \{(p, w) \in \sigma_\alpha : p \in D\}$.

Definition 4 (Compatibility of Runs). *Let $\mathcal{R}$ and $\mathcal{R}'$ be sets of runs, possibly from system models with different synchrony assumptions and failure models. Runs $\mathcal{R}'$ are compatible with runs $\mathcal{R}$ for processes in D, denoted by $\mathcal{R}' \preceq_D \mathcal{R}$, if $\forall \alpha \in \mathcal{R}' \exists \beta \in \mathcal{R} : \alpha \overset{D}{\sim} \beta$.*

If $\mathcal{R}$ and $\mathcal{R}'$ are sets of runs for the same protocol A, and in the same model $\mathcal{M}$, then both induce subcomplexes of a common protocol complex $\mathcal{A}$. We will call those subcomplexes $\overline{\mathcal{R}}$ and $\overline{\mathcal{R}'}$ respectively. We define $D\text{-skel}(\overline{\mathcal{R}'})$ as the subcomplex of $\overline{\mathcal{R}'}$ where all vertices correspond to processes of D with views from $\overline{\mathcal{R}'}$. Given these definitions, it is clear that $\mathcal{R}' \preceq_D \mathcal{R}$ if and only if $D\text{-skel}(\overline{\mathcal{R}'}) \subseteq D\text{-skel}(\overline{\mathcal{R}})$. Note that the applicability of Definition 3 is limited, as it only works for sets of runs from the same protocol in the same model. However, we will give a more general definition below, which provides some correspondence between runs of *different* protocols in *different* models: Herein, $\mathcal{M}$ and $\mathcal{M}'$ only need to be computationally compatible, but may otherwise differ in the number of processes, synchrony assumptions, failure models, etc.

Definition 5 (D-View embedding). *Let A and B be protocols with a non-empty set of common processes S, and $D \subseteq S$ with $s = |S|$, and let $\mathcal{A}$ and $\mathcal{B}$ be the protocol complexes corresponding to A's runs in model $\mathcal{M}$ (with $s + r$ processes, $r \geq 0$) and B's runs in model $\mathcal{M}'$ (with $s + k$ processes, $k \geq 0$), respectively. Consider sets of runs $\mathcal{R}$ and $\mathcal{R}'$ from protocol A in model $\mathcal{M}$ and B in model $\mathcal{M}'$ respectively, and the corresponding subcomplexes $\overline{\mathcal{R}} \subseteq \mathcal{A}$ and $\overline{\mathcal{R}'} \subseteq \mathcal{B}$. We say that $\overline{\mathcal{R}'}$ is D-view embedded in $\overline{\mathcal{R}}$, if for every $(p, w_s, m_1, \dots, m_k) \in D\text{-skel}(\overline{\mathcal{R}'})$ with w_s denoting the message buffers for the processes in S the following holds for every $1 \leq i \leq k$: (i) $p_i \notin S \Rightarrow m_i = \bot$; (ii) There exists $(p, w_s, m'_1, \dots m'_r) \in V(\overline{\mathcal{R}})$ such that*

$$m'_j = \begin{cases} m_j & \text{if } p_j \in S, \\ \bot & \text{if } p_j \notin S; \end{cases}$$

(iii) $\mu : D\text{-skel}(\overline{\mathcal{R}'}) \to \overline{\mathcal{R}}$ defined by $(p, w_s, m_1, \dots, m_k) \mapsto (p, w_s, m'_1, \dots, m'_r)$ is a simplicial map. Note that μ is an embedding of $D\text{-skel}(\overline{\mathcal{R}'})$, i.e. an injective simplicial map.

If both sets of runs come from the same protocol ($B = A$) and the same synchrony model ($\mathcal{M} = \mathcal{M}'$), then the embedding is given by the inclusion $\iota : D\text{-skel}(\overline{\mathcal{R}'}) \to \overline{\mathcal{R}}$ with $\iota(p, w) = (p, w)$. This matches with the previous observation that $\mathcal{R}' \preceq_D \mathcal{R}$ if and only if $D\text{-skel}(\overline{\mathcal{R}'}) \subseteq D\text{-skel}(\overline{\mathcal{R}})$ in this case. More generally, we can formulate compatibility of runs in terms of D-view embeddings.

Lemma 2 (Compatible runs are D-view embedded). *Let $\mathcal{R}$ and $\mathcal{R}'$ be sets of runs from algorithms A and B in model $\mathcal{M}$ and $\mathcal{M}'$ respectively. Let S be the set of common processes for $\mathcal{R}$ and $\mathcal{R}'$ and $D \subseteq S$. Then $\mathcal{R}' \preceq_D \mathcal{R} \Leftrightarrow \overline{\mathcal{R}'}$ is D-view embedded in $\overline{\mathcal{R}}$.*

The following Definition 6 is crucial for expressing the consequences of partitioning in our topological setting. Essentially, it says that it is reflected by a splitting of the decision map.

Definition 6 (Decision map split). *Let $\mathcal{A}$ be the protocol complex for a given algorithm A on a model $\mathcal{M} = \langle \Pi \rangle$ and $\mathcal{A}'$ a non-empty subcomplex of $\mathcal{A}$. Let $D \subseteq \Pi$ be a set of processes in $\mathcal{M}$, $\overline{D} = \Pi \backslash D$ and $B = A_{|D}$ the restriction of algorithm A to D in a given model $\mathcal{M}' = \langle D \rangle$ with possibly different synchrony assumptions and failure models, resulting in protocol complex $\mathcal{B}$. We say that D splits the decision map of A at $\mathcal{A}'$ with respect to $\mathcal{M}'$ if $\mathcal{B}$ is D-view embedded in $\mathcal{A}'$ and*

$$\mu_{|\mathcal{A}'}(p, w) = [\mu_D * \mu_{|\overline{D}}](p, w) := \begin{cases} \mu_D(p, w) & \text{if } p \in D, \\ \mu_{|\overline{D}}(p, w) & \text{if } p \in \overline{D}. \end{cases}$$

Herein, $\mu_{|\mathcal{A}'}$ is the decision map μ of $\mathcal{A}$ restricted to $\mathcal{A}'$, μ_D is the decision map for restricted algorithm B at the extended complex $\mathcal{A}_D$, and $\mu_{|\bar{D}}$ is the decision map μ restricted to $\bar{D}$-skel$(\mathcal{A}')$.

An illustration of decision map splitting for a fixed $D = \{r, q\}, \bar{D} = \{p\}$ can be found in Fig. 1b. Note that the subcomplex for D (at the bottom) is the full 1-round subdivided complex for 2 processes, whereas the subcomplex for $\bar{D}$ represents a process that only hears from itself. Figure 1a shows the analogous splitting in the full one-round complex.

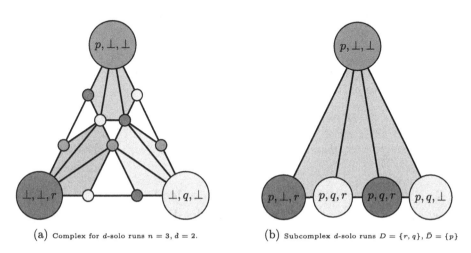

(a) Complex for d-solo runs $n = 3, d = 2$.   (b) Subcomplex d-solo runs $D = \{r, q\}, \bar{D} = \{p\}$

Fig. 1. 1-round protocol complex of d-solo runs for $n = 3, d = 2$: (a) full complex, (b) zoom-in for $D = \{r, q\}, \bar{D} = \{p\}$. Nodes represent local process states (views), with the color encoding the respective process and with the 3-tuple encoding (m_p, m_q, m_r).

The following Lemma 3 shows that decision map splitting is equivalent to the extended complex of the restricted algorithm being equal to the corresponding D-skeleton:

Lemma 3 (Decision map splitting condition). *Let $\mathcal{A}$ be the protocol complex for a given algorithm A in a model $\mathcal{M} = \langle \Pi \rangle$ and $\mathcal{A}'$ a non-empty subcomplex of $\mathcal{A}$. Let $D \subseteq \Pi$ be a set of processes in $\mathcal{M}$, $B = A_{|D}$ a restriction of algorithm A to D in a model $\mathcal{M}'$ with possibly different synchrony assumptions and failure assumptions, $\mathcal{A}_D$ the extended complex of B with respect to Π and μ be the decision map for $\mathcal{A}$, then D splits μ at $\mathcal{A}'$ with respect to $\mathcal{M}' \Leftrightarrow \mathcal{A}_D = D$-skel$(\mathcal{A}')$.*

We will show in the next section that decision map splitting is equivalent to finding a partition of Π into $D, \bar{D}$ and a set of runs where D decides independently from $\bar{D}$. However, notice that $\bar{D}$ could use information from D to decide.

4   Topological BRS Theorem

Since we have already established an equivalence between topological conditions and run compatibility, we can proceed to state a slightly more general version of the BRS theorem from a topological perspective. The BRS theorem requires that conditions (A)–(D), as stated in Theorem 1 below, hold, in order to guarantee the k-set agreement impossibility. In the following lemmas, we will state topological properties that are slightly weaker than the original conditions (A)–(D).

Let $\mathcal{M} = \langle \Pi \rangle$ be a model and A an algorithm that runs in $\mathcal{M}$. Assume that each $p \in \Pi$ starts with an input value taken from a set of at least k different values.[2] Also assume that there is a distinguished set $D \subseteq \Pi$, and a partition of $\Pi \backslash D = \overline{D}$ given by $D_1, \ldots, D_{k-1}$. Let $\{v_1, \ldots, v_{k-1}\}$ be a fixed set of different values. Let $\mathcal{M}_A$ denote the runs of algorithm A in model $\mathcal{M}$.

Theorem 1 (Original BRS theorem [9]). *We consider the following runs of algorithm A in model $\mathcal{M}$ and a restricted model $\mathcal{M}' = \langle D \rangle$:*

dec-D *Any $p_j \in D$ receives no messages from any process in $\overline{D}$ until every process in D has decided.*
dec-$\overline{D}$ *For every D_i, there is some $q \in D_i$ that decides v_i, which was proposed by some $p \in \overline{D}$.*

$\mathcal{R}_{(D)}$ *denotes the set of runs from $\mathcal{M}$ where* **dec-D** *holds,* $\mathcal{R}_{(D,\overline{D})}$ *denotes the set of runs from $\mathcal{M}$ where both* **dec-D** *and* **dec-$\overline{D}$** *hold. If all of the following conditions (A)* $\mathcal{R}_{(D)}$ *is nonempty, (B)* $\mathcal{R}_{(D)} \preceq_D \mathcal{R}_{(D,\overline{D})}$, *(C) Consensus is not solvable in $\mathcal{M}'$, (D)* $\mathcal{M}'_{A_{|D}} \preceq_D \mathcal{M}_A$, *hold, then A does not solve k-set agreement.*

We generalize the statement of the BRS theorem by slightly relaxing conditions (A)–(D):

Lemma 4 (Condition equivalence). *Let $\mathcal{M}$, $\mathcal{M}'$, A, D, $\overline{D}$ and D_i be as defined for the BRS theorem. Then, (A)-(D) imply (A') $\mathcal{R}_{(D)}$ is nonempty, (B') Consensus is not solvable in $\mathcal{M}'$, (C') $\mathcal{M}'_{A_{|D}} \preceq_D \mathcal{R}_{(D,\overline{D})}$.*

The following Lemma 5 is the key technical lemma for the topological equivalence of a slightly stronger version of the BRS theorem established in Theorem 2 below.

Lemma 5 (BRS equivalence). *Let $\mathcal{M}$, $\mathcal{M}'$, A, D, $\overline{D}$ and D_i be as defined for the BRS theorem. Then (A')–(C') are equivalent to the following :*

(1) There exists a non empty subcomplex $\mathcal{A}'$ of $\mathcal{A}$ such that $D\text{-skel}(\mathcal{A}') = \mathcal{A}_D$ (the extended complex for $A_{|D}$ with respect to $\mathcal{M}$).
(2) For each D_i, the decision map $\mu_{|\mathcal{A}'}$ maps every view from D_i into a decision configuration that includes v_i as a decision value.

---

[2] Note that the original BRS theorem actually assumed that every process starts with a unique value.

(3) Each v_i is the input value for some process $p \in \overline{D}$ at subcomplex $\mathcal{A}'$.
(4) Consensus is not solvable in $\mathcal{M}' = \langle D \rangle$.

The following Theorem 2 provides the main result of this section, the topological version of a generalization of the BRS theorem.

Theorem 2 (Decision split theorem). *Let $\mathcal{M} = \langle \Pi \rangle$ be a model and A an algorithm that runs in $\mathcal{M}$. Let $\mathcal{A}$ be the protocol complex of A, $\mathcal{M}' = \langle D \rangle$ be a model with D a subset of Π, and μ the decision map for $\mathcal{A}$. Assume that the following conditions hold:*

(a) There exists a non-empty subcomplex $\mathcal{A}'$, such that D splits μ at $\mathcal{A}'$ with respect to $\mathcal{M}'$.
(b) Consensus is not solvable in $\mathcal{M}'$.
(c) Processes in $\overline{D} = \Pi \backslash D$ always decide at least $k-1$ input values from $\overline{D}$ for runs in $\mathcal{A}'$.

Then, A does not solve k-set agreement.

We conclude this section with some corollaries of Theorem 2.

Corollary 1. *Let $\mathcal{M}$, $\mathcal{M}'$, A, D, $\overline{D}$ and D_i be as defined for the original BRS theorem (Theorem 1). If conditions (1)–(4) given in Lemma 5 hold, then A does not solve k-set agreement.*

Corollary 2. *Let $\mathcal{M}$, $\mathcal{M}'$, A, D, $\overline{D}$ and D_i be as defined for the original BRS theorem (Theorem 1). If conditions (A')–(C') in Lemma 4 hold, then A does not solve k-set agreement.*

5   Partition Compatibility in Shared Memory and Set Agreement

In the previous sections, we developed a topological version of the BRS theorem [9], which allows to reduce the impossibility of k-set agreement to the impossibility of consensus in a wide variety of message-passing systems. In this section, we will transfer some of the resulting insights to the standard *asynchronous shared memory* (ASM) model. We thus consider a set of processes $\Pi = \{p_1, \ldots, p_n\}$ and a shared snapshot memory $M = (e, m_1, \ldots, m_n)$, where e is a buffer for global shared variables [not used in our protocols], and each m_i corresponds to the local memory portion of process p_i in the snapshot memory. To simplify our reasoning, we will assume that the protocols are full information immediate snapshot layered protocols, which does not change the computability power of our algorithms [17,20].

In this model, each process executes a predefined number r of asynchronous rounds, called layers. Each layer i consists of concurrent write-read snapshots. A write-read snapshot at layer i consists of writing the full view (the complete local state) of a process into its corresponding part of the memory, and immediately after writing, taking a snapshot of the views from processes at the same layer i. The initial view of a process consists only of its input value; therefore, during the first round, each process only writes its input value to the shared memory. Note that each process' current view contains the history of previous views, so we need not be concerned with overwriting previous views in the shared memory. The final view of a process in an r-round protocol is its view after round r, and the protocol's decision map μ maps the process's final view to some output value.

Definition 7 (Views). *The view for a given process p at round k is defined as follows: If $k = 0$, the view consists of a tuple (p, s, v), where p is the process id, s is the initial local state of p and v is the input value for p. If $k > 0$, the view consists of a tuple $(p, s, v_1, \ldots, v_n)$, where each v_j corresponds to either the view of process p_j at the end of round $k - 1$ if the write-read execution for round k happened before or at the same time as the write-read execution for process p, or else $\perp$, which represents that p finished its write-read in round k before p_j.*

Note that this definition of the processes' views is extremely useful, since it has a nice combinatorial structure: Herlihy and Shavit showed in [20] that there is an isomorphism between the standard chromatic subdivision of the input complex and the protocol complex for a general 1-layer immediate snapshot protocol. Since each layer i is only determined by the previous layers and the scheduling of layer i, it follows by induction that a $k+1$ layered protocol complex corresponds to the $k + 1$-th chromatic subdivision of the input complex.

We start our considerations by using the BRS theorem to show that d-set agreement cannot be implemented in the *d-solo model* introduced in [18] if at most d processes may crash. In the d-solo model for asynchronous shared memory, up to d processes may run solo, i.e., have no *other* process in their view in every round of a run. The wait-free 1-solo model is equivalent to asynchronous read-write shared memory, and any d-solo model can be simulated in the 1-solo model. Since d-set agreement for $d < n$ cannot be implemented in the wait-free read-write model [20], it cannot be implemented in the wait-free d-solo model either. The following Theorem 3 shows that this does not change if one strengthens the model by allowing only up to a single crash.

Theorem 3 (d-set agreement impossibility in d-solo model). *In the d-solo model, it is impossible to solve d-set agreement if just a single process may crash, not even with a colored, i.e., non-anonymous, algorithm.*

The remaining part of this section is devoted to the very different result of translating the insights gained from the topological version of the BRS theorem to the way of how partitioning is reflected in protocol complexes. More specifically, we show that set agreement can always be solved in ASM systems with *partition-compatible runs*, which are runs that allow some processes to hide their information from others. Our key insight is that partition-compatible runs "pierce" a hole at the center of the protocol complex, which allows the border to be retracted continuously to the center, thereby allowing to solve set agreement.

One implication of this result is that partitioning arguments cannot be used to prove the set-agreement impossibility in general: Assume that there is such a proof, which necessarily relies on some set of partition-compatible runs. After all, any partitioning argument rules out runs where every process has a complete view of all other processes in all iterations. Since we can construct a correct set agreement protocol for this set of runs, however, we have a contradiction.

The central idea of partitioning arguments, which exploit limited communication between sets of processes, stimulated the notion of partition-compatible views and (sets of) runs:

Definition 8 (Partition compatibility). *A view $(p_i, s, v_1, \ldots, v_n)$ of a process p_i at the end of some round $k \geq 1$ is called partition-compatible, if p_i did not get information from some p_j during round k, i.e., when $v_j = \perp$. A set of runs S is partition-compatible if, for every run $\alpha \in S$, there exists some participating process p_i and a round k at the end of which p_i's view is partition-compatible.*

Note that it is the presence of the complete-view run, i.e., the presence of the complete-view simplex in the corresponding protocol complex, that makes a set of runs *not* partition compatible. This, once made, obvious observation gave us the idea for the following simple set agreement protocol:

We define the following 1-layer protocol P for the immediate snapshot model with a set of processes $\Pi = \{p_0, \ldots, p_n\}$.

Definition 9 (1-layer protocol). *Let $p_i \in \Pi = \{p_0, \ldots, p_n\}$ be a process and $(m_0, \ldots, m_n)$ its view. Since we are considering the iterated immediate snapshot model, the protocol is determined just by the number of communication rounds (1 in this case) and the decision map. We define*

$$\mu(p_i, m_0, \ldots, m_n) = \begin{cases} m_i & \text{if } \perp = m_j \text{ for some } j \in \{1, \ldots, n\}, \\ m_{(i+1) \bmod n+1} & \text{otherwise.} \end{cases}$$

Obviously, since μ always chooses the input value for p_i unless all other input values have been observed, μ satisfies the validity condition. In fact, we can prove the following result:

Lemma 6 (Correctness of the 1-round protocol). *Let S be a set of partition compatible runs for a 1-round immediate snapshot protocol. Then, μ solves set agreement in S.*

An immediate consequence of Lemma 6 is that a partitioning argument cannot be used for showing n-set agreement impossibility for 1-round immediate snapshot protocols. However, in order to show that a partitioning argument cannot be used for a general shared memory protocol, we need to show this result for any number of layers. In order to do so, we define a k-round set agreement protocol for any value of k:

Definition 10 (k-layer protocol). *Consider a general k-layered immediate snapshot protocol. Let $p_i \in \Pi$ be a process, and $(m_0, \ldots, m_n)$ its final view. We* denote $L^\ell(m_0, \ldots, m_n)$ as the view for p_i at layer ℓ. Alternatively, if α is a run of an r-layered protocol and $s < r$ then L^s_α denotes the s-layered protocol run induced by α. We say that a view $v = (m_0, \ldots, m_n)$ of a process at a layer s is incomplete if either $\bot \in v$ or if there exists $\ell < s$ and $0 \leq r \leq n$ such that $L^\ell(m_r)$ is an incomplete view; recall that m_r is p_r's view in round $s - 1$ or $\bot$. Let

$$\mu^1 = \mu,$$

$$\mu^{k+1}(p_i, m_0, \ldots, m_n) = \begin{cases} \mu^k(p_i, L^k(m_0, \ldots, m_n)) & \text{if } (m_0, \ldots, m_n) \text{ is incomplete,} \\ \mu^k(p_{i+1}, L^k(m_0, \ldots, m_n)) & \text{otherwise.} \end{cases}$$

We can prove the following result:

Lemma 7 (Correctness of the k-round protocol). *Let S be a set of partition compatible runs for a k-round immediate snapshot protocol. Then, the decision map μ^k solves n-set agreement in S.*

We can thus state the main result of this section:

Theorem 4. *Let S be any set of partition compatible runs in the iterated immediate snapshot model. Then, there exists a protocol P that solves set agreement for any run in S.*

6   Conclusions

We developed a topological version of a generalization of the BRS theorem. Our findings reveal that partitioning is reflected by a "color splitting" of the algorithm's decision map, which separates the sub-complexes representing the partitioned processes. We used these insights to show that the impossibility of wait-free set agreement in the layered immediate snapshot model cannot be proved using partitioning arguments: For any set of partition compatible runs it is possible to construct a simple protocol that solves set agreement.

Extending the latter to k-set agreement and investigating the applicability of the BRS theorem to alternative shared memory systems remain as open questions.

References

1. Afek, Y., Gafni, E., Rajsbaum, S., Raynal, M., Travers, C.: The k-simultaneous consensus problem. Distrib. Comput. **22**(3), 185–195 (2010). https://doi.org/10.1007/s00446-009-0090-8
2. Alistarh, D., Aspnes, J., Ellen, F., Gelashvili, R., Zhu, L.: Why extension-based proofs fail. CoRR abs/1811.01421 (2018)
3. Alistarh, D., Aspnes, J., Ellen, F., Gelashvili, R., Zhu, L.: Why extension-based proofs fail. In: Proceedings of the 51st Annual ACM SIGACT Symposium on Theory of Computing, STOC 2019, Phoenix, AZ, USA, 23–26 June 2019, pp. 986–996 (2019). https://doi.org/10.1145/3313276.3316407

4. Alistarh, D., Gilbert, S., Guerraoui, R., Travers, C.: Brief announcement: new bounds for partially synchronous set agreement. In: Lynch, N.A., Shvartsman, A.A. (eds.) DISC 2010. LNCS, vol. 6343, pp. 404–405. Springer, Heidelberg (2010). https://doi.org/10.1007/978-3-642-15763-9_40
5. Attiya, H., Bar-Noy, A., Dolev, D.: Sharing memory robustly in message-passing systems. J. ACM **42**(1), 124–142 (1995). https://doi.org/10.1145/200836.200869
6. Attiya, H., Castañeda, A.: A non-topological proof for the impossibility of k-set agreement. Theor. Comput. Sci. **512**, 41–48 (2013)
7. Attiya, H., Paz, A.: Counting-based impossibility proofs for renaming and set agreement. In: Aguilera, M.K. (ed.) DISC 2012. LNCS, vol. 7611, pp. 356–370. Springer, Heidelberg (2012). https://doi.org/10.1007/978-3-642-33651-5_25
8. Biely, M., Robinson, P., Schmid, U.: Weak synchrony models and failure detectors for message passing (k-)set agreement. In: Abdelzaher, T., Raynal, M., Santoro, N. (eds.) OPODIS 2009. LNCS, vol. 5923, pp. 285–299. Springer, Heidelberg (2009). https://doi.org/10.1007/978-3-642-10877-8_23
9. Biely, M., Robinson, P., Schmid, U.: Easy impossibility proofs for k-set agreement in message passing systems. In: Fernàndez Anta, A., Lipari, G., Roy, M. (eds.) OPODIS 2011. LNCS, vol. 7109, pp. 299–312. Springer, Heidelberg (2011). https://doi.org/10.1007/978-3-642-25873-2_21
10. Bonnet, F., Raynal, M.: On the road to the weakest failure detector for k-set agreement in message-passing systems. Theor. Comput. Sci. **412**(33), 4273–4284 (2011). https://doi.org/10.1016/j.tcs.2010.11.007
11. Borowsky, E., Gafni, E.: Generalized FLP impossibility result for t-resilient asynchronous computations. In: STOC 1993: Proceedings of the 25th Annual ACM Symposium on Theory of Computing, pp. 91–100. ACM, New York (1993). https://doi.org/10.1145/167088.167119
12. Bouzid, Z., Travers, C.: (anti-$\Omega^x \times \Sigma_z$)–based k-set agreement algorithms. In: Lu, C., Masuzawa, T., Mosbah, M. (eds.) OPODIS 2010. LNCS, vol. 6490, pp. 189–204. Springer, Heidelberg (2010). https://doi.org/10.1007/978-3-642-17653-1_16
13. Brewer, E.A.: Towards robust distributed systems (abstract). In: Proceedings of the Nineteenth Annual ACM Symposium on Principles of Distributed Computing, PODC 2000. ACM, New York (2000). https://doi.org/10.1145/343477.343502
14. Fich, F., Ruppert, E.: Hundreds of impossibility results for distributed computing. Distrib. Comput. **16**, 121–163 (2003). https://doi.org/10.1007/s00446-003-0091-y
15. Fischer, M.J., Lynch, N.A., Paterson, M.S.: Impossibility of distributed consensus with one faulty process. J. ACM **32**(2), 374–382 (1985)
16. Gilbert, S., Lynch, N.: Brewer's conjecture and the feasibility of consistent, available, partition-tolerant web services. SIGACT News **33**(2), 51–59 (2002). https://doi.org/10.1145/564585.564601
17. Herlihy, M., Kozlov, D.N., Rajsbaum, S.: Distributed Computing Through Combinatorial Topology. Morgan Kaufmann, Burlington (2013). https://store.elsevier.com/product.jsp?isbn=9780124045781
18. Herlihy, M., Rajsbaum, S., Raynal, M., Stainer, J.: From wait-free to arbitrary concurrent solo executions in colorless distributed computing. Theor. Comput. Sci. **683**, 1–21 (2017). https://doi.org/10.1016/j.tcs.2017.04.007
19. Herlihy, M., Rajsbaum, S., Tuttle, M.R.: An overview of synchronous message-passing and topology. Electron. Notes Theor. Comput. Sci. **39**(2), 1–17 (2000). https://doi.org/10.1016/S1571-0661(05)01148-5
20. Herlihy, M., Shavit, N.: The topological structure of asynchronous computability. J. ACM **46**(6), 858–923 (1999). https://doi.org/10.1145/331524.331529

21. de Prisco, R., Malkhi, D., Reiter, M.: On k-set consensus problems in asynchronous systems. IEEE Trans. Parallel Distrib. Syst. **12**(1), 7–21 (2001). https://doi.org/10.1109/71.899936

22. Rincon, H., Winkler, K., Schmid, U., Rajsbaum, S.: A topological view of partitioning arguments: reducing k-set agreement to consensus. Technical report TUW-281149, TU Wien (2019). https://publik.tuwien.ac.at/files/publik_281149.pdf

23. Saks, M., Zaharoglou, F.: Wait-free k-set agreement is impossible: the topology of public knowledge. SIAM J. Comput. **29**(5), 1449–1483 (2000). https://doi.org/10.1137/S0097539796307698

Logarithmic Expected-Time Leader Election in Population Protocol Model

Yuichi Sudo[1(✉)], Fukuhito Ooshita[2], Taisuke Izumi[3], Hirotsugu Kakugawa[4], and Toshimitsu Masuzawa[1]

[1] Osaka University, Suita, Japan
y-sudou@ist.osaka-u.ac.jp
[2] Nara Institute of Science and Technology, Ikoma, Japan
[3] Nagoya Institute of Technology, Nagoya, Japan
[4] Ryukoku University, Otsu, Japan

Abstract. In this paper, we present the *first* leader election protocol in the population protocol model that stabilizes within $O(\log n)$ parallel time in expectation with $O(\log n)$ states per agent, where n is the number of agents. Given a rough knowledge m of the population size n such that $m \geq \log_2 n$ and $m = O(\log n)$, the proposed protocol guarantees that exactly one leader is elected and the unique leader is kept forever thereafter.

1 Introduction

We consider the *population protocol* (PP) model [1] in this paper. A network called *population* consists of a large number of finite-state automata, called *agents*. Agents make *interactions* (i.e., pairwise communication) each other by which they update their states. The interactions are opportunistic, that is, they are unpredictable. Agents are strongly anonymous: they do not have identifiers and they cannot distinguish their neighbors with the same states. As with the majority of studies on population protocols [1–11], we assume that the network of agents is a complete graph and that the scheduler selects an interacting pair of agents at each step uniformly at random.

In this paper, we focus on leader election problem, which is one of the most fundamental and well studied problems in the PP model. The leader election problem requires that starting from a specific initial configuration, a population reaches a configuration in which exactly one leader exists and the population keeps that unique leader thereafter.

There have been many works which study the leader election problem in the PP model (Tables 1 and 2). Angluin et al. [1] gave the first leader election protocol, which stabilizes in $O(n)$ parallel time in expectation and uses only constant space of each agent, where n is the number of agents and "parallel

This work was supported by JSPS KAKENHI Grant Numbers 17K19977, 18K18000, 19H04085, and 19K11826 and JST SICORP Grant Number JPMJSC1606.

M. Ghaffari et al. (Eds.): SSS 2019, LNCS 11914, pp. 323–337, 2019.
https://doi.org/10.1007/978-3-030-34992-9_26

time" means the number of steps in an execution divided by n. If we stick to constant space, this linear parallel time is optimal; Doty and Soloveichik [8] showed that any constant space protocol requires linear parallel time to elect a unique leader. Alistarh and Gelashvili [2] made a breakthrough in 2015; they achieve poly-logarithmic stabilization time ($O(\log^3 n)$ parallel time) by increasing the number of states from $O(1)$ to only $O(\log^3 n)$. Thereafter, the stabilization time has been improved by many studies [4–7,12]. Gąsieniec et al. [6] gave a state-of-art protocol that stabilizes in $O(\log n \cdot \log\log n)$ parallel time with only $O(\log\log n)$ states. Its space complexity is optimal; Alistarh et al. [3] shows that any leader election algorithm with $o(n/(\text{polylog } n))$ parallel time requires $\Omega(\log\log n)$ states. Michail et al. [7] gave a protocol with $O(\log n)$ parallel time but with a linear number of states. Those protocols with non-constant number of states [2–6,12] are not *uniform*, that is, they require some rough knowledge of n. For example, in the protocol of [5], an $\Theta(\log\log n)$ value must be hard-coded to set the maximum value of one variable (named l in that paper). One can find detailed information about the leader election in the PP model in two survey papers [13,14].

The stabilization time of [7] is optimal; any leader election algorithm requires $\Omega(\log n)$ parallel time even if it uses any large number of states and assumes the exact knowledge of population size n [9]. At the beginning of an execution, all the agents are in the same initial state specified by a protocol. Therefore, simple analysis on Coupon Collector's problem shows that we cannot achieve $o(\log n)$ parallel stabilization time if an agent in the initial state is a leader. The lower bound of [9] shows that we cannot achieve $o(\log n)$ parallel time even if we define the initial state such that all the agents are non-leaders initially.

Our Contribution. In this paper, we present the *first* time-optimal leader election protocol P_{LL} with sub-polynomial number of states. Specifically, P_{LL} stabilizes in $O(\log n)$ parallel time and uses only $O(\log n)$ states per agent. Compared to a state of the art protocol [6], P_{LL} achieves shorter (and best possible) stabilization time but uses larger space of each agent. Compared to [7], P_{LL} achieves drastically small space while maintaining the same (and optimal) stabilization time. The protocol P_{LL} is non-uniform as with the existing non-constant space protocols; it requires a rough knowledge m of n such that $m \geq \log_2 n$ and $m = \Theta(\log n)$.

We give P_{LL} as an asymmetric protocol in the main part of this paper *only for simplicity* of presentation and analysis of stabilization time. Actually, we can change P_{LL} to a symmetric protocol, which we discuss in Sect. 5. In particular, that section proposes the first implementation of totally independent and fair (*i.e.*, unbiased) coin flips in the symmetric version of the PP model. Although the implementation of coin flips in [3] is almost independent and fair, the totally independent and fair coin clips achieved in this paper can contribute to a simple analysis in a variety kind of protocols in the PP model.

2   Preliminaries

A *population* is a network consisting of *agents*. We denote the set of all the agents by V and let $n = |V|$. We assume that a population is complete graph, thus every pair of agents (u, v) can interact, where u serves as the *initiator* and v serves as the *responder* of the interaction. Throughout this paper, we use the phrase "with high probability" to denote "with probability $1 - O(n^{-1})$".

A *protocol* $P(Q, s_{init}, T, Y, \pi_{out})$ consists of a finite set Q of states, an initial state $s_{init} \in Q$, a transition function $T : Q \times Q \to Q \times Q$, a finite set Y of output symbols, and an output function $\pi_{out} : Q \to Y$. Every agent is in state s_{init} when an execution of protocol P begins. When two agents interact, T determines their next states according to their current states. The *output* of an agent is determined by π_{out}: the output of an agent in state q is $\pi_{out}(q)$. As with all papers listed in Table 1 except for the one of [1], we assume that a rough knowledge of n is available. Specifically, we assume that an integer m such that $m \geq \log_2 n$ and $m = \Theta(\log n)$ is given, thus we can design $P(Q, s_{init}, T, Y, \pi_{out})$ using this input m, *i.e.*, the parameters Q, s_{init}, T, Y, and π_{out} can depend on m.

A *configuration* is a mapping $C : V \to Q$ that specifies the states of all the agents. We define $C_{init,P}$ as the configuration of P where every agent is in state s_{init}. We say that a configuration C changes to C' by the interaction $e = (u, v)$, denoted by $C \xrightarrow{e} C'$, if $(C'(u), C'(v)) = T(C(u), C(v))$ and $C'(w) = C(w)$ for all $w \in V \setminus \{u, v\}$.

A *schedule* $\gamma = \gamma_0, \gamma_1, \cdots = (u_0, v_0), (u_1, v_1), \ldots$ is a sequence of interactions. A schedule determines which interaction occurs at each *step*, *i.e.*, interaction γ_t happens at step t. In particular, we consider a *uniform random scheduler* $\Gamma = \Gamma_0, \Gamma_1, \ldots$ where each Γ_t $(t \geq 0)$ is a random variable that specifies the interaction (u_t, v_t) at *step* t and satisfies $\Pr(\Gamma_t = (u, v)) = \frac{1}{n(n-1)}$ for any distinct $u, v \in V$. All interactions $\Gamma_0, \Gamma_1, \ldots$ are independent of each other. Given an initial configuration C_0, the *execution* of protocol P is uniquely defined as $\Xi_P(C_0, \Gamma) = C_0, C_1, \ldots$ such that $C_t \xrightarrow{\Gamma_t} C_{t+1}$ for all $t \geq 0$. We say that agent

Table 1. Leader election protocols. (Stabilization time is shown in terms of parallel time and in expectation.)

	States	Stabilization time
[1]	$O(1)$	$O(n)$
[2]	$O(\log^3 n)$	$O(\log^3 n)$
[3]	$O(\log^2 n)$	$O(\log^{5.3} n \cdot \log \log n)$
[4]	$O(\log n)$	$O(\log^2 n)$
[5]	$O(\log \log n)$	$O(\log^2 n)$
[6]	$O(\log \log n)$	$O(\log n \cdot \log \log n)$
[7]	$O(n)$	$O(\log n)$
This work	$O(\log n)$	$O(\log n)$

Table 2. Lower bounds for leader election (Stabilization time is shown in terms of parallel time and in expectation.)

	States	Stabilization time
[8]	$O(1)$	$\Omega(n)$
[3]	$< 1/2 \log \log n$	$\Omega(n/(\text{polylog } n))$
[9]	any large	$\Omega(\log n)$

$v \in V$ *participates* in Γ_t if v is either the initiator or the responder of Γ_t. We say that a configuration C of protocol P is reachable if the initial configuration $C_{\text{init},P}$ changes to C by some finite sequence of interactions $\gamma_0, \gamma_1, \ldots, \gamma_k$. We define $\mathcal{C}_{\text{all}}(P)$ as the set of all reachable configurations of P.

The leader election problem requires that every agent should output L or F which means "leader" or "follower" respectively. Let $\mathcal{S}_P$ be the set of the configurations such that, for any configuration $C \in \mathcal{S}_P$, exactly one agent outputs L (*i.e.*, is a leader) in C and no agent changes its output in execution $\Xi_P(C, \gamma)$ for any schedule γ. We say that a protocol P solves the leader election if execution $\Xi_P(C_{\text{init},P}, \Gamma)$ reaches a configuration in $\mathcal{S}_P$ with probability 1. For any leader election protocol P, we define the expected stabilization time of P as the expected number of steps during which execution $\Xi_P(C_{\text{init},P}, \Gamma)$ reaches a configuration in $\mathcal{S}_P$, divided by the number of agents n. The division by n is needed because we evaluate the stabilization time in terms of parallel time.

We write the natural logarithm of x as $\ln x$ and the logarithm of x with base 2 as $\lg x$. We do not indicate the base of logarithm in an asymptotical expression such as $O(\log n)$. By an abuse of notation, we will identify an interaction (u, v) with the set $\{u, v\}$ whenever convenient.

In the proposed protocol, we often use the technique called *one-way epidemic* [15]. See the full paper [16] for the formal definition of one-way epidemic and rigorous analysis on the number of steps required to finish one-way epidemic, especially when the epidemic is executed by a sub-population (*i.e.*, a part of the whole population).

3   Key Ideas of Logarithmic Leader Election

In this section, we give key ideas of the proposed protocol P_{LL}. Each agent v keeps an output variable $v.\texttt{leader} \in \{\textit{false}, \textit{true}\}$. An agent outputs L when the value of $\texttt{leader}$ is *true* and it outputs F when it is *false*. An execution of P_{LL} can be regarded as a competition by agents. At the beginning of the execution, every agent has $\texttt{leader} = \textit{true}$, that is, all agents are leaders. Throughout the execution, every leader tries to remain a leader and tries to make all other leaders followers so that it becomes the unique leader in the population. The competition consists of three modules *QuickElimination()*, *Tournament()*, and *BackUp()*, which are executed in this order. These three modules guarantee the following properties:

QuickElimination(): An execution of this module takes $O(\log n)$ parallel time in expectation. For any $i \geq 2$, exactly i leaders survive an execution of *QuickElimination*() with probability at most 2^{1-i}. The execution never eliminates all leaders, i.e., at least one leader always survives.

Tournament(): An execution of this module takes $O(\log n)$ parallel time in expectation. By an execution of *Tournament*(), which starts with $i \geq 2$ leaders, the unique leader is elected with probability at least $1 - O(i/\log n)$. This lower bound of probability is independent of an execution of the previous module *QuickElimination*(). The execution never eliminates all leaders, i.e., at least one leader always survives.

BackUp(): An execution of this component elects a unique leader within $O(\log^2 n)$ parallel time in expectation.

From above, it holds that the number of leaders is exactly one with probability at least $1 - \sum_{i=2}^{n} O\left(i/(2^{i-1}\log n)\right) = 1 - O(1/\log n)$ after executions of *QuickElimination*() and *Tournament*() finish. Therefore, combined with *BackUp*(), protocol P_{LL} elects a unique leader within $(1 - O(1/\log n)) \cdot \log n + O(1/\log n) \cdot O(\log^2 n) = O(\log n)$ parallel time in expectation.

In the remainder of this subsection, we briefly give key ideas to design the three modules satisfying the above guarantees. We will present a way to implement the following ideas with $O(\log n)$ states per agent in the next subsection (Sect. 4). In this subsection, keep in mind only that these ideas are easily implemented with poly-logarithmic number of states per agent, that is, with a constant number of variables with $O(\log \log n)$ bits. For the following description of the key ideas, we assume a kind of global synchronization, for example, we assume that each agent begins an execution of *Tournament*() after *all* agents finish necessary operations of *QuickElimination*(). We also present a way to implement such a synchronization in Sect. 4.

3.1 Key Idea for *QuickElimination*()

The goal of this module is to reduce the number of leaders such that, for any $i \geq 2$, the resulting number of leaders is exactly i with probability at most 2^{1-i} while guaranteeing that not all leaders are eliminated. This module is based on almost the same idea as *the lottery protocol* in [3]. The protocol P_{LL} achieves much faster stabilization time than the lottery protocol thanks to tighter analysis on the number of surviving leaders, which we will see below, and the combination with the other two modules.

First, consider the following game:

- (i) Each agent in V executes a sequence of independent fair coin flips, each of which results in head with probability $1/2$ and tail with probability $1/2$, until it observes tail for the first time,
- (ii) Let s_v be the number of heads that v observes in the above coin flips and let $s_{\max} = \max_{v \in V} s_v$,
- (iii) The agents v with $s_v = s_{\max}$ are winners and the other agents are losers.

Let $i \geq 2$ and $j \geq 0$. Consider the situation that exactly i agents observe that their first j coin flips result in head and define $p_{i,j}$ as the probability that all the i agents win the game in the end starting from this situation. Starting from this situation, if all the i agents observe tail in their $j + 1$-st coin flips then exactly i agents win the game with probability 1; if all the i agents observe head in their $j + 1$-st coin flips then exactly i agents win with probability $p_{i,j+1}$; Otherwise, the number of winners of the game is less than i with probability 1. Therefore, we have $p_{i,j} = 2^{-i} + 2^{-i} \cdot p_{i,j+1}$. Since we have $p_{i,j} = p_{i,j+1}$ thanks to memoryless property of this game, solving this equality gives $p_{i,j} = 1/(2^i - 1) \leq 2^{1-i}$. Let k_i be the minimum integer j such that exactly i agents observe that all of their first j coin flips result in head. We define $k_n = 0$ for simplicity. Then, for any $i \geq 0$, we have

$$\Pr(|\{v \in V \mid s_v = s_{\max}\}| = i) = \sum_{j=0}^{\infty} \Pr(k_i = j) \cdot p_{i,j} \leq 2^{1-i} \sum_{j=0}^{\infty} \Pr(k_i = j) \leq 2^{1-i}.$$

Module *QuickElimination()* simulates this game in the population protocol model. Every time an agent v has an interaction, we regard the interaction as the coin flip by v. If v is an initiator at the interaction, we regard the result of the coin flip as head; Otherwise we regard it as tail. The correctness of this simulation for coin flips comes from the definition of the uniform random scheduler: at each step, an interaction where v is an initiator happens with probability $1/n$ and an interaction where v is a responder also happens with probability $1/n$. Strictly speaking, this simple simulation of coin flips does not guarantee independence of coin flips by u and v for any distinct $u, v \in V$. However, the actual P_{LL} defined in Sect. 4 completely simulates independent coin flips of leaders and we will explain it in Sect. 4. Each agent v computes and stores s_v on variable $v.\mathtt{level}_Q$ by counting the number of interactions that it participates in as an initiator until it interacts as a responder for the first time. After every agent v computes s_v on $v.\mathtt{level}_Q$, the maximum value of $\mathtt{level}_Q$, i.e., $s_{\max}$, is propagated from agent to agent via *one-way epidemic* [15], that is,

- each agent memorizes the largest value of $\mathtt{level}_Q$ it has observed, and
- the larger value is propagated to the agent with smaller value at every interaction.

It is proven in [15] that all agents obtain the largest value within $O(\log n)$ parallel time with high probability by this simple propagation. If agent v knows $s_v < s_{\max}$, v changes $v.\mathtt{leader}$ from *true* to *false*, that is, v becomes a follower. Thus, when one-way epidemic of $s_{\max}$ finishes, only the agents v satisfying $s_v = s_{\max}$ are leaders. From the above discussion, for any $i \geq 2$, the number of such surviving leaders is exactly i with probability at most 2^{1-i}. On the other hand, there is at least one agent v with $s_v = s_{\max}$, thus this module never eliminates all leaders. A logarithmic number of states is sufficient for $\mathtt{level}_Q$ because each agent v gets more than $c \lg n$ consecutive heads with probability at most n^{-c} for any $c \geq 1$.

3.2   Key Idea for *Tournament*()

Starting from a configuration where the number of leaders is i, the goal of *Tournament*() is to reduce the number of leaders from i to one with probability $1 - O(i/\log n)$ while guaranteeing that not all leaders are eliminated. The idea of this component is simple. As with the *QuickElimination*(), we use coin flips in *Tournament*(). Every leader v maintains variable v.rand. Initially, v.rand $= 0$. Every time it has an interaction, it updates v.rand by v.rand $\leftarrow 2v$.rand $+ j$ where j indicates whether v is a responder in the interaction or not, *i.e.*, $j = 0$ if v is a initiator and $j = 1$ if v is a responder. This operation stops when v encounters $\lceil \log_2 m \rceil = O(\log \log n)$ interactions. Thus, when all the i leaders encounter at least $\lceil \log_2 m \rceil$ interactions, for every leader v, v.rand is a random variable uniformly chosen from $\{0, 1, \ldots, 2^{\lceil \log_2 m \rceil} - 1\}$. Although u.rand and v.rand are not independent of each other for any distinct leader u and v, we will present a way to remove any dependence between u.rand and v.rand in Sect. 4. As with *QuickElimination*(), the maximum value **rand** is propagated to the whole population via one-way epidemic within $O(\log n)$ parallel time with high probability and only leaders with the maximum value remain leaders in the end of *Tournament*().

Let $v_1, v_2, \ldots, v_i$ be the i leaders that survive *QuickElimination*(), $r_1, r_2, \ldots, r_i$ be the resulting value of v_i.rand, and $r_{\max}(j) = \max(r_1, r_2, \ldots, r_j)$ for any $j = 1, 2, \ldots, i$. Clearly, the number of leaders at the end of *Tournament*() is exactly one if $r_{j+1} \neq r_{\max}(j)$ holds for all $j = 1, 2, \ldots, i - 1$. By the union bound and independence between $r_1, r_2, \ldots, r_i$, this holds with probability at least $1 - \sum_{j=1}^{i-1} 2^{-\lceil \log_2 m \rceil} \geq 1 - i/m \geq 1 - i/(\lg n)$. On the other hand, an execution of *Tournament*() never eliminates all leaders since there is always at least one leader v_j that satisfies $r_j = r_{\max}(i)$.

3.3   Key Idea for *BackUp*()

The goal of *BackUp*() is to elect a unique leader within $O(\log^2 n)$ parallel time in expectation. We must guarantee this expected time regardless of the number of the agents that survive both *QuickElimination*() and *Tournament*() and remain leaders at the beginning of an execution of *BackUp*(). We can only assume that at least one leader exists at the beginning of the execution. We use coin flips also for *BackUp*(). Every leader v maintains v.level$_B$. Initially, v.level$_B = 0$. Every leader v repeats the following procedure until v.level$_B$ reaches $5m$ or v becomes a follower.

- Make a coin flip. If the result is head (i.e., v participates in an interaction as an initiator), v increments v.level$_B$ by one. If the result is tail, v does nothing.
- Wait for sufficiently long but logarithmic parallel time so that the maximum level$_B$ propagates to the whole population via one-way epidemic. If it observes larger value in the epidemic, it becomes a follower, that is, it executes v.leader $\leftarrow$ *false*. Furthermore, if v interacts with another leader with

the same level during this period and v is a responder in the interaction, v becomes a follower.

Let j be an arbitrary integer such that $1 \leq i \leq 5m$. Consider the first time that $\mathtt{level}_B$ of some leader, say v, reaches j. Let $V' \subseteq V$ be the set of leaders at that time. By the definition of the above procedure, every $u \in V'$ other than v satisfies $u.\mathtt{level}_B < j$, and u makes a coin flip at most once with high probability until the maximum value j is propagated from v to u. If the result of the one coin flip is tail, u becomes a follower. Therefore, with probability at least $1/2 - O(n^{-1}) > 1/3$, no less than half of leaders in $V' \setminus v$ becomes followers, that is, the number of leaders decreases to at most $1 + \lfloor |V'|/2 \rfloor$. Chernoff bound guarantees that the number of leaders becomes one with high probability until $v.\mathtt{level}_B$ for every leader v reaches $5m$. Even if multiple leaders survive at that time, we have simple election mechanism to elect a unique leader; when two leaders with the same level interacts with each other, one of them becomes a follower. This simple election mechanism elects a unique leader within $O(n)$ parallel time in expectation. Therefore, the total expected parallel time to elect a unique leader is $O(m \log n) + O(n^{-1}) \cdot O(n) = O(\log^2 n)$.

4   Implementation of Logarithmic Leader Election

In this section, we present detailed description of the proposed protocol P_{LL}. The key ideas presented in the previous subsection achieve $O(\log n)$ stabilization time if it is implemented correctly. For implementation, they need some kind of global synchronization and *independent* coin flips. Furthermore, a naive implementation of the key ideas requires a poly-logarithmic number of states (*i.e.,* $O(\log^c n)$ states for $c > 1$) per agent while our goal is to achieve $O(\log n)$ states per agent. We present how to address these issues in this section.

Table 3. Variables of P_{LL}

Groups	Variables	Initial values
All agents	$\mathtt{leader} \in \{false, true\}$	$true$
	$\mathtt{tick} \in \{false, true\}$	$false$
	$\mathtt{status} \in \{X, A, B\}$	X
	$\mathtt{epoch}, \mathtt{init} \in \{1, 2, 3, 4\}$	1
	$\mathtt{color} \in \{0, 1, 2\}$	0
V_B	$\mathtt{count} \in \{0, 1, \ldots, c_{max} - 1\}$	Undefined
$V_A \cap V_1$	$\mathtt{level}_Q \in \{0, 1, \ldots, l_{max}\}$	Undefined
	$\mathtt{done} \in \{false, true\}$	Undefined
$V_A \cap (V_2 \cup V_3)$	$\mathtt{rand} \in \{0, 1, \ldots, 2^\Phi - 1\}$	Undefined
	$\mathtt{index} \in \{0, 1, \ldots, \Phi - 1\}$	Undefined
$V_A \cap V_4$	$\mathtt{level}_B \in \{0, 1, \ldots, l_{max}\}$	Undefined

All variables of P_{LL} are listed in Table 3. All agents manage six variables `leader`, `tick`, `status`, `epoch`, `init`, and `color`. To implement the key ideas above with $O(\log n)$ states, we divide the population into multiple subpopulations or *groups*, as in [6], where agents in different groups manage different variables in addition to the above six variables. In the remainder of this paper, we refer the above six variables by *common variables* and other variables by *additional variables*. The population is divided to six groups based on two common variables `status` $\in \{X, A, B\}$ and `epoch` $\in \{1, 2, 3, 4\}$, that is, V_X, V_B, $V_A \cap V_1$, $V_A \cap (V_2 \cup V_3)$, $V_A \cap V_4$ where we denote $V_Z = \{v \in V \mid v.\mathtt{status} = Z\}$ for $Z \in \{X, A, B\}$ and $V_i = \{v \in V \mid v.\mathtt{epoch} = i\}$ for $i \in \{1, 2, 3, 4\}$. We have no additional variables for agents in group V_X, one additional variable `count` $\in \{0, 1, \ldots, c_{\max} - 1\}$ for agents in V_B where $c_{\max} = 41m$, two additional variables $\mathtt{level}_Q \in \{0, 1, \ldots, l_{\max}\}$ and `done` $\in \{false, true\}$ for agents in $V_A \cap V_1$ where $l_{\max} = 5m$, two additional variables `rand` $\in \{0, 1, \ldots, 2^\Phi - 1\}$ and `index` $\in \{0, 1, \ldots, \Phi - 1\}$ for agents in $V_A \cap (V_2 \cup V_3)$ where $\Phi = \lceil \frac{2}{3} \lg m \rceil$, and one additional variable $\mathtt{level}_B \in \{0, 1, \ldots, l_{\max}\}$ for agents in $V_A \cap V_4$. Agents in any group have only $O(\log n)$ states. This is because every common variable has constant size domain, every group other than $V_A \cap (V_2 \cup V_3)$ has at most one non-constant additional variable and any of such variables can take $O(\log n)$ values, and an agent in $V_A \cap (V_2 \cup V_3)$ has two additional variables `rand` and `index` and the combination of the two variables can take $2^\Phi \cdot \Phi = O(m^{2/3} \log m) \subset O(\log n)$ values. Therefore, the number of states per agent used by P_{LL} is $O(\log n)$.

Lemma 1. *The number of states per agent used by P_{LL} is $O(\log n)$.*

Independently of the six groups defined above, we define another groups V_L and V_F based on a common variable `leader`; V_L (resp., V_F) is the set of agents $v \in V$ such that $v.\mathtt{leader} = true$ (resp., $v.\mathtt{leader} = false$). We introduce these two groups only for simplicity of notation.

The pseudocode of P_{LL} is given in Algorithm 1, which has four modules *CountUp()*, *QuickElimination()*, *Tournament()*, and *BackUp()*. Due to the lack of space, we omit the pseudocodes of *QuickElimination()*, *Tournament()*, and *BackUp()*, while the pseudocode of *CountUp* is presented in Algorithm 2. See the full paper [16] for the omitted pseudocodes. The main function of P_{LL} (Algorithm 1) consists of four parts. The first part (Lines 1–6) assigns status A or B to each agent. The second part (Lines 7–10) manages variable `epoch` using module *CountUp()*. Initially, $v.\mathtt{epoch} = 1$ holds, that is, $v \in V_1$ holds for all $v \in V$. In an execution of P_{LL}, $v.\mathtt{epoch}$ never decreases and increases by one every sufficiently large logarithmic parallel time in expectation until it reaches 4 as we will explain later. In the third part (Lines 11–15), we initialize additional variables when an agent increases its epoch. Each agent v has a common variable `init`, which is set to 1 initially. Whenever $v.\mathtt{epoch}$ increases, $v.\mathtt{epoch} > v.\mathtt{init}$ must hold, then v initialize additional variables according to v's group and executes $v.\mathtt{init} \leftarrow v.\mathtt{epoch}$. For example, when the `epoch` of agent $v \in V_A$ changes from 3 to 4 *i.e.*, v moves from group $V_A \cap V_3$ to $V_A \cap V_4$, it initializes an additional variable $a_i.\mathtt{level}_B$ to 0 (Line 13). Additional variables for groups V_B and $V_A \cap V_1$

Algorithm 1. P_{LL}

Notations:

$l_{\max} = 5m$, $c_{\max} = 41m$, $\Phi = \lceil \frac{2}{3} \lg m \rceil$

$V_Z = \{v \in V \mid v.\text{status} = Z\}$ for $Z \in \{X, A, B\}$

$V_i = \{v \in V \mid v.\text{epoch} = i\}$ for $i \in \{1, \ldots, 4\}$

Output function π_{out}:

if $v.\text{leader} = true$ holds, then the output of agent v is L, otherwise F.

Interaction between initiator a_0 and responder a_1:

1: **if** $a_0, a_1 \in V_X$ **then**
2:     $(a_0.\text{status}, a_0.\text{level}_Q, a_0.\text{done}, a_0.\text{leader}) \leftarrow (A, 0, false, true)$
3:     $(a_1.\text{status}, a_1.\text{count}, a_1.\text{leader}) \leftarrow (B, 0, false)$
4: **else if** $\exists i \in \{0, 1\} : a_i \in V_X \wedge a_{1-i} \notin V_X$ **then**
5:     $(a_i.\text{status}, a_i.\text{level}_Q, a_i.\text{done}, a_i.\text{leader}) \leftarrow (A, 0, true, false)$
6: **end if**

7: $a_0.\text{tick} \leftarrow a_1.\text{tick} \leftarrow false$
8: $CountUp()$
9: **for all** $i \in \{0, 1\}$ such that $a_i.\text{tick}$ **do** $a_i.\text{epoch} = \max(a_i.\text{epoch} + 1, 4)$ **endfor**
10: $a_0.\text{epoch} \leftarrow a_1.\text{epoch} \leftarrow \max(a_0.\text{epoch}, a_1.\text{epoch})$

11: **for all** $i \in \{0, 1\}$ such that $a_i.\text{epoch} > a_i.\text{init}$ **do**   // Initialize variables for each group
12:     **if** $a_i \in V_A \cap (V_2 \cup V_3)$ **then** $(a_i.\text{rand}, a_i.\text{index}) \leftarrow (0, 0)$ **endif**
13:     **if** $a_i \in V_A \cap V_4$ **then** $a_i.\text{level}_B \leftarrow 0$ **endif**
14:     $a_i.\text{init} \leftarrow a_i.\text{epoch}$
15: **end for**

16: **if** $a_0, a_1 \in V_1$ **then**
17:     Execute $QuickElimination()$
18: **else if** $a_0, a_1 \in V_2 \vee a_0, a_1 \in V_3$ **then**
19:     Execute $Tournament()$
20: **else if** $a_0, a_1 \in V_4$ **then**
21:     Execute $BackUp()$
22: **end if**

---

are initialized not in this part but in the first part as we will explain in Sect. 4.1. In the fourth part (Lines 16–22), agents execute modules based on the values of their **epoch**. Specifically, agents execute $QuickElimination()$, $Tournament()$, and $BackUp()$ while they are in V_1, $V_2 \cup V_3$, and V_4 respectively.

In the key idea depicted in Sect. 3.2, each leader v makes fair coin flips exactly $\lceil \lg m \rceil = \Theta(\log \log n)$ times. However, this requires $\Omega(\log n \cdot \log \log n)$ states per agent because this procedure requires not only variable $v.\text{rand}$ that stores the results of those flips but also variable $v.\text{index}$ to memorize how many times v already made coin flips. Therefore, in an execution of $Tournament()$, each agent makes fair coin flips only $\Phi = \lceil \frac{2}{3} \lg m \rceil$ times, and we execute this module $Tournament()$ twice. That is why we assign two epochs (*i.e.*, the second and the third epochs) to $Tournament()$.

In the remainder of this section, we explain how to assign status to agents, how to synchronize the population by $CountUp()$, and how to implement independent coin flips in three modules $QuickElimination()$, $Tournament()$, and $BackUp()$.

Algorithm 2. $CountUp()$

Interaction between initiator a_0 and responder a_1:

23: **for all** $i \in \{0, 1\}$ such that $a_i \in V_B$ **do**
24:     $a_i.\text{count} \leftarrow a_i.\text{count} + 1 \pmod{c_{\max}}$
25:     **if** $a_i.\text{count} = 0$ **then**
26:         $a_i.\text{color} \leftarrow a_i.\text{color} + 1 \pmod 3$
27:         $a_i.\text{tick} \leftarrow true$
28:     **end if**
29: **end for**
30: **if** $\exists i \in \{0, 1\} : a_{1-i}.\text{color} = a_i.\text{color} + 1 \pmod 3$ **then**
31:     $a_i.\text{color} \leftarrow a_{1-i}.\text{color}$
32:     $a_i.\text{tick} \leftarrow true$
33:     **if** $a_i \in V_B$ **then** $a_i.\text{count} \leftarrow 0$ **endif**
34: **end if**

4.1 Assignment of Status

At the beginning of an execution, all agents are in V_X, that is, the statuses of all agents are the "initial" status X. Every agent is given status A or B at its first interaction where A means "leader candidate" and B means "timer agent". As we will explain later, the unique leader is elected from V_A and agents in V_B are mainly used to synchronize the population with their count-up timers.

Agents determine their status, A or B, by the following simple way. When two agents in V_X meet, the initiator and the responder are given status A and B, respectively (Line 2–3). The initiator initializes its additional variable level_Q and **done** to 0 and *false* respectively and remains a leader (Line 2) while the responder initializes its additional variable **count** to 0 and becomes a follower by **leader** $\leftarrow$ *false* (Line 3). When an agent in V_X meets an agent in V_A or V_B, it gets status A but it becomes a follower. It also initializes its additional variable level_Q and **done** to 0 and *true* respectively (Line 5). For agent v, assigning *true* to $v.\text{done}$ means that v never joins a game with coin flips in $QuickElimination()$.

No agent changes its status once it gets status A or B, and no follower becomes a leader in an execution of P_{LL}. Therefore, we have the following lemma.

Lemma 2. *In an execution of P_{LL}, $|V_A| \geq n/2$, $|V_F| \geq n/2$, and $|V_B| \geq 1$ always hold after every agent gets status A or B.*

Proof. Consider any configuration in $\mathcal{C}_{\text{all}}(P_{LL})$ where every agent has status A or B. Let x (resp., y and z) be the the number of agents which get status A (resp., B and A) by Line 2 (resp., Line 3 and Line 5). We have $x = y \leq n/2$ by the definition of P_{LL}, which gives $|V_A| = x + z = n - y \geq n/2$. Moreover,

$|V_L| \leq x \leq n/2$ holds because the number of leaders is monotonically non-increasing in an execution of P_{LL}. The first interaction of the execution assigns one agent with status V_B, hence $|V_B| \geq 1$ holds.   □

4.2 Synchronization and Epochs

When a unique leader exists in the population, we can synchronize the population by *phase clocks* with constant space per agent [15]. Recently, in [5] and [6], it is proven that even when we cannot assume the existence of the unique leader, Phase clocks can be used for synchronization if we are allowed to use $O(\log \log n)$ states per agent. However, this synchronization with $O(\log \log n)$ states without the unique leader involves complicated analysis. Since we use $O(\log n)$ states for another modules, we achieve synchronization in simpler way with $O(\log n)$ states per agent.

The pseudocode of our synchronization is shown in Algorithm 2. We use common variables color $\in \{0, 1, 2\}$ in all agents and an additional variable count $\in \{0, 1, \ldots, c_{\max} - 1\}$ for agents in group V_B. Initially, all agents have the same color, namely, 0. The color of an agent is incremented by modulo 3 when the agent changes its color. We say that the agent *gets a new color* when this event happens. Roughly speaking, our goal is to guarantee that

- (i) whenever one agent gets a new color (*e.g.*, changes its color from 0 to 1), the new color spreads to the whole population within $O(\log n)$ parallel time with high probability,
- (ii) thereafter, all agents keep the same color for sufficiently long but $\Theta(\log n)$ parallel time with high probability.

Specifically, "sufficiently long but $\Theta(\log n)$ time" in (ii) means sufficiently long period such that any $O(\log n)$ parallel time operations in *QuickElimination()*, *Tournament()*, and *BackUp()*, such as one-way epidemic of some value, finishes with high probability during the period.

At every interaction, module *CountUp()* is invoked (Line 8) and variables color and count can be changed only in this module. In *CountUp()*, every agent in V_B increments its count by one modulo $c_{\max}$ (Line 24). For every $v \in V_B$, if this incrementation changes v.count from $c_{\max} - 1$ to 0, v gets a new color by incrementing v.color by one modulo 3 (Line 26). Once one agent gets a new color, the new color spreads to the whole population via one-way epidemic in the whole population. Specifically, if agents u and v satisfying u.color $= v$.color$+1$ (mod 3) meets, v executes v.color $\leftarrow u$.color and resets its count to 0 (Line 31–33). One-way epidemic requires $\Theta(\log n)$ parallel time and each agent requires $\Theta(c_{\max})$ parallel time to count up $c_{\max}$ times (*i.e.*, encounter $c_{\max}$ interactions). Therefore, this synchronization guarantees the above requirements (i) and (ii) if we give $c_{\max}$ a sufficiently large $\Theta(\log n)$ value (actually, $c_{\max} = 41m$ in this paper). See the full paper [16] for detailed analysis.

Every time an agent v gets a new color, it raises a tick flag, i.e., assigns v.tick $\leftarrow$ *true* (Lines 27 and 32). This common variable v.tick is used only for simplicity of the pseudocode and it does not affect the transition at v's next interaction (v.tick is reset to *false* in Line 7), unlike the other variables. When v.tick is raised, v.epoch increases by one unless it has already reached 4 (Line 9). After two agents u and v execute Lines 7–9 at an interaction, u.epoch $= v$.epoch usually holds. However, this equation does not hold when synchronization fails. For this case, we substitute $\max(u$.epoch, v.epoch) into u.epoch and v.epoch in Line 10. Variable v.tick is also used to determine when each leader makes a coin flip in module *BackUp*() because each leader must wait for sufficiently long logarithmic parallel time between any two consecutive coin flips, as explained in Sect. 3.3.

As mentioned above, every agent gets a new color in every sufficiently large $\Theta(\log n)$ parallel time with high probability. This means that, for every $v \in V$, v.tick is raised and v.epoch increases by one with high probability in every sufficiently large $\Theta(\log n)$ parallel time until v.epoch reaches 4. If this synchronization fails, *e.g.*, some agent gets a color 1 without keeping color 0 for $\Theta(\log n)$ parallel time, the modules *QuickElimination*() and *Tournament*() may not work correctly. However, starting from any configuration after a synchronization fails arbitrarily, module *CountUp*() and Lines 7–10 guarantee that all agents proceeds to the forth epoch within $O(\log n)$ parallel time in expectation, and thereafter *BackUp*() guarantees that exactly one leader is elected within $O(n)$ parallel time in expectation. Hence, P_{LL} guarantees that a unique leader is elected with probability 1. The above $O(n)$ parallel time never prevent us from achieving stabilization time of $O(\log n)$ parallel time in expectation because (i) synchronization fails at each phase with probability $O(1/n)$, (ii) our protocol P_{LL} uses $O(\log n)$ phases in total (*i.e.*, one phase for *QuickElimination*(), two phases for *Tournament*(), and $l_{\max}$ phases for *BackUp*()), and hence (iii) the synchronization fails and damages the progress of an execution of P_{LL} with probability $O(\log n/n)$ in total.

4.3   Independent Coin Flips

In the three modules *QuickElimination*(), *Tournament*(), and *BackUp*(), only leaders make coin flips. We implement independent coin flips in a simple way: A leader makes a coin flip only if it interacts with a follower. In other words, when a leader v wants to make a coin flip in an interaction with agent u, it does make a coin flip if u is a follower, but it does not otherwise. After all agents get their status A or B, $|V_F| \geq n/2$ always holds by Lemma 2. Therefore, the frequency of coin-flipping decreases only by half, thus we can asymptotically ignore the impact of this slowdown on the stabilization time of P_{LL}.

Therefore, we can say that protocol P_{LL} correctly implements the key ideas in Sect. 3 with $O(\log n)$ states per agent. Thus, we obtain the following lemma. (See the full paper [16] for the complete proof.)

Theorem 1. *Let* $\Xi = \Xi_{P_{LL}}(C_{\mathrm{init}, P_{LL}}, \Gamma) = C_0, C_1, \dots$. *Execution* Ξ *reaches a configuration in* $\mathcal{S}_{\mathcal{P}_{\mathcal{L}\mathcal{L}}}{}^1$ *within* $O(\log n)$ *parallel time in expectation.*

---

[1] Recall that $\mathcal{S}_{\mathcal{P}_{\mathcal{L}\mathcal{L}}}$ is the set of configurations such that, for any configuration $C \in \mathcal{S}_{\mathcal{P}_{\mathcal{L}\mathcal{L}}}$, exactly one agent outputs L (*i.e.*, is a leader) in C and no agent changes its output in execution $\Xi_{P_{LL}}(C, \gamma)$ for any schedule γ.

Proof (Proof Sketch). When an execution of P_{LL} elects a unique leader, a configuration at that time must belong to $\mathcal{S}_{\mathcal{P}_{\mathcal{LL}}}$. This is because the number of leaders are monotonically non-increasing and no interaction brings a configuration with no leader in an execution of P_{LL}. Modules *QuickElimination()* and *Tournament()* guarantee that exactly one leader is elected within $O(\log n)$ parallel time with probability $1 - O(1/\log n)$. Even if these two modules fail, module *BackUp()* elects exactly one leader within $O(\log^2 n)$ parallel time with probability $1 - O(\log n/n)$. Note that this error probability $O(\log n/n)$ comes from the synchronization mechanism: synchronization fails during these $O(\log^2 n)$ parallel time with probability $O(\log n/n)$. Even if module *BackUp()* does not elect exactly one leader within $O(\log^2 n)$ parallel time, it elects exactly one leader within $O(n)$ parallel time in expectation. Therefore, we conclude that the stabilization time of P_{LL} is at most $O(\log n) + O(1/\log n) \cdot O(\log^2 n) + O(1/n) \cdot O(n) = O(\log n)$ parallel time.   □

5   Discussion Towards Symmetric Transitions

In the field of PP model, several works are devoted to design a *symmetric protocol*. A protocol $P(Q, s_{\text{init}}, T, Y, \pi_{\text{out}})$ is symmetric if its transition function T satisfies $(p', q') = T(p, q) \Leftrightarrow (q', p') = T(q, p)$ for any $p, q, p', q' \in Q$. In other words, a symmetric protocol is a protocol that does not utilize the roles of the two agents at an interaction, initiator and responder. Suppose that two agents have an interaction and their states changes from p, q to p', q', respectively. In a symmetric protocol, $p = q \Rightarrow p' = q'$ always hold. This property is important for some applications such as chemical reaction networks.

The proposed protocol P_{LL} described above is not symmetric, however, we can make it symmetric by the following strategy. Protocol P_{LL} performs asymmetric actions only for assignment of status and flipping fair and independent coins, described in Sects. 4.1 and 4.3, respectively. To assign the agents their statuses by symmetric transitions, we only have to add additional status Y and make the following three rules: $X \times X \to Y \times Y$, $Y \times Y \to X \times X$, $X \times Y \to A \times B$. Furthermore, similarly to the original rules of P_{LL}, when an agent v with status X or Y meets an agent with status A or B, v gets status A but it becomes a follower. This modification does not make any harmful influence on the analysis of stabilization time, at least asymptotically. Coin flips are dealt with in the same way. We assign a *coin status* J, K, F_0, or F_1 to each follower. Every time a leader v becomes a follower, initial status J is assigned to v. Thereafter, when two followers meet, they change their coin statuses according to the following rules: $J \times J \to K \times K$, $K \times K \to J \times J$, $J \times K \to F_0 \times F_1$. These rules guarantee that the numbers of the followers with state F_0 and F_1 are always equal. Therefore, a leader can make a fair and independent coin flip every time it meets a follower whose coin state is not J or K.

References

1. Angluin, D., Aspnes, J., Diamadi, Z., Fischer, M.J., Peralta, R.: Computation in networks of passively mobile finite-state sensors. Distrib. Comput. **18**(4), 235–253 (2006)
2. Alistarh, D., Gelashvili, R.: Polylogarithmic-time leader election in population protocols. In: Halldórsson, M.M., Iwama, K., Kobayashi, N., Speckmann, B. (eds.) ICALP 2015. LNCS, vol. 9135, pp. 479–491. Springer, Heidelberg (2015). https://doi.org/10.1007/978-3-662-47666-6_38
3. Alistarh, D., Aspnes, J., Eisenstat, D., Gelashvili, R., Rivest, R.L.: Time-space trade-offs in population protocols. In: Proceedings of the Twenty-Eighth Annual ACM-SIAM Symposium on Discrete Algorithms, pp. 2560–2579. SIAM (2017)
4. Alistarh, D., Aspnes, J., Gelashvili, R.: Space-optimal majority in population protocols. In: Proceedings of the Twenty-Ninth Annual ACM-SIAM Symposium on Discrete Algorithms, pp. 2221–2239. SIAM (2018)
5. Gąsieniec, L., Staehowiak, G.: Fast space optimal leader election in population protocols. In: Proceedings of the Twenty-Ninth Annual ACM-SIAM Symposium on Discrete Algorithms, pp. 2653–2667. SIAM (2018)
6. Gąsieniec, L., Stachowiak, G., Uznański, P.: Almost logarithmic-time space optimal leader election in population protocols. arXiv preprint arXiv: 1802.06867 (2018)
7. Michail, O., Spirakis, P.G., Theofilatos, M.: Simple and fast approximate counting and leader election in populations. In: Izumi, T., Kuznetsov, P. (eds.) SSS 2018. LNCS, vol. 11201, pp. 154–169. Springer, Cham (2018). https://doi.org/10.1007/978-3-030-03232-6_11
8. Doty, D., Soloveichik, D.: Stable leader election in population protocols requires linear time. Distrib. Comput. **31**(4), 257–271 (2018)
9. Sudo, Y., Masuzawa, T.: Leader election requires logarithmic time in population protocols. arXiv preprint arXiv:1906.11121 (2019)
10. Sudo, Y., Nakamura, J., Yamauchi, Y., Ooshita, F., Kakugawa, H., Masuzawa, T.: Loosely-stabilizing leader election in a population protocol model. Theor. Comput. Sci. **444**, 100–112 (2012)
11. Sudo, Y., Ooshita, F., Kakugawa, H., Masuzawa, T., Datta, A.K., Larmore, L.L.: Loosely-stabilizing leader election with polylogarithmic convergence time. In: 22nd International Conference on Principles of Distributed Systems, OPODIS 2018, pp. 30:1–30:16 (2018)
12. Bilke, A., Cooper, C., Elsässer, R., Radzik, T.: Brief announcement: population protocols for leader election and exact majority with $o(log^2 n)$ states and $o(log^2 n)$ convergence time. In: Proceedings of the 38th ACM Symposium on Principles of Distributed Computing, pp. 451–453. Springer (2017)
13. Alistarh, D., Gelashvili, R.: Recent algorithmic advances in population protocols. ACM SIGACT News **49**(3), 63–73 (2018)
14. Elsässer, R., Radzik, T.: Recent results in population protocols for exact majority and leaderelection. Bull. EATCS **3**(126), 1–34 (2018)
15. Angluin, D., Aspnes, J., Eisenstat, D.: Fast computation by population protocols with a leader. Distrib. Comput. **21**(3), 183–199 (2008)
16. Sudo, Y., Ooshita, F., Izumi, T., Kakugawa, H., Masuzawa, T.: Logarithmic expected-time leader election in population protocol model. arXiv preprint arXiv:1812.11309 (2018)

A Self-stabilizing 1-Maximal Independent Set Algorithm

Hideyuki Tanaka[1(✉)], Yuichi Sudo[1], Hirotsugu Kakugawa[2],
Toshimitsu Masuzawa[1], and Ajoy K. Datta[3]

[1] Osaka University, Suita, Japan
tanaka.hideyuki@ist.osaka-u.ac.jp
[2] Ryukoku University, Otsu, Japan
[3] University of Nevada, Las Vegas, USA

Abstract. We consider the 1-maximal independent set (1-MIS) problem: given a graph $G = (V, E)$, our goal is to find an 1-maximal independent set (1-MIS) of a given network G, that is, a maximal independent set (MIS) $S \subset V$ of G such that $S \cup \{v, w\} \setminus \{u\}$ is not an independent set for any nodes $u \in S$, and $v, w \notin S$ ($v \neq w$). We give a silent, self-stabilizing, and asynchronous distributed algorithm to construct 1-MIS on a network of any topology. We assume the processes have unique identifiers and the scheduler is unfair and distributed. The time complexity, *i.e.*, the number of rounds to reach a legitimate configuration in the worst case, of the proposed algorithm is $O(nD)$, where n is the number of processes in the network and D is the diameter of the network. We use a composition technique called *loop composition* [Datta et al. 2017] to iterate the same procedure consistently, which results in a small space complexity, $O(\log n)$ bits per process.

1 Introduction

Nowadays, distributed systems generally consist of *numerous* computers (or processes) where the processes collaboratively solve a problem with communicating each other. Because of the huge scale, distributed systems are prone to have faults of their components. Therefore, it is important to design a algorithm, which correctly works even if some of the processes are crashed, the topology of a network changes, and/or the memory of some processes are corrupted arbitrarily.

Self-stabilization [5] is a promising technique to achieve high fault tolerance. An execution of a self-stabilizing algorithm is guaranteed to reach a legitimate configuration eventually (*Convergence property*), which satisfies the specification of a given problem and keeps the legitimacy thereafter (*Closure property*). These two properties of self-stabilization make distributed systems tolerate any number and any kind of transient faults in a sense that the system can recover and attain

This work was supported by JSPS KAKENHI Grant Numbers 17K19977, 18K18000, 19H04085, and 19K11826 and JST SICORP Grant Number JPMJSC1606.
A. K. Datta passed away on May 26, 2019. Rest in Peace, Ajoy.

M. Ghaffari et al. (Eds.): SSS 2019, LNCS 11914, pp. 338–353, 2019.
https://doi.org/10.1007/978-3-030-34992-9_27

the desired behavior from any illegitimate configuration that those faults may cause.

In this paper, we consider the 1-maximal independent set problems which is the variant of the maximal independent set problem. Given a graph (or a network) $G = (V, E)$, a set $S \subseteq V$ of nodes (or processes) is independent if any two nodes in S are not neighbors. Finding a large independent set of a given graph is important for many applications in distributed systems, for example, clustering in wireless networks (See [1] in detail). However, finding the *maximum* independent set is NP-hard [11]. Therefore, many studies in the literature give a solution to find a *maximal* independent set (MIS), *i.e.*, an independent set such that no proper superset of it is independent. Unfortunately, the maximality of an independent set does not always guarantee a large cardinality of the set. For example, every star graph has an MIS consisting of only one node. Therefore, we consider a stronger maximality called *1-maximality*, which Bollobás et al. [2] introduced[1]. An MIS $S \subseteq V$ is 1-maximal if $S \cup \{v, w\} \setminus \{u\}$ is not independent for any $u \in S$, and $v, w \notin S$ ($v \neq w$). The 1-maximality offers better solution in many cases. For example, the star graph of n nodes has exactly one 1-maximal independent set (1-MIS), whose size is $n - 1$.

1.1   Related Work

The maximal independent set problem is one of the most fundamental problems in graph theory and the field of distributed computing, thus it has been studied in many literature. Table 1 summarizes recent results on self-stabilizing MIS algorithms, where n and D denote the number of processes and the diameter of the network, respectively. In 1995, Shukla et al. [10] gave a self-stabilizing MIS algorithm for any anonymous network. Their algorithm assumes the central scheduler, *i.e.*, exactly one process executes an atomic action at each step. Its worst-case convergence from any configuration, to a configuration where an MIS is constructed requires $O(n)$ steps and it uses only $O(1)$ bits per process. Ikeda et al. [7] gave a different self-stabilizing MIS algorithm. Their algorithm assumes the existence of the process-identifiers, however, it works correctly under the distributed scheduler, which can activate any number of enabled processes simultaneously at each step. The space complexity is still $O(1)$, but the convergence increases to $O(n^2)$ steps. Turau [11] gave a self-stabilizing MIS algorithm with improved convergence time to $O(n)$ in the same settings.

Shi et al. [9] gave the first self-stabilizing 1-MIS algorithm. It assumes that the network topology is a tree and assumes the central scheduler. Its convergence time is $O(n^2)$ steps and the space complexity is $O(1)$ bits per process. Namba [8] gave a self-stabilizing 1-MIS algorithm for any arbitrary graph with assuming the central scheduler. No analysis was presented for convergence time in terms of the number of steps, but its convergence time is $O(n^2)$ (asynchronous) rounds. To construct 1-MIS, his algorithm runs n sub-algorithms in parallel, thus it uses $O(n \log n)$ bits of memory space per process.

---

[1] They actually introduce more general maximality, k-maximality for any $k \geq 1$.

It is worthwhile to mention that deterministic construction for MIS (and thus 1-MIS) is impossible in an anonymous network of an arbitrary topology with the distributed scheduler, due to the lack of ability to break symmetry.

Table 1. Self-stabilizing maximal and 1-maximal independent set algorithms. n denotes the number of processes, D denotes the diameter of the network.

	Problem	Topology	ID	Scheduler	Convergence time		Space
					Steps	Rounds	
[10]	MIS	Any	Unavailable	Central	$O(n)$	$O(n)$	$O(1)$ bits
[7]	MIS	Any	Available	Distributed	$O(n^2)$	$O(n)$	$O(1)$ bits
[11]	MIS	Any	Available	Distributed	$O(n)$	$O(n)$	$O(1)$ bits
[9]	1-MIS	Tree	Unavailable	Central	$O(n^2)$	$O(n)$	$O(1)$ bits
[8]	1-MIS	Any	Available	Central	–	$O(n^2)$	$O(n \log n)$ bits
Proposed	1-MIS	Any	Available	Distributed	–	$O(nD)$	$O(\log n)$ bits

1.2 Our Contributions

We give a silent self-stabilizing 1-MIS algorithm under a distributed scheduler for any arbitrary network. We assume the existence of process-identifiers. Its convergence time is $O(nD)$ rounds, where D is the diameter of the network, while the space complexity is $O(\log n)$ per process. We use a composition technique called *loop composition*, which Datta et al. [3] introduced recently. This technique enables the processes to iterate the same subalgorithm an unbounded number of times consistently until an 1-MIS is constructed, which results in a smaller space complexity, $O(\log n)$ bits per process. To the best of our knowledge, the loop composition technique is utilized only for the k-grouping problem [3] although it seem applicable to many problems. Thus, our result shows the applicability by providing the second success case of the loop composition in the literature.

2   Preliminaries

A undirected network $G = (V, E)$ consisting of process set V and link set E is given. We denote the number of processes and the diameter of G by n and D, respectively. We assume $n \geq 2$. We assume that the network G is connected without loss of generality; if G is not connected, it suffices to construct an 1-MIS for each component of G. Each process v has a unique identifier $v.id$ chosen from a set ID of non-negative integers where $|ID| = O(poly(n))$. Let N_v denote the neighbors of a process v, i.e., $N_v = \{u \in V \mid \{u, v\} \in E\}$. We call the processes in N_v *v-neighbors*. By an abuse of notation, we will identify each process with its identifier, and vice versa, whenever convenient. We call a member of ID a *false identifier* if it is not the identifier of any process in V.

We use the locally shared memory model [5]. A process is modeled by a finite state machine. The state of a process is defined by the values of its variables. A process can read the variables of its own and its neighbors' variables simultaneously, but can update only its own variables. A distributed algorithm defines behavior of each process v by a finite set of actions of the following from: $<label><guard> \longrightarrow <statement>$. The *label* of each action is a number used for reference. The *guard* is a predicate on the variables and identifiers of v and it's neighbors. The *statement* updates the state (or variables) of v. An action can be executed only if it is *enabled*, *i.e.*, its guard evaluates to true, and a process is *enabled* if at least one of its actions is enabled. The evaluation of guard and the execution of the corresponding statement are presumed to take place in one atomic step. For simplicity, we use notation "$v.x \longleftarrow \chi(v)$" to represent an action "$v.x \neq \chi(v) \longrightarrow v.x \leftarrow \chi(v)$" for any variable x and any function $\chi(v)$. Thus, the action "$v.x \longleftarrow \chi(v)$" is enabled if and only if $v.x \neq \chi(v)$. We also use a symbol $\bot$ to represent "null value" and define $\min \emptyset = \bot$ and $\min\{a, \bot\} = a$.

A *configuration* of the network is an n-dimensional vector consisting of a state for each process. We denote by $\gamma(v).x$ the value of variable x of process v in configuration γ. Each transition from a configuration to another, called a *step* of the algorithm, is driven by a *daemon*. We assume the *distributed daemon* in this paper; at each step, the distributed daemon selects one or more enabled processes to execute an action. If a selected process has two or more enabled actions, it executes the action with the smallest label number. We write $\gamma \mapsto_{\mathcal{A}} \gamma'$ if configuration γ can change to γ' by one step of algorithm $\mathcal{A}$. We define an *execution* of algorithm $\mathcal{A}$ to be a sequence of configurations $\gamma_0, \gamma_1, \cdots$ such that $\gamma_i \mapsto_{\mathcal{A}} \gamma_{i+1}$ for all $i \geq 0$. We assume the daemon to be *weakly-fair*, meaning that a continuously enabled process must be selected eventually.

A self-stabilization algorithm ensures that an execution eventually recovers a correct configuration even if it is started from any configuration, *i.e.*, each process may start from any state. An execution is *maximal* if it is infinite, or it terminates at a *final* configuration, *i.e.*, a configuration where no process is enabled. We say that a configuration γ of $\mathcal{A}$ is *safe* for $\mathcal{L}$ if every execution $\gamma_0, \gamma_1, \ldots$ of $\mathcal{A}$ starting from γ (*i.e.*, $\gamma_0 = \gamma$) always satisfies $\mathcal{L}$, that is, $\mathcal{L}(\gamma_i)$ holds for all $i \geq 0$. Algorithm $\mathcal{A}$ is said to be *self-stabilizing* for $\mathcal{L}$ if there exists a set $\mathcal{C}$ of configurations of $\mathcal{A}$ such that every configuration in $\mathcal{C}$ is safe for $\mathcal{L}$ and every maximal execution $\gamma_0, \gamma_1, \ldots$ of $\mathcal{A}$ reaches a configuration in $\mathcal{C}$, *i.e.*, $\gamma_i \in \mathcal{C}$ holds for some $i \geq 0$. We also say that $\mathcal{A}$ is *silent* if every execution of $\mathcal{A}$ is finite. Thus, a silent algorithm $\mathcal{A}$ is self-stabilizing for predicate $\mathcal{L}$ if and only if every final configuration satisfies $\mathcal{L}$.

We sometimes regard a predicate on configurations as the set of configurations. For example, we write $\gamma \in \mathcal{L}_1 \cap \mathcal{L}_2$ when a configuration γ satisfies both predicates $\mathcal{L}_1$ and $\mathcal{L}_2$.

We measure time complexity of an execution in *rounds* [6]. We say that process v is *neutralized* at step $\gamma_i \mapsto \gamma_{i+1}$ if v is enabled at γ_i and not at γ_{i+1}. We define the first round of an execution $\varrho = \gamma_0, \gamma_1, \ldots$ to be the minimum prefix $\gamma_0 \ldots \gamma_s$ during which every process enabled at γ_0 executes an action or is neutralized. The second round of ϱ to be the first round of the execution

$\gamma_s, \gamma_{s+1}, \ldots$ and so forth. We evaluate the number of rounds of ϱ, denoted by $R(\varrho)$, as the execution *time* of ϱ.

2.1 Problem Specification

We specify the 1-maximal independent set problem. A set $S \subseteq V$ of processes is called a *independent set* (or just IS) of G if no two processes in S are neighbors in G, that is, $\forall u, v \in S : \{u, v\} \notin E$. An independent set of G is called a *maximal independent set* (or just MIS) of G if it is not a proper subset of any other independent set of G. A maximal independent set S of G is called an *1-maximal independent set* (or just 1-MIS) of G if we cannot increase cardinality of S without violating the *independent* property by removing one process and adding two or more processes, that is, for any process $v \in S$ and any distinct processes $u, w \notin S$, set $S \cup \{u, w\} \setminus \{v\}$ is not an independent set. We assume that each process v has a variable $v.\text{mis} \in \{\textbf{true}, \textbf{false}\}$. We define predicate $\mathcal{L}_{1\text{MIS}}$ on configurations as follows: $\mathcal{L}_{1\text{MIS}}(\gamma) = \textbf{true}$ holds if and only if, in configuration γ, $\{v \in V \mid v.\text{mis} = \textbf{true}\}$ is an 1-maximal independent set. Our goal is to give a silent and self-stabilizing algorithm for $\mathcal{L}_{1\text{MIS}}$.

3 Loop Compotision

We use the loop composition [3] to design a silent self-stabilizing 1-MIS algorithm. The loop composition is a technique to iterate a given algorithm repeatedly in a harmonious way. To utilize the loop composition, we must design two algorithms $\mathcal{A}$ and $\mathcal{P}$ and a predicate E for a given predicate $\mathcal{L}$. Algorithm $\mathcal{A}$ is a base algorithm that we aim to execute repeatedly. It must satisfy the three requirements, *shiftable convergence*, *loop convergence*, and *correctness*. These requirements are defined in the sequel. Predicate $E : V \mapsto \{\textbf{false}, \textbf{true}\}$ is an locally checkable error-detecting predicate. We say that a configuration γ is *erroneous* for E if $E(v)$ holds for some $v \in V$ in γ. Otherwise, we say that γ is *non-erroneous* for E. Algorithm $\mathcal{P}$ is a silent self-stabilizing algorithm that brings the system to a non-erroneous configuration for E, starting from any configuration. Then, we obtain a composite algorithm $\textbf{Loop}(\mathcal{A}, E, \mathcal{P})$ [3], which is a silent self-stabilizing for $\mathcal{L}$. Very roughly speaking, $\textbf{Loop}(\mathcal{A}, E, \mathcal{P})$ shows the following behavior. Recall that a configuration is final for $\mathcal{A}$ if and only if no action of $\mathcal{A}$ is enabled in any process.

```
1: repeat
2:     if the current configuration is erroneous then
3:         Execute P, which bring the system to a non-erroneous configuration.
4:     else
5:         Execute A, which bring the system to a final configuration for A
6:         Copy the outputs of A to the inputs of A
7:     end if
8: until The current configuration is final and the inputs and the outputs of A
       are the same.
```

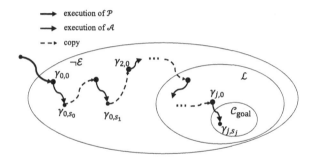

Fig. 1. An execution of $\mathbf{Loop}(\mathcal{A}, E, \mathcal{P})$

In what follows, we describe the conditions for $\mathcal{A}$ and $\mathcal{P}$ and explain the meaning of *copying* from the output of $\mathcal{A}$ to the input of $\mathcal{A}$. We define $O_{\mathcal{A}}$ (resp. $O_{\mathcal{P}}$) as the set of variables of $\mathcal{A}$ (resp.$\mathcal{P}$) whose values can be updated by actions of $\mathcal{A}$ (resp.$\mathcal{P}$), and $I_{\mathcal{A}}$ (resp. $I_{\mathcal{P}}$) as the set of variables $\mathcal{A}$ (resp.$\mathcal{P}$) whose values are never updated and only read by actions of $\mathcal{A}$ (resp. $\mathcal{P}$). We assume $O_{\mathcal{A}} \cap O_{\mathcal{P}} = \emptyset$ and $I_{\mathcal{P}} = \emptyset$. The error detecting predicate $E(v)$ is evaluated by process $v \in V$, and its evaluation depends on variables in $I_{\mathcal{A}} \cup O_{\mathcal{P}}$ of the v-neighbors and v itself. Let $\mathcal{E}$ be a predicate on configurations such that $\mathcal{E}(\gamma)$ holds if and only if $\bigvee_{v \in V} E(v)$ holds in configuration γ. We assume that algorithm $\mathcal{A}$ has a *copying variable* $\overline{x} \in I_{\mathcal{A}}$ for every variable $x \in O_{\mathcal{A}}$. We define γ^{copy} as the configuration obtained by replacing the value of $v.\overline{x}$ with the value of $v.x$ for every process v and every variable $x \in O_{\mathcal{A}}$ in configuration γ. We define predicate $\mathcal{C}_{\text{goal}}(\mathcal{A}, E)$ as follows: configuration γ satisfies $\mathcal{C}_{\text{goal}}(\mathcal{A}, E)$ if and only if $\gamma \in \neg\mathcal{E}$, $\gamma^{copy} = \gamma$, and no action of $\mathcal{A}$ is enabled in any process. We must design $\mathcal{A}$ to satisfy the following three requirements for some $R_{\mathcal{A}}$ and $L_{\mathcal{A}}$:

Shiftable Convergence. Every maximal execution of $\mathcal{A}$ that starts from a configuration in $\neg\mathcal{E}$ terminates at a configuration γ such that $\gamma^{copy} \in \neg\mathcal{E}$.

Loop Convergence. If $\varrho_0, \varrho_1 \ldots$ is an infinite sequence of maximal executions of $\mathcal{A}$ where $\varrho_i = \gamma_{i,0}, \gamma_{i,1}, \ldots, \gamma_{i,s_i}$, $\gamma_{0,0} \in \neg\mathcal{E}$, and $\gamma_{i+1,0} = \gamma_{i,s_i}^{copy}$ for each $i \geq 0$, then $\gamma_{j,s_j} \in \mathcal{C}_{\text{goal}}(\mathcal{A}, E)$ and $R(\varrho_0) + R(\varrho_1) + \ldots R(\varrho_j) \leq R_{\mathcal{A}}$ hold for some $j < L_{\mathcal{A}}$, and

Correctness. $\gamma \in \mathcal{C}_{\text{goal}}(\mathcal{A}, E) \Rightarrow \gamma \in \mathcal{L}$ holds for every configuration γ.

Two integers $L_{\mathcal{A}}$ and $R_{\mathcal{A}}$ are an upper bound on the number of interactions of $\mathcal{A}$'s executions and an upper bound on the total number of rounds of those (iterated) executions in $\mathcal{A}$. We also must design $\mathcal{P}$ such that every maximal execution of $\mathcal{P}$ terminates at a configuration in $\neg\mathcal{E}$ within $T_{\mathcal{P}}$ rounds.

If we design $\mathcal{A}$, E, and $\mathcal{P}$ that satisfy the above conditions, the composited algorithm $\mathbf{Loop}(\mathcal{A}, E, \mathcal{P})$ presented in [3] has the property described in the following theorem.

Theorem 1 ($[3,4]^2$). *Algorithm* **Loop**($\mathcal{A}, E, \mathcal{P}$) *is a silent and self-stabilizing for predicate* $\mathcal{L}$. *Every execution of* **Loop**($\mathcal{A}, E, \mathcal{P}$) *terminates within* $O(n + T_{\mathcal{P}} + R_{\mathcal{A}} + L_{\mathcal{A}}D)$ *rounds. Its space complexity is* $O(S_{\mathcal{A}} + S_{\mathcal{P}} + \log n)$ *bits per process, where* $S_{\mathcal{A}}$ *(resp.* $S_{\mathcal{P}}$*) is space complexity of* $\mathcal{A}$ *(resp.* $\mathcal{P}$*) in bits per process.*

We briefly describe the behavior of **Loop**($\mathcal{A}, E, \mathcal{P}$) in the rest of this section. In an execution of **Loop**($\mathcal{A}, E, \mathcal{P}$), the processes simulates an execution of $\mathcal{A}$ or an execution of $\mathcal{P}$. Starting from any configuration, the processes eventually agree with which algorithm they should simulate. When they detect that the current configuration does not satisfy $\neg\mathcal{E}$, they simulate an execution of $\mathcal{P}$, by which a configuration in $\neg\mathcal{E}$ is reached. (See the (leftmost) thick arrow in Fig. 1). If the current configuration satisfies $\neg\mathcal{E}$, then the processes simulate $\mathcal{A}$. (See the solid arrows in Fig. 1). Whenever a simulated execution of $\mathcal{A}$ terminates, the processes checks whether or not they already reaches a configuration in $\mathcal{C}_{\text{goal}}(\mathcal{A}, E)$. If it does, they will do nothing thereafter. Otherwise, they copy the values of all input variables to the corresponding copy variables, which results in transition of the current configuration. (See the dashed arrows in Fig. 1). After that they simulate a new execution of $\mathcal{A}$ again. From the property of the shiftable convergence and the loop convergence of $\mathcal{A}$, they eventually reach a configuration in $\mathcal{C}_{\text{goal}}(\mathcal{A}, E)$ and they terminate (although they cannot detect termination). Thereafter, legitimate predicate $\mathcal{L}$ is always satisfied thanks to the correctness property of $\mathcal{A}$.

4   Self-stabilizing 1-MIS Algorithm

In this section, we design a silent self-stabilizing algorithm for 1-MIS using the loop composition method described in the previous section. Specifically, we give a base algorithm *Inc* and an initialization algorithm *Init* and an error detecting predicate E_{MIS} such that **Loop**(*Inc*, E_{MIS}, *Init*) is a silent and self-stabilizing algorithm for $\mathcal{L}_{1\text{MIS}}$. The space complexity of **Loop**(*Inc*, E_{MIS}, *Init*) is $O(\log n)$ bits per process and the worst case time complexity is $O(nD)$ rounds.

Every process v maintains a Boolean variable $v.\text{mis} \in \{\textbf{false}, \textbf{true}\}$ and the corresponding copying variable $v.\overline{\text{mis}} \in \{\textbf{false}, \textbf{true}\}$. A variable $v.\overline{\text{mis}}$ is an output variable of *Init* and an input variable of *Inc*. A variable $v.\text{mis}$ is not accessed by *Init*. It is updated only by *Inc*. We define two sets $S_I = \{v \in V \mid v.\overline{\text{mis}}\}$ and $S_O = \{v \in V \mid v.\text{mis}\}$. Define $\mathcal{L}_{\text{input}}$ as the predicate on configurations such that $\mathcal{L}_{\text{input}}(\gamma) = \textbf{true}$ if and only if S_I is an MIS in a configuration γ.

Our goal is to design *Inc*, E_{MIS}, and *Init* such that;

– If S_I is not an MIS, *i.e.*, the current configuration deviates from $\mathcal{L}_{\text{input}}$, then at least one process v must detect the deviation with error detecting predicate $E_{\text{MIS}}(v)$, that is, $\mathcal{L}_{\text{input}}(\gamma)$ holds if and only if $\neg \bigvee_{v \in V} E_{\text{MIS}}(v)$ holds in a configuration γ,

---

[2] Loop composition **Loop**($\mathcal{A}, E, \mathcal{P}$) was originally given in [3], and its time complexity was slightly improved by [4].

- Every maximal execution of *Init* starting from any configuration terminates within $O(n)$ rounds at a configuration in $\mathcal{L}_{\text{input}}$,
- Every maximal execution ϱ of *Inc* starting from a configuration in $\mathcal{L}_{\text{input}}$ where S_I is not an 1-MIS terminates at a configuration where S_O is an MIS such that $|S_O| \geq |S_I| + 1$, within $O(\epsilon + 1)$ rounds where ϵ is $|S_O| - |S_I|$ in the final configuration of ϱ, and
- Every maximal execution ϱ of *Inc* starting from a configuration in $\mathcal{L}_{\text{input}}$ where S_I is an 1-MIS terminates within $O(1)$ rounds at a configuration where $S_O = S_I$.

Table 2. *Init*

[**Actions of process** v]
$\text{I}_1: \quad v.\overline{\text{mis}} \longleftarrow (\forall w \in N_v : \neg w.\overline{\text{mis}} \vee (v.id < w.id))$

Note that the predicate $\mathcal{L}_{\text{input}}$ corresponds to $\neg \mathcal{E}$ in the previous section. If the above conditions hold, then *Inc*, E_{MIS}, and *Init* satisfy all the conditions of the loop composition for $\mathcal{L} = \mathcal{L}_{\text{1MIS}}$, $T_{Init} = O(n)$, $R_{Inc} = O(n)$, $L_{Inc} = n$, thus **Loop**(*Inc*, E_{MIS}, *Init*) is a silent and self-stabilizing algorithm for $\mathcal{L}_{\text{1MIS}}$ and its time complexity is $O(n + T_{Init} + R_{inc} + L_{inc} \cdot D) = O(nD)$ rounds.

We give *Init* and E_{MIS} in Sect. 4.1 and give *Inc* in Sect. 4.2.

4.1   Error Detecting Predicate E_{MIS} and Algorithm *Init*

First, we give the error detecting predicate E_{MIS} as follows:

$$E_{\text{MIS}}(v) \equiv (v.\overline{\text{mis}} \wedge \exists u \in N_v : u.\overline{\text{mis}}) \vee (\neg v.\overline{\text{mis}} \wedge \forall w \in N_v : \neg w.\overline{\text{mis}}).$$

Lemma 1. *For any configuration* γ, $\mathcal{L}_{\text{input}}(\gamma)$ *holds if and only if* $\neg \bigvee_{v \in V} E_{\text{MIS}}(v)$ *holds in* γ.

Proof. Note that $\neg E_{\text{MIS}}(v) \equiv (v.\overline{\text{mis}} \Rightarrow \forall u \in N_v : \neg u.\overline{\text{mis}}) \wedge (\neg v.\overline{\text{mis}} \Rightarrow \exists w \in N_v : w.\overline{\text{mis}})$. Suppose that $\neg \bigvee_{v \in V} E_{\text{MIS}}(v)$ holds in a configuration γ, that is, we have $\neg E_{\text{MIS}}(v)$ for all $v \in V$ in γ. Then, every process in S_I has no neighbor in S_I and every process in $V \setminus S_I$ has at least one process in S_I. Hence, S_I is an MIS in γ and $\mathcal{L}_{\text{input}}(\gamma)$ holds. Suppose the other case, that is, $\bigvee_{v \in V} E_{\text{MIS}}(v)$ in γ. In this case, some process in S_I has a neighbor in S_I or some process in $V \setminus S_I$ has no process in S_I. Hence, S_I is not an MIS in γ and $\mathcal{L}_{\text{input}}(\gamma)$ does not hold.

□

Next, we give an algorithm *Init*. The goal of this algorithm is to bring the network to a configuration where S_I is an MIS within $O(n)$ rounds starting from any configuration. The algorithm *Init* consists of only one action I_1 as given in

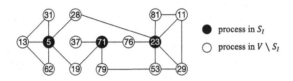

Fig. 2. Example of S_I

Fig. 3. Example of S_O for Fig. 2

Table 2. In this algorithm, a process with a smaller identifier has a higher priority to be a member of S_I. If a process v finds that there is no neighbor with a smaller identifier in S_I, then v becomes a member of S_I, that is, v executes $v.\overline{\text{mis}} \leftarrow \textbf{true}$. Otherwise, that is, if there exists a process with a smaller identifier in $N_v \cap S_I$, v executes $v.\overline{\text{mis}} \leftarrow \textbf{false}$.

Lemma 2. *Every maximal execution of Init starting from any configuration terminates within $O(n)$ rounds at a configuration in $\mathcal{L}_{\text{input}}$.*

Proof. First, we claim that any final configuration of *Init* satisfies $\mathcal{L}_{\text{input}}$. By definition of notation "$v.x \longleftarrow \chi(v)$" (See Sect. 2), a process v is enabled if and only if $v.\overline{\text{mis}} \not\equiv (\forall w \in N_v : \neg w.\overline{\text{mis}} \vee (v.id < w.id)))$. Therefore, in a final configuration γ, where no process is enabled, every process v in S_I has no neighbor in S_I and every process v in $V \setminus S_I$ has at least one neighbor in S_I. Thus, any final configuration satisfies $\mathcal{L}_{\text{input}}$.

Next, we prove that every maximal execution terminates within $O(n)$ rounds. Let $v_1, v_2, \ldots, v_n$ be the processes in V such that $v_1.id < v_2.id < \cdots < v_n.id$. The guard of I_1 in process v, *i.e.*, $v.\overline{\text{mis}} \not\equiv (\forall w \in N_v : \neg w.\overline{\text{mis}} \vee (v.id < w.id)))$, depends only on $v.\overline{\text{mis}}$ and $w.\overline{\text{mis}}$ such that $w.id \leq v.id$. Therefore, v_1 becomes disabled in the first round of any maximal execution of *Init* and never becomes enabled thereafter. Similarly, v_i becomes disabled within one round after all neighboring processes with smaller identifiers than v_i are disabled. Thus, any maximal execution terminates within $O(n)$ rounds. $\qquad\square$

4.2   Algorithm *Inc*

We give an algorithm *Inc* in this section. This algorithm assumes that S_I is an MIS of G. The goal of this algorithm is to bring the network to a configuration where S_O is an MIS such that $|S_O| \geq |S_I| + 1$ if S_I is not an 1-MIS. Otherwise the goal is to reach a configuration where $S_O = S_I$ holds. Note that if S_I is not an 1-MIS, S_O such that $|S_O| \geq |S_I| + 1$ holds necessarily exists. For example, see Fig. 2 where $S_I = \{5, 23, 71\}$ is an MIS of G.

4.2.1   Key Idea

In this subsection, we give a key idea to find an MIS S_O such that $S_O = S_I$ if S_I is an 1-MIS of G, otherwise $|S_O| \geq |S_I| + 1$. Implementation of this idea as a distributed algorithm will be described in Sect. 4.2.2.

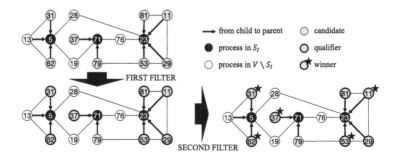

Fig. 4. The first and the second filters

First, we define the parent-child relationship on processes: If a process $v \in V \setminus S_I$ has exactly one neighbor u in S_I, *i.e.*, $N_v \cap S_I = \{u\}$, we say that u is a *parent* of v and v is a *child* of u. That only constructs tree whose height is 1. As we will see later, each process v memorizes the identifier of its parent in v.parent if v has a parent. Otherwise, we assign v.parent $= \bot$. Define C_u as the set of children of process u, that is, $C_u = \{v \in N_u \mid v.\text{parent} = u\}$. Let u and v be any two processes such that u is a parent of v, *i.e.*, v.parent $= u$. If $|C_u| - |C_u \setminus N_v| \geq 2$, we call v a *candidate*. In other words, v is a candidate if and only if it has a parent and this parent has another child in $V \setminus N_v$. Let us see an example (See Fig. 4). Process 31 is a candidate because its parent, process 5, has another child, process 62, which is not a neighbor of process 31. On the other hand, process 13 is not a candidate because all other children of process 5 are neighbors of Process 13. We use the word "candidate" because we can increase the size of MIS by adding some candidates and removing their parent. For example, we obtain larger MIS than S_I if we remove process 5 and add the two non-neighboring candidates among its children, processes 31 and 62. However, we cannot add all candidates to obtain a larger MIS. This is because some of the candidates may be neighbors and thus have conflicts to join the independent set. For example, we cannot add both processes 53 and 79, and we cannot add both processes 11 and 81.

Next, we perform two distinct filers to avoid conflicts. We call a process that survives the first filter (resp. the second filter) a *qualifier* (resp. a *winner*). If a candidate v does not have any neighboring candidate w such that w.parent $<$ v.parent, v is a qualifier. Otherwise, v is not a qualifier. For example, in the example of Fig. 4, process 79 is not a qualifier because the identifier of its parent is 71 and one of its neighbors, process 53, has a parent with identifier $23 < 71$. The other candidates are qualifiers in this example. Any two qualifiers that have distinct parents are not neighbors, thus do not have a conflict while some two qualifiers with the same parent may have a conflict. We chose winners among the qualifiers in the second filter by comparing their identifiers in the same way as algorithm *Init*. The winners are decided by each parent such that it has two or more qualifiers. Let $q_1, q_2, \ldots, q_s$ be the qualifiers of G in the ascending order, *i.e.*, $q_1 < q_2 < \ldots q_s$. Define the set W of qualifiers recursively as follows; $q_1 \in W$, and for each $i \geq 2$, $q_i \in W$ if and only if there is no q_i-neighbor $q_j \in W$ such that $j < i$. The qualifiers in the resulting W are winners and the other qualifiers are non-winners. In the example of Fig. 4, Processes 11, 31, 37, 53, and 62 are winners. We have the following two lemmas about winners.

Lemma 3. *There is at least one process that has two or more winners in its children if S_I is not an 1-MIS of G.*

Proof. Let u be the process with the minimum identifier such that it has a candidate among its children. Such u must exist if S_I is not an 1-MIS. Let c_1 be the candidate in C_u with the smallest identifier. By definition of the candidates, there exists one or more other candidates in C_u that are not c_1-neighbors. Let c_2 be the candidate with the smallest identifier among them. Both c_1 and c_2 survive the first and the second filters thanks to the minimality of u's identifier and the absence of a candidate in C_u that makes c_1 or c_2 drop in the second filter. Hence, there exist at least two winners in u's children.   □

Lemma 4. *There exists no winner if S_I is an 1-MIS of G.*

Proof. Assume for contradiction that S_I is an 1-MIS of G and there is a candidate $v \in V$. Let u be the parent of v. By definition of a candidate, u has at least one candidate other than v, say w, in its children, which is not a neighbor of v. Since v and w are candidates, $S_I \cap (N_v \cup N_w) = \{u\}$ holds. Therefore, $S_O = S_I \cup \{v, w\} \setminus \{u\}$ is an independent set and $|S_O| = |S_I| + 1$, which contradicts the assumption that S_I is an 1-MIS of G.   □

Finally, we choose $S_O = S_A(S_I) \cup S_B(S_I) \cup S_C(S_I)$ where $S_A(S_I)$, $S_B(S_I)$, and $S_C(S_I)$ are the sets of processes that we will define in the following. These sets depend on S_I, but we always omit S_I from their notations, *i.e.*, just write S_A, S_B, and S_C, whenever it is clear from the contest. Define S_A as the set of all processes u in S_I such that u has less than two winners in its children. Define S_B as the set of all winners v such that $v.\text{parent} \notin S_A$. By definition, $S_A \cup S_B$ is an independent set of G. Moreover, $|S_A \cup S_B| \geq |S_I| + |S_B|/2 \geq |S_I| + 1$ holds if S_I is not an 1-MIS. This is because (i) $S_B \neq \emptyset$ holds by Lemma 3, and (ii) each process in $S_I \setminus S_A$ has at least two winners in its children, thus $|S_B| \geq 2|S_I \setminus S_A|$ holds. On the other hand, $S_A = S_I$ and $S_B = \emptyset$ holds if S_I is an 1-MIS, by Lemma 4. Note that $S_A \cup S_B$ is an independent set but may not be an MIS of G (if S_I is not an 1-MIS). In the example of Fig. 4, $S_A \cup S_B = \{11, 31, 53, 62, 71\}$ is not an MIS because $S_A \cup S_B \cup \{28\}$ is also an independent set of G. We define the set S_C as the maximal set of non-winners such that $p_i \in S_C$ if and only if there exists no p_i-neighbor in $S_A \cup S_B$ and no p_i-neighbor with a smaller identifier than p_i in S_C The set $S_A \cup S_B \cup S_C$ is an independent set since each $p_i \in S_C$ does not have a neighbor in $S_A \cup S_B \cup S_C$. Furthermore, $S_A \cup S_B \cup S_C$ is an MIS of G because otherwise there must be a non-winner $v \in V \setminus S_I$ such that there exists no v-neighbor in $S_A \cup S_B \cup S_C$, but it implies $v \in S_C$ by definition of S_C, a contradiction. Therefore, we have the following three lemmas.

Lemma 5. *The set $S_A \cup S_B \cup S_C$ is an MIS of G if S_I is an MIS of G.*

Lemma 6. $|S_A \cup S_B \cup S_C| \geq |S_I| + |S_B|/2 + |S_C| \geq |S_I| + 1$ *if S_I is an MIS but not an 1-MIS of G.*

Lemma 7. $S_A \cup S_B \cup S_C = S_I$ *and* $S_B = S_C = \emptyset$ *if S_I is an 1-MIS of G.*

Thus we achieve our goal by letting $S_O = S_A \cup S_B \cup S_C$. If S_I is not an 1-MIS, $|S_O| \geq |S_I| + 1$. Otherwise, $S_O = S_I$. In the example of Fig. 4, $S_A = \{71\}$, $S_B = \{11, 31, 53, 62\}$, $S_C = \{28\}$, and hence $|S_O| = 6 > 3 = |S_I|$ (See Fig. 3).

Table 3. *Inc*

[Actions of process v]		
M$_1$:	v.parent	$\longleftarrow$ *Parent*(v)
M$_2$:	v.numchild	$\longleftarrow |C_v|$
M$_3$:	v.cand	$\longleftarrow$ *Cand*(v)
M$_4$:	v.qualifier	$\longleftarrow$ *Qualifier*(v)
M$_5$:	v.winner	$\longleftarrow$ *Winner*(v)
M$_6$:	v.mis	$\longleftarrow InS_A(v) \vee InS_B(v) \vee InS_C(v)$

Table 4. Functions of *Inc*

$$Parent(v) = \begin{cases} \min\{w.id \mid w.\overline{\text{mis}}\} & \text{if } |\{w \in N_v \mid w.\overline{\text{mis}}\}| = 1 \\ \bot & \textbf{otherwise} \end{cases}$$

$$C_v = \{w \in N_v \mid w.\text{parent} = v\}$$

$$Cand(v) \equiv v.\text{parent} \neq \bot$$
$$\wedge \, (v.\text{parent}).\text{numchild} - |\{w \in N_v \mid v.\text{parent} = w.\text{parent}\}| \geq 2$$

$$Qualifier(v) \equiv v.\text{cand} \wedge (\forall w \in N_v : w.\text{cand} \Rightarrow v.\text{parent} \leq w.\text{parent})$$

$$Winner(v) \equiv v.\text{qualifier} \wedge (\forall w \in N_v : v.id < w.id \vee \neg w.\text{winner})$$

$$InS_A(v) \equiv v.\overline{\text{mis}} \wedge |\{w \in C_v \mid w.\text{winner}\}| \leq 1$$

$$InS_B(v) \equiv \neg v.\overline{\text{mis}} \wedge v.\text{winner} \wedge \neg (v.\text{parent}).\text{mis}$$

$$InS_C(v) \equiv \neg v.\overline{\text{mis}} \wedge \neg v.\text{winner} \wedge \begin{pmatrix} \forall w \in N_v \text{ s.t. } w.\text{mis} : \\ \neg \left(w.\overline{\text{mis}} \vee w.\text{winner} \vee w.id < v.id \right) \end{pmatrix}$$

4.2.2 Distributed Implementation

Each process v has six variables v.parent $\in ID$, v.numchild $\in \{0, 1, 2, \ldots, |N_v|\}$, v.cand $\in \{\text{false}, \text{true}\}$, v.qualifier $\in \{\text{false}, \text{true}\}$, v.winner $\in \{\text{false}, \text{true}\}$, and v.mis $\in \{\text{false}, \text{true}\}$, and the corresponding copying variables for the six variables. Each child v, *i.e.*, a process in $V \setminus S_I$ that has exactly one neighbor in S_I, stores the identifier of its parent on v.parent. Each process u in S_I store the number of its children, *i.e.*, $|C_v|$, on u.numchild. Since $|ID| = O(poly(n))$, two variables parent and numchild require $O(\log n)$ bits per each process. The Boolean variable v.cand, v.qualifier, and v.winner represent whether or not a process v is a candidate, a qualifier, and a winner, respectively.

Actions of *Inc* are given in Table 3 and the functions used in Table 3 are given in Table 4. We use a hierarchical composition to design *Inc*. Actions $M_1, M_2, \ldots, M_6$ maintain variables parent, numchild, cand, qualifier, winner, and mis, respectively. We say that an action M_i converges if all of $M_1, M_2, \ldots, M_i$ are disabled in all the processes. Generally, action M_i ($i \geq 2$) refers the variables maintained by $M_1, M_2, \ldots, M_{i-1}$. Therefore, before M_{i-1} converges, some of those variables in some process may not be correct, thus action M_i does not compute the correct value of the variable it maintains. However, after M_{i-1} converges, those variables maintained by $M_1, M_2, \ldots, M_{i-1}$ are correct in all the processes, thus action M_i can use the correct values of those variables.

Actions M_1, M_2, M_3, and M_4 are simple and straightforward. By these actions, each process v computes its parent (if it has), the number of its children, whether or not it is a candidate, and whether or not it is a qualifier, respectively. Actions M_5 simulates the second filter. By M_5, a qualifier v sets v.winner $= \text{true}$ (become a winner) if and only if every $w \in N_v$ such that w.winner $= \text{true}$ has larger identifier than v. The second filter implemented by Action M_5 obviously computes the correct value of winner, *i.e.*, v.winner holds if and only if $v \in W$. However, it sometimes requires more than a constant number of rounds. In the example shown in Fig. 4, process 53 may become the first winner because there is no winner among its neighbors in the configuration shown in the figure. However, process 29 may become a winner later, and then, process 53 will get back to a non-winner since process 29 has a smaller identifier. Process 29 also must get back to a non-winner because process 11 is the smallest qualifier among its neighbors and eventually becomes a winner. After that, process 53 will become a winner again because now it has no winner among its neighbors. Eventually, an execution of *Inc* reaches the configuration shown in Fig. 4. As we will see later, the flipping behavior like this example may require $\Theta(k)$ rounds in the worst case when some process in S_I has k winners among its children after this filter converges.

A variable v.mis is maintained by Action M_6. Our goal is to set v.mis such that $S_O = S_A \cup S_B \cup S_C$ holds. After M_5 converges, every process $v \in S_I$ computes v.mis correctly within one round by M_6; every $v \in S_I$ executes v.mis $\leftarrow \text{true}$ if and only if $v \in S_A$, *i.e.*, there is no more than one winner among v's children. Every $v \in W$ computes v.mis correctly in the next round;

it executes $v.\mathtt{mis} \leftarrow \mathbf{true}$ if and only if $v \in S_B$, i.e., $(v.\mathtt{parent}).\mathtt{mis} = \mathbf{false}$. Thereafter, every process $v \notin S_I \cup W$ computes $v.\mathtt{mis}$ correctly; it executes $v.\mathtt{mis} \leftarrow \mathbf{true}$ if and only if $v \in S_C$, i.e., there is no v's neighbor w such that $w \in S_A \cup S_B$ or $w \in S_C \wedge w.id < v.id$. Since the last computation (for $v \notin S_I \cup W$) is recursive, it requires $O(|S_C|)$ rounds for the same reason as the computation of M_5.

Lemma 8. *Every maximal execution ϱ of Inc starting from a configuration in $\mathcal{L}_{\text{input}}$ terminates at a configuration where $S_O = S_A \cup S_B \cup S_C$, within $O(1 + |S_B| + |S_C|)$ rounds.*

Proof. By definition of the six actions of *Inc*, $S_O = S_A \cup S_B \cup S_C$ holds when ϱ terminates. Therefore, it suffices to show that ϱ terminates within $O(1+|S_B|+|S_C|)$ rounds. Actions M_1, M_2, M_3, and M_4 converges within $O(1)$ rounds because Action M_i $(1 \le i \le 4)$ refers only variables that are maintained by the actions $M_1 \ldots, M_{i-1}$. The same does not hold for Actions M_5 because it is recursive in a sense that a qualifier v may refer a variable $\mathtt{winner}$ of some neighbors to compute its own $\mathtt{winner}$. In the following, for any process v and any variable x, we say that $v.x$ converges at a point of a maximal execution if v does not change the value of $v.x$ after the point. Let u be any process in S_I and W_u be the set of winners among w's children, i.e., $W_u = W \cap N_v$. Let $w_{u,1}, w_{u,2}, \ldots, w_{u,|W_u|}$ be the processes in W_u in the ascending order of their identifiers. After M_4 converged, $w_{u,1}.\mathtt{winner}$ converges to $\mathbf{true}$ within one round and $v.\mathtt{winner}$ of all its neighbors v converges to $\mathbf{false}$ within the next round. After that, $w_{u,2}.\mathtt{winner}$ converges within one round since $w_{u,2}$ has a smaller identifier than any qualifier in $N_{w_{u,2}} \setminus N_{w_{u,1}}$. Generally, $w_{u,i}.\mathtt{winner}$ converges to $\mathbf{true}$ within $O(i)$ rounds after M_4 converged. Thus, Action M_5 converges within $O(\max_{u \in S_I} |W_u|) = O(|S_B|)$ rounds. After M_5 converges, M_6 converges within $O(|S_C|)$ rounds for the same reason. □

In what follows, we show that algorithm *Inc* satisfies three requirements, shiftable convergence, loop convergence, and correctness.

Lemma 9 (Shiftable Convergence). *Every maximal execution ϱ of Inc starting from a configuration in $\mathcal{L}_{\text{input}}$ terminates at a configuration γ such that $\gamma^{\text{copy}} \in \mathcal{L}_{\text{input}}$.*

Proof. In the final configuration γ, $S_O = S_A \cup S_B \cup S_C$ holds by Lemma 8. This set S_O is an MIS of G by Lemma 5. Clearly, S_O in γ equals to S_I in γ^{copy}. Then, S_I is an MIS of G in γ^{copy}, which yields $\gamma^{\text{copy}} \in \mathcal{L}_{\text{input}}$. □

Recall that, for any configuration of *Inc*, $\gamma \in \mathcal{C}_{\text{goal}}(Inc, E_{\text{MIS}})$ means that $\gamma \in \mathcal{L}_{\text{input}}$, $\gamma^{copy} = \gamma$, and no action of *Inc* is enabled in any process.

Lemma 10 (Loop Convergence). *If $\varrho_0, \varrho_1 \ldots$ is an infinite sequence of maximal executions of $\mathcal{A}$ where $\varrho_i = \gamma_{i,0}, \gamma_{i,1}, \ldots, \gamma_{i,s_i}$, $\gamma_{0,0} \in \mathcal{L}_{\text{input}}$, and $\gamma_{i+1,0} = \gamma_{i,s_i}^{\text{copy}}$ for each $i \ge 0$, then $\gamma_{j,s_j} \in \mathcal{C}_{\text{goal}}(Inc, E_{\text{MIS}})$ and $R(\varrho_0)+R(\varrho_1)+\ldots R(\varrho_j) = O(n)$ hold for some $j \le n$.*

Proof. By Lemma 9, the initial configuration of every execution ϱ_i satisfies $\gamma_{i,0} \in \mathcal{L}_{\text{input}}$ by induction on $i \geq 0$. Let $S_{I,0}, S_{I,1}, \ldots$ be S_I in the initial configurations $\gamma_{0,0}, \gamma_{1,0}, \ldots$ of $\varrho_0, \varrho_1, \ldots$, respectively. Let $S_{O,0}, S_{O,1}, \ldots$ be S_O in the final configurations $\gamma_{0,s_0}, \gamma_{1,s_1}, \ldots$ of $\varrho_0, \varrho_1, \ldots$, respectively. By definition, $S_{I,i+1} = S_{O,i}$ holds for all $i \geq 0$. By Lemmas 6 and 8, $|S_{I,i}| < |S_{I,i+1}|$ holds unless $S_{I,i}$ is an 1-MIS of G. Since $|S_{I,i}| < n$ holds for all i and $|S_{I,0}| \geq 1$, there exists $j' < n$ such that $S_{I,j'} = S_{I,j'+1} = S_{I,j'+2} = \ldots$, *i.e.*, $S_{I,k}$ is the same for all $k \geq j'$ by Lemmas 7 and 8. We consider the minimum such j' in what follows. Algorithm *Inc* refers only variable $\overline{\texttt{mis}}$ among the six copying variables. Therefore, letting $v \in V$ be any process and x be any output variable, $v.x$ is the same in $\gamma_{j',s_{j'}}$ and $\gamma_{j'+1,s_{j'+1}}$. Furthermore, $v.x$ in $\gamma_{j',s_{j'}}$ equals to $v.\bar{x}$ in $\gamma_{j'+1,s_{j'+1}}$ because $\gamma_{j',s_{j'}}^{\text{copy}} = \gamma_{j'+1,0}$. and $v.\bar{x}$ never changes in execution $\varrho_{j'+1}$. Therefore, $\gamma_{j'+1,s_{j'+1}}^{\text{copy}} = \gamma_{j'+1,s_{j'+1}}$, which yields $\gamma_{j'+1,s_{j'+1}} \in \mathcal{C}_{\text{goal}}(Inc, E_{\text{MIS}})$. Lemmas 6, 7, and 8 gives $R(\varrho_0) + R(\varrho_1) + \ldots R(\varrho_{j'+1}) = O(n)$. □

Lemma 11 (Correctness). *Every configuration* $\gamma \in \mathcal{C}_{\text{goal}}(Inc, E_{\text{MIS}})$ *satisfies* $\mathcal{L}_{1\text{MIS}}$.

Proof. We have $S_I = S_O$ because $\gamma \in \mathcal{C}_{\text{goal}}(Inc, E_{\text{MIS}})$. Assume for contradiction that $S_O(= S_I)$ is not an 1-MIS of G in γ. Since γ is a final configuration (*i.e.*, no process is enabled), $S_O = S_A(S_I) \cup S_B(S_I) \cup S_C(S_I)$ holds by Lemma 8. Therefore, $|S_O| \geq |S_I| + 1$ holds by the assumption and Lemma 6, which yields $S_O \neq S_I$, a contradiction. □

Theorem 2. *Algorithm* **Loop**$(Inc, E_{\text{MIS}}, Init)$ *is silent and self-stabilizing for* $\mathcal{L}_{1\text{MIS}}$. *Every maximal execution of* **Loop**$(Inc, E_{\text{MIS}}, Init)$ *starting from any configuration terminates within* $O(nD)$ *rounds. Algorithm* **Loop**$(Inc, E_{\text{MIS}}, Init)$ *uses* $O(\log n)$ *bits per process.*

Proof. Immediately follows from Theorem 1 and Lemmas 1, 2, 9, 10, and 11 because $O(n + T_{Init} + R_{inc} + L_{inc} \cdot D) = O(nD)$. □

References

1. Awerbuch, B., Luby, M., Goldberg, A.V., Plotkin, S.A.: Network decomposition and locality in distributed computation. In: 30th Annual Symposium on Foundations of Computer Science, pp. 364–369. IEEE (1989)
2. Bollobás, B., Cockayne, E.J., Mynhardt, C.M.: On generalised minimal domination parameters for paths. In: Annals of Discrete Mathematics, vol. 48, pp. 89–97. Elsevier (1991)
3. Datta, A.K., Larmore, L.L., Masuzawa, T., Sudo, Y.: A self-stabilizing minimal k-grouping algorithm. In: Proceedings of the 18th International Conference on Distributed Computing and Networking, p. 3. ACM (2017)
4. Datta, A.K., Larmore, L.L., Masuzawa, T., Sudo, Y.: A self-stabilizing minimal k-grouping algorithm. arXiv preprint arXiv: 1907.10803 (2019)
5. Dijkstra, E.W.: Self-stabilizing systems in spite of distributed control. Commun. ACM **17**(11), 643–644 (1974)

6. Dolev, S.: Self-stabilization. MIT Press, Cambridge (2000)
7. Ikeda, M., Kamei, S., Kakugawa, H.: A space-optimal self-stabilizing algorithm for the maximal independent set problem. In: The Third International Conference on Parallel and Distributed Computing, Applications and Technologies (PDCAT), pp. 70–74 (2002)
8. Namba, E.: A hierarchical self-stabilizing 1-MIS algorithm. Master's thesis, Osaka University (2017). (in Japanese)
9. Shi, Z., Goddard, W., Hedetniemi, S.T.: An anonymous self-stabilizing algorithm for 1-maximal independent set in trees. Inf. Process. Lett. **91**(2), 77–83 (2004)
10. Shukla, S.K., Rosenkrantz, D.J., Ravi, S.S., et al.: Observations on self-stabilizing graph algorithms for anonymous networks. In: Proceedings of the Second Workshop on Self-stabilizing Systems, vol. 7, p. 15 (1995)
11. Turau, V.: Linear self-stabilizing algorithms for the independent and dominating set problems using an unfair distributed scheduler. Inf. Process. Lett. **103**(3), 88–93 (2007)

Black Hole Search Despite Byzantine Agents

Masashi Tsuchida, Fukuhito Ooshita$^{(\boxtimes)}$, and Michiko Inoue

Nara Institute of Science and Technology, Ikoma, Japan
{tsuchida.masashi.td8,f-oosita,kounoe}@is.naist.jp

Abstract. We study the black hole search problem of k mobile agents in synchronous Byzantine environments. The goal of the black hole search problem is to detect a black hole node, which deletes all agents visiting the node without any trace. We assume that the graph topology is arbitrary, each agent has a unique ID, and at most f_u strongly Byzantine agents exist. Under these assumptions, we propose an algorithm that detects a black hole node in $O(f_u n)$ rounds when $k \geq 2f_u + 2$ holds, where n is the number of nodes. We also show that it is impossible to solve the black hole search problem when $k \leq 2c_u + 1$ holds, where c_u is an upper bound of the number of crash agents. Since a crash fault is a special case of a Byzantine fault, the above result also applies to strongly Byzantine agents. This implies that our proposed algorithm is optimal in terms of the number of tolerable faulty agents. To the best of our knowledge, this is the first work to address the black hole search problem with Byzantine agents.

Keywords: Mobile agent · Black hole search problem · Synchronous network · Byzantine fault

1 Introduction

In recent years, a distributed system with multiple computers (nodes) has become larger. Since nodes communicate with each other, the communication complexity increases in huge distributed systems. This makes it complicated to design distributed systems because developers must maintain a huge number of nodes and treat massive data communication among them. To alleviate these problems, mobile agents have attracted attention as a new paradigm [1]. Agents are software programs that can autonomously move from node to node and execute various tasks in distributed systems. By using agents, nodes do not need to communicate with each other because agents themselves can collect and analyze data by moving around distributed systems. Therefore, we can simplify design of distributed systems by using agents. In addition, agents can efficiently

This work was supported in part by JSPS KAKENHI Grant Number 18K11167 and JST SICORP Grant Number JPMJSC1806.

M. Ghaffari et al. (Eds.): SSS 2019, LNCS 11914, pp. 354–367, 2019.
https://doi.org/10.1007/978-3-030-34992-9_28

execute tasks by cooperating with other agents. Hence many researchers study algorithms to realize cooperation among multiple agents.

Since a node exists as a physical entity, it may crash. In addition, as the number of nodes in distributed systems increases, the possibility that some faulty nodes exist also increases. To realize continuous availability, it is necessary to cope with faults such as node crashes. Several works address the black hole search problem of finding faulty nodes. In this problem, we assume that a faulty node erases all visited agents without any trace, and we call this node a black hole node. In addition, some of agents might be faulty because they move on the distributed systems and might be affected by several noise. Faulty agents may behave arbitrarily without following the algorithm. We call such agents Byzantine agents.

As mentioned above, agents and nodes may become faulty and hence it is important to design fault-tolerant algorithms. Many works assume that faults occur in either agents or nodes in the mobile agent environment. However, it is possible that faults of agents and nodes occur at the same time. In this paper, we consider distributed systems such that at most one black hole node exists and, at the same time, multiple Byzantine agents exist. Under this assumption, we propose an algorithm that detects a black hole node regardless of the behavior of Byzantine agents if the black hole exists. In addition, our proposed algorithm is optimal in terms of the number of tolerable faulty agents. To the best of our knowledge, this is the first work to address the black hole search problem in Byzantine environments.

Table 1. Searching for a black hole node in graphs (F-to-F is "face-to-face", Pure T is "pure token", Enhanced T is "enhanced token", n is the number of nodes, f_u is the upper bound of the number of Byzantine agents).

	Synchronicity	Graph	Network knowledge	Starting location	Comm. models	Byzantine agents	# agents	Time to solve
[4]	Sync.	Ring	Map	Common	F-to-F	No	≥ 2	$O(n \log n)$[a]
[8]	Sync.	Arbitrary	Map	Common	F-to-F	No	2	$3\frac{3}{8} \cdot$OPT
[2]	Sync.	Torus	Sense of direction	Arbitrary	Pure T & F-to-F	No	≥ 3	-
[5]	Async.	Ring	Map	Common	Enhanced T	No	≥ 2	$\Theta(n \log n)$[b]
[6]	Async.	Arbitrary	Map	Common	Pure T	No	≥ 2	$\Theta(n \log n)$[b]
Proposed	Sync.	Arbitrary	Map	Common	F-to-F	Yes	$\geq 2f_u + 2$	$O(f_u n)$

[a]It represents the number of rounds required to solve the problem.
[b]It represents the number of movements of the agent to solve the problem.

1.1 Related Works

The black hole search problem has been widely studied in literature [2,4–6, 8]. Table 1 summarizes some of the results. In the face-to-face model, agents communicate only when they stay at the same node [3]. In the pure token model, a token indicates 1-bit information, and each agent can carry a constant number of tokens and place them on nodes of the network [2,6]. In the enhanced token model, a token indicates 1-bit information similarly to the pure token model, but agents can place it in the middle of a node or on a port. For example, in [5], tokens indicate that the node connected to the end of an edge is a black hole.

The purpose of these works is to clarify the solvability of the black hole search problem in various environments, and to clarify the optimal cost if it can be solved. Many results for the black hole search problem have already been shown and they are surveyed by Peng et al. [9].

For synchronous networks, many deterministic algorithms to search for a black hole node have been proposed [2,4,8]. Dobrev et al. [4] devise *Cautious Walk*, which is a general technique to specify the location of the black hole node with only two agents. In the cautious walk, two agents cooperatively identify that nodes are safe or not. To check an adjacent node by two agents, first one agent a_i moves to the node and the other agent a_j waits for a_i to come back. If a_i comes back in the next round, the adjacent node is identified to be safe, and otherwise a_j finds that the node is a black hole node. Since the cautious walk is an effective technique to locate a black hole node, many related researches also use the cautious walk to reduce complexities including the number of agent moves. Klasing et al. [8] prove that the black hole search problem is NP-hard with respect to minimizing the number of rounds to solve the problem in arbitrary graphs. In addition, they propose a $3\frac{3}{8}$-approximation algorithm, showing the first non-trivial approximation ratio upper bound for this problem. The algorithm in [8] follows an intuitive approach of exploring the network graph via a spanning tree and they prove that this approach cannot lead to an approximation ratio better than $\frac{3}{2}$. Chalopin et al. [2] study oriented tori under the following assumptions: agents start from different nodes, and the communication model uses both the pure token model and the face-to-face model. They show that three agents, each with two movable tokens, are necessary and sufficient to solve the problem in any oriented torus.

For asynchronous networks, the cautious walk does not work well. This is because it is impossible to distinguish whether an agent disappears by a black hole node or it stacks in an asynchronous link [10]. The only way to detect a black hole node in an asynchronous network is to explore the entire network [7]. Dobrev et al. [5] show that the black hole search problem can be solved by at least two agents without knowledge of k in asynchronous ring networks with enhanced tokens when there is exactly one black hole node, where k is the number of agents. Dobrev et al. introduce an algorithm that allows agents to solve the problem with $\Theta(n \log n)$ moves, where n is the number of nodes. Flochini et al. [6] show that the black hole search problem can be solved in any asynchronous network with $\Theta(n \log n)$ moves if there are two agents with a map and a token when there is exactly one black hole node.

The main topics of related works are solvability and cost optimization in various environments. In literature no work consider Byzantine agents for the black hole search problem.

1.2   Our Contributions

In this work, we focus on the black hole search problem considering faults of agents. We assume that all agents start from the same node and that each agent knows the graph topology and the number of agents. In addition, we assume that agents can communicate only when they stay at the same node. We first prove that no algorithm can correctly detect a black hole node with $c_u + 1$ correct agents, where c_u is an upper bound of the number of crash agents. Since a crash fault is a special case of a Byzantine fault, the above result also applies to strongly Byzantine agents. Here, a strongly Byzantine agent is an agent that can behave arbitrarily, including changing its ID. Therefore, we can also say that, if at most f_u Byzantine agents exist, $2f_u + 2$ or more agents are necessary to locate a black hole node.

Next, we propose an algorithm to locate a black hole node with $2f_u + 2$ agents when at most f_u Byzantine agents exist. From the above impossibility result, our algorithm is optimal in terms of the number of tolerable faulty agents. To the best of our knowledge, this is the first work to address the black hole search problem in Byzantine environments.

2   Preliminaries

2.1   A Distributed System

A distributed system is modeled by a connected undirected graph $G = (V, E)$, where V is a set of nodes and E is a set of edges. The set of nodes is denoted by $V = \{v_0, v_1, \cdots, v_{n-1}\}$ and the number of nodes is denoted by $n = |V|$. A set of adjacent nodes of node v_i is denoted by $N_{v_i} = \{v_j | (v_i, v_j) \in E\}$. The degree of node v_i is defined as $d(v_i) = |N_{v_i}|$. Each edge is labeled locally by function $\lambda_{v_i} : \{(v_i, v_j) | v_j \in N_{v_i}\} \rightarrow \{1, 2, \cdots, d(v_i)\}$ such that $\lambda_{v_i}(v_i, v_j) \neq \lambda_{v_i}(v_i, v_p)$ holds for $v_j \neq v_p$. We say $\lambda_{v_i}(v_i, v_j)$ is a port number (or port) of edge (v_i, v_j) on node v_i.

Multiple agents exist in a distributed system. The number of agents is denoted by k, and a set of agents is denoted by $A = \{a_0, a_1, \cdots, a_{k-1}\}$. Each agent has a unique ID, and the ID of agent a_i is denoted by ID_i. In addition, each agent is equipped with a graph map of G.

In this paper, we show an impossibility result that, if the number of correct agents is less than $c_u + 2$, agents cannot detect a black hole node even if agents can leave information on a node. For this reason, we define a model such that each node has a whiteboard, which is an area where agents can leave some information.

In this paper, we assume that at most one black hole node exists in the distributed system. A black hole node has a stationary process of deleting any agent arriving at the node without any trace. Deleted agents cannot be recognized by other agents.

Each agent is modeled as a state machine (S, δ). The first element S is a set of agent states, where each agent state is determined by values of variables in its memory. The second element δ is the state transition function that decides the behavior of an agent. The input of δ is IDs and agent states of all agents on the current node, the content of the whiteboard on the current node, and the incoming port number. The output of δ is the next agent state, the next content of the whiteboard, decision of the next movement (stay or leave) and the outgoing port number if the agent leaves. Here, we define two special states $s_{fault}, s_{terminal} \in S$ and transitions related to these states. State s_{fault} is a state in which agents cannot move to other nodes and cannot change other states. When an agent gets crashed, it changes to s_{fault}. We will discuss this transition later. When an agent arrives at the black hole node, it changes to s_{fault}. In addition, no agent can observe the state and existence of the agent with s_{fault}. State $s_{terminal}$ is a terminal state in which agents decide to terminate. If a_i changes to $s_{terminal}$, it never changes its state after that.

A global configuration C is defined as a 3-tuple $C = (M, P, W)$. The first element M is a k-tuple $M = (m_0, m_1, \cdots, m_{k-1})$, where m_i is a state of agent $a_i(0 \le i \le k - 1)$. The second element P is a k-tuple $P = (p_0, p_1, \cdots, p_{k-1})$, where p_i is the position of agent $a_i(0 \le i \le k - 1)$. Here, we define that positions of agents with s_{fault} to be $null$. The third element W is a n-tuple $W = (w_0, w_1, \cdots, w_{n-1})$, where w_i is a state of the whiteboard of node $v_i(0 \le i \le n - 1)$.

We denote by $\mathcal{C}$ the set of all possible configurations. In initial configuration $C_0 \in \mathcal{C}$, all agents have the designated initial state $s_{initial} \in S$ and placed at a starting node $v_s(\in V)$. In addition, states of whiteboards on all nodes are $ws_{empty} \in WS$ in initial configuration. State ws_{empty} is a state in which no information is written on the whiteboard.

2.2 Fault Models of Agents

In this work, we consider two types of faults of agents: a Byzantine fault and a crash fault.

We call an agent which has a Byzantine fault as a Byzantine agent. Each Byzantine agent behaves arbitrarily without following the algorithm. In addition, we assume that Byzantine agents do not necessarily disappear even when they visit the black hole node, but they can disappear anywhere on their own. This implies that they can simulate disappearance by the black hole on any node. We consider Byzantine agents only in Sect. 4. In Sect. 4, we assume that at most f_u Byzantine agents exist in the distributed system, and it satisfies $k \ge 2f_u + 2$. We assume that all agents know f_u.

A crash fault makes an agent stop during the algorithm execution, and we call such an agent a crash agent. A crash agent a_c may stop operations in arbitrary timing and it never works after that. We define that an agent a_c transitions to s_{fault} when it stops operating. In addition, no agent can observe the state and existence of a_c after a_c stops operations. We consider crash agents only in Sect. 3. In Sect. 3, we assume that at most c_u crash agents exist in the distributed system and all agents know c_u.

We define correct agents as agents that always follow the algorithm. Note that faulty agents cannot be distinguished from correct agents as long as they correctly execute an algorithm.

2.3   Synchronicity

All agents start the algorithm at the same time, and then operate in sequential synchronous rounds. Each correct agent a_i executes the following operations during the r-th round ($r = 0, 1, 2, \ldots$):

1. At the beginning of the r-th round, a_i obtains a snapshot of its current node v. The snapshot includes a state of a whiteboard and states of all agents on v. Note that all agents on v obtain the same snapshot.
2. Based on the snapshot, a_i computes new states of the whiteboard and itself, and decides whether it stays or moves. In order to cope with cases where multiple agents write informations on the whiteboard at the same time, we define the state transition of the whiteboard. We define the update function of the whiteboard w_k. Each agent outputs a next content of the whiteboard in r-th round. The whiteboard w_k executes the update function with these contents as arguments and transits the state until $r + 1$-th round. Hence, the whiteboard w_k becomes the information obtained by merging the output of agents in $r + 1$-th round.
3. If a_i decides to move, it completes the movement before the beginning of the $(r + 1)$-th round.

On the other hand, Byzantine agents can visit any number of nodes and update states of whiteboards on the nodes during one round. An execution $E = C_0, C_1, \ldots$ is an infinite sequence of configurations such that C_i is a configuration at the beginning of the r-th round.

2.4   Black Hole Search Problem

The goal of the black hole search problem is to satisfy the following conditions in any execution.

- If a black hole node exists in the graph, (1) at least one correct agent remains surviving, and (2) all surviving correct agents report the location of the black hole node and terminate.
- If no black hole node exists in the graph, all correct agents report non-existence of a black hole node and terminate.

To evaluate the performance of the algorithm, we consider the maximum number of rounds required for a correct agent to terminate.

3   Impossibility with $c_u + 1$ Correct Agents

In this paper, we propose an algorithm that can detect a black hole node regardless of the behavior of Byzantine agents. In this section, we consider an environment in which some agents may crash. This is a weaker fault model than a Byzantine fault. We prove that, when at most c_u crash agents exist, $k \geq 2c_u + 2$ is necessary to solve the black hole search problem. That is, we prove that no algorithm can correctly detect a black hole node with $c_u + 1$ correct agents.

Theorem 1. *When c_u crash agents exist, no deterministic algorithm exists for the black hole search problem with $k \leq 2c_u + 1$ agents.*

Proof. We prove this theorem by contradiction. Assume that there exists a deterministic algorithm Alg that can solve the black hole search problem with $2c_u + 1$ agents for an arbitrary graph. This implies that $c_u + 1$ correct agents exist. We consider graph G that has a starting node v_s satisfying $d(v_s) = 2$. Let v_a and v_b be the two adjacent nodes of v_s. We consider two executions $E = C_0, C_1, \ldots$ and $E' = C'_0, C'_1, \ldots$ as executions on G. In execution E, we assume that v_a is a black hole node and that first c_u agents moving to v_b crash when they arrive at v_b. Similarly, in execution E', we assume that v_b is a black hole node and that first c_u agents moving to v_a crash when they arrive at v_a. If multiple agents move to v_b (resp., v_a) at the same time and the total number of such agents exceeds c_u in E (resp., E'), some of them crash so that the number of crash agents become c_u.

Let the t-th (resp., t'-th) round be the round in E (resp., E') such that all correct agents terminate before the end of the round. Let the h-th (resp., h'-th) round be the last round in E (resp., E') such that $h \leq t$ (resp., $h' \leq t'$) holds, at most c_u agents have moved to v_a before C_h (resp., C'_h), and at most c_u agents have moved to v_b before C_h (resp., C'_h). Let $h^* = \min(h, h')$. By induction, we prove the proposition that, for any ℓ $(0 \leq \ell \leq h^*)$, $C_\ell = C'_\ell$ holds and all agents exist only on v_s or *null* in C_ℓ. For the base case, clearly $C_0 = C'_0$ holds and all agents exist on starting node v_s in C_0.

For inductive steps, assume that, for some $\ell < h^*$, $C_\ell = C'_\ell$ holds and all agents exist only on v_s or *null* in C_ℓ. Let us consider the behaviors of the ℓ-th round. Since Alg is deterministic, agents make the same behaviors in E and E'. In addition, since all agents exist only on v_s or *null*, only agents on v_s can move. Clearly, for the state of the whiteboard on v_s and states of agents that do not move during the ℓ-th round, the states in $C_{\ell+1}$ are the same as in $C'_{\ell+1}$. Consider agents that move from v_s to v_a during the ℓ-th round. In E, since v_a is a black hole node, they change to state s_{fault}. In E', from the definition of E' and h^*, they crash and change to state s_{fault}. Similarly, consider agents that move from v_s to v_b during the ℓ-th round. In E, from the definition of E and h^*, they crash and change to s_{fault}. In E', since v_b is a black hole node, they change to s_{fault}. Hence all agents that move from v_s to v_a or v_b change to s_{fault} in $C_{\ell+1}$ and $C'_{\ell+1}$. By definition, the position of these agents are *null* . Therefore, $C_{\ell+1} = C'_{\ell+1}$ holds and all agents exist only on v_s or *null* in $C_{\ell+1}$.

From the above proposition, we have $C_{h^*} = C'_{h^*}$. Here, in C_{h^*} and C'_{h^*}, some correct agents have not yet terminated. This is because otherwise correct agents report the same node as the black hole node in E and E' despite the fact that black hole nodes are different.

Hence, since $h < t$ and $h' < t'$ hold, we can consider C_{h^*+1}. From the definition of h^*, during the h^*-th round, the number of agents moving to v_a or the number of agents moving to v_b exceeds c_u. If the number of agents moving to v_a exceeds c_u, agents cannot solve the black hole search problem in E because all correct agents visit black hole node v_a. If the number of agents moving to v_b exceeds c_u, agents cannot solve the black hole search problem in E' because all correct agents visit black hole node v_b. This is a contradiction.

4   Algorithm with $f_u + 2$ Correct Agents

In this section, we propose an algorithm that solves the black hole search problem. Here, we assume at most f_u Byzantine agents exist and each agent knows the graph topology and k. Since there exists no algorithm to search for a black hole node with $k \leq 2f_u+1$ agents from Theorem 1, we assume $k \geq 2f_u+2$ holds. Our algorithm solves the black hole search problem with $f_u + 2$ correct agents. In other words, our algorithm is optimal in terms of the number of tolerable faulty agents.

4.1   Overview

First, we give an overview of our algorithm. This algorithm achieves the black hole search problem in synchronous networks even if Byzantine agents exist. In order to simplify the explanation, at first, we explain the case where Byzantine agents do not exist. All agents start the algorithm from node v_s, and mark v_s as an explored node and all other nodes as unexplored nodes. The agents try to visit all nodes by using the DFS (depth-first search) traversal. Since all agents know the topology of the network, they can try to visit in the same order. The agents decide the next destination node v_d from neighboring nodes of the current node v. If v_d is an explored node, they just move to v_d. Otherwise, by using the cautious walk [4], they determine whether v_d is a safe node or a black hole node. The cautious walk is a method to search for a black hole node by agents paired on the same node. Let us assume that two agents a_i and a_j are paired. First, a_i stays in v and a_j once visits v_d. After that, a_j returns to v. At this time, if v_d is a safe node, a_i can see a_j on v. On the other hand, if v_d is a black hole node, a_i cannot see a_j on v. Thus, a_i can determine whether v_d is a safe node or a black hole node. If v_d is a black hole node, all living agents detect the black hole node v_d and then terminate. If v_d is a safe node, all agents mark v_d as an explored node and continue the DFS traversal. They repeat the same operations until they visit all nodes. If they visit all nodes, they detect non-existence of a black hole node and then terminate.

However, if Byzantine agents exist, the cautious walk does not work well. When a Byzantine agent visits a safe node v_d from v, it may not return to v to make other agents believe that v_d is a black hole node. On the other hand, when a Byzantine agent visits a black hole node v_d from v, it may return to v to make other agents believe that v_d is a safe node. These behaviors make information by cautious walk unreliable, which implies that the black hole cannot be correctly detected. To resolve this problem, we modify the cautious walk to work even if Byzantine agents exist. In the modified cautious walk, at least $f_u + 1$ agents visit the destination node v_d. More concretely, agents execute the cautious walk for one node until one of the following conditions holds; (1) $f_u + 1$ agents return to v from v_d, or (2) $f_u + 1$ agents do not return to v from v_d. If $f_u + 1$ agents return, v_d is a safe node because at least one correct agent returns. If $f_u + 1$ agents do not return, v_d is a black hole because at least one correct agent disappears.

4.2   Details

The pseudo-code of the algorithm is given in Algorithm 1. Each agent knows the number of agents k and the upper bound of the number of Byzantine agents f_u. Each agent a_i manages local variables $a_i.snap1$, $a_i.snap2$, $a_i.return$, $a_i.alive$ and $a_i.expid$. We explain the variables $a_i.snap1$, $a_i.snap2$, $a_i.return$ and $a_i.alive$ later.

Recall that, in each round, an agent obtains the snapshot, executes the local calculation, and then, possibly leaves the node (and reaches to the next node before the next round). In one round, each agent executes the operations until it leaves (lines 4, 11 and 29), waits (line 13), or terminates (line 24 and 32).

Each agent a_i stays at node v_s in the initial configuration. After a_i starts the algorithm, a_i visits all unexplored nodes to detect a black hole node. Since all agents know the topology of the graph, they visit all nodes in the same order by using the DFS traversal. As a specific method, at first, each agent simulates DFS for its own map, and decides the order of visiting n nodes. We omit the details of the DFS traversal from the pseudo-code. First, a_i decides the destination node v_d from neighboring nodes of the current node v. If the destination node is explored, a_i moves to v_d. Otherwise, a_i determines whether the destination node is a safe node or a black hole node.

In lines 6 to 30 of the pseudo-code, each agent checks the destination node v_d. Agent a_i stores the result of function $snapcount()$ in $a_i.snap1$. Variable $a_i.snap1$ is an array which stores the number of agents for each agent ID. Function $snapcount()$ is the function which obtains the number of agents for each agent ID from the current snapshot. Note that, multiple agents with the same ID may exist since Byzantine agents can freely change their IDs. We show the pseudo-code of $snapcount()$ in Algorithm 2. In order to simplify the pseudo-code, we write $count[x]$ for all x values, but actually agents hold only the values of $count[]$ for existing IDs.

After that, a_i repeats the following operations until the condition of line 23 or 25 is satisfied. As we explain later, a_i uses conditions of lines 23 and 25 to judge whether the destination node is a safe node or a black hole node. First, a_i selects the ID of the explorer agents and sets this ID to $a_i.expid$ (line 8). The selected agents are the agents with the smallest ID in $a_i.snap1$. Note that, since Byzantine agents can change their IDs, multiple agents may have the selected

Algorithm 1. Pseudo-code of agent a_i. The node v indicates the node which a_i is staying.

```
 1: while there is an unexplored node in the graph do
 2:    Decide the destination node
 3:    if the destination node is explored then
 4:       Go to the destination node
 5:    else
 6:       a_i.snap1 = snapcount()
 7:       while true do
 8:          a_i.expid = min(id : a_i.snap1[id] ≥ 1)
 9:          a_i.snap1[a_i.expid] = 0
10:          if a_i.expid == ID_i then
11:             Go to the destination node and then come back to v
12:          else
13:             Wait for two rounds
14:          end if
15:          a_i.snap2 = snapcount()
16:          a_i.return = 0, a_i.alive = 0
17:          for all m such that a_i.snap2[m] ≥ 1 do
18:             a_i.alive+ = a_i.snap2[m]
19:             if m ≤ a_i.expid then
20:                a_i.return+ = a_i.snap2[m]
21:             end if
22:          end for
23:          if k − a_i.alive ≥ f_u + 1 then
24:             Write the location of the black hole node to the map and terminate
25:          else if a_i.return ≥ f_u + 1 then
26:             break; //the destination node is correct
27:          end if
28:       end while
29:       Go to the destination node
30:    end if
31: end while
32: Report that there is no black hole node and terminate
```

ID. After that, a_i sets 0 to $a_i.snap1[a_i.expid]$ to record that $a_i.expid$ has already been selected. If $a_i.expid == ID_i$ holds, a_i is an explorer. Otherwise, a_i is an observer. If a_i is an explorer, a_i moves to the destination node and then comes back to v. If a_i is an observer, a_i waits for two rounds to see explorers again.

After exploring or waiting, a_i stores the result of function $snapcount()$ in $a_i.snap2$. Then a_i computes $a_i.alive$ and $a_i.return$ from $a_i.snap2$. Variable $a_i.alive$ represents the number of living agents, and variable $a_i.return$ represents the number of agents that visit v_d and then return to v. If v_d is a safe node, correct agents always come back to v and thus they are always counted in $a_i.alive$. In line 20, a_i computes $a_i.return$ by counting the number of agents whose ID is $a_i.expid$ or smaller because agents with smaller IDs have become explorers before. We use $a_i.alive$ for the black hole detection and $a_i.return$ for

Algorithm 2. function : $snapcount()$

1: **for all** x **do**
2:     $count[x] = 0$
3: **end for**
4: **for all** a_j such that a_j exists at v **do**
5:     $count[ID_j] += 1$
6: **end for**
7: $return\ count$

the safe node detection. If $k - a_i.alive \geq a_i.f_u + 1$ holds, v_d is a black hole node because the condition holds only if at least one correct agent is not included in $a_i.alive$ (lines 23 to 24). On the other hand, if $a_i.return \geq f_u + 1$ holds, v_d is a safe node because the condition holds only if at least one correct agent is included in $a_i.return$ (line 25). If v_d is a black hole node, the condition does not hold because correct agents cannot return to v and they are not included in $a_i.return$. If a_i determines that v_d is a safe node, a_i leaves the loop and moves to the destination node. If neither line 23 nor 25 is satisfied, each agent continues exploring for v_d until it is satisfied.

When a_i visits all unexplored nodes and does not detect a black hole node, the algorithm detects that no black hole node exists.

4.3   Proof of Correctness

Lemma 1. *Consider a configuration such that all correct agents exist on a safe node v and start to search for its adjacent node v_d. If v_d is a safe node, all correct agents determine that v_d is a safe node.*

Proof. We consider a correct agent a_i staying at v. Since all correct agents behave at the same time, they get the same snapshot of v in line 6. Thus, in line 8, each correct agent a_i calculates the same ID of an explorer and sets it to $a_i.expid$. Furthermore, all correct agents get the same snapshot again in line 15. Therefore, in line 18, each correct agent a_i has the same value in $a_i.alive$. The algorithm continues to send agents to v_d until (1) $k - a_i.alive \geq f_u + 1$ is satisfied or (2) $a_i.return \geq f_u + 1$ is satisfied. We prove that condition 1 does not hold and that condition 2 holds when at least $f_u + 1$ correct agents go to v_d and come back to v.

Since correct agents always return to v after visiting v_d, all correct agents exist on v when each agent a_i obtains $a_i.snap2$. Consequently, at least all correct agents are counted in $a_i.alive$. Therefore, $k - a_i.alive \geq f_u + 1$ is not satisfied because at most f_u Byzantine agents exist.

Let us consider the case where $x (\leq f_u)$ agents are missing from $a_i.snap1$ at v. That means, x agents are Byzantine agents and $k - x$ agents exists at v. Byzantine agents can be caunted $a_i.return$ by joining v, but it is not enough to satisfy $a_i, return \geq f_u + 1$. In addition, the $k - a_i.alive \geq f_u + 1$ cannot be satisfied because at most x agents can exist on different nodes from v without following the algorithm.

Hence, condition 1 is not satisfied, and agents move to v_d until condition 2 is satisfied. Since correct agents always return to v after visiting v_d and are counted in $a_i.return$, eventually $a_i.return \geq f_u + 1$ is satisfied. Therefore, the lemma holds. $\square$

Lemma 2. *Consider a configuration such that all correct agents exist on a safe node v and start to search for its adjacent node v_d. If v_d is a black hole node, at least one correct agent determines that v_d is a black hole node.*

Proof. We consider a correct agent a_i staying at v. Since all correct agents behave at the same time, they get the same snapshot of v. Thus, in line 8, each correct agent a_i calculates the same ID of an explorer and sets it to $a_i.expid$. Furthermore, all correct agents get the same snapshot again in line 15. Therefore, in line 18, each correct agent a_i has the same value in $a_i.alive$. The algorithm continues to send agents to v_d until (1) $k - a_i.alive \geq f_u + 1$ is satisfied or (2) $a_i.return \geq f_u + 1$ is satisfied. Note that, since only agents with ID $a_i.expid$ move to v_d at the same time, at most one correct agent moves to v_d at the same time. We prove that condition 2 does not hold and that condition 1 holds before $f_u + 2$ correct agents go to v_d.

Once correct agents visit v_d, they cannot return to v and hence they are not counted in $a_i.return$ or $a_i.alive$. Since there are at most f_u Byzantine agents, even if all Byzantine agents are included in $a_i.return$, condition 2 is never satisfied. Let us consider condition 1. When $f_u + 1$ correct agents visit v_d, $k - a_i.alive \geq f_u + 1$ holds because all of them are not counted in $a_i.alive$. Since at most one correct agent goes to v_d at the same time, at least one agent is alive when $k - a.alive \geq f_u + 1$ holds. Therefore, the lemma holds. $\square$

Theorem 2. *The algorithm solves the black hole search problem with $O(f_u n)$ rounds.*

Proof. By the assumption, all correct agents start the algorithm from safe node v_s. From Lemmas 1 and 2, when all correct agents exist on a safe node and start to search for its adjacent node v_d, they can correctly determine whether v_d is a safe node or a black hole node. Thus, as long as all the correct agents search for a safe node, they can continue searching. If they search for a black hole node, they can detect a black hole node. Therefore, the algorithm solves the black hole search problem.

Let us consider the time complexity. For each unexplored node, at most $2f_u + 1$ agents visit to determine whether it is a safe node or a black hole node. This requires at most two rounds for each agent, and hence the algorithm requires at most $2(2f_u + 1)$ rounds to search for each node. Since all agents know the topology of the graph, they can visit all nodes by using the DFS traversal. Since the length of the DFS traversal is $O(n)$, the algorithm solves the black hole search problem with $O(f_u n)$ rounds. $\square$

5   Conclusions

We consider the black hole search problem in synchronous Byzantine environments. In this work, we first showed the impossibility with $c_u + 1$ correct agents, where c_u is an upper bound of the number of crash agents. Since a crash fault is a special case of a Byzantine fault, the above result also applies to strongly Byzantine agents. Next, we proposed an algorithm to locate a black hole node with $2f_u + 2$ agents with $O(f_u n)$ rounds when at most f_u Byzantine agents exist in synchronous networks (n is the number of nodes). From the above impossibility result, our algorithm is optimal in terms of the number of tolerable faulty agents. To the best of our knowledge, this is the first work to address the black hole search problem in Byzantine environments.

A future task is to search for a black hole node in asynchronous Byzantine environments. The cautious walk cannot work in asynchronous networks because it is impossible to distinguish whether an agent disappeared in a black hole node or is stuck in a slow edge of the network. Hence we need to devise a technique to make correct agents collaboratively visit all nodes except for a black hole node even if Byzantine agents try to send correct agents to the black hole node.

References

1. Cao, J., Das, S.K.: Mobile Agents in Networking and Distributed Computing. Wiley, Hoboken (2012)
2. Chalopin, J., Das, S., Labourel, A., Markou, E.: Black hole search with finite automata scattered in a synchronous torus. In: Peleg, D. (ed.) DISC 2011. LNCS, vol. 6950, pp. 432–446. Springer, Heidelberg (2011). https://doi.org/10.1007/978-3-642-24100-0_41
3. Cooper, C., Klasing, R., Radzik, T.: Searching for black-hole faults in a network using multiple agents. In: Shvartsman, M.M.A.A. (ed.) OPODIS 2006. LNCS, vol. 4305, pp. 320–332. Springer, Heidelberg (2006). https://doi.org/10.1007/11945529_23
4. Dobrev, S., Flocchini, P., Prencipe, G., Santoro, N.: Mobile search for a black hole in an anonymous ring. Algorithmica 48(1), 67–90 (2007)
5. Dobrev, S., Královič, R., Santoro, N., Shi, W.: Black hole search in asynchronous rings using tokens. In: Calamoneri, T., Finocchi, I., Italiano, G.F. (eds.) CIAC 2006. LNCS, vol. 3998, pp. 139–150. Springer, Heidelberg (2006). https://doi.org/10.1007/11758471_16
6. Flocchini, P., Ilcinkas, D., Santoro, N.: Ping pong in dangerous graphs: optimal black hole search with pure tokens. In: Taubenfeld, G. (ed.) DISC 2008. LNCS, vol. 5218, pp. 227–241. Springer, Heidelberg (2008). https://doi.org/10.1007/978-3-540-87779-0_16
7. Flocchini, P., Santoro, N.: Distributed security algorithms by mobile agents. In: Chaudhuri, S., Das, S.R., Paul, H.S., Tirthapura, S. (eds.) ICDCN 2006. LNCS, vol. 4308, pp. 1–14. Springer, Heidelberg (2006). https://doi.org/10.1007/11947950_1
8. Klasing, R., Markou, E., Radzik, T., Sarracco, F.: Hardness and approximation results for black hole search in arbitrary networks. Theoret. Comput. Sci. 384(2–3), 201–221 (2007)

9. Peng, M., Shi, W., Corriveau, J.P., Pazzi, R., Wang, Y.: Black hole search in computer networks: state-of-the-art, challenges and future directions. J. Parallel Distrib. Comput. **88**, 1–15 (2016)
10. Shi, W., Garcia-Alfaro, J., Corriveau, J.P.: Searching for a black hole in interconnected networks using mobile agents and tokens. J. Parallel Distrib. Comput. **74**(1), 1945–1958 (2014)

Self-adjusting Linear Networks

Chen Avin[1] , Ingo van Duijn[2(✉)], and Stefan Schmid[3]

[1] School of Electrical and Computer Engineering,
Ben Gurion University of the Negev, Beersheba, Israel
avin@cse.bgu.ac.il
[2] Department of Computer Science, Aalborg University, Aalborg, Denmark
ingo@cs.aau.dk
[3] Faculty of Computer Science, University of Vienna, Vienna, Austria
stefan_schmid@univie.ac.at

Abstract. Emerging networked systems become increasingly flexible, reconfigurable, and "self-*". This introduces an opportunity to adjust networked systems in a demand-aware manner, leveraging spatial and temporal locality in the workload for *online* optimizations. However, it also introduces a tradeoff: while more frequent adjustments can improve performance, they also entail higher reconfiguration costs. This paper initiates the formal study of *list* networks which self-adjust to the demand in an online manner, striking a balance between the benefits and costs of reconfigurations. We show that the underlying algorithmic problem can be seen as a distributed generalization of the classic dynamic list update problem known from self-adjusting datastructures: in a network, requests can occur between *node pairs*. This distributed version turns out to be significantly harder than the classical problem it generalizes. Our main results are a $\Omega(\log n)$ lower bound on the competitive ratio, and a (distributed) online algorithm that is $\mathcal{O}(\log n)$-competitive if the communication requests are issued according to a *linear order*.

Keywords: Self-adjusting datastructures · Competitive analysis · Distributed algorithms · Communication networks

1 Introduction

Communication networks are becoming increasingly flexible, along three main dimensions: routing (enabler: software-defined networking), embedding (enabler: virtualization), and topology (enabler: reconfigurable optical technologies, for example [17]). In particular, the possibility to quickly reconfigure communication networks, e.g., by migrating (virtualized) communication endpoints [9] or by reconfiguring the (optical) topology [12], allows these networks to become *demand-aware*: i.e., to adapt to the traffic pattern they serve, in an online and "self-*" manner. In particular, in a *self-adjusting* network, frequently communicating node pairs can be moved *topologically closer*, saving communication costs (e.g., bandwidth, energy) and improving performance (e.g., latency, throughput).

M. Ghaffari et al. (Eds.): SSS 2019, LNCS 11914, pp. 368–382, 2019.
https://doi.org/10.1007/978-3-030-34992-9_29

However, today, we still do not have a good understanding yet of the algorithmic problems underlying self-adjusting networks. The design of such algorithms faces several challenges. As the demand is often not known ahead of time, *online* algorithms are required to react to changes in the workload in a clever way; ideally, such online algorithms are "competitive" even when compared to an optimal offline algorithm which knows the demand ahead of time. Furthermore, online algorithms need to strike a balance between the benefits of adjustments (i.e., improved performance and/or reduced costs) and their costs (i.e., frequent adjustments can temporarily harm consistency and/or performance, or come at energy costs).

The vision of self-adjusting networks is reminiscent of self-adjusting datastructures such as *self-adjusting lists* and *splay trees*, which optimize themselves toward the workload. In particular, the *dynamic list update problem*, introduced already in the 1980s by Sleator and Tarjan in their seminal work [23], asks for an online algorithm to reconfigure an unordered linked list datastructure, such that a sequence of lookup requests is served optimally and at minimal reconfiguration costs (i.e., pointer rotations). It is well-known that a simple *move-to-front* strategy, which immediately promotes each accessed element to the front of the list, is *dynamically optimal*, that is, has a constant competitive ratio.

This paper initiates the study of a most basic self-adjusting linear *network*, which can be seen as a *distributed* variant of the dynamic list update problem, generalizing the datastructure problem to networks: while datastructures serve requests originating from the front of the list (the "root") to access data items, networks serve *communication* requests between *pairs of nodes*. The objective is to move nodes which currently communicate frequently, closer to each other, while accounting for reconfiguration costs.

1.1 Related Work

One important area of related work arises in the context of the dynamic list update problem. Since the groundbreaking work by Sleator and Tarjan on amortized analysis and self-adjusting datastructures [23], researchers have also explored many interesting variants of self-adjusting datastructures, also using randomized algorithms [21] or lookaheads [1,3], or offline algorithms [5,20]. The deterministic Move-To-Front (MTF) algorithm is known to optimally solve the standard formulation of the list update problem: it is 2-competitive [23], which matches the lower bound [4]. To the best of our knowledge, the competitive ratio in the randomized setting (against an oblivious adversary) is still an open problem: the best upper bound so far is 1.6 [3], and the best lower bound 1.5 [24]. The randomized algorithm [3] makes an initial random choice between two known algorithms that have different worst-case request sequences, relying on the BIT [21] and TIMESTAMP [2] algorithms.

We also note that the self-adjusting linear network design problem can be considered a special case of general online problems such as the online Metrical Task System (MTS) problems. However, given the exponential number of possible configurations, the competitive ratio of generic MTS algorithms will be high if applied to our more specific problems (at least according to the existing bounds). Furthermore, we note that in case of list request graphs, the problem can also be seen as a learning problem and hence related to bandits theory [13].

In terms of reconfigurable networks, there exist several static [8,11] and dynamic [16,19,22] algorithms for bounded-degree networks, as well as hybrid variants [15] which combine static and reconfigurable links. However, these solutions do not apply to the list and do not provide performance guarantees over time (with the notable exception of [16] in a different model); the latter also applies to recent work on node migration models on the grid [7].

The paper closest to ours is by Olver et al. [18] who introduced the Itinerant List Update (ILU) problem: a relaxation of the classic dynamic list update problem in which the pointer no longer has to return to a home location after each request. The authors show an $\Omega(\log n)$ lower bound on the randomized competitive ratio and also present an offline polynomial-time algorithm and prove that it achieves an approximation ratio of $O(\log^2 n)$. In contrast, we in our paper focus on online algorithms and request graphs forming a list (or grid). In fact, we show that the lower bound $\Omega(\log n)$ even holds in this case, at least for deterministic algorithms. We also present an online algorithm which matches this bound in our model.

1.2   Formal Model

We initiate the study of pairwise communication problems in a dynamic network reconfiguration model, using the following notation:

- Let $d_G(u, v)$ denote the *(hop) distance* between u and v in a graph G.
- A *communication request* is a pair of communicating nodes from a set V.
- A *configuration* of V in a graph N (the host network) is an injection of V into the vertices of N; $C_{V \hookrightarrow N}$ denotes the set of all such configurations.
- A configuration $h \in C_{V \hookrightarrow N}$ is said to *serve* a communication request $(u, v) \in V \times V$ at cost $d_N(h(u), h(v))$.
- A finite *communication sequence* $\sigma = (\sigma_0, \sigma_1, \ldots, \sigma_m)$ is served by a sequence of configurations $h_0, h_1, \ldots, h_m \in C_{V \hookrightarrow N}$.
- The cost of serving σ is the sum of serving each σ_i in h_i plus the reconfiguration cost between subsequent configurations h_i, h_{i+1}.
- The reconfiguration cost between h_i, h_{i+1} is the number of *migrations* necessary to change from h_i to h_{i+1}; a migration swaps the images of two nodes u and v under h.
- $E_i = \{\sigma_1, \ldots, \sigma_i\}$ denotes the first i requests of σ interpreted as a set of edges on V, and $R(\sigma) = (V, E_m)$ denotes the *request graph* of σ.

In particular, we study the problem of designing a self-adjusting *linear network*: a network whose topologoy forms a d-dimensional grid. We are particularly interested in the 1-dimensional grid in this paper, the line:

Definition 1 (Distributed List Update). *Let* V, h, *and* σ *be as before, with*

$$N = (\{1, \ldots, n\}, \{(1, 2), (2, 3), \ldots, (n - 1, n)\})$$

representing a list graph. The cost of serving a $\sigma_i = (u, v) \in \sigma$ is given by $|h(u) - h(v)|$, i.e. the distance between u and v on N. Migrations can only occur between nodes configured on adjacent vertices in N.

Recall that the cost incurred by an algorithm A on σ is the sum of communication and reconfiguration costs. In the realm of online algorithms and competitive analysis, we compare an online algorithm ON to an offline algorithm OFF which has complete knowledge of σ ahead of time. We want to devise online algorithms ON which minimize the competitive ratio ρ:

$$\rho = \max_{\sigma} \frac{\text{cost}(ON(\sigma))}{\text{cost}(OFF(\sigma))}$$

As a first step, we in this paper consider the DISTRIBUTED LIST UPDATE problem for the case where the request graph $R(\sigma)$ has constant *graph bandwidth*: i.e. graphs for which there is a configuration in a line network such that any request can be served at constant cost. We refer to such a request graph as *linear demand.*

1.3 Contributions

This paper initiates the study of a most basic self-adjusting network, a line, which optimizes itself toward the dynamically changing linear demand, while amortizing reconfiguration cost. The underlying algorithmic problem is natural and motivated by emerging reconfigurable communication networks (e.g., based on virtual machine migration or novel optical technologies [10,17]). The problem can also be seen as a distributed version of the fundamental dynamic list update problem. Our first result is a negative one: we show that unlike the classic dynamic list update problem, which admits for constant-competitive online algorithms, there is an $\Omega(\log n)$ lower bound on the competitive ratio of any deterministic online algorithm for the distributed problem variant. Our second main contribution is a (distributed) online algorithm which is $\mathcal{O}(\log n)$-competitive for long enough sequences, given that the communication patterns exhibit linear demand.

1.4 Organization

The remainder of this paper is organized as follows. In Sect. 2, we put the problem and its challenges into perspective with respect to the list update problem. We then first derive the lower bound in Sect. 3 and present our algorithm and upper bound in Sect. 4. We conclude in Sect. 5.

2   From List Update to Distributed List Update

To provide an intuition of the challenges involved in designing online algorithms for distributed list update problems and to put the problem into perspective, we first revisit the classic list update problem and then discuss why similar techniques fail if applied to communicating node *pairs*, i.e., where requests not only come from the front of the list.

The *(dynamic) list update problem* [23] introduced by Sleater and Tarjan over 30 years ago is one of the most fundamental and oldest online problems: Given a set of n elements stored in a linked list, how to update the list over time such that it optimally serves a request sequence $\tau = (\tau_1, \tau_2, \ldots)$ where for each i, $\tau_i \in V$ is an arbitrary element stored in the list? The cost incurred by an algorithm is the sum of the access costs (i.e. scanning from the *front* of the list to the accessed element) and the number of *swaps* (switching two neighboring elements in the list). As accesses to the list elements start at the front of the list, it makes sense to amortize high access costs by moving frequently accessed elements closer to the front of the list. In fact, the well-known *Move-To-Front* (MTF) algorithm even moves an accessed element to the front *immediately*, and is known to be *constant competitive*: its cost is at most a factor 2 (or some other constant, depending on the cost model) worse than that of an optimal offline algorithm which knows the entire sequence τ ahead of time [23]. Throughout the literature, slightly different cost models have been used for the list update problem, though they only differ by a constant factor. Generally, a *cursor* is located at the head of the list at each request. Then, the algorithm can perform two operations, each operation incurring unit cost. (i) *Move* the cursor to the left, or to the right, one position; the element in the new position is referred to as *touched*. (ii) *Swap* the element at the cursor with the element one position to the left or right; the cursor also moves.

In the DISTRIBUTED LIST UPDATE problem, upon a request $\sigma_i = (s_i, t_i)$, the cursor is placed at s_i instead of the head of the list, and t_i needs to be looked up. To demonstrate the significance of this difference, we first present a paraphrased version of the proof by Tarjan and Sleator showing the dynamic optimality of MTF. After that, we showcase a simple access sequence differentiating the two problems.

2.1   An Expositional Proof for the Optimality of MTF

While the potential argument used to show dynamic optimality of the move-to-front strategy for the list access problem yields a very elegant and succinct proof [23], it lacks intuition which makes it difficult to generalise the argument. The key idea in the potential argument is to compare the execution of MTF to the execution of an arbitrary algorithm A. The algorithm is fixed for the analysis, but any valid algorithm can be used, e.g. the optimal offline algorithm. The state (represented by a list) of MTF and A are juxtaposed at every access, comparing how the order of elements in both lists differ. In fact, it is sufficient to only consider the relative order of two arbitrary but fixed elements, call them u and v. Consider the order of u and v in the state of A before it performs the ith access. If this order is the same as in MTF *before* it performs the ith acces, let $b_i = 0$ and otherwise $b_i = 1$.

Similarly, if their relative order is the same in MTF *after* its *ith* access, let $a_i = 0$ and otherwise $a_i = 1$. This describes an inversion sequence $b_1 a_1 b_2 a_2 \ldots b_m a_m$. Figure 1 illustrates this for MTF and an arbitrarily chosen algorithm A on a sequence $\tau = 6, 3, 1, 3, 6$, with the inversions of 1 and 6 described by the sequence 01111011100.

Suppose that $\tau_i \in \{u, v\}$ and that MTF touches u and v while accessing τ_i. The proof by Tarjan and Sleator boils down to three observations.

Observation 1. *MTF inverts u and v relative to A by accessing τ_i, i.e. $b_i \neq a_i$.*

Observation 2. *If $b_i = 0$, MTF and A agree on the order of u and v before τ_i. Since MTF touches both, A also touches both in order to access τ_i.*

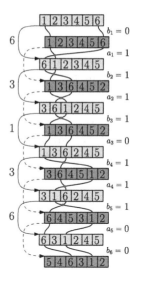

Fig. 1. MTF (yellow) and A (blue) on $\tau = 6, 3, 1, 3, 6$ (Color figure online)

Observation 3. *For $b_i = 1$, let $j < i$ be the largest index such that $b_j = 0$ or $a_j = 0$ (note that j exists because $b_1 = 0$). When $a_j = 0$, and thus $b_{j+1} = 1$, A inverts u and v and therefore must have touched both. When $b_j = 0$, and thus $a_j = 1$, MTF inverts u and v and one of them is τ_j. By Observation 2, if $b_j = 0$ and MTF touches u and v to access τ_j, then A does as well.*

The last observation is essentially the amortised argument rephrased as a charging argument. We can now easily prove the dynamic optimality of MTF.

Theorem 1 (Tarjan & Sleator). MTF *is 4-competitive.*

Proof. We prove that for all $\tau_i = v$ where MTF touches u, there is a move by A touching u. MTF first moves the cursor to τ_i, and then swaps τ_i to the front. Along the way it touches u twice, once with a move and once with a swap, incurring a cost of 2.

For $b_i = 0$ (resp. $b_i = 1$), we use Observation 2 (resp. 3) to charge the cost to A touching u while accessing τ_i (resp. τ_j). By Observation 1, $b_i \neq a_i$, and thus for any $\tau_k \in \{u, v\}$ with $i < k$, the largest index $j' < k$ with $b_{j'} = 0$ or $a_{j'} = 0$ must be at least i, and therefore $j < i \leq j'$. This guarantees that MTF charges at most a cost of 4 to one move of A. Since all the cost incurred by MTF is charged to some move of A, the claim follows.   □

In the original work by Tarjan and Sleator, MTF is shown to be 2-competitive. This is because their cost model allows accessed elements to be moved to the front 'for free'. If we allow this as well, the cursor touches u only once to access v, resulting in a factor 2.

2.2 The Challenge of Distributed List Update

Generalizing dynamic list update to DIS-
TRIBUTED LIST UPDATE introduces a number
of challenges which render the problem more
difficult. First, the natural inversion argument
no longer works: a reference point such as the
front of the list is missing in the distributed
setting. This makes it harder to relate algorithms to each other and hence also to define
a potential. Second, for general request graphs
$R(\sigma)$, an online algorithm needs to be able to
essentially "recognize" certain patterns over
time.

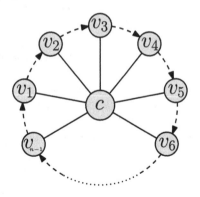

Fig. 2. A star graph used to
construct a cyclic sequence of
requests $\sigma_c = (c, v_1), (c, v_2), \ldots,$
$(c, v_{n-1}), (c, v_1), \ldots$

Regarding the latter, consider the set of
nodes $V = \{v_1, ..., v_n\}$ and let τ_c be a cyclic
sequence: for all $\tau_i, \tau_{i+1} \in \tau_c$ with $\tau_i = v_j$ and
$\tau_{i+1} = v_k$ it holds that $j + 1 = k(\mathrm{mod}\ n - 1)$.
From this we construct a similar sequence σ_c
for DISTRIBUTED LIST UPDATE on the set of nodes $V \cup \{c\}$, with $\sigma_i = (c, \tau_i)$.
This yields a star graph $R(\sigma_c)$ as denoted in Fig. 2. An offline algorithm can
efficiently serve the cyclic order by first embedding the elements in the order
$v_1, ...v_k$, and then moving the element c one position further after every request.
If the cost of embedding the initial order is dominated by serving all requests,
then the amortized cost is $\mathcal{O}(1)$ per request (per cycle there are $n - 1$ moves of
cost $\mathcal{O}(1)$ and once c is moved a distance n). However, in the list update model,
any sequence cycling through all elements is a worst-case sequence with $\Omega(n)$
per request. This demonstrates that a "dynamic cursor" can mean a factor n
difference in cost. What the sequence σ_c also demonstrates, is that aggregating
elements around a highly communicative node is suboptimal; in the particular
case of σ_c, it is this central node that needs to be moved.

Another pattern is a request sequence σ that forms a connected path in the
request graph $R(\sigma)$. When restricted to only these patterns, DISTRIBUTED LIST
UPDATE corresponds to the *Itinerant List Update Problem* (ILU) studied in [18].
In this work it is shown that deriving non-trivial upper bounds on the competitive ratio already seems notoriously hard (even offline approximation factors
are relatively high). Note that the star example can be expressed as a path,
i.e. $\sigma'_c = (c, v_1), (v_1, c), (c, v_2), (v_2, c), (c, v_3), \ldots$, demonstrating the significance
of understanding simple request patterns for DISTRIBUTED LIST UPDATE. This
is partly why in this paper we focus on request graphs with a linear demand.

3 A Lower Bound

This section derives a lower bound on the competitive ratio of any algorithm for
DISTRIBUTED LIST UPDATE.

Theorem 2. *The competitive ratio* $\rho = \max_\sigma \frac{cost(ON(\sigma))}{cost(OFF(\sigma))}$ *for* DISTRIBUTED
LIST UPDATE, *with* $|\sigma| = \Omega(n^2)$, *is at least* $\Omega(\log n)$. *This bound holds for
arbitrarily long sequences, but if* $|\sigma| = \mathcal{O}(n^2)$, *it even holds if the request graph
is a list graph.*

To prove this, we consider an arbitrary online algorithm *ON* for DISTRIBUTED
LIST UPDATE. The main idea is to have an adaptive online adversary construct a
sequence σ_{ON} that depends on the algorithm *ON*. The adversary constructs σ_{ON}
so that the resulting request graph $R(\sigma_{ON})$ is a list graph. Because an offline
algorithm knows $R(\sigma_{ON})$ in advance, it can immediately configure it and serve
all requests at optimal cost of 1; since $|\sigma| = \Omega(n^2)$, the configuration cost of
$\mathcal{O}(n^2)$ is negligible. We show that the online algorithm is forced to essentially
reconfigure its layout $\log n$ times, resulting in the desired ratio. To facilitate our
analysis, we use the same notion of the *distortion* of an embedding as is used in
the Minimum Linear Arrangement (MLA) [14] problem.

Definition 2. *Given a request graph* $G = (V, E)$ *with* $E \subseteq V \times V$, *let* $E^+ =
\{(u, v) \mid d_G(u, v) < \infty\}$ *denote the transitive closure of* E.
For $h \in C_{V \hookrightarrow N}$, *let* $d_h(E)$ *denote the distortion of* E, *which is defined as:*

$$d_h(E) = \sum_{(u,v) \in E^+} d_h(u, v)$$

By summing over edges in E^+ (instead of E), the cost of a badly embedded
edge $e \in E$ is essentially multiplied by the number of paths in E that contain e.
This means that the distortion of an embedding of a list is worse if the badly
embedded edges occur in the middle of the list, see Fig. 3a. To build σ_{ON}, the
adversary gradually commits to the edges of $R(\sigma_{ON})$. Having already requested
$\sigma_1, \ldots, \sigma_i$, then depending on the distortion the adversary:

Option 1: picks $\sigma_{i+1} = \arg\max_{(u,v) \in E_i} d_h(u, v)$.
Option 2: reveals a new batch of edges $M \subset V \times V$.

From these two options, the adversary's strategy becomes clear; Option 1 forces
the highest possible cost to *ON* based on E_i and h, and Option 2 introduces new
communication edges to force an increase in distortion. What is left to show is
how the value of $d_h(E_i)$ comes into play, and which edges the adversary commits
to. The adversary reveals at most $n - 1$ edges (since the final request graph is
a list), and they will be revealed in batches of size $n/2$, $n/4$, $n/8$, etc., resulting
in $\log n$ batches. After each batch, for *ON* to remain optimal it must permute
its layout at cost $\Omega(n^2)$, totaling a cost of $\Omega(n^2 \log n)$ for all batches combined.
To ensure that $R(\sigma_{ON})$ is a list graph, the partial request graph E_i (i.e., the
set of revealed edges) always comprises a set of disjoint *sublists*. Therefore, the
adversary only reveals edges that concatenate two sublists in E_i. Initially E_i is
empty and the corresponding sublists are all singleton sets of $u \in V$.

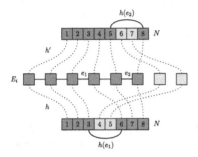

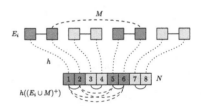

(a) Two embeddings h and h' of a set of edges. Even though both embeddings embed only a single edge suboptimally, the distortion of $d_h(E)$ is bigger than $d_{h'}(E)$ because more paths in E_i cross e_1 than e_2.

(b) A visualization of $d_h(E_i \cup M)$: the list graph N, E_i (solid) and M (dashed) are sets of edges, configured on N by h (dotted). The sum of length of the configured edges $h((E_i \cup M)^+)$ is the distortion $d_h(E_i \cup M)$.

Fig. 3. Illustrations of distortion.

To help decide which edges to reveal, we use the distortion to associate a cost to batches of edges that the adversary can commit to. Let $M \subseteq V \times V \setminus E_i$ be any set of edges such that the graph $(V, E_i \cup M)$ comprises a set of disjoint sublists. For a configuration h of ON, the set M induces a distortion of $d_h(E_i \cup M)$, as shown in Fig. 3b. We show that for any embedding that ON chooses, the adversary can find a set M so that the distortion is large.

Lemma 1. *Let N be a list graph, and $E \subseteq V \times V$ a set of edges so that the graph $G = (V, E)$ induces k disjoint sublists. For every $h \in C_{V \hookrightarrow N}$, there exists a set $M \subseteq V \times V$ of at most $k/2$ edges such that $d_h(E \cup M) = \Omega(\frac{n^3}{k})$ and $(V, E \cup M)$ comprises a set of disjoint lists.*

To prove this lemma, we use the following fact (with proof in the full paper [6]):

Theorem 3. *Let $x_1, \ldots, x_k$ and $y_1, \ldots, y_k$ be sequences of k nonnegative numbers, and let x (resp. y) denote $\sum_{i=1}^{k} x_i$. Let the weight of an involution[1] over the indices $1, \ldots, k$ be defined as $w(f) = \sum_{i=1}^{k} x_i y_{f(i)}$.*
The average weight over all involutions is $\Omega(\frac{xy}{k})$.

Proof (Lemma 1). Let $L_1, ..., L_k \subseteq E$ be the sublists in G. For all pairs (i, j), let (L_i, L_j) denote any edge so that $L_i \cup L_j \cup \{(L_i, L_j)\} = L_i \oplus L_j$ is connected. For any involution f on the sublists we have:

$$2d_h(E \cup \{(L_i, L_{f(i)}) \mid i \neq f(i)\}) \geq \sum_{i=1}^{k} d_h(L_i \oplus L_{f(i)}). \tag{1}$$

---

[1] A function f such that $f(f(x)) = x$ for all x.

The factor 2 is necessary because for i such that $i \neq f(i)$, the term $d_h(L_i \oplus L_{f(i)})$ appears twice in the sum. Now partition N into three sublists: a left part $X = \{1, \dots, \lceil n/3 \rceil\}$, a right part $Y = \{\lfloor 2n/3 \rfloor, \dots, n\}$, and the centre part $C = N \setminus (X \cup Y)$. Let $h_X(L_i)$ (resp. $h_Y(L_i)$) denote the number of elements of L_i that h maps onto X (resp. Y). Every two vertices u, v so that $h(u) \in X$ and $h(v) \in Y$ are by construction at least $|C| = \Theta(n)$ apart on N, and therefore we can lower bound $d_h(L_i \oplus L_j)$ by:

$$d_h(L_i \oplus L_j) \geq |C| \cdot h_X(L_i) h_Y(L_j) \tag{2}$$

For an involution f drawn uniformly at random, Theorem 3 gives us a bound on the expected value of the following:

$$\mathbf{E}\left(\sum_{i=1}^{k} h_X(L_i) h_Y(L_{f(i)}) \right) = \Omega\left(\frac{\lceil n/3 \rceil^2}{k} \right) \tag{3}$$

Therefore, there exists an involution f for which we have:

$$2d_h(E \cup \{(L_i, L_{f(i)}) \mid i \neq f(i)\}) \overset{(1)}{\geq} \sum_{i=1}^{k} d_h(L_i \oplus L_{f(i)})$$

$$\overset{(2)}{\geq} |C| \cdot \sum_{i=1}^{k} h_X(L_i) h_Y(L_{f(i)})$$

$$\overset{(3)}{=} \Theta(n) \cdot \Omega(n^2/k) = \Omega\left(\frac{n^3}{k} \right)$$

Since this holds for any choice of (L_i, L_j), we can pick them so that $(V, E \cup \{(L_i, L_{f(i)}) \mid i \neq f(i)\})$ comprises a set of disjoint lists.   □

This lemma (and the proof) reveals how the adversary commits to a new batch of edges in Option 2 (essentially a random matching will do). Observe that the number of edges is at most half the number of sublists in E_i. In the worst case we have to assume it is exactly half, and thus that the number of sublists is halved after every new batch of edges is selected. Next we show the precondition for the adversary to opt for Option 1, including a lower bound on the corresponding cost imposed on ON.

Lemma 2. *Let N be a list graph, $h \in C_{V \hookrightarrow N}$ a configuration, and $E \subseteq V \times V$ a set of edges so that the graph $G = (V, E)$ has n/ℓ disjoint sublists of size ℓ. If $d_h(E) = \Omega(\ell n^2)$, then there exists an edge $(u, v) \in E$ such that $d_h(u, v) = \Omega(n/\ell)$.*

Proof. There are at most $n/\ell \cdot \binom{\ell}{2} = \mathcal{O}(\ell n)$ distinct simple paths in G, meaning that the average distortion of these paths is $\frac{\Omega(\ell n^2)}{\mathcal{O}(\ell n)} = \Omega(n)$. The highest distortion is at least the average, and every path in G has length at most ℓ. On this path, there must exist an edge with distortion $\Omega(n/\ell)$, since if all edges have a distortion of $o(n/\ell)$, the total would be $o(n)$.   □

Combined, Lemmas 1 and 2 imply that the adversary can either request an edge at cost $\Omega(n/\ell)$, or increase the distortion to $\Omega(\ell n^2)$ by revealing a new batch of edges. The final ingredient is a lower bound on how much cost the adversary can impose on ON in between these batches.

Lemma 3. *Let N be a list graph, $E \subset V \times V$ a set of communication edges. If $h, h' \in C_{V \hookrightarrow N}$ are two embeddings that differ only in the order of two adjacent elements u and v, then $d_h(E) \leq d_{h'}(E) + 2\ell$, where ℓ is the size of the largest sublist in E.*

Proof. Consider all simple paths in E that end in u. At most ℓ paths ending in u (or v) are reduced by 1, and therefore $d_h(E) - d_{h'}(E) \leq 2\ell$. $\qquad\square$

Combining the previous lemmata, we can prove the main technical result.

Lemma 4. *For every online algorithm A, there is a sequence σ_{ON} of length $\mathcal{O}(\varepsilon n^{1+\varepsilon} \log n)$ such that $cost(ON(\sigma_{ON})) = \Omega(\varepsilon n^2 \log n)$, for $0 < \varepsilon \leq 1$. Furthermore, the resulting request graph $R(\sigma_{ON})$ is a list graph.*

Proof. W.l.o.g. assume that $n = 2^p$ for some integer p. This implies that the number of edges in every new batch is a power of 2; consequently, the sublists in any set E_i of revealed edges have size $2^k = \ell$ for some integer k.

Consider the situation right after a batch of edges is revealed, where all sublists have size ℓ. By Lemma 1 this implies that the distortion is $\Omega(\ell n^2)$. Let $\sigma = \sigma_i, \sigma_{i+1}, ..., \sigma_{i+\ell n}$ be the requests obtained by repeatedly requesting the edge in E_i with largest distortion. There are two situations:

- Throughout serving σ, the distortion is always at least $\Omega(\ell n^2)$. Then by Lemma 2 each σ_j, $i \leq j \leq i + \ell n$ incurred a cost of $\Omega(n/\ell)$, at total cost $\Omega(n^2)$.
- By serving σ, ON halves the distortion, thus reducing it by at least $\Omega(\ell n^2)$. Then, since by Lemma 3 every swap reduces the distortion by at most 2ℓ, ON must have used at least $\Omega(n^2)$ swaps.

This argument holds for each batch of edges revealed. The adversary stops when the sublists have size $2^{\varepsilon \log n}$, yielding a sequence σ_{ON} with

$$|\sigma_{ON}| = \sum_{\ell \in \{2^0, ..., 2^{\varepsilon \log n}\}} \ell n = \mathcal{O}(n^{1+\varepsilon})$$

and $cost(\sigma_{ON}) = \Omega(\varepsilon n^2 \log n)$. By Lemma 2, the adversary only requests edges that are introduced using the matching from Lemma 1. Any edge introduced by the latter Lemma concatenates two already existing sublists, hence $R(\sigma_{ON})$ is a list graph. $\qquad\square$

To wrap up the proof for Theorem 2, we conclude by showing that for any online algorithm ON, the sequence σ_{ON} can be solved in $\mathcal{O}(n^2)$ by an optimal offline algorithm.

Proof (Proof of Theorem 2). Let ON be any online algorithm solving DISTRIBUTED LIST UPDATE. Apply Lemma 4 with $\varepsilon = 1/2$, yielding $\text{cost}(ON(\sigma_{ON})) = \Omega(n^2 \log n)$. Since σ_{ON} is a list graph, an offline algorithm can embed this graph at (worst case optimal) cost $\Theta(n^2)$, and serve every request at optimal cost $\mathcal{O}(1)$. This yields $\text{cost}(OFF(\sigma_{ON})) = \Theta(n^2)$, and thus

$$\rho = \frac{\text{cost}(ON(\sigma))}{\text{cost}(OFF(\sigma))} = \Omega(\log n)$$

In order to make this bound hold for arbitrary long sequences, we slightly modify the adversary. After every $\mathcal{O}(n^2)$ requests it serves, it can reconfigure to a new list at cost $\mathcal{O}(n^2)$, and repeat the argument to force cost of $\Omega(n^2 \log n)$ to ON for the subsequent $\mathcal{O}(n^2)$ requests.

Remark. We can generalise the model for DISTRIBUTED LIST UPDATE to include cases where both the request graph and the host graph G are a d-dimensional grid, for constant d; we dub this problem DISTRIBUTED GRID UPDATE. On a request (u, v), the cursor is placed at u and the request is served when it touches v. The same operations are allowed: **moving** the cursor, or **swapping** with on of its 2^d neighbors (also moving the cursor).

Lemma 5. *For every online algorithm* ON *for* DISTRIBUTED GRID UPDATE, *there is a sequence* σ_{ON} *of length* $\mathcal{O}(\varepsilon n^{1+\varepsilon} \log n)$ *such that* $\text{cost}(ON(\sigma_{ON})) = \Omega(\varepsilon n^{1+1/d} \log n)$, *for* $0 < \varepsilon \leq 1$. *The resulting request graph* $R(\sigma_{ON})$ *is a d-dimensional grid graph.*

The proof of Lemma 5 is essentially identical to that of Lemma 4. An overview of the necessary modifications are given in the full paper [6].

4   An Upper Bound

This section presents a $\mathcal{O}(\log n)$-competitive online algorithm for DISTRIBUTED LIST UPDATE. Our main technical lemma shows that the total cost spent on learning the optimal embedding never exceeds $\mathcal{O}(n^2 \log n)$. We propose a simple greedy algorithm that identifies a *locally optimal* embedding, and always moves towards this embedding. Observe that a set of k sublists can be embedded perfectly on a line graph in at most $2^k k!$ ways (they are permuted in some order, and every list has at most two orientations). Given a configuration $h \in C_{V \hookrightarrow N}$ of the lists, we define the locally optimal embedding to be an optimal embedding one that takes the fewest number of reconfigurations to reach, starting at h. Formally, if $opt(E)$ is the set of optimal embeddings of a set edges, then the h-optimal embedding of E is

$$h[E] = \underset{h' \in opt(E)}{\arg \min} \sum_{v \in V} |h(v) - h'(v)|$$

With such a configuration we associate the cost:

$$\Phi_h[E] = \sum_{v \in V} |h(v) - h[E](v)|$$

Let GREAD be the algorithm (it GREedily ADjoins sublists), that upon seeing a new edge σ_i, *immediately* moves to the embedding $h[E_i \cup \{\sigma_{i+1}\}]$.

For each E_i, let $\mathcal{V}(E_i)$ be the connected components of (V, E_i), so that $\mathcal{V}_\sigma = \cup_{1 \leq i \leq m} \mathcal{V}(E_i)$ is the set of all sublists induced by σ. This naturally defines a binary tree $T_\sigma = (\mathcal{V}_\sigma, E_\sigma)$: for every first occurence σ_i of $(u, w) \in E_m$ connecting two sublists U, W in $R(E_i)$, there are two corresponding edges $(U, U \cup W)$ and $(W, U \cup W)$ in E_σ. For every $\sigma_i \in E_m$, GREAD incurs some cost for reconfiguring, and the following lemma bounds this cost.

Lemma 6. *Let h be an optimal embedding of E_i, and let σ_{i+1} be an edge connecting two sublists U and W of E_i. It holds that*

$$\Phi_h[E_i \cup \{\sigma_{i+1}\}] \leq n \cdot \min(|U|, |W|)$$

Proof. Since E_i is optimally embedded by h, we simply need to move the smaller of U and W into its correct location so that $E_i \cup \{\sigma_{i+1}\}$ is optimally embedded. This requires every element in the smaller list to be moved at most n locations, therefore $\Phi_h[E_i \cup \{\sigma_{i+1}\}] \leq n \min(|U|, |W|)$.

For a node $U \in \mathcal{V}_\sigma$, let left(U) and right(U) denote U's left and right child respectively. Further, let $w(U)$ denote the number of nodes in the subtree rooted at U. Observe that for any binary tree with nodes N, it holds that

$$\sum_{v \in N} \min(w(\text{left}(v)), w(\text{right}(v))) \leq |N| \log |N|$$

Theorem 4. *For any σ, with $|\sigma| = m$, such that $|E_m| = k$ and $R(\sigma)$ is a list,*

$$cost(\text{GREAD}(\sigma)) = \mathcal{O}(m + nk \log k)$$

Proof. Let h_i denote the configuration after request σ_1, and let h_0 denote the trivial optimal initial embedding. Then the total cost of GREAD is the sum of reconfiguring after every σ_i plus accessing every request at cost 1:

$$cost(\text{GREAD}(\sigma)) - m = \sum_{i=0}^{m} \Phi_{h_i}[E_i \cup \{\sigma_{i+1}\}]$$

$$\leq \sum_{U \in \mathcal{V}_\sigma} n \min(w(\text{left}(U)), w(\text{right}(U)))$$

$$\leq nk \log k$$

As a corollary, it is not hard to show that GREAD achieves optimal $\log n$ competitiveness for the worst case sequence constructed in Sect. 3. Additionally, in the full paper [6] we show a distributed implementation of this algorithm using message passing.

5   Conclusion

We presented a first and asymptotically tight, i.e., $\Theta(\log n)$-competitive online algorithm for self-adjusting reconfigurable linear networks with linear demand. Both our lower and upper bounds are non-trivial, and we believe that our work opens several interesting directions for future research. In particular, it would be very interesting to shed light on the competitive ratio achievable in more general network topologies, and to study randomized algorithms.

References

1. Albers, S.: A competitive analysis of the list update problem with lookahead. Theoret. Comput. Sci. **197**(1–2), 95–109 (1998)
2. Albers, S.: Improved randomized on-line algorithms for the list update problem. SIAM J. Comput. **27**(3), 682–693 (1998)
3. Albers, S., Von Stengel, B., Werchner, R.: A combined bit and timestamp algorithm for the list update problem. Inf. Process. Lett. **56**(3), 135–139 (1995)
4. Albers, S., Westbrook, J.: Self-organizing data structures. In: Fiat, A., Woeginger, G.J. (eds.) Online Algorithms. LNCS, vol. 1442, pp. 13–51. Springer, Heidelberg (1998). https://doi.org/10.1007/BFb0029563
5. Ambühl, C.: Offline list update is NP-hard. In: Paterson, M.S. (ed.) ESA 2000. LNCS, vol. 1879, pp. 42–51. Springer, Heidelberg (2000). https://doi.org/10.1007/3-540-45253-2_5
6. Avin, C., van Duijn, I., Schmid, S.: Self-adjusting linear networks. arXiv preprint arXiv:1905.02472 (2019)
7. Avin, C., Haeupler, B., Lotker, Z., Scheideler, C., Schmid, S.: Locally self-adjusting tree networks. In: 2013 IEEE 27th International Symposium on Parallel and Distributed Processing, pp. 395–406. IEEE (2013)
8. Avin, C., Hercules, A., Loukas, A., Schmid, S.: Towards communication-aware robust topologies. ArXiv Technical Report (2017)
9. Avin, C., Loukas, A., Pacut, M., Schmid, S.: Online balanced repartitioning. In: Gavoille, C., Ilcinkas, D. (eds.) DISC 2016. LNCS, vol. 9888, pp. 243–256. Springer, Heidelberg (2016). https://doi.org/10.1007/978-3-662-53426-7_18
10. Avin, C., Mondal, K., Schmid, S.: Demand-aware network designs of bounded degree. In: Proceedings International Symposium on Distributed Computing (DISC) (2017)
11. Avin, C., Mondal, K., Schmid, S.: Demand-aware network design with minimal congestion and route lengths. In: Proceedings of IEEE INFOCOM (2019)
12. Avin, C., Schmid, S.: Toward demand-aware networking: a theory for self-adjusting networks. In: ACM SIGCOMM Computer Communication Review (CCR) (2018)
13. Bubeck, S., Cesa-Bianchi, N., et al.: Regret analysis of stochastic and nonstochastic multi-armed bandit problems. Found. Trends® Mach. Learn. **5**(1), 1–122 (2012)
14. Díaz, J., Petit, J., Serna, M.: A survey of graph layout problems. ACM Comput. Surv. (CSUR) **34**(3), 313–356 (2002)
15. Fenz, T., Foerster, K.T., Schmid, S., Villedieu, A.: Efficient non-segregated routing for reconfigurable demand-aware networks. In: Proceedings of IFIP Networking (2019)

16. Huq, S., Ghosh, S.: Locally self-adjusting skip graphs. In: Proceedings of IEEE 37th International Conference on Distributed Computing Systems (ICDCS), pp. 805–815 (2017)
17. Ghobadi, M., et al.: Projector: agile reconfigurable data center interconnect. In: Proceedings of ACM SIGCOMM, pp. 216–229 (2016)
18. Olver, N., Pruhs, K., Schewior, K., Sitters, R., Stougie, L.: The itinerant list update problem. In: 13th Workshop on Models and Algorithms for Planning and Scheduling Problems, p. 163 (2017)
19. Peres, B., Souza, O., Goussevskaia, O., Schmid, S., Avin, C.: Distributed self-adjusting tree networks. In: Proceedings of IEEE INFOCOM (2019)
20. Reingold, N., Westbrook, J.: Off-line algorithms for the list update problem. Inf. Process. Lett. **60**(2), 75–80 (1996)
21. Reingold, N., Westbrook, J., Sleator, D.D.: Randomized competitive algorithms for the list update problem. Algorithmica **11**(1), 15–32 (1994)
22. Schmid, S., Avin, C., Scheideler, C., Borokhovich, M., Haeupler, B., Lotker, Z.: Splaynet: towards locally self-adjusting networks. IEEE/ACM Trans. Netw. (ToN) **24**, 1421–1433 (2016)
23. Sleator, D.D., Tarjan, R.E.: Amortized efficiency of list update and paging rules. Commun. ACM **28**(2), 202–208 (1985)
24. Teia, B.: A lower bound for randomized list update algorithms. Inf. Process. Lett. **47**(1), 5–9 (1993)

Author Index

Printed in the United States
By Bookmasters